D0764201

LINEAR ALGEBRA

LINEAR ALGEBRA

CARROLL WILDE
Naval Postgraduate School
Monterey, California

Addison-Wesley Publishing Company

Reading, Massachusetts • Menlo Park, California • New York
Don Mills, Ontario • Wokingham, England • Amsterdam • Sydney • Bonn
Singapore • Tokyo • Madrid • Bogotá • Santiago • San Juan

Sponsoring Editor: Thomas N. Taylor
Production Supervisor: Herbert Nolan
Editorial and Production Services: Cobb/Dunlop Publisher Services, Inc.
Text Design: Design Five
Illustrator: V.A.P. Group, Ltd.
Art Consultant: Loretta M. Bailey
Manufacturing Supervisor: Roy Logan
Cover Designer: Marshall Henrichs

Library of Congress Cataloging in Publication Data

Wilde, Carroll O.
Linear Algebra

Includes index.
1. Algebras, Linear. I. Title.
QA184.W525 1987 512'.5 87-1313
ISBN 0-201-13089-0

Copyright ©1988 by Addison-Wesley Publishing Company, Inc.
All rights reserved. No part of this publication may be reproduced, stored in a retrieval system, or transmitted, in any form or by any means, electronic, mechanical, photocopying, recording, or otherwise, without the prior written permission of the publisher. Printed in the United States of America. Published simultaneously in Canada.

ABCDEFGHIJ-HA-8987

To my wife,
Roxanne M. Wilde

In loving memory of my parents,
Carrie B. and Edwin F. Wilde, Sr.

PREFACE

Linear algebra provides a foundation for much important mathematical theory as well as a vehicle for solving many useful applied problems. Intended for a one-semester introduction to linear algebra, this book seeks to convey an understanding of fundamental principles and techniques, and a sense of their value in modern applications. High school algebra should be adequate preparation. Some optional calculus-based material is included for readers with this additional background.

Features

Promotion of harmony between theory and application. When carefully coupled, mathematical theory and physical applications can have a synergistic effect for the beginning student of linear algebra. We have tried to meet basic needs of readers who are specializing in mathematics and those with primary interests in applied fields, and have striven to provide complementary insights to both groups. [See page 289]

An application section in each chapter. This feature is designed to facilitate a significant integration of applications with the theory while simultaneously enabling an uninterrupted flow of concept within each chapter. These sections are designated as optional (o) to allow additional flexibility in the orientation of the courses it serves. [See pages 48–55]

Detailed and careful explanations. In all presentations, clarity of exposition has high priority. Where appropriate, notes are included to explain steps in procedures or reasoning. The goal here is optimal support for the most basic level of learning, *comprehension*. Proofs of most theorems are included. [See pages 176–183]

An unusually large number of exercises. Each section contains a very large set of exercises graded from routine to difficult. The "drill and practice" exercises are aimed at *retention and recall* of information. The more challenging problems are aimed at a higher level of learning, *the creative use of knowledge*. [See pages 199–203]

Emphasis on geometric interpretation. The interplay between geometry and linear algebra is used to develop conceptual insight as well as intuitive feeling for the algebraic concepts and methods. Many figures are included to help provide the geometric emphasis. [See pages 1–5]

Introduction of concept via examples. Wherever appropriate, examples are given *prior* to the formal presentation of a new concept. This device is used to motivate and to help the reader develop the ability to advance from the concrete to more general mathematical principles. [See pages 345–346]

Careful discussions of computing. Some computational aspects of linear algebra are discussed in the application sections. We try to give the reader a sense of the role of computing in applied work and the delicate nature of some problems in this area, without detracting from the main development. We also convey the idea that computing for linear algebra warrants further study in numerical analysis. [See pages 55–56]

A special section on vectors. A section on vector algebra is given to prepare for a smooth presentation of matrix algebra. This section also provides groundwork for the subsequent study of the core material on vector spaces. [See pages 61–74]

Introduction to linear programming. The final chapter is an optional introduction to linear programming. This material is intended to stimulate interest in further study of a topic that has both theoretical and practical value, in keeping with the general spirit of embracing both directions. [See pages 395–430]

Computer package. This book is accompanied by a diskette with software for a microcomputer designed to support the learning of linear algebra. The programs on the diskette are user-friendly, and are intended to shift the emphasis away from arithmetic and toward the concepts and methods of linear algebra. This software is designed to enhance the learning of linear algebra rather than to solve large problems.

Organization

The contents and their organization are relatively standard, facilitating adaptation to many existing course syllabi. Each chapter is primarily concerned with the basic object of study, a system of linear algebraic equations. **Chapter 1** provides the fundamental means for solving linear systems and for studying them, Gaussian elimination. Applications include interpolation of curves and surfaces, partial fraction expansions, and heat transfer. **Chapter 2** is devoted to algebraic operations on matrices and their relation to linear systems. The study includes the LU decomposition of a matrix, the Leslie population model, genetics, and finite Markov chains. The determinant function in **Chapter 3** provides a useful condition for the existence of solutions of linear systems. The determinant is used in finding equations of curves, areas of plane figures, and volumes of solids.

Chapter 4 brings deeper insight into linear systems through an axiomatic approach. We define a vector space in terms of properties of plane and space vectors, and derive some consequences for linear systems. This chapter contains a general summary of the theory of linear systems. **Chapter 5** treats the interplay between the algebraic and analytic structures of vector spaces. This study sheds more light on linear systems and provides a framework for least squares curve fitting.

Chapter 6 on linear transformations completes the preparations for the capstone study of eigenvalues and eigenvectors. We apply the ideas to problems in computer graphics and integral calculus. The eigenvalue problem of **Chapter 7** is the gateway to a host of traditional and modern mathematical studies. We study applications to quadratic forms, a population model, and differential equations. **Chapter 8** is an optional unit on linear programming designed to arouse interest in the practical area of constrained optimization.

A Note to the Student

This book is meant to be read "with a pencil." We hope you will participate *actively* in the development, filling in necessary details, raising questions, trying to anticipate what is coming, and trying to furnish proofs. The text and the exercises are structured to encourage and facilitate your active involvement. Treat each result as though it were your own, and soon it will be.

Acknowledgments

The author owes a large debt to many individuals who contributed to this book directly or indirectly. The reviewers, who are listed separately, provided extensive guidance and suggestions, and their advice resulted in numerous improvements in the manuscript. I am especially indebted to Richard W. Hamming for many stimulating conversations as well as critical reviews of the manuscript and other documents pertaining to the book. The focus group, whose participants are also listed separately, provided valuable early guidance and helped to determine the general direction of this study and relative emphasis of the topics. I had many conversations about linear algebra with Gordon E. Latta; in particular, he suggested the proof of Theorem 4.12. I am grateful to Frank D. Faulkner for his contributions, especially in the early stages of the work; he contributed many of the exercises in Sections 1.3 and 1.4. I appreciate the list of corrections and comments from M. B. Ulmer. Finally, I wish to thank George B. Thomas, Jr. and Ross L. Finney for permission to use material from the seventh edition of *Calculus and Analytic Geometry*. This material appears in Appendix A and in Appendix C of this book.

I wish to acknowledge the outstanding cooperation and support of the team at Addison-Wesley. I am especially grateful to Stephanie Botvin, Jeffrey Pepper, Chip Price, Leslie Richardson, Adeline Ruggles, and Tom Taylor. These individuals made a difficult job look easy and made much of the work I had to do a pleasure. I also acknowledge the support, encouragement, and patience shown by my wife, Roxanne, throughout many long months on this project.

I am solely responsible for any errors in this book. I would appreciate having any errors brought to my attention.

C. W.

REVIEWER LIST

Jean H. Bevis, Georgia State University
Charles K. Cook, Tri-State University
Richard Crittenden, University of Alabama—Birmingham
George J. Davis, Georgia State University
Bruce Edwards, University of Florida
Murray Eisenberg, University of Massachusetts—Amherst
Ralph P. Grimaldi, Rose-Hulman Institute of Technology
Christopher E. Hee, Eastern Michigan University
Robert Heller, Mississippi State University
Harold Hochstadt, Polytechnic University
Dale T. Hoffman, Bellvue Community College
W. A. McWorter, Ohio State University
Steven A. Pruess, Colorado School of Mines
W. Robert Stephenson, Iowa State University
Marvin Zeman, Southern Illinois University—Carbondale
Stanley M. Zoltek, George Mason University

Focus Group Participants

Gail Broome, Providence College
John Fogarty, University of Massachusetts—Amherst
Edward Hinson, University of New Hampshire
Michael Rosen, Brown University
John Smith, Boston College
Taffee Tanimoto, University of Massachusetts—Boston

CONTENTS

LINEAR ALGEBRA

1 Linear Systems and Gaussian Elimination

Solving linear systems *is a central activity in linear algebra. The* Gaussian elimination *method used in this text provides more than just solution data—it facilitates identification of important and useful properties of linear systems and their solution sets. In fact, Gaussian elimination is our primary means of* understanding *linear systems, and the knowledge gained from this method will facilitate the study of some fundamental concepts in Chapters 4 and 5.*

1.1 LINEAR SYSTEMS

We begin by defining linear systems and their solutions, and interpreting them geometrically. We give a procedure for solving certain *upper triangular* linear systems. This procedure constitutes one major stage of Gaussian elimination. In this section, some physical interpretations of linear systems are also described.

Linear Systems and Their Solutions

Any real numbers a, b, and c, such that a and b are not both zero, correspond to a line in the rectangular Cartesian plane: a point $P: (x, y)$ in the plane lies on the line if and only if

$$ax + by = c.$$

In this relation, x and y are called **variables**. The relation itself is called a **linear equation in x and y**. The line is said to be **determined by the equation**, and the points on the line are said to **satisfy the equation**.

Any real a, b, c, and d, such that a, b, and c are not all zero, correspond to a plane in space in a similar way. A point $P: (x, y, z)$ lies on the plane if and only if

$$ax + by + cz = d.$$

This relation is a **linear equation in the variables x, y, and z**. The plane is **determined by the equation**, and the points on the plane **satisfy the equation**.

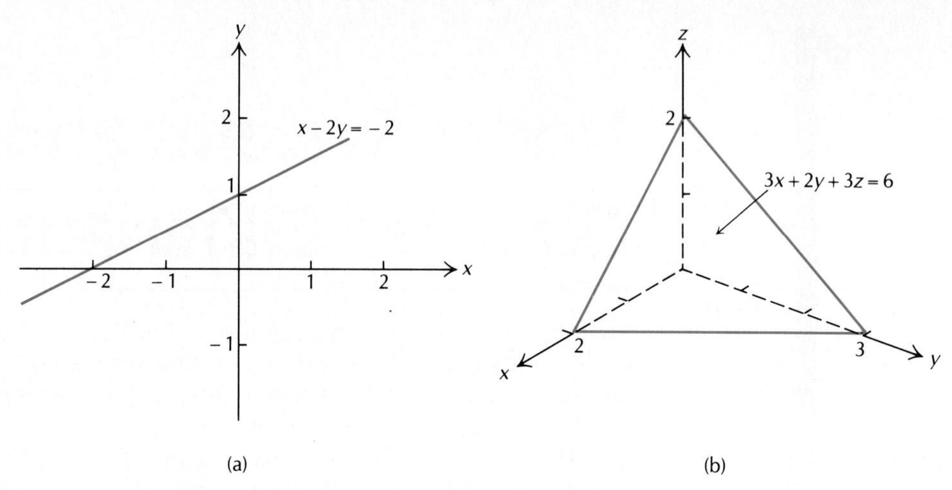

Figure 1.1 Linear equations: (a) in the plane; (b) in space.

Linear equations are illustrated in Fig. 1.1. Appendix B contains a more detailed explanation of linear equations in two and three variables.

More generally, an expression of the form

$$a_1x_1 + \cdots + a_nx_n = b$$

is called a **linear equation in $\mathbf{x_1, \ldots, x_n}$**. The lefthand side of this expression is called a **linear form in $\mathbf{x_1, \ldots, x_n}$**.

A set of one or more linear equations *in the same variables* is called a **linear system**. For instance,

$$\begin{aligned} x - y &= -2 \\ -x + 2y &= 5 \end{aligned}$$

is a linear system of two equations in x and y. Similarly,

$$\begin{aligned} x + y + z &= 6 \\ x + y &= 3 \\ 4x + y + z &= 9 \end{aligned}$$

is a linear system of three equations in x, y, and z, where we understand the second equation as

$$x + y + 0z = 3.$$

Some configurations of two equations in x and y are illustrated in Fig. 1.2. Several possibilities for three equations in x, y, and z are in Fig. 1.3.

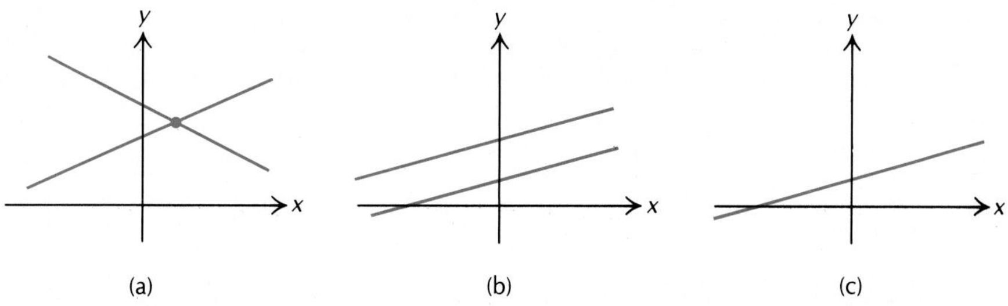

Figure 1.2 The possible configurations for the lines determined by two linear equations in x and y. (a) The lines intersect at a unique solution point. (b) The lines are parallel so the system has no solutions. (c) The lines coincide, and each point on this line is a solution of the system.

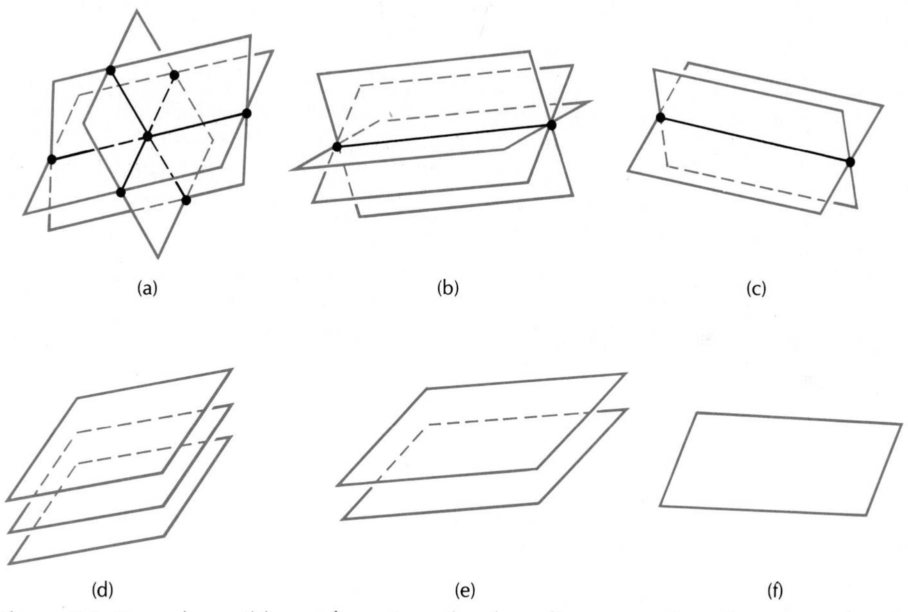

Figure 1.3 Several possible configurations for three linear equations in x, y, and z: (a) planes intersect at a unique solution point; (b) planes intersect in a line of solutions; (c) two planes coincide and intersect the third in a line of solutions; (d) three planes parallel, so no solution; (e) two planes coincide and are parallel to the third, so no solution; (f) three planes coincide in a plane of solutions.

DEFINITION 1.1

A **solution of a linear system** is a set of values for the variables that satisfy each equation in the system. To **solve** a linear system means to find *all* its solutions.

Example 1

Show that the assignment of values $x = 1$ and $y = 3$ constitutes a solution of the linear system

$$\begin{aligned} x - y &= -2 \\ -x + 2y &= 5. \end{aligned}$$

Solution

Substituting the given values for x and y into both equations gives

$$\begin{aligned} 1 - 3 &= -2 \\ -1 + 2(3) &= 5. \end{aligned}$$

Since both statements are true, the given values satisfy both equations, and hence constitute a solution of the given system by Definition 1.1. □

In Example 1, the given assignment of values to the variables was verified as a solution of the given system. No indication was given on how the values were found or whether other solutions exist. Both of these issues will be addressed in Section 1.2. For the present, the geometric interpretation in Fig. 1.4 indicates a unique solution for this particular system, hence that it is solved, by Definition 1.1.

Cartesian coordinates provide a convenient way to denote solutions of linear systems. If the variables are kept in the same order in each equation, then the values that form a solution of the system are the Cartesian coordinates of a point, and the assignment of values is indicated by the order of the variables. For

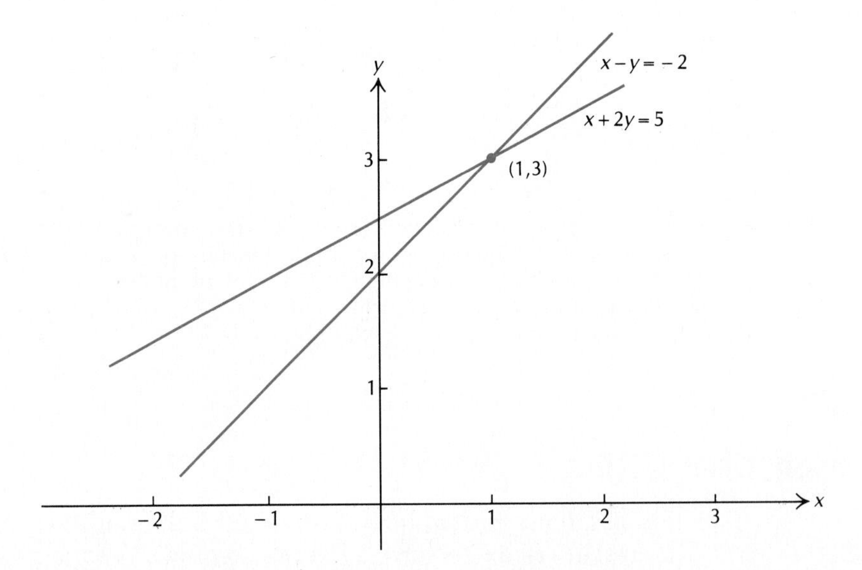

Figure 1.4 The linear system in Example 1.

the system in Example 1, the *ordered pair* (1, 3) denotes the solution obtained by assigning 1 to x and 3 to y. The first coordinate 1 is assigned to the first variable x and the second coordinate 3 is assigned to the second variable y. For a linear system in x, y, z, the *ordered triple* (r, s, t) denotes the solution obtained by assigning r to x, s to y, and t to z. For a linear system in $x_1, \ldots, x_n$, the ordered *n-tuple* $(t_1, \ldots, t_n)$ denotes the solution obtained by assigning t_i to x_i for $i = 1, \ldots, n$.

Example 2

Show that (1, 2, 3) is, and (1, 5, 0) is not, a solution of the linear system

$$\begin{aligned} x + y + z &= 6 \\ x + y \phantom{{}+z} &= 3 \\ 4x + y + z &= 9. \end{aligned}$$

Solution

Since the assignment $x = 1$, $y = 2$, and $z = 3$ satisfies all three equations, (1, 2, 3) is a solution of the given system, by Definition 1.1. The assignment $x = 1$, $y = 5$, and $z = 0$ does not satisfy the second equation (although it does satisfy the first and the third), so (1, 5, 0) is not a solution of the given system, again by Definition 1.1. □

The linear system in Example 2 is illustrated in Fig. 1.5. The graph shows the planes determined by the equations when they intersect at the point (1, 2, 3).

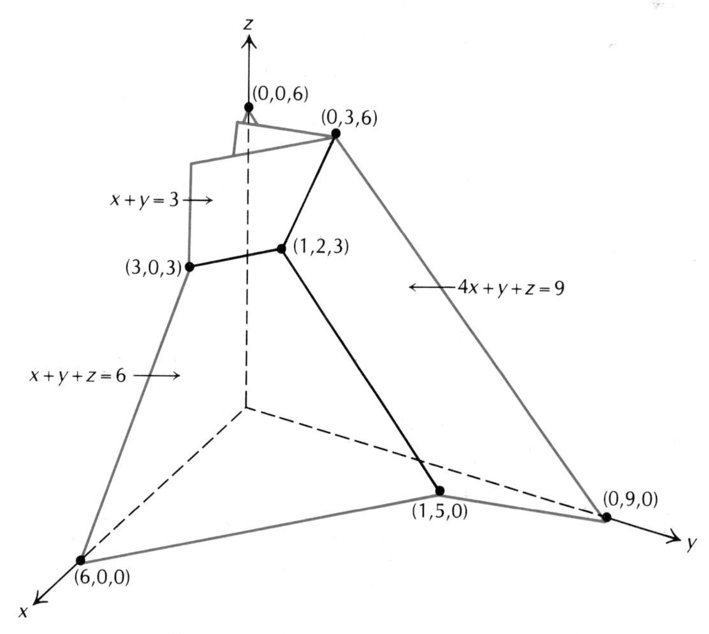

Figure 1.5 The linear system in Example 2.

The point $(1, 5, 0)$ lies on the intersection of the first and third planes, but not on the second.

Solution of Upper Triangular Systems

Some linear systems have a simplified form enabling us to find solutions directly.

Example 3

Solve the linear system

$$\begin{aligned} 2x - y + 3z &= 5 \\ 2y - z &= 4 \\ 2z &= 4. \end{aligned}$$

Solution

The third equation determines z as $z = 2$. Substituting 2 for z in the second equation gives

$$2y - 2 = 4.$$

Solving this equation yields $y = 3$. Substituting 2 for z and 3 for y in the first equation produces

$$2x - 3 + 3(2) = 5,$$

from which $x = 1$. Thus, the given system has one solution, the ordered triple $(1, 3, 2)$. □

The second and third equations in Example 3 can also be written as

$$\begin{aligned} 0x + 2y - z &= 4 \\ 0x + 0y + 2z &= 4, \end{aligned}$$

to reflect the same set of variables in all equations. Thus, the statement "$z = 2$" in Example 3 implies many solutions of the third equation: any ordered triple of the form $(x, y, 2)$. The subsequent statement "$y = 3$" includes as solutions of the second and third equations any ordered triple of the form $(x, 3, 2)$. For ease of presentation, we normally use only abbreviated expressions as in Example 3.

The linear system in Example 3 is called *upper triangular* because of the shape indicated by the colored shading. The triangle is the upper half of a square, above one diagonal.

Any upper triangular system with every coefficient on the diagonal nonzero has a unique solution. To see why, start at the bottom and work upward. The last equation has a unique solution for the last variable. We may substitute this value into the next-to-last equation and then solve the equation for the corresponding variable. The overall result is a unique solution for the last two variables. This procedure may be continued through the system. At the final step, the first equation is solved for the first variable. The overall result is a unique solution for the system. A formal proof requires mathematical induction, which is described in Appendix A.

Applications of Linear Systems

Linear systems arise in diverse mathematical and physical environments. In some physical problems, linear systems are mathematical models resulting directly from an application of some principle or from an analysis of the problem. For instance, Section 1.5 contains linear systems arising from an application of Kirchhoff's laws to electronic circuits with multiple loops. The solutions of the systems yield the currents in these loops. In this same section, linear algebra is applied to a problem of heat transfer in a thin, insulated plate. In this case, the solution represents steady-state temperatures at points in the plate.

Even when the direct mathematical model for a physical problem is *not* a linear system, linear algebra is often applied to a related mathematical problem. For instance, linear algebra may be used for the interpolation of data points by a curve of prescribed type. If the points are satellite tracking data, the solution of the linear system may give the equation of an orbital path. Linear algebra may also be applied to partial fraction problems. In an application of partial fractions, the solution of a linear system may yield the equation of motion for a harmonic oscillator.

One reason why linear systems appear as mathematical models is that many basic concepts lead directly to linear forms. For instance, the total distance d traveled by three objects moving at constant rates r_1, r_2, and r_3 meters per second over straight-line paths through times t_1, t_2, and t_3 seconds, respectively, is

$$d = r_1t_1 + r_2t_2 + r_3t_3 \qquad \text{meters.}$$

Similarly, if principal amounts of p_1, p_2, p_3, and p_4 dollars are invested at simple interest rates r_1, r_2, r_3, and r_4, respectively, then the total interest for one year is

$$I = p_1r_1 + p_2r_2 + p_3r_3 + p_4r_4 \qquad \text{dollars.}$$

Linear systems may arise when conditions are imposed on linear forms such as these. Some examples may be found in the application sections of this book.

REVIEW CHECKLIST

1. Define linear form, linear equation, and linear system.
2. Define a solution of a linear system, and apply it in checking particular values.
3. Use ordered n-tuples for solutions of linear systems.
4. State what it means to "solve" a linear system.
5. Sketch graphs for linear systems in two or three variables.
6. Solve upper triangular linear systems with nonzero coefficients on the diagonal.
7. Describe informally why an upper triangular system with nonzero diagonal coefficients must have a unique solution.

EXERCISES

In Exercises 1–3, sketch the lines determined by the equations in a rectangular Cartesian coordinate system. Then use the graph to estimate the intersection of the lines, and check whether the values form a solution of the system.

1. $\begin{aligned} 2x + y &= 4 \\ x - 2y &= -3 \end{aligned}$

2. $\begin{aligned} x + y &= 3 \\ x - 3y &= 7 \end{aligned}$

3. $\begin{aligned} 2x - y &= 1 \\ 3x + y &= 4 \end{aligned}$

In Exercises 4–6, sketch the lines determined by the equations in the given systems. What do the graphs indicate concerning solutions of the systems?

4. $\begin{aligned} x - y &= 1 \\ x + y &= 5 \\ 2x - y &= 0 \end{aligned}$

5. $\begin{aligned} 3x - y &= 5 \\ 6x - 2y &= 3 \end{aligned}$

6. $\begin{aligned} x - y &= -1 \\ 3x - y &= 1 \\ 2x + y &= 4 \end{aligned}$

In Exercises 7–9, determine whether the given point is a soluiton of the linear system

$$\begin{aligned} 2x - y &= 1 \\ x + 3y &= 4. \end{aligned}$$

7. $(1,1)$

8. $(2,3)$

9. $(-2,2)$

In Exercises 10–12, determine whether the given point is a solution of the linear system

$$\begin{aligned} 2x - y + z &= 5 \\ x + y + 2z &= 1. \end{aligned}$$

10. $(1,-2,1)$

11. $(2,-1,0)$

12. $(4,2,-1)$

In Exercises 13–15, determine whether the given point is a solution of the linear system

$$\begin{aligned} x - 2y + z &= 0 \\ 2x \qquad - z &= 0 \\ 4y - 3z &= 0. \end{aligned}$$

13. $(0, 0, 0)$ **14.** $(2, 3, 4)$ **15.** $(-4, -6, -8)$

In Exercises 16–18, find the value(s) of a, if any exist, for which the given ordered triple is a solution of the system used in Exercises 10–12.

16. $(3, 1, a)$ **17.** $(4, 1, -2a)$ **18.** $(5, 1, -2a)$

In Exercises 19–21, solve the given linear system.

19. $$\begin{aligned} 2x + 3y &= 4 \\ 2y &= 4 \end{aligned}$$

20. $$\begin{aligned} x - y - z &= 4 \\ 2y + z &= 1 \\ 3z &= -9 \end{aligned}$$

21. $$\begin{aligned} 2x_1 + x_2 - 4x_3 + 2x_4 &= 2 \\ 3x_2 + x_3 + x_4 &= 6 \\ 2x_3 + 3x_4 &= -1 \\ x_4 &= -1 \end{aligned}$$

In Exercises 22–24, determine if the given system of equations is a linear system.

22. $$\begin{aligned} 2x - 3y &= 1 \\ x - y^2 &= 2 \end{aligned}$$

23. $$\begin{aligned} x + 3xy &= 2 \\ 2x - 4y &= 3 \end{aligned}$$

24. $$\begin{aligned} 3x - y &= 4 \\ 2x + 5y &= 6 \end{aligned}$$

25. If a mass of 4 grams is placed at -3 and a mass of 2 grams is placed at 1 on the x-axis, where must a mass of 5 grams be placed in order that the total moment of the system about the origin be zero? (The total moment is the sum of individual moments, each of which is a product of mass by position.)

26. (*Analytic geometry based*) If the x- and y-axes in a rectangular coordinate system are rotated counterclockwise through an angle θ, the formulas

$$x' = x\cos\theta + y\sin\theta$$

$$y' = -x\sin\theta + y'\cos\theta$$

from analytic geometry relate the xy-coordinates of any point to its $x'y'$-coordinates, as shown in the figure.

a. If $\theta = 30°$, find the $x'y'$-coordinates of the point whose xy-coordinates are $x = 2$, $y = 2$.

b. Check to verify by Definition 1.1 that for any x' and y', the ordered pair

$$(x, y) = (x'\cos\theta - y'\sin\theta, x'\sin\theta + y\cos\theta)$$

is a solution for the xy-coordinates of the point with the given $x'y'$-coordinates. (Note that the *form* of the expressions in the solution resembles the form of the

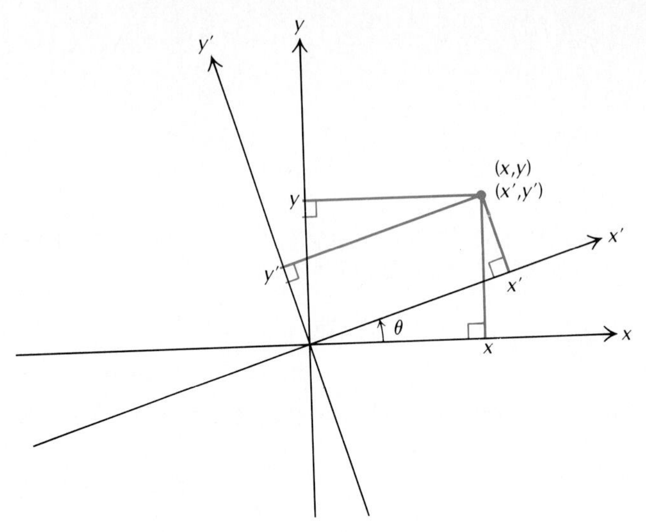

Rotation of axes for exercise 26.

original rotation formulas with θ replaced by $-\theta$. Is this result reasonable from the geometry of the problem?)

1.2 GAUSSIAN* ELIMINATION

Once we obtain a linear system from mathematical or physical considerations of the problem, solving the system is often the next step. The solution method we use, *Gaussian elimination*, is an orderly procedure that is used widely. This method, and variations of it, provide the principal algorithms for computer solution of linear systems.

Gaussian Elimination as a Two-Pass Procedure

The general form of an upper triangular system is

$$\begin{aligned} a_{11}x_1 + a_{12}x_2 + \cdots + a_{1n}x_n &= b_1 \\ a_{22}x_2 + \cdots + a_{2n}x_n &= b_2 \\ &\;\;\vdots \\ a_{nn}x_n &= b_n. \end{aligned}$$

* **Carl Friedrich Gauss** (1777–1855) made contributions in many areas of mathematics, including linear algebra, number theory, and geometry. His work also foreshadowed many important results in analysis and topology.

In this form, a **double subscript** locates each coefficient. The first subscript indicates the equation and the second indicates the variable. For instance, if $n \geq 4$, then a_{34} is the coefficient for the fourth variable in the third equation.

The procedure described in Section 1.1 for solving an upper triangular system is called **back substitution**. If each $a_{ii} \neq 0$ in an upper triangular system, then back substitution yields a unique solution. For such a system, Gaussian elimination consists of two distinct phases, called **passes**.

Pass 1. Construction of an upper triangular system whose solution is the same as that of the given system.

Pass 2. Solution of the upper triangular system by back substitution.

The goal of Pass 1 leads to the following important concept.

DEFINITION 1.2

Two linear systems are said to be **equivalent** if every solution of either system is also a solution of the other.

Figure 1.6 and the following example illustrate equivalent linear systems.

Example 1

The linear systems

$$\begin{aligned} 3x + 2y &= -4 \\ 2x - y &= 3 \end{aligned} \qquad \text{and} \qquad \begin{aligned} 3x + 2y &= -4 \\ 4x - 2y &= 6 \end{aligned}$$

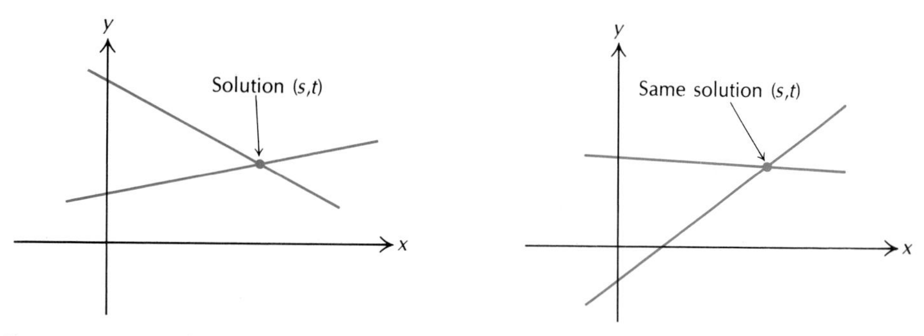

Figure 1.6 Equivalent systems: they may be distinct systems, but their solutions are the same.

are equivalent. Because the two first equations are the same, they have the same solutions. In addition, since

$$2x - y = 3 \qquad \text{if and only if} \qquad 2(2x - y) = 2(3),$$

the two second equations have the same solutions. Thus, the two systems have the same solutions, so they are equivalent by Definition 1.2. ☐

As noted, Pass 1 of Gaussian elimination is the construction of an upper triangular system that is equivalent to a given system. The upper triangular system is obtained through a sequence of arithmetic operations, one type of which was illustrated in Example 1. Theorem 1.1 describes the types of operations used in Gaussian elimination.

THEOREM 1.1

If operations of the following types are applied to a linear system, the resulting system is equivalent to the original.

Type 1. Multiply any equation by a nonzero constant.

Type 2. Interchange any two equations.

Type 3. Add a constant times one equation to another equation.

Proof For a Type 1 operation, let c be a nonzsero constant. Then, for the ith equation of the system,

$$a_{i1}x_1 + \cdots + a_{in}x_n = b_i$$

if and only if

$$c(a_{i1}x_1 + \cdots + a_{in}x_n) = cb_i,$$

by properties of multiplication of real numbers. Hence, a Type 1 operation procduces an equivalent system.

In a Type 2 operation, two equations are interchanged, and the solution set of the system is unchanged since the same set of equations must be satisfied.

For a Type 3 operation, if

$$a_{i1}x_1 + \cdots + a_{in}x_n = b_i \tag{1}$$

and

$$a_{j1}x_1 + \cdots + a_{jn}x_n = b_j, \tag{2}$$

adding "equals to equals" gives

$$(3) \qquad (ca_{i1} + a_{j1})x_1 + \cdots + (ca_{in} + a_{jn})x_n = cb_i + b_j.$$

Thus any solution of a system containing equations (1) and (2) is a solution of any system containing equations (1) and (3). Conversely, if equations (1) and (3) are satisfied, let c be any constant. From equation (1),

$$-ca_{i1}x_1 - \cdots - ca_{in}x_n = -cb_i,$$

and adding equals to equals again gives

$$a_{j1}x_1 + \cdots + a_{jn}x_n = b_j.$$

Thus any solution of a system containing equations (1) and (3) is a solution of any system containing equations (1) and (2). Since the remaining equations of any system are unaffected, a Type 3 operation also yields an equivalent system. □

How are these operations applied in Gaussian elimination?

Example 2

Solve the linear system

$$\begin{aligned} x - 2y &= -1 \\ 2x + y &= 8 \end{aligned}$$

by Gaussian elimination.

Solution

For the first pass, a Type 3 operation "eliminates" the coefficient of x in the second equation:

$$\begin{aligned} -2x + 4y &= 2 \\ 2x + y &= 8 \\ \hline 0x + 5y &= 10. \end{aligned}$$

Add -2 times the first equation to the second.

If the original second equation is replaced by this result, the new system

$$\begin{aligned} x - 2y &= -1 \\ 5y &= 10 \end{aligned}$$

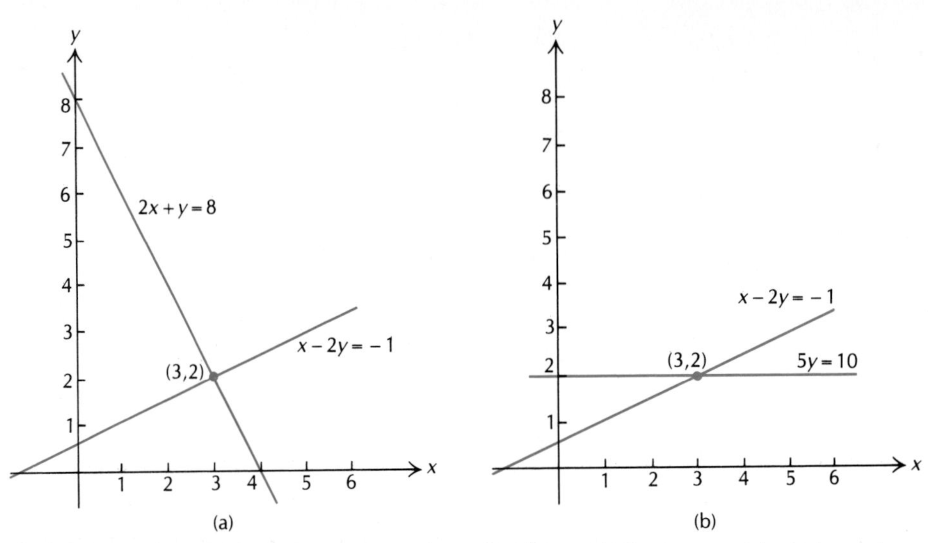

Figure 1.7 (a) The given system in Example 1. (b) The equivalent upper triangular system.

is in upper triangular form. The unique solution (3, 2) may be found by back substitution, the second pass of the elimination. Since the given system is equivalent to the upper triangular system by Theorem 1.1, the given system is solved.

Figure 1.7 illustrates the method used in Example 2. Figure 1.7(a) shows the given system and Fig. 1.7(b) shows the upper triangular system. Figure 1.7(b) also shows the equivalence of these systems—they have the same solution, (3, 2).

Example 3

Solve the linear system

$$\begin{aligned} 2x - 4y + 2z &= 2 \\ 3x - 6y + 4z &= 6 \\ x + y - 4z &= 1 \end{aligned}$$

by Gaussian elimination.

Solution

For the first pass, construct an equivalent upper triangular system by applying the operations described in Theorem 1.1.

$$\begin{aligned} 2x - 4y + 2z &= 2 \\ 3x - 6y + 4z &= 6 \\ x + y - 4z &= 1 \end{aligned}$$

Given.

$$\begin{aligned} x - 2y + z &= 1 \\ 3x - 6y + 4z &= 6 \\ x + y - 4z &= 1 \end{aligned}$$

Obtained by a Type 1 operation (the first equation was multiplied by 1/2).

$$\begin{aligned} x - 2y + z &= 1 \\ z &= 3 \\ x + y - 4z &= 1 \end{aligned}$$

Obtained by a Type 3 operation (-3 times the first equation was added to the second).

$$\begin{aligned} x - 2y + z &= 1 \\ z &= 3 \\ 3y - 5z &= 0 \end{aligned}$$

Obtained by a Type 3 operation (-1 times the first equation was added to the third).

$$\begin{aligned} x - 2y + z &= 1 \\ 3y - 5z &= 0 \\ z &= 3 \end{aligned}$$

Obtained by a Type 2 operation (the second and third equations were interchanged).

This upper triangular system is equivalent to the given system. Back substitution yields the unique solution $(8, 5, 3)$. □

Matrix Representation of Linear Systems

Just as the powers of 10 are normally omitted in decimal numbers and the position of a digit indicates its meaning, the variable names are omitted in solving linear systems. Instead, the coefficients a_{ij} and the constants b_i on the righthand side are arranged in a rectangular array and the position of each number indicates its meaning. Thus the system

$$\begin{aligned} x - 2y &= -1 \\ 2x + y &= 8 \end{aligned}$$

from Example 2 is represented by the array

$$\left[\begin{array}{rr|r} 1 & -2 & -1 \\ 2 & 1 & 8 \end{array}\right].$$

(The dashed vertical line shows where the sides of the equations are separated.) Such a rectangular array is called a *matrix*.

DEFINITION 1.3

A **matrix** is a rectangular array of objects arranged in rows and columns. The objects are the **entries** of the matrix.

In this text, matrix entries will be numbers or symbols representing numbers. When a linear system is represented by a matrix, each row corresponds to an equation. Thus solving a linear system involves operating on the rows of the matrix. For this reason, each operation in Pass 1 of Gaussian elimination (see Theorem 1.1) is called an **elementary row operation** on the corresponding matrix.

Elementary Row Operations

1. Interchange rows i and j.
2. Multiply row i by a nonzero constant c.
3. Add c times row i to row j, where $i \neq j$.

Special notation is used for elementary row operations. For instance,

$$R1 \leftrightarrow R3$$

denotes the interchange of rows 1 and 3,

$$R2 \leftarrow 4R2$$

denotes the multiplication of row 2 by row 4, and

$$R3 \leftarrow R3 + 3R2$$

denotes the addition of 3 times row 2 to row 3.

Example 4

Solve the linear system

$$\begin{aligned} x + \ \ y - z &= -2 \\ x + 2y + z &= \ \ 3 \\ 2x + 3y - z &= -1 \end{aligned}$$

by Gaussian elimination.

Solution

The variable x is eliminated from the second and third equations first. The operations are indicated to the right:

$$\left[\begin{array}{ccc|c} 1 & 1 & -1 & -2 \\ 1 & 2 & 1 & 3 \\ 2 & 3 & -1 & -1 \end{array}\right] \rightarrow \left[\begin{array}{ccc|c} 1 & 1 & -1 & -1 \\ 0 & 1 & 2 & 5 \\ 0 & 1 & 1 & 3 \end{array}\right] \quad \begin{array}{l} R2 \leftarrow R2 - R1 \\ R3 \leftarrow R3 - 2R1. \end{array}$$

Then y is eliminated from the third equation:

$$\left[\begin{array}{rrr|r} 1 & 1 & -1 & -1 \\ 0 & 1 & 2 & 5 \\ 0 & 1 & 1 & 3 \end{array}\right] \rightarrow \left[\begin{array}{rrr|r} 1 & 1 & -1 & -2 \\ 0 & 1 & 2 & 5 \\ 0 & 0 & -1 & -2 \end{array}\right] \qquad R3 \leftarrow R3 - R2.$$

This matrix represents the upper triangular system

$$\begin{aligned} x + y - \ z &= -2 \\ y + 2z &= \ \ 5 \\ -z &= -2. \end{aligned}$$

By Theorem 1.1, this upper triangular system is equivalent to the given system. The solution, $(-1,1,2)$, may be found by back substitution. ☐

Example 5

Solve the linear system

$$\begin{aligned} x_1 - 2x_2 + \ x_3 - \ x_4 &= -2 \\ -2x_1 + 5x_2 - \ x_3 + 4x_4 &= \ \ 1 \\ 3x_1 - 7x_2 + 4x_3 - 4x_4 &= -4 \\ 2x_1 - 3x_2 + 5x_3 - \ x_4 &= -2 \end{aligned}$$

by Gaussian elimination.

Solution

Since $a_{11} \neq 0$, x_1 may be eliminated from equations 2, 3, and 4. The operations are indicated to the right:

$$\left[\begin{array}{rrrr|r} 1 & -2 & 1 & -1 & -2 \\ -2 & 5 & -1 & 4 & 1 \\ 3 & -7 & 4 & -4 & -4 \\ 2 & -3 & 5 & -1 & -2 \end{array}\right]$$

$$\rightarrow \left[\begin{array}{rrrr|r} 1 & -2 & 1 & -1 & -2 \\ 0 & 1 & 1 & 2 & -3 \\ 0 & -1 & 1 & -1 & 2 \\ 0 & 1 & 3 & 1 & 2 \end{array}\right] \qquad \begin{array}{l} R2 \leftarrow R2 + 2R1 \\ R3 \leftarrow R3 - 3R1 \\ R4 \leftarrow R4 - 2R1 \end{array}$$

Now x_2 is eliminated from equations 3 and 4 as follows:

$$\left[\begin{array}{rrrr|r} 1 & -2 & 1 & -1 & -2 \\ 0 & 1 & 1 & 2 & -3 \\ 0 & -1 & 1 & -1 & 2 \\ 0 & 1 & 3 & 1 & 2 \end{array}\right]$$

$$\rightarrow \left[\begin{array}{rrrr|r} 1 & -2 & 1 & -1 & -2 \\ 0 & 1 & 1 & 2 & -3 \\ 0 & 0 & 2 & 1 & -1 \\ 0 & 0 & 2 & -1 & 5 \end{array}\right] \qquad \begin{array}{l} R3 \leftarrow R3 + R2 \\ R4 \leftarrow R4 - R2. \end{array}$$

Finally, eliminate x_3 from equation 4 by adding -1 times row 3 to row 4:

$$\left[\begin{array}{rrrr|r} 1 & -2 & 1 & -1 & -2 \\ 0 & 1 & 1 & 2 & -3 \\ 0 & 0 & 2 & 1 & -1 \\ 0 & 0 & 2 & -1 & 5 \end{array}\right]$$

$$\rightarrow \left[\begin{array}{rrrr|r} 1 & -2 & 1 & -1 & -2 \\ 0 & 1 & 1 & 2 & -3 \\ 0 & 0 & 2 & 1 & -1 \\ 0 & 0 & 0 & -2 & 6 \end{array}\right] \qquad R4 \leftarrow R4 - R3.$$

This matrix represents the upper triangular system

$$\begin{aligned} x_1 - 2x_2 + x_3 - x_4 &= -2 \\ x_2 + x_3 + 2x_4 &= -3 \\ 2x_3 + x_4 &= -1 \\ -2x_4 &= 6. \end{aligned}$$

This system is equivalent to the given system, and its unique solution, $(-2, 2, 1, -3)$, may be found by back substitution. □

DEFINITION 1.4

The matrix representing a linear system is called the **augmented matrix** of the system. The matrix formed from the augmented matrix by deleting the last column is called the **coefficient matrix**.

The procedure used in Examples 2–4 can now be more easily described.

Gaussian Elimination, Version 1.0

To solve a linear system of n equations in n variables, carry out the following procedure:

1. Form an equivalent upper triangular system through a sequence of elementary row operations (Pass 1).
2. If the first nonzero entry in row i appears in column i for $i = 1, \ldots, n$, then the original system has a unique solution. Find this solution by applying back substitution to the upper triangular system (Pass 2).

Linear systems with multiple solutions or no solutions are studied in Sections 1.3 and 1.4. In these studies, versions of Gaussian elimination are given for other types of linear systems.

Gauss-Jordan* Elimination

Instead of solving the upper triangular system by back substitution, we may continue to apply elementary row operations until each diagonal entry (at row i and column i) equals 1 and each remaining entry in the first n columns equals 0. When this form is reached, the solution appears in the final column. This procedure, known as **Gauss-Jordan elimination**, is illustrated in the following example.

Example 6

Solve the linear system

$$\begin{aligned} x + 2y - z &= 4 \\ -3x + 2y + z &= 2 \\ 3x + y - 2z &= 3 \end{aligned}$$

by Gauss-Jordan elimination.

* Camille Jordan (1838–1922) is known for a celebrated result in geometry, the famous "Jordan curve theorem" as well as for his contributions to linear algebra.

Solution

We begin as usual by finding an upper triangular form equivalent to the given system:

$$\left[\begin{array}{rrr|r} 1 & 2 & -1 & 4 \\ -3 & 2 & 1 & 2 \\ 3 & 1 & -2 & 3 \end{array}\right] \rightarrow \left[\begin{array}{rrr|r} 1 & 2 & -1 & 4 \\ 0 & 8 & -2 & 14 \\ 0 & -5 & 1 & -9 \end{array}\right] \qquad \begin{array}{l} R2 \leftarrow R2 + 3R1 \\ R3 \leftarrow R3 - 3R1 \end{array}$$

$$\rightarrow \left[\begin{array}{rrr|r} 1 & 2 & -1 & 4 \\ 0 & 4 & -1 & 7 \\ 0 & -20 & 4 & -36 \end{array}\right] \qquad \begin{array}{l} R2 \leftarrow (\tfrac{1}{2})R2 \\ R3 \leftarrow 4R3 \end{array}$$

$$\rightarrow \left[\begin{array}{rrr|r} 1 & 2 & -1 & 4 \\ 0 & 4 & -1 & 7 \\ 0 & 0 & -1 & -1 \end{array}\right] \qquad R3 \leftarrow R3 + 5R2.$$

Instead of applying back substitution, reduce the first nonzero entry in each row to 1, called the **leading 1** of that row, and eliminate the nonzero entries above these *leading 1's*:

$$\left[\begin{array}{rrr|r} 1 & 2 & -1 & 4 \\ 0 & 4 & -1 & 7 \\ 0 & 0 & -1 & -1 \end{array}\right] \rightarrow \left[\begin{array}{rrr|r} 1 & 2 & -1 & 4 \\ 0 & 4 & -1 & 7 \\ 0 & 0 & 1 & 1 \end{array}\right] \qquad R3 \leftarrow -R3$$

$$\rightarrow \left[\begin{array}{rrr|r} 1 & 2 & 0 & 5 \\ 0 & 4 & 0 & 8 \\ 0 & 0 & 1 & 1 \end{array}\right] \qquad \begin{array}{l} R1 \leftarrow R1 + R3 \\ R2 \leftarrow R2 + R3 \end{array}$$

$$\rightarrow \left[\begin{array}{rrr|r} 1 & 2 & 0 & 5 \\ 0 & 1 & 0 & 2 \\ 0 & 0 & 1 & 1 \end{array}\right] \qquad R2 \leftarrow (\tfrac{1}{4})R2$$

$$\rightarrow \left[\begin{array}{rrr|r} 1 & 0 & 0 & 1 \\ 0 & 1 & 0 & 2 \\ 0 & 0 & 1 & 1 \end{array}\right]. \qquad R1 \leftarrow R1 - 2R2$$

The solution (1, 2, 1) may be read from the final matrix. ☐

The Gauss-Jordan procedure is rarely used in practice because Gaussian elimination is more efficient. Why, then, should we be interested in Gauss-Jordan elimination? One reason lies in the *form* of the final matrix: *the first nonzero entry in each row is a 1, and this 1 is the only nonzero entry in its column.* We shall use this form to communicate important ideas concerning linear systems.

REVIEW CHECKLIST

1. Find the coefficient matrix and the augmented matrix of a linear system.

2. Identify and describe the three types of elementary row operations, and use the symbols for them.
3. Solve a given linear system by applying Gaussian elimination to the augmented matrix of the system.
4. Describe Gaussian elimination in general terms, clearly identifying the two major phases.
5. Describe the form of the final augmented matrix produced in the first phase of Gaussian elimination for a square system with a unique solution, using double subscripts.
6. Formulate a definition of equivalent linear systems and a statement of the basic theorem on equivalent systems.
7. State why the original augmented matrix and the final augmented matrix produced in the first phase of Gaussian elimination represent equivalent linear systems.
8. Solve a given linear system with a unique solution by applying Gauss-Jordan elimination to the augmented matrix.

EXERCISES

In Exercises 1–4, find the augmented matrix of the given system.

1. $\begin{aligned} x - 2y &= 4 \\ 2x + y &= -2 \end{aligned}$

2. $\begin{aligned} 2x - 3y &= 0 \\ x + y &= 0 \end{aligned}$

3. $\begin{aligned} x - 3y + z &= 2 \\ 2x \quad\quad - z &= 1 \\ 3x + y - 2z &= 3 \end{aligned}$

4. $\begin{aligned} x + 2y \quad\quad &= 0 \\ 3x + y - z &= 0 \\ 5x + 3y + 2z &= 0 \end{aligned}$

In Exercises 5–8, find the linear system represented by the given matrix.

5. $\left[\begin{array}{rr|r} 2 & 4 & 0 \\ 1 & -2 & 5 \end{array}\right]$

6. $\left[\begin{array}{rr|r} 1 & 2 & 0 \\ -3 & 4 & 0 \end{array}\right]$

7. $\left[\begin{array}{rrr|r} 1 & -2 & 4 & 4 \\ 2 & 3 & -1 & 0 \\ -2 & 0 & 5 & -1 \end{array}\right]$

8. $\left[\begin{array}{rrr|r} 1 & 0 & 2 & 0 \\ 3 & -1 & -1 & 0 \\ 1 & 2 & 2 & 0 \end{array}\right]$

In Exercises 9–23, solve the given linear systems by Gaussian elimination.

9. $\begin{aligned} x - 2y &= 1 \\ -x + 3y &= 0 \end{aligned}$

10. $\begin{aligned} x + 2y &= 3 \\ x - y &= -3 \end{aligned}$

11. $\begin{aligned} x - y &= 0 \\ 2x + y &= 6 \end{aligned}$

12. $\begin{aligned} 2x + 3y &= -1 \\ 4x - y &= 5 \end{aligned}$

13. $\begin{aligned} 2x + 3y &= 0 \\ 4x - y &= 0 \end{aligned}$

14. $\begin{aligned} 2x - y &= 4 \\ x + y &= -1 \end{aligned}$

15. $\begin{aligned} 1.2x - 0.8y &= 2.0 \\ -1.5x + 0.25y &= -4.0 \end{aligned}$

16. $\begin{aligned} 1.5x + 2.3y &= 4 \\ 0.7x - 4.1y &= 5.2 \end{aligned}$

17. $\begin{aligned} x + y + z &= 1 \\ x + 2y + 3z &= -1 \\ x + 4y + 9z &= -9 \end{aligned}$

18. $\begin{aligned} x + 2y - z &= 4 \\ -x - y + 2z &= -1 \\ 2x + y + 3z &= 7 \end{aligned}$

19.
$$\begin{aligned} x + y + 4z &= -4 \\ x + 3y - 2z &= 8 \\ 2x - y + 5z &= -2 \end{aligned}$$

20.
$$\begin{aligned} 2x - y + z &= 8 \\ x + y + 2z &= 4 \\ 3x - 3y - z &= 10 \end{aligned}$$

21.
$$\begin{aligned} x_1 - x_2 + x_3 - x_4 &= 2 \\ x_1 - x_3 + 2x_4 &= 0 \\ -x_1 + 2x_2 - 2x_3 + 7x_4 &= -7 \\ 2x_1 - x_2 - x_3 &= 3 \end{aligned}$$

22.
$$\begin{aligned} x_1 - 2x_2 + x_3 - x_4 &= 2 \\ -x_1 + 4x_2 - 2x_3 + 3x_4 &= -1 \\ 2x_1 + x_3 + 3x_4 &= 9 \\ -2x_1 + 6x_2 + 5x_4 &= 2 \end{aligned}$$

23.
$$\begin{aligned} x_1 - x_2 + 2x_3 + 2x_4 + x_5 &= 3 \\ 2x_1 + x_2 + 5x_3 + 2x_4 + 2x_5 &= 6 \\ -x_1 + 4x_2 - 6x_4 + x_5 &= -2 \\ -2x_1 - 4x_2 - 4x_3 - x_4 + x_5 &= -3 \\ 2x_1 + 4x_2 + 4x_3 + 7x_4 - x_5 &= 9 \end{aligned}$$

24–38. Solve the systems in Exercises 9–23 by Gauss-Jordan elimination.

In Exercises 39–42, find the value(s) of a for which the given linear system does not have a unique solution.

39.
$$\begin{aligned} x - 2y &= 5 \\ 3x + ay &= 1 \end{aligned}$$

40.
$$\begin{aligned} ax + 3y &= 1 \\ x - 9y &= 3 \end{aligned}$$

41.
$$\begin{aligned} x - y + 2z &= 3 \\ 2x + ay + 3z &= 1 \\ -3x + 3y + z &= 4 \end{aligned}$$

42.
$$\begin{aligned} ax + y - z &= 0 \\ x - 2y + 3z &= 1 \\ 2x - y + 2z &= 1 \end{aligned}$$

In Exercises 43–44, find the value(s) of a for which the given systems are equivalent.

43.
$$\begin{aligned} x + 2y &= 1 \\ 2x + 5y &= 1 \end{aligned} \qquad \begin{aligned} x + ay &= 4 \\ -x + 2y &= -5 \end{aligned}$$

44.
$$\begin{aligned} x + 2y - z &= 1 \\ 2x - y + z &= 6 \\ -2x + 2y + z &= 1 \end{aligned} \qquad \begin{aligned} x - y + z &= 4 \\ x + y - z &= 0 \\ 2x + y + az &= 2 \end{aligned}$$

In Exercises 45–48, solve the given system by Gaussian elimination, expressing the solution in terms of a, b, and c.

45.
$$\begin{aligned} x + 3y &= a \\ 2x + 2y &= b \end{aligned}$$

46.
$$\begin{aligned} 2x + y &= a \\ x + y &= b \end{aligned}$$

47.
$$\begin{aligned} x + y - z &= a \\ x + 2y - 2z &= b \\ 2x - y + 2z &= c \end{aligned}$$

48.
$$\begin{aligned} x - y &= a \\ -x + 2y - z &= b \\ 3x + 2z &= c \end{aligned}$$

49. If the coordinate axes in the Cartesian plane are rotated 60° counterclockwise, the coordinates of a point (x, y) are given in the $x'y'$ system by

$$\begin{aligned} x' &= (1/2)x + (\sqrt{3}/2)y \\ y' &= -(\sqrt{3}/2)x + (1/2)y. \end{aligned}$$

Use Gaussian elimination to solve this system for x and y in terms of x' and y'.

50. Solve the general equations of rotation

$$x' = (\cos\theta)x + (\sin\theta)y$$
$$y' = -(\sin\theta)x + (\cos\theta)y$$

for x and y by Gaussian elimination.

1.3 HOMOGENEOUS LINEAR SYSTEMS

Homogeneity is used extensively in the theory and the application of linear systems. In Chapters 4 and 5, we use this concept to study *linear dependence* and *independence*, *basis*, and *dimension*, all of which lie at the core of linear algebra. In Chapter 7, we use homogeneous systems to solve *eigenvalue* problems, which arise in many follow-on studies in theoretical and applied mathematics.

Homogeneous Systems and Nontrivial Solutions

For constants a and b, not both zero, the equation

$$ax + by = 0$$

determines a line through the origin in the plane. Similarly, for a, b, and c not all zero, the equation

$$ax + by + cz = 0$$

determines a plane through the origin in space. (Appendix B contains a brief review of lines and planes.) These equations illustrate the following general form for the ith equation in a linear system:

$$a_{i1}x_1 + \cdots + a_{in}x_n = 0.$$

DEFINITION 1.5

A linear system is said to be **homogeneous** if the constant on the right equals zero in every equation, otherwise the system is **nonhomogeneous**.

We interpret this definition geometrically for systems in two or three variables. In the plane, since any line determined by an equation of the form

$$ax + by = 0$$

passes through $(0, 0)$, any two such lines may intersect at the origin or they may coincide, but they cannot be parallel. The possible configurations for two lines in the plane are shown in Fig. 1.8.

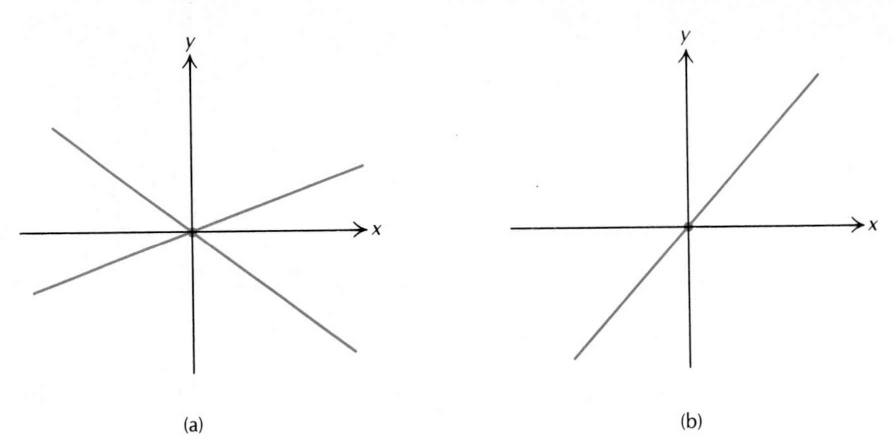

Figure 1.8 The possibilities for the solution of 2×2 homogeneous linear systems: (a) the only solution is $(0, 0)$; (b) the given lines coincide in a line of solutions.

In space, since the plane determined by an equation of the form

$$ax + by + cz = 0$$

passes through $(0, 0, 0)$, any three such planes may intersect at the origin or in a line, or they may coincide, but no two may be parallel. The possible configurations for three lines in space are shown in Fig. 1.9.

For any homogeneous system in n variables, the n-tuple $(0, 0, \ldots, 0)$ is called the **trivial solution**. As Figures 1.8 and 1.9 suggest, this may be the only solution of a given homogeneous system. One problem is to determine whether *nontrivial* solutions exist and another is to find them when they do.

Example 1

Solve the system

$$\begin{aligned} x - 2y + 2z &= 0 \\ 4x - 7y + 3z &= 0 \\ 2x - y + 2z &= 0 \end{aligned}$$

using Gaussian elimination.

Solution

Since every entry in the last column of the augmented matrix equals 0, only the coefficient matrix is needed.

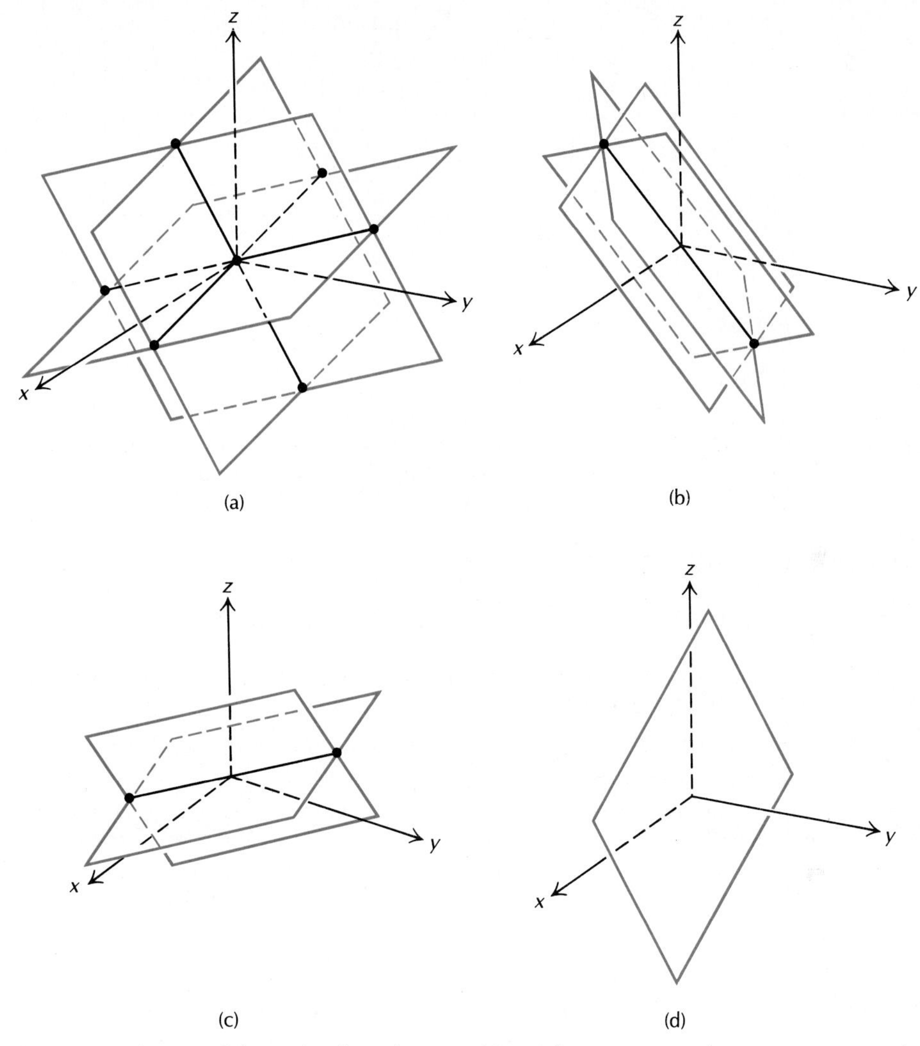

Figure 1.9 The possibilities for the solution of 3×3 homogeneous linear systems: (a) the only solution is $(0, 0, 0)$; (b) three distinct planes intersect in a line of solutions; (c) two planes coincide and intersect the third in a line of solutions; (d) the given planes coincide in a plane of solutions.

$$\begin{bmatrix} 1 & -2 & 2 \\ 4 & -7 & 3 \\ 2 & -1 & 2 \end{bmatrix} \rightarrow \begin{bmatrix} 1 & -2 & 2 \\ 0 & 1 & -5 \\ 0 & 3 & -2 \end{bmatrix} \quad \begin{matrix} \\ R2 \leftarrow R2 - 4R1 \\ R3 \leftarrow R3 - 2R1 \end{matrix}$$

$$\rightarrow \begin{bmatrix} 1 & -2 & 2 \\ 0 & 1 & -5 \\ 0 & 0 & 13 \end{bmatrix} \quad \begin{matrix} \\ \\ R3 \leftarrow R3 - 3R2. \end{matrix}$$

This matrix represents the equivalent upper triangular system

$$\begin{aligned} x - 2y + 2z &= 0 \\ y - 5z &= 0 \\ 13z &= 0, \end{aligned}$$

which has only the trivial solution $(0,0,0)$. $\square$

Example 2

Solve the homogeneous linear system

$$\begin{aligned} x + 2y - z &= 0 \\ x + 3y + 2z &= 0 \\ 3x + 8y + 3z &= 0 \end{aligned}$$

by Gaussian elimination.

Solution

With annotations for the elementary row operations omitted, the first pass of the elimination is given by

$$\begin{bmatrix} 1 & 2 & -1 \\ 1 & 3 & 2 \\ 3 & 8 & 3 \end{bmatrix} \to \begin{bmatrix} 1 & 2 & -1 \\ 0 & 1 & 3 \\ 0 & 2 & 6 \end{bmatrix} \to \begin{bmatrix} 1 & 2 & -1 \\ 0 & 1 & 3 \\ 0 & 0 & 0 \end{bmatrix}.$$

The final matrix represents the equivalent upper triangular system

$$\begin{aligned} x + 2y - z &= 0 \\ y + 3z &= 0. \end{aligned}$$

For any value of z, the second pass, back substitution, gives: $y = -3z$, and then $x = z - 2(-3z) = 7z$. The solution consists of all points of the form $(7z, -3z, z)$, which is the line through the origin and $(7, -3, 1)$. $\square$

Remarks on Example 2

1. We may solve the equation

$$y + 3z = 0$$

for z instead of y. This choice gives $z = -y/3$ and then $x = (-y/3) - 2y = -7y/3$. The solution of the given system is then described as the set of all points of the form $(-7y/3, y, -y/3)$. The solution is the same—only its description is changed.

2. A new arbitrary constant, such as t, may be used for the solution. If we let $z = t$ and solve for $y = -3t$ and $x = 7t$, the resulting form is $(7t, -3t, t)$. Again, the solution consists of the same points in space.
3. We described the solution as a "line" in space.

Appendix B contains background information in case an explanation is needed.

Example 3

Solve the homogeneous linear system

$$x + 3y - 3z = 0$$

$$x + 3y - 2z = 0$$

$$2x + 6y - 3z = 0$$

by Gaussian elimination.

Solution

The first pass results are:

$$\begin{bmatrix} 1 & 3 & -3 \\ 1 & 3 & -2 \\ 2 & 6 & -3 \end{bmatrix} \to \begin{bmatrix} 1 & 3 & -3 \\ 0 & 0 & 1 \\ 0 & 0 & 3 \end{bmatrix} \to \begin{bmatrix} 1 & 3 & -3 \\ 0 & 0 & 1 \\ 0 & 0 & 0 \end{bmatrix}.$$

This matrix represents the equivalent upper triangular system

$$x + 3y - 3z = 0$$

$$z = 0.$$

From the second equation, $z = 0$. Then for any value of y, the first equation yields $x = -3y$. The solution of the given system consists of all points $(-3y, y, 0)$, which is the line through the origin and $(-3, 1, 0)$. ☐

Example 4

Solve the homogeneous linear system

$$\begin{aligned} x + 4y - 3z &= 0 \\ 4x + 16y - 12z &= 0 \\ -3x - 12y + 9z &= 0 \end{aligned}$$

by Gaussian elimination.

Solution

The coefficient matrix is completely simplified in one step:

$$\begin{bmatrix} 1 & 4 & -3 \\ 4 & 16 & -12 \\ -3 & -12 & 9 \end{bmatrix} \to \begin{bmatrix} 1 & 4 & -3 \\ 0 & 0 & 0 \\ 0 & 0 & 0 \end{bmatrix}.$$

Only one equation remains,

$$x + 4y - 3z = 0,$$

and it determines the plane of solutions. An explicit form for any solution may be found as in Examples 2 and 3. For any z and y, solve for x to find

$$x = 3z - 4y.$$

Thus a solution is a point of the form $(3z - 4y, y, z)$. We may also describe this plane using three of its points. If $y = 0$ and $z = 1$, then $x = 3$. For $y = 1$ and $z = 0$, we have $x = -4$. The solution set is the plane through the points $(3, 0, 1)$, $(-4, 1, 0)$, and $(0, 0, 0)$. (Again, see Appendix B.) □

Row Echelon Form and the General Solution

Examples 1–4 illustrate the forms that result from the first pass of Gaussian elimination for 3×3 homogeneous systems. These forms are shown in Fig. 1.10.

In any matrix, a row in which every entry equals 0 is called a **zero row**. A row with at least one entry other than 0 is called a **nonzero row**. For the forms in Fig. 1.10, the first nonzero entry in any row has been set equal 1. This *normalization* corresponds to dividing both sides of an equation by a nonzero number. The following is a general description of these forms.

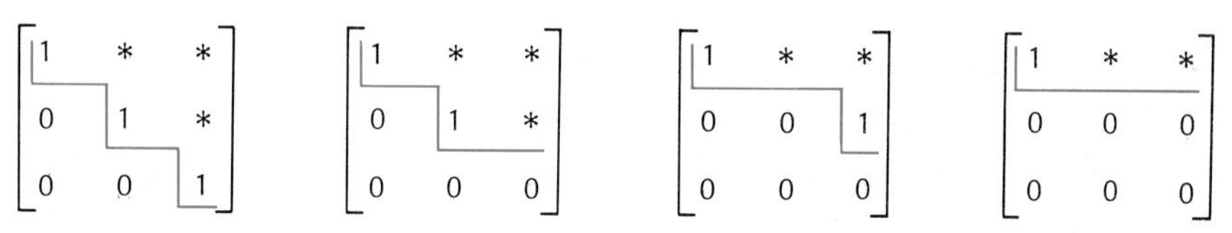

Figure 1.10 Possible forms of a 3×3 coefficient matrix ready for back substitution.

DEFINITION 1.6

A matrix is in **row echelon form** if:

1. every zero row appears below all nonzero rows;
2. the first nonzero entry in any row is a 1, the leading 1 of that row;
3. the leading 1 in row j is to the right of and below the leading 1 in row i whenever $i < j$.

A row echelon matrix is in **reduced row echelon form** if:

4. each leading 1 is the only nonzero entry in its column. (This is the form produced in Gauss-Jordan elimination.)

Example 5

Determine whether the following matrices are in row echelon, reduced row echelon form, or neither of these forms.

$$\begin{bmatrix} 1 & 1 & 2 & 2 & 0 \\ 0 & 0 & 1 & 0 & 1 \\ 0 & 0 & 0 & 1 & 3 \end{bmatrix}, \quad \begin{bmatrix} 1 & 1 & 0 & 0 & 0 \\ 0 & 0 & 0 & 1 & 0 \\ 0 & 0 & 0 & 0 & 0 \end{bmatrix},$$

$$\begin{bmatrix} 1 & 1 & 0 & 0 & 0 \\ 0 & 0 & 1 & 0 & 1 \\ 0 & 0 & 0 & 1 & 3 \end{bmatrix}, \quad \begin{bmatrix} 1 & 0 & 3 & 0 & 1 \\ 0 & 0 & 0 & 1 & 2 \\ 0 & 1 & 2 & 0 & 1 \end{bmatrix}.$$

Solution

The first three of these matrices are in row echelon form since they satisfy conditions (1)–(3) of Definition 1.6. The second and third matrices also satisfy condition (4), and hence are in reduced row echelon form. The fourth matrix is not in an echelon form, since it does not satisfy condition (3). ☐

Using echelon matrices, we can update the description of Gaussian elimination in Section 1.2 to include homogeneous systems.

Gaussian Elimination, Version 2.0

Begin with an $n \times n$ linear system represented by the coefficient matrix for a homogeneous system or by the augmented matrix for a nonhomogeneous system.

1. Form an equivalent row echelon system using elementary row operations (Pass 1).
2. Solve the row echelon system by back substitution (Pass 2). If the row echelon matrix has no zero rows, the solution is unique. If the system is homogeneous and the row echelon matrix has at least one zero row, each nonzero equation is solved for the variable of the leading 1, called the **leading variable**, in terms of the remaining variables.

Examples 1–4 illustrate Gaussian elimination. In Gauss-Jordan elimination, we continue on to the reduced row echelon matrix instead of carrying out a back substitution.

Example 6

Solve the system

$$\begin{aligned} x_1 + x_2 - x_3 - 3x_4 - x_5 &= 0 \\ 2x_1 + 3x_2 - 5x_3 - 4x_4 + x_5 &= 0 \\ x_1 \qquad + 2x_3 - 4x_4 - 3x_5 &= 0 \\ -x_1 - 3x_2 + 7x_3 + 2x_4 - 2x_5 &= 0 \\ x_1 + 2x_2 - 4x_3 \qquad + 3x_5 &= 0 \end{aligned}$$

by Gauss-Jordan elimination.

Solution

We apply elementary row operations to the coefficient matrix:

$$\begin{bmatrix} 1 & 1 & -1 & -3 & -1 \\ 2 & 3 & -5 & -4 & 1 \\ 1 & 0 & 2 & -4 & -3 \\ -1 & -3 & 7 & 2 & -2 \\ 1 & 2 & -4 & 0 & 3 \end{bmatrix}$$

$$\rightarrow \begin{bmatrix} 1 & 1 & -1 & -3 & -1 \\ 0 & 1 & -3 & 2 & 3 \\ 0 & -1 & 3 & -1 & -2 \\ 0 & -2 & 6 & -1 & -3 \\ 0 & 1 & -3 & 3 & 4 \end{bmatrix} \quad \begin{array}{l} \\ R2 \leftarrow R2 - 2R1 \\ R3 \leftarrow R3 - R1 \\ R4 \leftarrow R4 + R1 \\ R5 \leftarrow R5 - R1 \end{array}$$

$$\rightarrow \begin{bmatrix} 1 & 1 & -1 & -3 & -1 \\ 0 & 1 & -3 & 2 & 3 \\ 0 & 0 & 0 & 1 & 1 \\ 0 & 0 & 0 & 3 & 3 \\ 0 & 0 & 0 & 1 & 1 \end{bmatrix} \quad \begin{array}{l} \\ \\ R3 \leftarrow R3 + R2 \\ R4 \leftarrow R4 + 2R2 \\ R5 \leftarrow R5 - R2 \end{array}$$

$$\rightarrow \begin{bmatrix} 1 & 1 & -1 & 0 & 2 \\ 0 & 1 & -3 & 0 & 1 \\ 0 & 0 & 0 & 1 & 1 \\ 0 & 0 & 0 & 0 & 0 \\ 0 & 0 & 0 & 0 & 0 \end{bmatrix} \quad \begin{array}{l} R1 \leftarrow R1 + 3R3 \\ R2 \leftarrow R2 - 2R3 \\ \\ R4 \leftarrow R4 - 3R3 \\ R5 \leftarrow R5 - R3 \end{array}$$

$$\rightarrow \begin{bmatrix} 1 & 0 & 2 & 0 & 1 \\ 0 & 1 & -3 & 0 & 1 \\ 0 & 0 & 0 & 1 & 1 \\ 0 & 0 & 0 & 0 & 0 \\ 0 & 0 & 0 & 0 & 0 \end{bmatrix} \quad \begin{array}{l} R1 \leftarrow R1 - R2 \\ \\ \\ \\ \\ \end{array}$$

Now solve, in order, for x_4, x_2, and x_1 in terms of x_5 and x_3:

$$x_4 = -x_5, \qquad x_2 = 3x_3 - x_5, \qquad x_1 = -2x_3 - x_5.$$

The solution consists of a points of the form

$$(-2x_3 - x_5, 3x_3 - x_5, x_3, -x_5, x_5). \quad \square$$

As illustrated in Example 6, a *reduced* row echelon matrix is used for the back substitution in Gauss-Jordan elimination. This form is a *characteristic* of the given system. That is, for a given linear system there is only one equivalent reduced row echelon system with the variables arranged in the original order. We shall not prove this assertion, but we illustrate it in the following example.

Example 7

Starting with the same matrix, the reductions

$$\begin{bmatrix} 1 & 2 & 1 \\ -1 & 3 & 4 \end{bmatrix} \rightarrow \begin{bmatrix} 1 & 2 & 1 \\ 0 & 5 & 5 \end{bmatrix} \rightarrow \begin{bmatrix} 1 & 2 & 1 \\ 0 & 1 & 1 \end{bmatrix}$$

and

$$\begin{bmatrix} 1 & 2 & 1 \\ -1 & 3 & 4 \end{bmatrix} \to \begin{bmatrix} -1 & 3 & 4 \\ 1 & 2 & 1 \end{bmatrix} \to \begin{bmatrix} -1 & 3 & 4 \\ 0 & 5 & 5 \end{bmatrix} \to \begin{bmatrix} 1 & -3 & -4 \\ 0 & 1 & 1 \end{bmatrix}$$

yield *distinct* row echelon matrices. In either case, continuing to reduced row echelon form yields the matrix

$$\begin{bmatrix} 1 & 0 & -1 \\ 0 & 1 & 1 \end{bmatrix}.$$

If we started with the coefficient matrix of a homogeneous system in x, y, and z, the solution $(z, -z, z)$ may be obtained from any of these echelon forms. If we started with the augmented matrix of a nonhomogeneous system in x and y, the solution $(-1, 1)$ may be obtained from any of the same echelon forms. □

We can formulate some important results based on these examples. The ideas are presented now to facilitate the immediate study of homogeneous systems. A justification will follow from the study of vector spaces in Chapters 4 and 5.

An $n \times n$ homogeneous linear system may have *only the trivial solution.* If so, any row echelon matrix equivalent to the coefficient matrix has a "staircase" form: to go from one leading 1 to the next, move down one row and right one column. In this case, no zero rows appear in any echelon form. In fact, the bottom row has a 1 in column n as its only nonzero entry. The general *row echelon form* and the *reduced row echelon form* are, respectively,

$$\begin{bmatrix} 1 & a_{12} & a_{13} & \cdots & a_{1n} \\ 0 & 1 & a_{23} & \cdots & a_{2n} \\ 0 & 0 & 1 & \cdots & a_{3n} \\ \vdots & \vdots & \vdots & & \vdots \\ 0 & 0 & 0 & \cdots & 1 \end{bmatrix} \quad \text{and} \quad \begin{bmatrix} 1 & 0 & 0 & \cdots & 0 \\ 0 & 1 & 0 & \cdots & 0 \\ 0 & 0 & 1 & \cdots & 0 \\ \vdots & \vdots & \vdots & & \vdots \\ 0 & 0 & 0 & \cdots & 1 \end{bmatrix}.$$

If an $n \times n$ homogeneous linear system has nontrivial solutions, then at least one zero row appears at the bottom of any equivalent row echelon matrix. The number of *nonzero* rows, r, is the same for any equivalent row echelon matrix. The number r is a characteristic of the given system, and it has a special name. This concept is defined for any matrix, whether it is regarded as a coefficient matrix or as an augmented matrix.

DEFINITION 1.7

The **row rank** of a matrix is the number of nonzero rows in any equivalent row echelon matrix.

For homogeneous linear systems, the significance of the row rank of the coefficient matrix is given in the following summary.

Solution Sets of Homogeneous Linear Systems

Let n be the number of variables in a homogeneous system and r be the row rank of the coefficient matrix. If $r = n$, the system has only the trivial solution. If $r < n$, then we may solve the r nonzero equations in any equivalent row echelon system for the leading variables in terms of the remaining $n - r$ variables. These $n - r$ variables then become arbitrary constants, called **parameters**, for the solution set. The solution of the given system consists of all n-tuples of the resulting parametric form.

If a given homogeneous system has m equations and n variables, where $m < n$, then the row rank r *must* be less than n, since $r \leq m < n$, hence the system has nontrivial solutions. Such systems are included routinely in the exercises.

We close this section with an instructive geometric illustration concerning row rank.

Example 8

Determine whether the points $(1, 2, 1)$, $(-2, 4, 1)$, $(8, -8, -1)$ lie in a single plane through the origin.

Solution

The equation of a plane through the origin in space has the form

$$ax + by + cz = 0,$$

where a, b, and c are not all zero. The given points all lie in one such plane if and only if the coordinates of each satisfy a single equation of this form, i.e., if and only if the homogeneous system

$$\begin{aligned} a + 2b + c &= 0 \\ -2a + 4b + c &= 0 \\ 8a - 8b - c &= 0 \end{aligned}$$

has a nontrivial solution. We need only bring the coefficient matrix to row echelon form:

$$\begin{bmatrix} 1 & 2 & 1 \\ -2 & 4 & 1 \\ 8 & -8 & -1 \end{bmatrix} \to \begin{bmatrix} 1 & 2 & 1 \\ 0 & 8 & 3 \\ 0 & -24 & -9 \end{bmatrix} \to \begin{bmatrix} 1 & 2 & 1 \\ 0 & 8 & 3 \\ 0 & 0 & 0 \end{bmatrix}.$$

Since the row rank is 2 and the number of unknowns is 3, the linear system for a, b, and c has a nontrivial solution, so the given points *do* lie in a single plane through the origin. In fact, there is exactly one such plane, and its equation,

$$2x + 3y - 8z = 0,$$

may be obtained by solving the row echelon equivalent of the original system. You may check to verify that the given points all lie on this plane. □

REVIEW CHECKLIST

1. Solve homogeneous linear systems by Gaussian elimination or by Gauss-Jordan elimination.
2. Define a row echelon matrix and a reduced row echelon matrix, and determine whether a given matrix is in either of these forms.
3. Define the row rank of a matrix, and find the row rank of a given matrix.
4. Formulate a description of Gaussian elimination to include homogeneous systems and systems with unique solutions.
5. Describe general solution characteristics for a homogeneous system in terms of the row rank of the coefficient matrix, the leading variables in a row echelon form, and parameters for the solution.

EXERCISES

In Exercises 1–20, solve the given linear system by Gaussian elimination.

1. $\begin{aligned} x - 4y &= 0 \\ 2x - 8y &= 0 \end{aligned}$

2. $\begin{aligned} x - 3y &= 0 \\ 2x + 4y &= 0 \end{aligned}$

3. $\begin{aligned} 3x + 2y &= 0 \\ 6x + 4y &= 0 \end{aligned}$

4. $\begin{aligned} 2x + 3y - z &= 0 \\ 10x + 15y - 5z &= 0 \end{aligned}$

5. $\begin{aligned} x_1 + 2x_2 - 3x_3 + 3x_4 &= 0 \\ 4x_1 + 8x_2 - 12x_3 + 12x_4 &= 0 \end{aligned}$

6. $\begin{aligned} x - 3y + 2z &= 0 \\ 2x - 4y - 4z &= 0 \end{aligned}$

7. $\begin{aligned} 3x - y - z &= 0 \\ x + 2y + z &= 0 \end{aligned}$

8. $\begin{aligned} x - 3y + 2z &= 0 \\ x + y + z &= 0 \\ 5x - 3y + 7z &= 0 \end{aligned}$

9. $\begin{aligned} 2x - y + 3z &= 0 \\ 4x - 2y + 6z &= 0 \\ 6x - 2y + 7z &= 0 \end{aligned}$

10. $\begin{aligned} x + 2y + 2z &= 0 \\ 2x + 3y + z &= 0 \\ x \qquad - 4z &= 0 \end{aligned}$

11. $\begin{aligned} x + 3y + 4z &= 0 \\ 2x + 7y + 7z &= 0 \\ 2x + 5y + 6z &= 0 \end{aligned}$

12. $\begin{aligned} x - 3y + 3z &= 0 \\ 2x - 8y + 7z &= 0 \\ 3x - 5y + 7z &= 0 \end{aligned}$

13. $\begin{aligned} 3x - 2y + z &= 0 \\ x + 4y + 2z &= 0 \\ 7x \qquad + 4z &= 0 \end{aligned}$

14. $\begin{aligned} x_1 - 2x_2 + 2x_3 + x_4 &= 0 \\ 2x_1 - 5x_2 + 2x_3 + 3x_4 &= 0 \\ -3x_1 + 7x_2 - 4x_3 + 2x_4 &= 0 \end{aligned}$

15. $\begin{aligned} 2x_1 + x_2 - x_3 - 2x_4 &= 0 \\ 3x_1 - 2x_2 - 2x_3 + x_4 &= 0 \\ 5x_1 + 6x_2 - 2x_3 - 9x_4 &= 0 \end{aligned}$

16. $\begin{aligned} x_1 + 2x_2 + x_3 - 3x_4 &= 0 \\ 2x_1 + 5x_2 + x_3 - 3x_4 &= 0 \\ -x_1 + x_2 - 3x_3 + 10x_4 &= 0 \\ 3x_1 + 8x_2 + x_3 - 3x_4 &= 0 \end{aligned}$

17. $\begin{aligned} x_1 + 3x_2 + x_3 + 2x_4 &= 0 \\ x_1 + 2x_2 + 4x_3 \qquad &= 0 \\ x_1 + 5x_2 - 5x_3 + 6x_4 &= 0 \\ 2x_1 + 3x_2 + 11x_3 - 4x_4 &= 0 \end{aligned}$

18. $\begin{aligned} x_1 + 2x_2 - x_3 + x_4 &= 0 \\ 4x_1 + 8x_2 - 4x_3 + 4x_4 &= 0 \\ 3x_1 + 6x_2 - 3x_3 + 3x_4 &= 0 \\ 2x_1 + 4x_2 - 2x_3 + 2x_4 &= 0 \end{aligned}$

19. $\begin{aligned} x_1 \qquad + x_3 + 2x_4 &= 0 \\ 3x_1 + x_2 + x_3 + 7x_4 &= 0 \\ 2x_1 - x_2 + 4x_3 + 6x_4 &= 0 \\ x_1 + 2x_2 - 2x_3 + 8x_4 &= 0 \end{aligned}$

20. $\begin{aligned} x_1 \qquad + 2x_3 + x_4 + x_5 &= 0 \\ x_2 - x_3 - 3x_4 - 4x_5 &= 0 \\ 2x_1 + x_2 + 3x_3 - x_4 - 2x_5 &= 0 \\ 2x_1 + 3x_2 + 5x_3 - 4x_4 - 6x_5 &= 0 \\ 2x_1 - x_2 + x_3 + 2x_4 + 2x_5 &= 0 \end{aligned}$

21–40. Solve the systems in Exercises 1–20 by Gauss-Jordan elimination.

In Exercises 41–42, determine whether the intersection of the planes determined by the given equations is a point or a line without solving the system.

41. $\begin{aligned} x + 2y - z &= 0 \\ 3x + 3y - 2z &= 0 \\ 2x + 2y - 3z &= 0 \end{aligned}$

42. $\begin{aligned} 2x - y + 3z &= 0 \\ 3x + 2y + z &= 0 \\ 4x - 3y + 2z &= 0 \end{aligned}$

In Exercises 43–44, determine if the given points lie on a single plane through the origin.

43. $(1,1,2), (-3,3,4), (-2,12,16)$

44. $(-1,1,-4), (3,5,2), (1,15,-10)$

In Exercises 45–48, find the row rank of the given matrix.

45. $\begin{bmatrix} 1 & -2 & 1 \\ 3 & 1 & 4 \\ 2 & -9 & 1 \end{bmatrix}$

46. $\begin{bmatrix} 1 & 4 & -1 & 2 \\ 3 & 1 & 2 & 4 \\ 5 & -5 & 10 & 8 \end{bmatrix}$

47. $\begin{bmatrix} 1 & -2 & 2 & 0 \\ 4 & -1 & 5 & 4 \\ 2 & 3 & 1 & 4 \\ 0 & 7 & -3 & 4 \end{bmatrix}$

48. $\begin{bmatrix} 1 & 2 & -1 & 3 \\ 4 & 3 & -1 & 10 \\ 2 & -1 & 1 & 4 \\ 3 & 1 & 0 & 7 \end{bmatrix}$

In Exercises 49–52, find two distinct row echelon matrices, each of which is equivalent to the given matrix.

49. $\begin{bmatrix} 2 & 1 \\ 1 & 2 \end{bmatrix}$

50. $\begin{bmatrix} 2 & 1 & 3 \\ 1 & 2 & -3 \end{bmatrix}$

51. $\begin{bmatrix} 2 & -1 & -1 \\ 1 & 1 & 2 \\ 3 & -3 & -1 \end{bmatrix}$

52. $\begin{bmatrix} 2 & 1 & -1 \\ 1 & 1 & 2 \\ 4 & 1 & -5 \end{bmatrix}$

53–56. For the matrices in Exercises 49–52, find the equivalent reduced row echelon matrix.

57. Find an equation for the plane through the origin and the points $(1,1,1)$, $(2,-1,3)$.

In Exercises 58–60, find the value(s) of a for which the given system has nontrivial solutions.

58. $\begin{aligned} x - y + 2z &= 0 \\ x + 2y + az &= 0 \\ 3x - 2y + 4z &= 0 \end{aligned}$

59. $\begin{aligned} ax - 3y + z &= 0 \\ 2x + y + z &= 0 \\ 3x + 2y - 2z &= 0 \end{aligned}$

60. $\begin{aligned} (1-a)x + 2y &= 0 \\ 2x + (4-a)y &= 0 \end{aligned}$

61. Let p be a function of the form

$$p(x) = a_0 + a_1x + a_2x^2 + a_3x^3.$$

Show that if 1, 2, 3, and 4 are roots of the equation $p(x) = 0$, then $a_0 = a_1 = a_2 = a_3 = 0$.

1.4 GENERAL LINEAR SYSTEMS

Nonhomogeneous linear systems with either no solutions or many solutions sometimes arise in applications of linear algebra. For instance, in studies of physical systems, measurement and modeling errors may lead to linear systems with no (exact) solution. Interpolation of data sets, which is studied in Chapter 5, may involve systems of these types. When a nonhomogeneous system has more than one solution, its solution set is similar to that of a homogeneous system and is found in much the same way. Our final version of Gaussian elimination is included in this section.

Consistency

In order to emphasize solution techniques, we have so far limited our study to linear systems that have solutions. In general, a nonhomogeneous system may have no solutions. For instance, this occurs when the equations in a 2×2 system determine parallel lines. A 2×2 system with no solutions has the form

$$\begin{aligned} ax + by &= c \\ kax + kby &= d, \end{aligned}$$

where $d \neq kc$. In this case, the first two entries in the two rows are proportional, but the last entries are not in the same proportion. We may also apply an elementary row operation:

$$\left[\begin{array}{cc|c} a & b & c \\ ka & kb & d \end{array}\right] \rightarrow \left[\begin{array}{cc|c} a & b & c \\ 0 & 0 & d - kc \end{array}\right].$$

The second row in the final matrix represents the equation

$$0x + 0y = d - kc,$$

which has no solution since $d - kc \neq 0$.

The following example illustrates the application of Gaussian elimination to a particular 3×3 system with no solution.

Example 1

Solve the linear system

$$x + 2y + 2z = 1$$

$$2x + 5y + 2z = 4$$

$$3x + 8y + 2z = 8$$

by Gaussian elimination.

Solution

The results of the first pass are:

$$\left[\begin{array}{ccc|c} 1 & 2 & 2 & 1 \\ 2 & 5 & 2 & 4 \\ 3 & 8 & 2 & 8 \end{array}\right] \rightarrow \left[\begin{array}{ccc|c} 1 & 2 & 2 & 1 \\ 0 & 1 & -2 & 2 \\ 0 & 2 & -4 & 5 \end{array}\right] \rightarrow \left[\begin{array}{ccc|c} 1 & 2 & 2 & 1 \\ 0 & 1 & -2 & 2 \\ 0 & 0 & 0 & 1 \end{array}\right].$$

The bottom row of the row echelon matrix represents the equation

$$0x + 0y + 0z = 1.$$

Since this equation has no solutions, the given system also has no solutions. □

The geometry of Example 1 is worth noting. The solution of the first two equations

$$x + 2y + 2z = 1$$

$$2x + 5y + 2z = 4$$

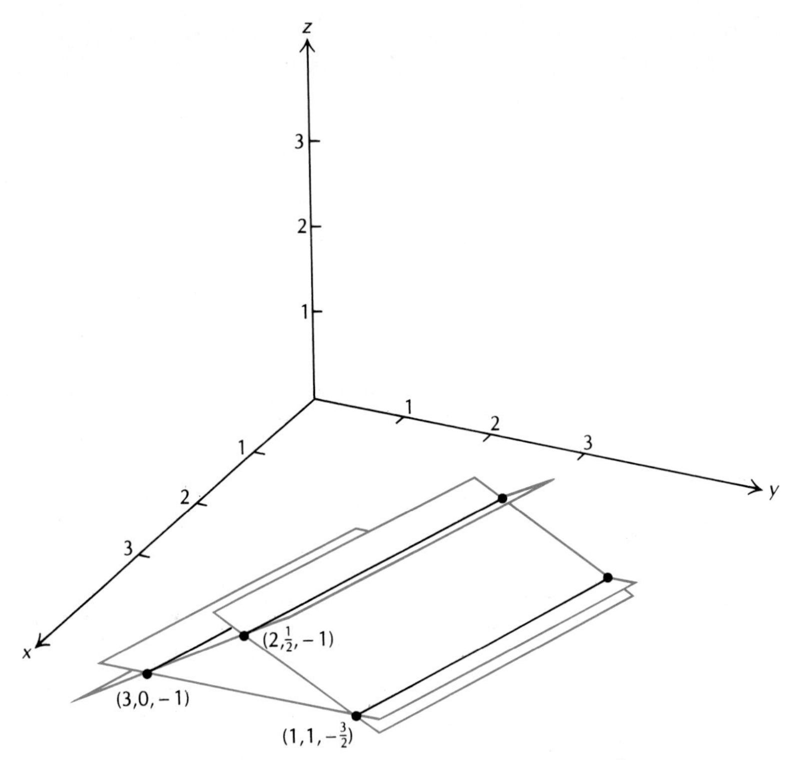

Figure 1.11 The inconsistent system of Example 1.

may be obtained from the first two rows of the final matrix. The solution is the line through the points $(-3,2,0)$ and $(3,0,-1)$. Similarly, the solution of the first and third equations is the line through $(-4,5/2,0)$ and $(2,1/2,-1)$, and the solution of the second and third is the line through $(-2,2,-1)$ and $(1,1,-3/2)$. These three lines are mutually parallel, as indicated in Fig. 1.11. (See Appendix B for information on parallel lines in space.)

DEFINITION 1.8

A linear system is said to be **consistent** if it has at least one solution, otherwise it is **inconsistent**.

Any homogeneous system is consistent because it has at least the trivial solution. Inconsistency of a 3×3 nonhomogeneous system may be related to any of several geometric possibilities. The planes determined by the equations may be mutually parallel, as illustrated in Fig. 1.12a. Two of the planes may coincide and be parallel to the third, as in Fig. 1.12b. Two of the planes may be parallel and both intersect the third, as in Fig. 1.12c. The planes may also intersect pairwise in mutually parallel lines, as illustrated in Fig. 1.12d and in Example 1.

The inconsistency in the nonhomogeneous 3×3 system in Example 1 becomes apparent from the row echelon form of the augmented matrix. The first three entries in the bottom row of this matrix equal 0 and the last entry equals 1. This example illustrates the following test.

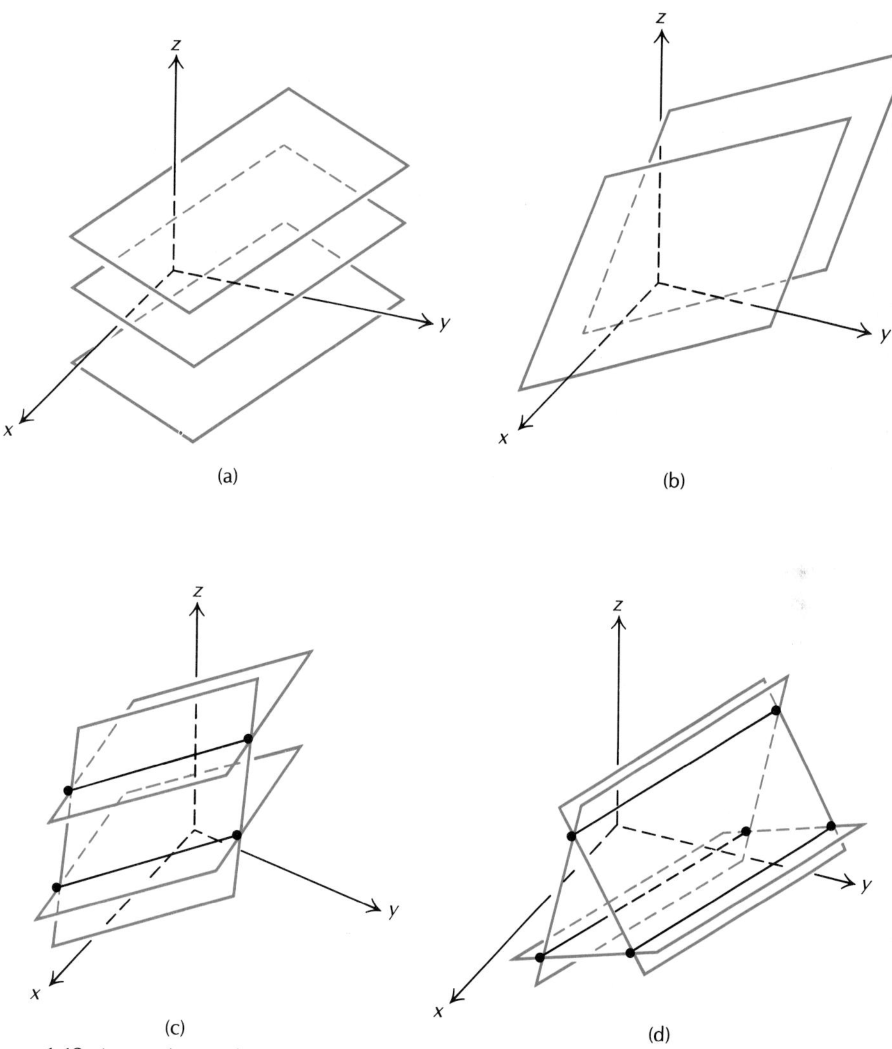

Figure 1.12 Inconsistent 3×3 linear systems: (a) three mutually parallel planes; (b) two planes coincide and are parallel to the third; (c) two planes are parallel and intersect the third; (d) each pair of planes intersects in a line parallel to the third plane.

Consistency Test

> To test an $m \times n$ system (m equations, n variables) for consistency, bring the augmented matrix to row echelon form. If this matrix has a row in which the first n entries equal zero and the $(n + 1)$st entry equals one, then the system is inconsistent. If the row echelon matrix has no such row, then the system is consistent.

In hand calculations, an inconsistency may be detected before a row echelon form is reached. For instance, the bottom two rows of the second matrix in Example 1 represent the system

$$y - 2z = 2$$

$$2y - 4z = 5,$$

which has no solution. For a system with n variables, an early discovery of two rows that are not proportional but whose first n entries are proportional saves at least one elementary row operation.

Consistency of a linear system can be determined from the row ranks of the coefficient and augmented matrices. The following theorem is a corollary to the consistency test.

THEOREM 1.2

A linear system is consistent if and only if the row rank of the coefficient matrix equals the row rank of the augmented matrix.

In some applications, we may impose extra conditions on the variables in order to "smooth out" measurement or modeling errors. Consistency is an especially important issue for $m \times n$ systems with $m > n$. Figure 1.13 illustrates two 3×2 linear systems, one of which is consistent and the other inconsistent.

Linear systems with more equations than unknowns may also be solved by Gaussian elimination.

Example 2

Solve the 3×2 linear system

$$x + 2y = 1$$

$$2x - y = 12$$

$$3x + 2y = 12.$$

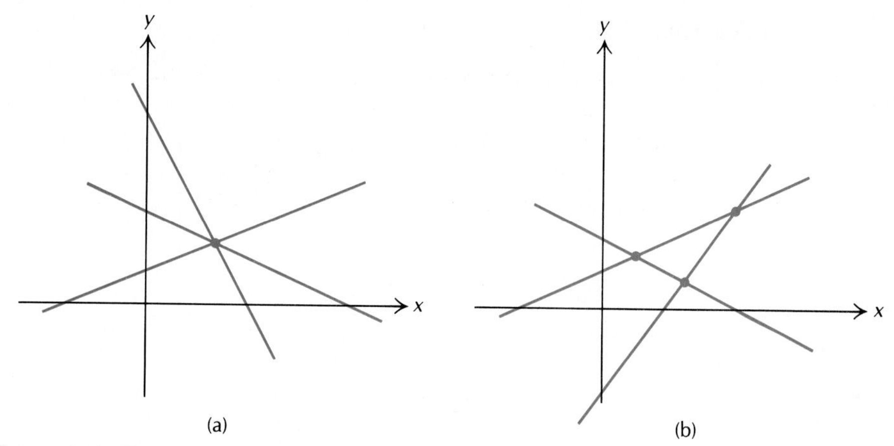

Figure 1.13 Systems of three equations in two variables: (a) a consistent system; (b) an inconsistent system.

Solution

Bring the augmented matrix to row echelon form:

$$\left[\begin{array}{rr|r} 1 & 2 & 1 \\ 2 & -1 & 12 \\ 3 & 2 & 12 \end{array}\right] \to \left[\begin{array}{rr|r} 1 & 2 & 1 \\ 0 & -5 & 10 \\ 0 & -4 & 9 \end{array}\right] \to \left[\begin{array}{rr|r} 1 & 2 & 1 \\ 0 & 1 & -2 \\ 0 & 0 & 1 \end{array}\right].$$

Since the rank of the coefficient matrix is 2 and the rank of the augmented matrix is 3, the system is inconsistent by Theorem 1.2. This system illustrates the configuration in Fig. 1.13b. □

General Solution Concepts

The procedure for solving nonhomogeneous linear systems with more than one solution is similar to the procedure used for homogeneous systems.

Example 3

Solve the linear system

$$\begin{aligned} x + 3y - 2z &= 1 \\ 2x + 3y - z &= -1 \\ x - 3y + 4z &= -5. \end{aligned}$$

Solution

The first pass of Gaussian elimination gives

$$\left[\begin{array}{rrr|r} 1 & 3 & -2 & 1 \\ 2 & 3 & -1 & -1 \\ 1 & -3 & 4 & -5 \end{array}\right] \rightarrow \left[\begin{array}{rrr|r} 1 & 3 & -2 & 1 \\ 0 & -3 & 3 & -3 \\ 0 & -6 & 6 & -6 \end{array}\right] \rightarrow \left[\begin{array}{rrr|r} 1 & 3 & -2 & 1 \\ 0 & 1 & -1 & 1 \\ 0 & 0 & 0 & 0 \end{array}\right].$$

Since the augmented matrix and the coefficient matrix both have row rank 2, the system is consistent. For any z, we may solve the second equation in the row echelon system to find $y = z + 1$, and then solve the first to find $x = -z - 2$. The solution consists of all ordered triples of the form $(-z - 2, z + 1, z)$. □

Examples 1–3 illustrate the use of Gaussian elimination to determine whether a given system is consistent and to solve it when it is. As Example 2 also illustrates, the row rank of any $m \times n$ coefficient matrix is at most n, even if $m > n$. This is because in the move from any leading 1 to the next in a row echelon matrix, each step down one row corresponds to at least one step right. With earlier examples, we have illustrated enough possibilities to prepare for our final version of Gaussian elimination.

Gaussian Elimination, Version 3.0

Begin with an $m \times n$ linear system represented by its coefficient matrix if the system is homogeneous, or its augmented matrix if the system is nonhomogeneous.

1. Form an equivalent row echelon system using elementary row operations (Pass 1).
2. Find the row ranks r and r' of the coefficient matrix and the augmented matrix, respectively. If $r' > r$, then the system has no solution (i.e., it is inconsistent).
3. If $r = r'$, solve the row echelon system for the r leading variables by back substitution (Pass 2). The solution is unique if $r = n$. If $r < n$, then the remaining $n - r$ variables appear as parameters in the general form of a solution.

We illustrate the general algorithm with a 5×4 system.

Example 4

Solve the linear system

$$
\begin{aligned}
x_1 + 2x_2 - x_3 \phantom{{}- 2x_4} &= 1 \\
x_1 + 3x_2 - x_3 - 2x_4 &= 4 \\
-x_1 \phantom{{}+ 3x_2} + x_3 - 4x_4 &= 5 \\
2x_1 + 3x_2 - 2x_3 + 2x_4 &= -1 \\
3x_1 + 4x_2 - 3x_3 + 4x_4 &= -3.
\end{aligned}
$$

Solution

Pass 1 of Gaussian elimination yields

$$
\left[\begin{array}{rrrr|r}
1 & 2 & -1 & 0 & 1 \\
1 & 3 & -1 & -2 & 4 \\
-1 & 0 & 1 & -4 & 5 \\
2 & 3 & -2 & 2 & -1 \\
3 & 4 & -3 & 4 & -3
\end{array}\right]
\to
\left[\begin{array}{rrrr|r}
1 & 2 & -1 & 0 & 1 \\
0 & 1 & 0 & -2 & 3 \\
0 & 2 & 0 & -4 & 6 \\
0 & -1 & 0 & 2 & -3 \\
0 & -2 & 0 & 4 & -6
\end{array}\right]
$$

$$
\to
\left[\begin{array}{rrrr|r}
1 & 2 & -1 & 0 & 1 \\
0 & 1 & 0 & -2 & 3 \\
0 & 0 & 0 & 0 & 0 \\
0 & 0 & 0 & 0 & 0 \\
0 & 0 & 0 & 0 & 0
\end{array}\right].
$$

Since the row ranks of the coefficient matrix and augmented matrix are equal, the system is consistent. For any x_4, solve the second equation of the final system to find $x_2 = 2x_4 + 3$. With this expression for x_2 and with x_3 arbitrary, solve the first equation to find $x_1 = x_3 - 4x_4 - 5$. The solution consists of all points of the form

$$(x_3 - 4x_4 - 5, 2x_4 + 3, x_3, x_4). \quad \square$$

When a linear system has more than one solution, we may always obtain a parametric description of the solution set by solving an equivalent row echelon system for the leading variables in terms of the remaining variables, as in Example 4. Solution sets may also be expressed in terms of specific solution points, as illustrated in the following example.

Example 5

Suppose the row echelon matrix

$$
\begin{bmatrix}
1 & 2 & 4 & -1 & 1 \\
0 & 1 & -1 & 2 & -3
\end{bmatrix}
$$

is equivalent to the coefficient matrix of a homogeneous system in $x_1, \ldots, x_5$. Then for any x_3, x_4, and x_5, we may solve for the leading variables to obtain the parametric form

$$(-6x_3 + 5x_4 - 7x_5, x_3 - 2x_4 + 3x_5, x_3, x_4, x_5)$$

of any solution.

We may also find the general solution of the given system by first finding certain particular solutions. For instance,

$$\begin{array}{llll} \text{for} & x_3 = 1,\ x_4 = 0,\ x_5 = 0, & \text{we have} & x_2 = 1,\ x_1 = -6, \\ \text{for} & x_3 = 0,\ x_4 = 1,\ x_5 = 0, & \text{we have} & x_2 = -2,\ x_1 = 5, \\ \text{for} & x_3 = 0,\ x_4 = 0,\ x_5 = 1, & \text{we have} & x_2 = 3,\ x_1 = -7. \end{array}$$

Thus, the 5-tuples

$$(-6,1,1,0,0),\ (5,-2,0,1,0),\ \text{and}\ (-7,3,0,0,1)$$

are solutions of the given system. These particular solutions determine the general solution in the following way. Multiplying the individual coordinates of these three points by x_3, x_4, and x_5, respectively, gives

$$(-6x_3, x_3, x_3, 0, 0), \quad (5x_4, -2x_4, 0, x_4, 0), \quad \text{and} \quad (-7x_5, 3x_5, 0, 0, x_5).$$

These 5-tuples are also solutions of the given system, and adding the individual coordinates of these points yields the general solution. These observations will be clarified in the study of vector spaces in Chapters 4 and 5. □

In solving linear systems, we approach homogeneous and nonhomogeneous systems in slightly different ways. You can see the differences in the diagram in Fig. 1.14.

This completes the introduction to linear systems. The basic solution method, Gaussian elimination, applies to *all* linear systems in theory, but other methods may be used in practice for improved efficiency or accuracy. Section 1.5 gives some idea of computational aspects of linear algebra, although a more thorough understanding requires additional study, such as a course in numerical analysis.

REVIEW CHECKLIST

1. State definitions of consistent and inconsistent linear systems.
2. Describe the consistency test involving a particular type of row in a row echelon equivalent to a given system.

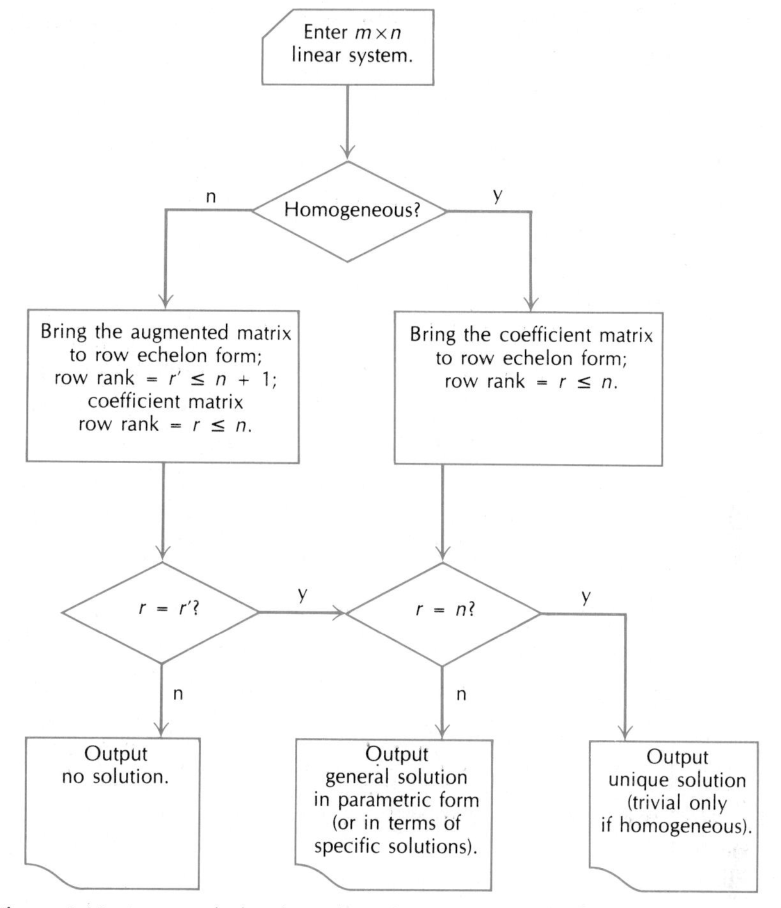

Figure 1.14 A general plan for solving linear systems by Gaussian elimination.

3. Relate consistency to row rank of the coefficient and augmented matrices of a system.
4. Determine whether a given linear system is consistent.
5. Formulate a comprehensive description of Gaussian elimination.
6. Find general solutions of homogeneous and nonhomogeneous systems.
7. Outline a general macro-approach to solving linear systems, as diagrammed in Fig. 1.14.

EXERCISES

In Exercises 1–2, determine by inspection whether the given system is consistent.

1. $$\begin{aligned} 2x - 4y + z &= 3 \\ 6x - 12y + 3z &= 9 \end{aligned}$$

2. $$\begin{aligned} x - 2y + 2z &= 4 \\ -5x + 10y - 10z &= 10 \end{aligned}$$

In Exercises 3–25, solve the given system by Gaussian elimination.

3. $$\begin{aligned} x + 2y - z &= 1 \\ 3x + y + z &= 1 \\ -x + 4y + 2z &= 1 \end{aligned}$$

4. $$\begin{aligned} x + y - z &= 1 \\ 2x - y + 2z &= 1 \\ 2x + 5y - 6z &= 1 \end{aligned}$$

5. $$\begin{aligned} x + 3y + 2z &= 4 \\ 3x - y - 4z &= 2 \\ 2x + y - z &= 3 \end{aligned}$$

6. $$\begin{aligned} x + 3y + 2z &= 0 \\ 3x - y - 4z &= 0 \\ 2x + y - z &= 0 \end{aligned}$$

7. $$\begin{aligned} x - 2y - 3z &= -1 \\ x - y - 2z &= 1 \\ 2x - 5y - 6z &= 3 \end{aligned}$$

8. $$\begin{aligned} 2x + 5y &= 5 \\ 4x - y &= 7 \\ 3x + 2y &= 6 \end{aligned}$$

9. $$\begin{aligned} 2x + 5y &= 5 \\ 4x - y &= 4 \\ 3x + 2y &= 5 \end{aligned}$$

10. $$\begin{aligned} 2x + 5y &= 5 \\ 4x - y &= 7 \\ 3x + 2y &= 6 \\ x - 3y &= -1 \end{aligned}$$

11. $$\begin{aligned} x + y - 2z &= 0 \\ x + 2y &= 0 \\ x - y - 6z &= 0 \\ 3x + 4y - 4z &= 0 \end{aligned}$$

12. $$\begin{aligned} x + 3y + 2z &= 4 \\ 3x - y - 4z &= 2 \\ 2x - y - z &= 2 \end{aligned}$$

13. $$\begin{aligned} x + 3y + 2z &= 4 \\ 3x - y - 4z &= 2 \\ 2x - y - z &= 3 \end{aligned}$$

14. $$\begin{aligned} x + 3y + 2z &= 4 \\ 3x - y - 4z &= 2 \\ 2x - y - z &= 5 \end{aligned}$$

15. $$\begin{aligned} 2x - 4y + z &= 4 \\ x - 2y + 2z &= -1 \\ x - 2y + z &= 1 \end{aligned}$$

16. $$\begin{aligned} x_1 + 2x_2 + x_3 - 4x_4 &= -1 \\ 2x_1 + 4x_2 + 3x_3 + 7x_4 &= 1 \\ x_1 + 2x_2 + 3x_3 + x_4 &= 1 \end{aligned}$$

17. $$\begin{aligned} x_1 + 2x_2 - x_3 + 3x_4 &= 2 \\ 2x_1 + 5x_2 - 2x_3 + 7x_4 &= 5 \\ -x_1 + x_2 + x_3 &= 1 \end{aligned}$$

18. $$\begin{aligned} x_1 + 3x_2 + x_3 - x_4 &= 1 \\ 2x_1 + 5x_2 + x_3 - 5x_4 &= 0 \\ x_1 + x_2 - x_3 - x_4 &= -4 \end{aligned}$$

19. $$\begin{aligned} 2x_1 + x_2 - x_3 + x_4 &= 3 \\ x_1 + 2x_2 + x_3 + 2x_4 &= 2 \\ 4x_1 - x_2 - 5x_3 - x_4 &= 5 \end{aligned}$$

20. $$\begin{aligned} x + 2y + 3z &= 4 \\ 3x + 4y - z &= 2 \\ 2x + 3y + z &= 3 \\ x - 2z &= -1 \end{aligned}$$

21. $$\begin{aligned} x + 2y + 3z &= 4 \\ 3x + 4y - z &= 2 \\ 2x + 3y + z &= 3 \\ x + y - 2z &= -1 \end{aligned}$$

22. $$\begin{aligned} 2x_1 - 4x_2 + x_3 - 2x_4 &= 0 \\ x_1 + x_2 + 2x_3 - x_4 &= 0 \\ x_1 - x_2 + x_3 - x_4 &= 0 \\ x_1 - 3x_2 - x_4 &= 0 \end{aligned}$$

23. $$\begin{aligned} 2x_1 - 4x_2 + x_3 - 2x_4 &= 0 \\ x_1 + x_2 + 2x_3 - x_4 &= 0 \\ x_1 - x_2 + x_3 - x_4 &= 0 \\ x_1 - 3x_2 + x_3 - x_4 &= 0 \end{aligned}$$

24. $$\begin{aligned} 2x_1 - 4x_2 + x_3 - 2x_4 &= 0 \\ x_1 + x_2 + 2x_3 - x_4 &= 0 \\ x_1 - x_2 + x_3 + x_4 &= 0 \\ x_1 - 3x_2 + x_3 - x_4 &= 0 \end{aligned}$$

25. $$\begin{aligned} 2x_1 - 4x_2 + x_3 - 2x_4 &= 1 \\ x_1 + x_2 + 2x_3 - x_4 &= 2 \\ x_1 - x_2 + x_3 - x_4 &= 1 \\ x_1 - 3x_2 - x_4 &= 0 \end{aligned}$$

In Exercises 26–27, sketch a graph of the given system. Use the graph to determine whether the system is consistent and if so, estimate the solution from the graph.

26. $$\begin{aligned} x + y &= 5 \\ 2x - y &= 1 \\ x - 2y &= -1 \end{aligned}$$

27. $$\begin{aligned} 2x - y &= 0 \\ x - 2y &= -4 \\ x + 2y &= 4 \end{aligned}$$

In Exercises 28–29, determine whether the point $(-1, 4, 2)$ is on the plane through the origin and the given points.

28. $(1, 2, 3)$, $(2, -1, 4)$

29. $(3, 1, 4)$, $(2, 5, 6)$

In Exercises 30–31, determine whether the plane $2x + y - 3z = 0$ contains the line of intersection of the given planes.

30. $$\begin{aligned} x + y + z &= 0 \\ x - y - 3z &= 0 \end{aligned}$$

31. $$\begin{aligned} x - y + 2z &= 0 \\ 2x - y - z &= 0 \end{aligned}$$

In Exercises 32–33, find the value(s) of a for which the given system is consistent.

32. $$\begin{aligned} x + y - z &= 2 \\ 3x + 2y + 3z &= 1 \\ 2x + y + 4z &= a \end{aligned}$$

33. $$\begin{aligned} 2x + y - z &= 0 \\ x + ay + 3z &= 2 \\ 2x + y + z &= 4 \end{aligned}$$

In Exercises 34–35, find the value(s) of a and b for which the given system is consistent.

34. $$\begin{aligned} x + 2y + az &= 2 \\ -x + y + 3z &= 0 \\ 2x + y + 3z &= b \end{aligned}$$

35. $$\begin{aligned} 2x + y + 2z &= a \\ 3x + z &= b \\ 2x - 5y - 6z &= 4 \end{aligned}$$

36. Without finding the equation of any plane involved, decide whether the point $(1, 2, 1)$ lies on the plane determined by the points $(2, 3, 2)$, $(4, 3, 0)$, $(2, 5, 6)$.

In Exercises 37–38, let $p(x) = 1 - 2x + x^2$ and $q(x) = 2 - x + x^2$ be polynomials. Determine whether the given polynomial $r(x)$ may be expressed in the form $r(x) = cp(x) + kq(x)$ for some values of c and k.

37. $r(x) = 1 + x + x^2$

38. $r(x) = 2 + x - x^2$

39. Refer to the linear system of Example 3, page 41. Show that $(-2, 1, 0)$ is a solution of the given system. Show also that $(-z, z, z)$ is the general solution of the associated homogeneous system

$$\begin{aligned} x + 3y - 2z &= 0 \\ 2x + 3y - z &= 0 \\ x - 3y + 4z &= 0. \end{aligned}$$

Hence, observe that each coordinate in the general solution found in Example 3 may be obtained as the sum of the corresponding coordinates of $(-2, 1, 0)$ and $(-z, z, z)$.

(o)1.5 APPLICATIONS AND COMPUTING

Interpolation refers to the problem of finding an equation of a prescribed type to be satisfied by some given points. The problem arises, for instance, when we record observations of an event and we need to estimate values between the recorded data points. When the desired equation involves variables and unknown constants, we obtain a system of equations for the constants by substituting the given values for the variables. Often we must first determine a suitable form of the unknown equation.

Example 1

Find an equation for the parabola having a vertical axis of symmetry and passing through the points $(-1, 7)$, $(1, -1)$, and $(2, -2)$.

Solution

A parabola of this type is determined by an equation of the form

$$y = Ax^2 + Bx + C$$

(see Appendix B). We substitute the given values into this equation to obtain the linear system

$$\begin{aligned} A - B + C &= 7 \\ A + B + C &= -1 \\ 4A + 2B + C &= -2. \end{aligned}$$

Solve this system by Gaussian elimination:

$$\left[\begin{array}{ccc|c} 1 & -1 & 1 & 7 \\ 1 & 1 & 1 & -1 \\ 4 & 2 & 1 & -2 \end{array}\right] \to \left[\begin{array}{ccc|c} 1 & -1 & 1 & 7 \\ 0 & 2 & 0 & -8 \\ 0 & 6 & -3 & -30 \end{array}\right] \to \left[\begin{array}{ccc|c} 1 & -1 & 1 & 7 \\ 0 & 1 & 0 & -4 \\ 0 & 0 & -3 & -6 \end{array}\right].$$

The solution $(1, -4, 2)$ may be found by back substitution. This solution yields the equation

$$y = x^2 - 4x + 2. \quad \square$$

What happens if we try to pass a parabola through three collinear points? A problem of this type is given in Exercise 6.

The form chosen for an unknown interpolating equation is crucial, and there is often some latitude in the choice. For instance, for the parabola in Example 1, we may use the form

$$y - k = a(x - h)^2$$

and solve for the constants a, h, and k.

Example 2

Find an equation for the plane through the points $(-1,-1,1)$, $(2,5,1)$, and $(3,-1,-1)$ in space.

Solution

The equation of a plane in space may be written in the form

$$Ax + By + Cz + D = 0,$$

where A, B, and C are not all zero. Substitute the given values for the variables to obtain the system

$$\begin{aligned} -A - B + C + D &= 0 \\ 2A + 5B + C + D &= 0 \\ 3A - B - C + D &= 0. \end{aligned}$$

This system may be solved by Gaussian elimination:

$$\begin{bmatrix} -1 & -1 & 1 & 1 \\ 2 & 5 & 1 & 1 \\ 3 & -1 & -1 & 1 \end{bmatrix} \to \begin{bmatrix} -1 & -1 & 1 & 1 \\ 0 & 3 & 3 & 3 \\ 0 & -4 & 2 & 4 \end{bmatrix} \to \begin{bmatrix} -1 & -1 & 1 & 1 \\ 0 & 1 & 1 & 1 \\ 0 & 0 & 6 & 8 \end{bmatrix}.$$

Back substitution yields the solution $(-2D/3, D/3, -4D/3, D)$. Any value of D other than 0 may be used. For instance, $D = -3$ yields

$$2x - y + 4z - 3 = 0. \quad \square$$

If we use the form in Example 2 to interpolate four points by a plane, the resulting linear system has a nontrivial solution if and only if the points are coplanar. If two distinct points are given, the solution of the linear system represents the family of all planes containing the line determined by the given points. If one point is given, the linear system consists of a single equation representing the family of all planes through the point. Some of these ideas are illustrated in Exercises 15–17.

Combining fractions over a common denominator is a familiar and useful procedure. Sometimes we need to reverse this procedure and express a single fraction as a sum of simpler fractions. Such sums are called **partial fraction** expansions.

We use partial fraction expansions in calculus as a technique of integration. We also use them in solving differential equations by transform methods. In both cases, we expand a fraction of the form $p(x)/q(x)$, where $p(x)$ and $q(x)$ are polynomial functions and the degree of $p(x)$ is less than that of $q(x)$. Readers who have used various ad hoc methods for such problems may well appreciate the beauty of this example.

Example 3

Find a partial fraction expansion for the quotient

$$\frac{2x - 1}{(x+1)^2(x^2+1)}.$$

Solution

One possible form for the expansion is

$$\frac{2x - 1}{(x+1)^2(x^2+1)} = \frac{Ax + B}{x^2 + 1} + \frac{Cx + D}{(x+1)^2}.$$

(How did we find this form? The denominator of the given fraction is the product of $(x + 1)^2$ by $x^2 + 1$. A denominator of this form results when we add fractions with these individual denominators. The numerator of each term on the right is a general polynomial of degree one less than the degree of the corresponding denominator. The sum of these terms yields the given denominator, which is of degree 4, and the numerator has the general form of a polynomial of degree at most 3.)

Combining the righthand side over the least common denominator $(x + 1)^2(x^2 + 1)$ gives the numerator

$$(Ax + B)(x^2 + 2x + 1) + (Cx + D)(x^2 + 1).$$

Expand and rearrange the terms to obtain

$$(A + C)x^3 + (2A + B + D)x^2 + (A + 2B + C)x + (B + D).$$

Since this polynomial must equal the numerator on the lefthand side for all x, we equate coefficients of like powers of x:

$$\begin{aligned}
A \qquad + C \qquad &= 0 \qquad (\text{for } x^3)\\
2A + B \qquad + D &= 0 \qquad (\text{for } x^2)\\
A + 2B + C \qquad &= 2 \qquad (\text{for } x^1 = x)\\
B \qquad + D &= -1 \qquad (\text{for } x^0 = 1).
\end{aligned}$$

The solution of this system, $(1/2, 1, -1/2, -2)$, yields the partial fraction expansion

$$\frac{2x - 1}{(x+1)^2(x^2+1)} = \frac{1}{2}\frac{x + 2}{x^2 + 1} - \frac{1}{2}\frac{x + 4}{(x+1)^2}. \qquad \square$$

For some partial fraction problems, linear algebra methods are not the best. Other methods are quicker when, for example, the given denominator can be expressed as a product of distinct linear factors, such as $(x-a)(x-b)$ with $a \neq b$. When shortcut methods fail, linear algebra methods may always be applied.

Linear algebra can be applied to *heat transfer* problems. If the temperature at the lateral boundary of an insulated, thin metal plate is held constant, the temperature of the plate will settle down to a steady-state distribution that depends on the boundary temperatures. To estimate this steady-state distribution, we place a rectangular grid on the plate and approximate the temperature at the grid points.

Example 4

Figure 1.15 shows a square plate with interior grid points P_1, P_2, P_3, and P_4. The diagram also gives the temperature on each edge of the plate and shows eight undesignated grid points on the edges. Assume that the temperature t_i at each P_i equals the average temperature at four neighboring grid points. Find t_1, t_2, t_3, and t_4.

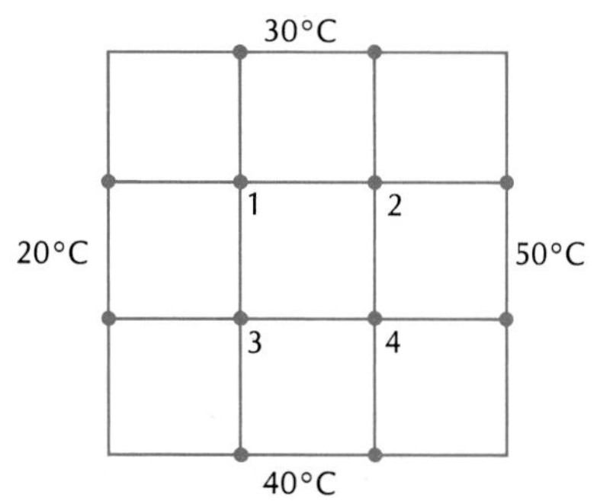

Figure 1.15 Square metal plate with grid for Example 3.

Solution

By the required assumption,

$$
\begin{aligned}
t_1 &= (20 + 30 + t_2 + t_3)/4 \\
t_2 &= (t_1 + 30 + 50 + t_4)/4 \\
t_3 &= (20 + t_1 + t_4 + 40)/4 \\
t_4 &= (t_3 + t_2 + 50 + 40)/4
\end{aligned}
$$

Applying Gaussian elimination (after rearranging each equation into the usual format) gives:

$$\left[\begin{array}{rrrr|r} 4 & -1 & -1 & 0 & 50 \\ -1 & 4 & 0 & -1 & 80 \\ -1 & 0 & 4 & -1 & 60 \\ 0 & -1 & -1 & 4 & 90 \end{array}\right] \rightarrow \left[\begin{array}{rrrr|r} -1 & 0 & 4 & -1 & 60 \\ 0 & -1 & -1 & 4 & 90 \\ -1 & 4 & 0 & -1 & 80 \\ 4 & -1 & -1 & 0 & 50 \end{array}\right]$$

$$\rightarrow \left[\begin{array}{rrrr|r} -1 & 0 & 4 & -1 & 60 \\ 0 & -1 & -1 & 4 & 90 \\ 0 & 4 & -4 & 0 & 20 \\ 0 & -1 & 15 & -4 & 290 \end{array}\right]$$

$$\rightarrow \left[\begin{array}{rrrr|r} 1 & 0 & -4 & 1 & -60 \\ 0 & 1 & 1 & -4 & -90 \\ 0 & 0 & -2 & 4 & 95 \\ 0 & 0 & 16 & -8 & 200 \end{array}\right]$$

$$\rightarrow \left[\begin{array}{rrrr|r} 1 & 0 & -4 & 1 & -60 \\ 0 & 1 & 1 & -4 & -90 \\ 0 & 0 & 1 & -2 & -47.5 \\ 0 & 0 & 0 & 1 & 40 \end{array}\right].$$

The solution for (t_1, t_2, t_3, t_4) is $(30, 37.5, 32.5, 40)$. ☐

Note two points in connection with Example 4. First, the approximation improves as the number of grid points increases, but this also increases the size of the linear system to be solved. Four points were used for ease of hand calculation. To determine the effect of adding grid points, we embedded the four points from Example 4 in a 5×5 interior grid. Using the same boundary temperatures, we solved the resulting 25×25 linear system on a computer. Rounded to two places, the temperatures at the original four points are 29.85, 37.50, 32.42, and 40.15.

There are other possible averaging schemes. In one, we use the temperatures at the eight nearest points (the horizontal, vertical, and diagonal neighbors). For the plate in Example 4, we also need prescribed temperatures at the corners of the plate. Exercise 20 illustrates this averaging scheme.

Linear algebra can be applied to *electrical network* analysis via Ohm's law and Kirchhoff's laws.*

* **Gustav Kirchhoff** (1824–1887) was a German physicist whose fundamental laws governing electrical circuits lead directly to linear systems.

Ohm's law *For a current I flowing through a resistance R, the voltage drop is given by V = IR.*

Kirchhoff's current law *The algebraic sum of the currents flowing into any junction is zero, i.e., the total current in must equal the total out.*

Kirchhoff's voltage law *The algebraic sum of the voltage drops around any closed path equals the algebraic sum of the applied voltages in that path.* (Here "algebraic" means relative to a direction chosen as positive.)

Example 5

Find the currents I_1, I_2, and I_3 for the network of Fig. 1.16. The units are ohms (o), volts (v), and amperes (a).

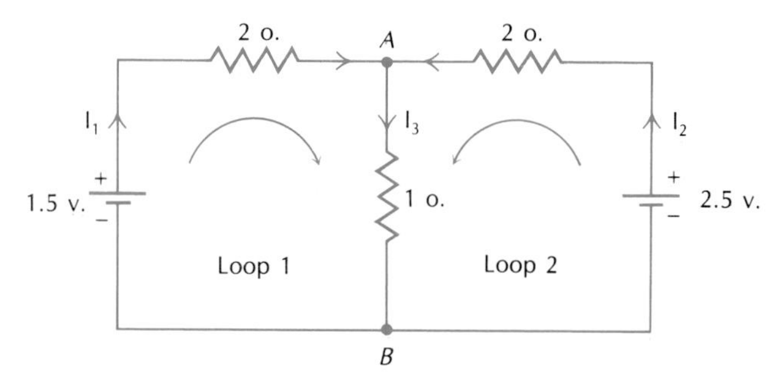

Figure 1.16 Ladder network for Example 5.

Solution

Kirchhoff's current law applied at either junction A or junction B yields

$$I_1 + I_2 - I_3 = 0.$$

Apply the voltage law to the closed paths labeled loop 1 and loop 2 in Fig. 1.17. The results are the second and third equations, respectively, in the linear system for the currents:

$$\begin{aligned} 2I_1 \qquad\quad + I_3 &= 1.5 \\ 2I_2 + I_3 &= 2.5. \end{aligned}$$

Gaussian elimination yields the currents $I_1 = .25$, $I_2 = .75$, and $I_3 = 1$, all in amperes. □

Example 6

Find the currents $I_1, \ldots, I_6$ for the network of Fig. 1.17.

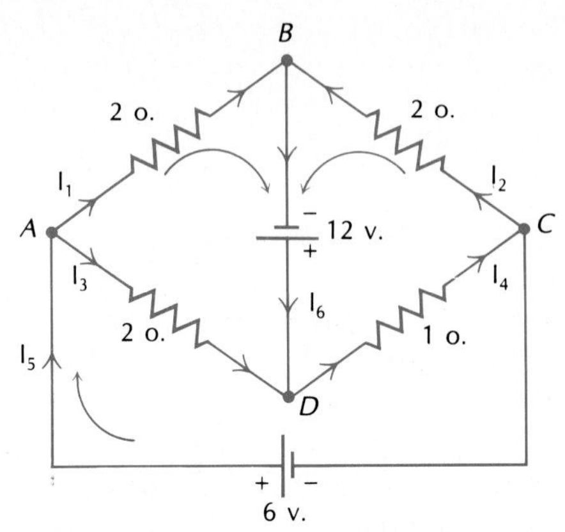

Figure 1.17 Bridge network for Example 6.

Solution

The current law applied at junctions A, B, and C, respectively, gives

$$I_1 + I_3 = I_5$$

$$I_1 + I_2 = I_6$$

$$I_2 + I_5 = I_4.$$

Apply the voltage to three of the closed paths in the network:

$$\begin{aligned} 2I_1 \quad\quad - 2I_3 \quad\quad &= 12 \quad \text{(path } ABDA\text{)} \\ 2I_3 + I_4 &= \ \ 6 \quad \text{(path } ADCA\text{)} \\ -2I_1 + 2I_2 + 2I_3 + I_4 &= \ \ 0 \quad \text{(path } ADCBA\text{).} \end{aligned}$$

Then solve by Gaussian elimination, using the augmented matrix

$$\left[\begin{array}{cccccc|c} 1 & 0 & 1 & 0 & -1 & 0 & 0 \\ 1 & 1 & 0 & 0 & 0 & -1 & 0 \\ 0 & 1 & 0 & -1 & 1 & 0 & 0 \\ 1 & 0 & -1 & 0 & 0 & 0 & 6 \\ 0 & 0 & 2 & 1 & 0 & 0 & 6 \\ -2 & 2 & 2 & 1 & 0 & 0 & 0 \end{array}\right].$$

The results for I_1, I_2, I_3, I_4, I_5, and I_6 are, respectively, 5.4, 2.4, -0.6, 7.2, 4.8, and 7.8 amperes. ☐

In Example 6 we ignored paths *ABCA*, *BDCB*, *BDACB*, *BDCAB*, and junction *D* in applying Kirchhoff's laws. Different choices of paths and junctions lead to different systems, but as long as a system has (at least) six equations and has row rank six, it will be equivalent to the one we used.

Computing

Hand calculation is adequate for simple linear systems. Machine calculation is used for systems that are difficult or inconvenient to solve by hand. Special care may be necessary to prevent errors in computing. We heartily recommend "Pitfalls of Computing" by G. E. Forsythe, an article from the November 1970 issue of the *American Mathematical Monthly*. Section 6 on linear systems is especially relevant. The next example illustrates roundoff error, a common problem in computational linear algebra. This example is essentially the same as one that appeared in the Forsythe article and had been treated earlier in the book, *Computer Solution of Linear Algebraic Systems* by G. E. Forsythe and C. B. Moler.

Example 7

Solve the linear system

$$\begin{aligned} .0001x + y &= 2 \\ x + y &= 1, \end{aligned}$$

rounding all work to three significant digits.

Solution

We try Gaussian elimination, rounding the result of each calculation:

$$\left[\begin{array}{cc|c} 0.0001 & 1.00 & 2.00 \\ 1.00 & 1.00 & 1.00 \end{array}\right] \rightarrow \left[\begin{array}{cc|c} 0.0001 & 1.00 & 2.00 \\ 0.00 & -10000 & -20000 \end{array}\right].$$

(When rounding to three significant digits, we keep a number of the form $d_1 \cdot d_2 d_3$ multiplied by a power of 10, where d_1, d_2, and d_3 are decimal digits, 0 to 9.) Back substitution yields $y = 2.00$, $x = 0.00$. This is a relatively poor result since the true solution, to 5 decimal places, is $(-1.00010, 2.00010)$. However, interchanging rows first gives

$$\left[\begin{array}{cc|c} 1.00 & 1.00 & 1.00 \\ 0.0001 & 1.00 & 2.00 \end{array}\right] \rightarrow \left[\begin{array}{cc|c} 1.00 & 1.00 & 1.00 \\ 0.00 & 1.00 & 2.00 \end{array}\right],$$

from which $y = 2.00$, $x = -1.00$. ▭

The difficulty in the back substitution in Example 7 arises from the relative sizes of the coefficients. The row interchange used in the second solution in Example 7 to circumvent this difficulty illustrates a technique called **pivoting**. In **partial pivoting** for an $m \times n$ system, before eliminating the nonzero entries below the current row i in the current column j, we locate an element among the entries $a_{ij}, \ldots, a_{mj}$ whose absolute value is maximum. If this element is in row k, where $k \neq i$, then an exchange of rows i and k precedes the elimination. In *full pivoting*, we use the maximum over all matrix entries from row i to row m and from column j to column n. Full pivoting is rarely used in practice.

Although pivoting solves the roundoff problem in many cases, it is by no means a panacea. For instance, if both equations in the linear system of Example 7 are multiplied by 10,000 and Gaussian elimination with partial pivoting is applied to the resulting system, roundoff error again produces a poor result. (This idea was also given in the Forsythe article.) Thus partial pivoting fails for this example.

Of course, computed values generally are not rounded to three significant digits. However, in any computer, *all* floating point calculations are rounded to a known number of significant digits, so examples of small systems with a roundoff error problem can always be constructed. Worse yet, since a computer may perform hundreds of thousands or even millions of arithmetic operations in solving a linear system, even relatively small roundoff errors in each operation may be built into large ones.

So what is the solution to the roundoff problem? No single technique solves the problem completely. Pivoting helps control roundoff error and pivoting coupled with double or higher precision is even more effective. Sometimes procedures other than Gaussian elimination should be used. Gaussian elimination is but one of a family of techniques called *direct methods*. There is another family of numerical methods called *iterative methods*. A study of numerical analysis is recommended for anyone seriously interested in computing for linear algebra.

In spite of difficulties, many linear algebra problems can be handled correctly and efficiently on a computer. Programs for solving linear systems may be written in FORTRAN, BASIC, Pascal, C, and other programming languages. Appendix D contains example programs that can be implemented conveniently on a microcomputer. Some excellent software packages are available commercially. One notable example is *PC-Matlab*, a powerful microcomputer version of the mainframe program *Matlab*. With proper care and caution, you can use computers successfully in studying and applying linear algebra.

REVIEW CHECKLIST

1. Interpolate given sets of points by curves in the plane or surfaces in space.
2. Find partial fraction expansions using linear algebra.
3. Find steady-state temperature distributions in heat transfer applications.
4. Set up and solve linear systems for the currents in simple electrical networks.

5. Set up and solve linear systems for other simple applications similar to those illustrated in the examples.
6. Discuss the roundoff error problem for computational linear algebra.

EXERCISES

In Exercises 1–3, find an equation for the line in the plane through the given points.

1. $(1,1), (2,3)$ **2.** $(3,-1), (2,4)$ **3.** $(-1,2), (1,5), (3,8)$

In Exercises 4–5, explain what happens to the resulting linear system when the points $(1,-1), (2,2), (3,-2)$ are interpolated by an equation of the given form.

4. $Ax + By = C$ **5.** $y = mx + b$

6. Explain what happens when the points $(1,-2), (2,0), (3,2)$ are interpolated by an equation of the form $y = Ax^2 + Bx + C$. (Compare with Example 1, page 48.)

In Exercises 7–8, find an equation of the form $y = Ax^2 + Bx + C$ that is satisfied by the given points.

7. $(-1,-3), (1,5), (2,6)$ **8.** $(1,2), (3,4), (9,6)$

In Exercises 9–10, find an equation for the circle through the given points.

9. $(0,-1), (2,1), (4,-1)$

10. $(1,2), (2,1), (2-\sqrt{2}/2, 2-\sqrt{2}/2)$

In Exercises 11–12, find an equation of the form $y = ax^3 + bx^2 + cx + d$ that is satisfied by the given points.

11. $(1,0), (-1,-4), (2,8), (-2,-8)$

12. $(-1,1), (1,1), (2,-5), (3,-3)$

13. Find an equation for the fourth degree polynomial through the points $(1,2), (-1,2), (2,5), (-2,17), (3,42)$.

14. Find an equation for the plane in space containing the point $(3,2,-1)$ and the line consisting of all points of the form $(1+t, 1+t, 2-2t)$, where t is a real number.

15. Find an equation representing the family of all planes in space containing the line in Exercise 14.

In Exercises 16–17, explain what happens when we try to find a plane in space containing the given points.

16. $(1,1,2), (2,2,0), (3,2,-1), (-1,3,2)$

17. $(1,1,2), (2,2,0), (3,2,-1), (-1,3,3)$

18. Find an equation of the form $z = ax^2 + by^2$ that is satisfied by the points $(1, 2, 1)$, $(2, -2, 1)$.

19. Find an equation for the sphere through the points $(2, -1, 2)$, $(1, 0, 2)$, $(1, -1, 3)$, $(0, -1, 2)$. Use the form

$$(x - a)^2 + (y - b)^2 + (z - c)^2 = r^2.$$

20. For the plate in Example 4, page 51, find the steady-state temperatures at the four interior grid points, assuming the temperature at each is the average of the temperatures at the eight nearest points. At each corner, use the average temperature at the two neighboring boundary grid points.

21–22. Find the steady-state temperatures at the interior grid points in the corresponding diagram using the averaging method of Example 4, page 51.

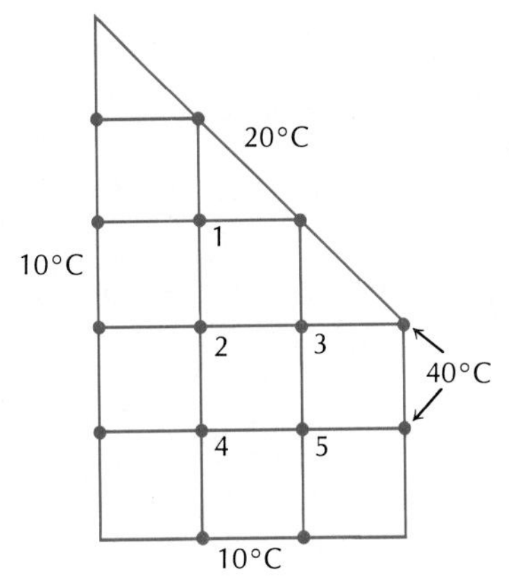

Metal plate with grid for Exercise 21.

25°C
15°C
5°C
1
10°C
2
3
4
10°C
5°C
5
10°C
20°C

Metal plate with grid for Exercise 22.

In Exercises 23–26, find the coefficients A, B, C, D in the indicated partial fraction expansion.

23. $$\frac{1}{(x^2 + 1)(x^2 + 4)} = \frac{Ax + B}{x^2 + 1} + \frac{Cx + D}{x^2 + 4}$$

24. $$\frac{3x^3 - x^2 + 6x - 4}{(x^2 + 1)(x^2 + 4)} = \frac{Ax + B}{x^2 + 1} + \frac{Cx + D}{x^2 + 4}$$

25. $$\frac{1}{(x - 1)^2(x + 1)^2} = \frac{Ax + B}{(x - 1)^2} + \frac{Cx + D}{(x + 1)^2}$$

26. $$\frac{3x^3 + 5x^2 + 17x + 2}{(x - 1)^2(x + 1)^2} = \frac{Ax + B}{(x - 1)^2} + \frac{Cx + D}{(x + 1)^2}$$

27–29. Find the currents $I_1, \ldots, I_5$ for the electrical network in the corresponding diagram.

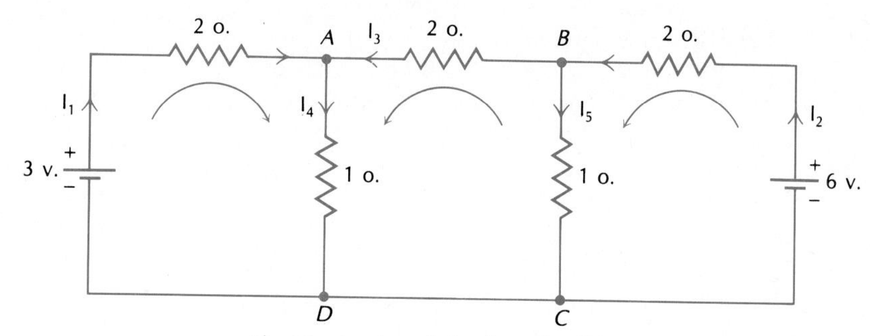

Electrical network for Exercise 27.

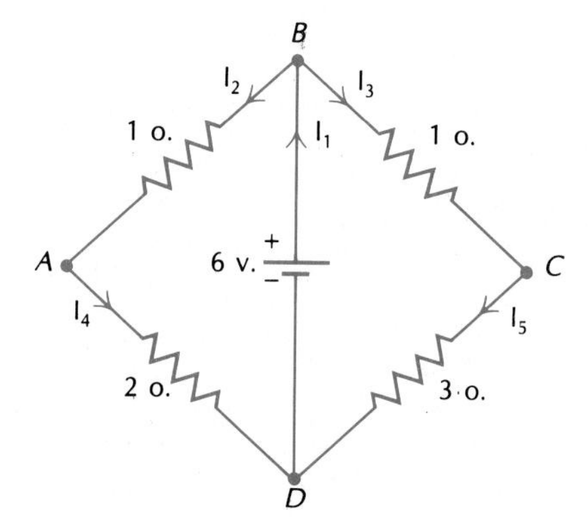

Electrical network for Exercise 28.

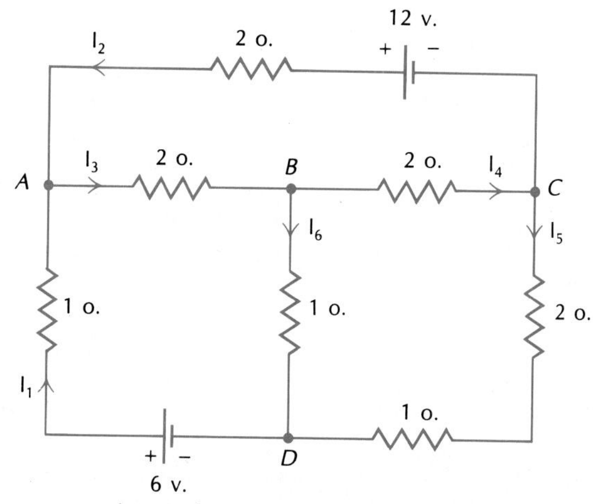

Electrical network for Exercise 29.

REVIEW EXERCISES

In Exercises 1–6, solve the given linear system by Gaussian elimination.

1. $$\begin{aligned} 4x + 2y - 3z &= 0 \\ 8x + 4y - 6z &= 0 \\ 2x + y - z &= 0 \end{aligned}$$

2. $$\begin{aligned} x - 3y + 2z &= 0 \\ 2x + y + 4z &= 0 \\ -x + 2y + 2z &= 0 \end{aligned}$$

3. $$\begin{aligned} 2x - y &= 0 \\ x + 4y + z &= 0 \\ 4x - 11y - 2z &= 0 \end{aligned}$$

4. $$\begin{aligned} 2x - y - z &= 1 \\ x + 3y - 2z &= 1 \\ 9x - y - 6z &= 1 \end{aligned}$$

5. $$\begin{aligned} x_1 - x_2 + 2x_3 &= 1 \\ 4x_2 - 3x_3 - 2x_4 &= 1 \\ 2x_1 + 2x_2 + x_3 - 2x_4 &= 3 \\ 7x_1 + 5x_2 + 5x_3 - 6x_4 &= 10 \end{aligned}$$

6. $$\begin{aligned} x_1 + x_2 + x_3 + x_4 &= 1 \\ 4x_1 + 4x_2 + 3x_3 + 3x_4 &= 1 \\ x_1 + 2x_2 + 4x_3 + x_4 &= 6 \\ 2x_1 + x_2 - 2x_3 + 3x_4 &= -4 \end{aligned}$$

In Exercises 7–10, solve the given system by Gauss-Jordan elimination.

7. $$\begin{aligned} x - 3y &= 5 \\ 2x - 9y &= 7 \end{aligned}$$

8. $$\begin{aligned} 2x + y &= 3 \\ 3x - 2y &= 1 \end{aligned}$$

9. $$\begin{aligned} x + 2y - 2z &= 0 \\ 2x + 3y - 4z &= 0 \\ -2x - 4y + 3z &= 0 \end{aligned}$$

10. $$\begin{aligned} 3x - 2y + 2z &= 3 \\ 3x - 2y + 4z &= 5 \\ x - y - 2z &= 1 \end{aligned}$$

11. Determine whether the following linear systems are equivalent.

$$\begin{aligned} x + y &= 3 \\ 2x + 3y &= 0 \end{aligned} \qquad \begin{aligned} x + 2y &= -3 \\ 3x + 4y &= 3 \end{aligned}$$

In Exercises 12–13, find the row rank of the given matrix.

12. $$\begin{bmatrix} 1 & -2 & 3 \\ 4 & -6 & 2 \\ 2 & 0 & 3 \end{bmatrix}$$

13. $$\begin{bmatrix} 1 & 3 & 5 \\ 3 & 5 & 5 \\ 0 & 2 & 5 \end{bmatrix}$$

14. Suppose the row rank of the coefficient matrix of a certain homogeneous linear system is 3. If the system has five variables, how many parameters appear in the general solution?

15. Suppose the row rank of the augmented matrix of a consistent nonhomogeneous linear system is 6. If the system has 10 variables, how many parameters appear in the general solution?

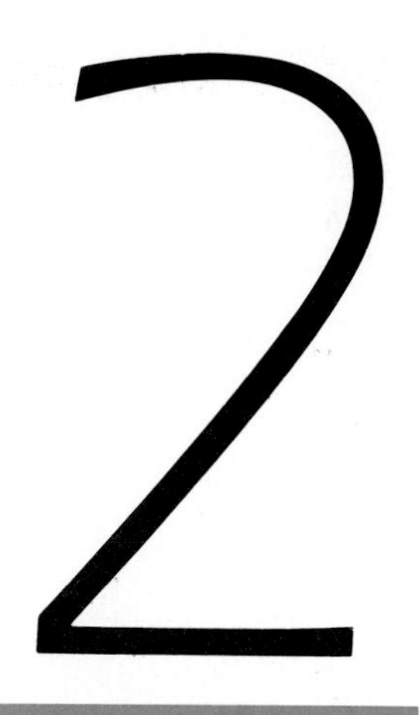

Matrix Algebra

By studying algebraic operations on matrices, we gain a new perspective on elementary row operations and enhance our understanding of Gaussian elimination and the ability to use it. The study includes matrix inversion, *a fundamental concept used to develop methods of solving linear systems and to communicate ideas concerning them. We also illustrate the application of matrix algebra to science and engineering problems.*

2.1 VECTOR ALGEBRA

Since the entries in a matrix are arranged in rows and columns, we often work with the individual rows and columns in performing the operations of matrix algebra. In fact, since *matrix multiplication* will be defined in terms of row and column operations, our study begins with these simpler operations and their properties. The concepts developed in this section are especially important. They are used throughout this chapter, extensively in Chapters 4 and 5, and widely in the more general field of mathematics.

Sums and Scalar Multiples

In geometry and in basic physics, a **vector** is defined as a quantity representing a *magnitude* and a *direction.* A vector may be interpreted geometrically as a line segment directed from an *initial point* to a *terminal point*. The orientation of this "arrow" represents the direction and its length represents the magnitude. Physically, a vector models a quantity such as velocity, acceleration, momentum, force, or pressure.

When vectors are used in a rectangular Cartesian coordinate system, algebra as well as geometry may be applied. In the Cartesian plane, any nonzero point P may be regarded as the terminal point of a vector with initial point at the origin (see Fig. 2.1). Since the point P corresponds uniquely to an ordered pair of numbers (x, y), so does any nonzero vector. For completeness, the ordered pair $(0, 0)$ is regarded as a "vector" having length zero and no direction. Therefore,

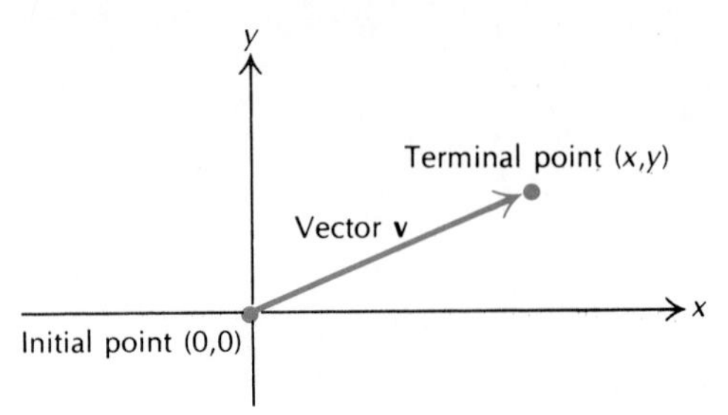

Figure 2.1 A plane vector.

we *redefine* a **plane vector** as an ordered pair of real numbers:

$$\mathbf{v} = (x, y).$$

(Note the use of boldface type for vectors.)

In a similar way, any point in space may be regarded as the terminal point of a vector with initial point at the origin (see Fig. 2.2). Since points in space correspond uniquely to ordered triples of real numbers, we redefine a **space vector** as an ordered triple of real numbers:

$$\mathbf{v} = (x, y, z).$$

Example 1

The vector $\mathbf{v} = (3, 4)$ may be represented geometrically as the directed line segment from the origin in the plane to the point P: $(3, 4)$. This is the point located 3 units to the right of the origin and 4 units up (Fig. 2.3). By the Pythagorean theorem, the length of this segment is $\sqrt{3^2 + 4^2} = 5$. We use the same notation $(3, 4)$ for either the point or the directed line segment. The context of any particular discussion determines which of these two meanings is intended. □

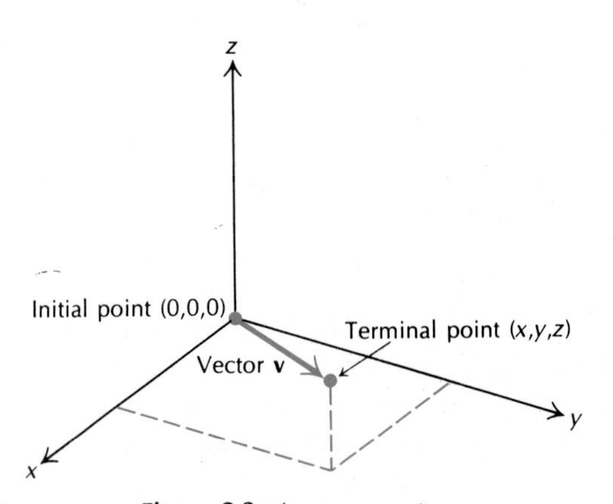

Figure 2.2 A space vector.

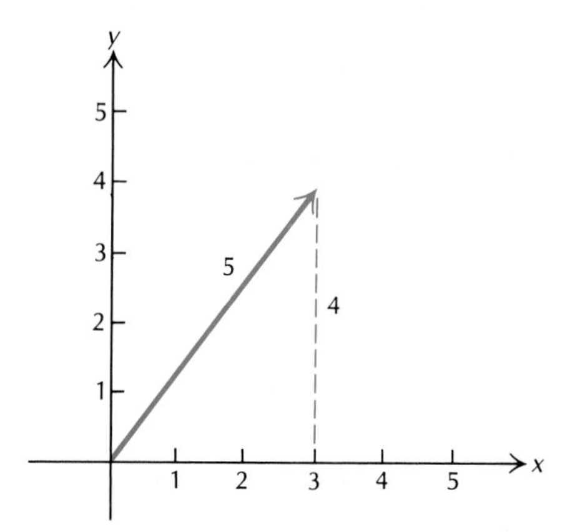

Figure 2.3 The vector $\mathbf{v} = (3, 4)$ in the Cartesian plane.

A vector need not have its initial point at the origin. For instance, if the vector represented by the ordered pair $(1, 2)$ is moved so that its initial point is located at the point $(1, -1)$, then the terminal point is $(2, 1)$. These directed segments have the same magnitude and direction.

The special cases of ordered pairs in the plane and ordered triples in space illustrate the general form used in the application of algebra to vectors. The idea is formalized in a definition of *one of the most useful basic concepts in mathematics*.

DEFINITION 2.1

A **vector** is an ordered n-tuple: $\mathbf{v} = (x_1, \ldots, x_n)$. The numbers $x_1, \ldots, x_n$ are called the **components** of $\mathbf{v}$. The **norm**, **magnitude**, or **length**, of $\mathbf{v}$ is given by

$$\|\mathbf{v}\| = \sqrt{x_1^2 + \cdots + x_n^2}.$$

A *component* of a vector is a *coordinate* of the corresponding point. In Chapter 7, vectors of *complex numbers* (Appendix C) are used in several examples. Throughout the remainder of this book, vectors have real components.

Example 2

Find the norm of each of the vectors

$$\mathbf{u} = (1, -2, 0, 2) \quad \text{and} \quad \mathbf{v} = (2, 5, -1, 3, 3).$$

Solution

By Definition 2.1,

$$\|\mathbf{u}\| = \sqrt{1^2 + (-2)^2 + 0^2 + 2^2} = 3$$

and

$$\|\mathbf{v}\| = \sqrt{2^2 + 5^2 + (-1)^2 + 3^2 + 3^2} = 4\sqrt{3}. \quad \square$$

Since vectors are ordered n-tuples and, in turn, ordered n-tuples represent solutions of linear systems, vectors have immediate relevance to linear algebra. This association is only the beginning. Vectors play a much more fundamental role in linear systems via *vector algebra*, the study of *vector addition* and *vector multiplication by a scalar*. Vector addition extends the idea of the *resultant* of two vectors in the plane or in space to any two n-tuples. Multiplication by a scalar extends the idea of changing scale, or size, in a similar way. These operations are illustrated for plane vectors in Fig. 2.4.

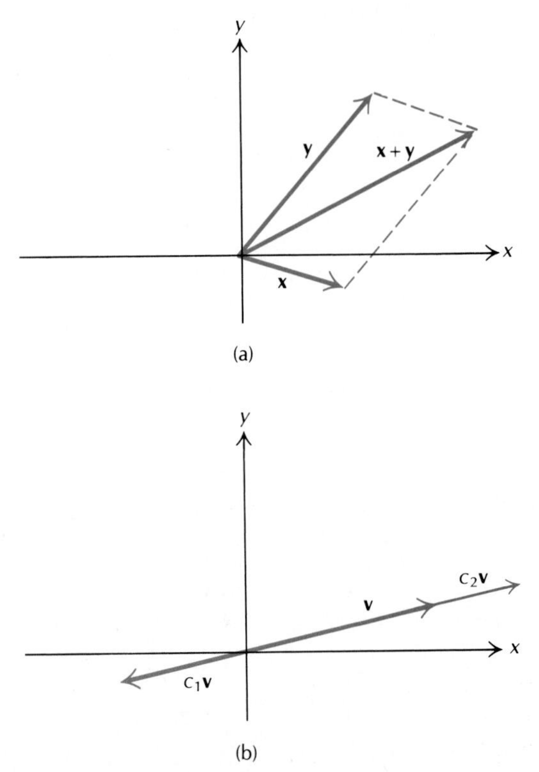

Figure 2.4 (a) The resultant of $\mathbf{x}$ and $\mathbf{y}$ formed by completing a parallelogram. (b) Two scalar multiples of a vector $\mathbf{v}$, $c_1\mathbf{v}$ and $c_2\mathbf{v}$. Here $-1 < c_1 < 0$ and $c_2 > 1$.

For general n-tuples, the same letter (such as x or y) is often used for both a vector and its components. The vector is indicated by boldface type, the components by subscripts.

DEFINITION 2.2

Let $\mathbf{x} = (x_1, \ldots, x_n)$ and $\mathbf{y} = (y_1, \ldots, y_n)$ be vectors with the same number of components. We say that **x equals y**, and write

$$\mathbf{x} = \mathbf{y},$$

if $x_i = y_i$ for $i = 1, \ldots, n$. The **sum of x and y** is given by

$$\mathbf{x} + \mathbf{y} = (x_1 + y_1, \ldots, x_n + y_n).$$

Any real number c is called a **scalar**, and the vector

$$c\mathbf{x} = (cx_1, \ldots, cx_n)$$

is called a **scalar multiple of x**.

Example 3

For $\mathbf{x} = (2, -1, 4, 5)$ and $\mathbf{y} = (0, 3, -1, -4)$, we have

$$\mathbf{x} + \mathbf{y} = (2 + 0, -1 + 3, 4 + (-1), 5 + (-4)) = (2, 2, 3, 1),$$

$$2\mathbf{x} = (2(2), 2(-1), 2(4), 2(5)) = (4, -2, 8, 10),$$

$$(1/2)\mathbf{x} = (1, -1/2, 2, 5/2),$$

$$(-3)\mathbf{x} = (-6, 3, -12, -15),$$

and

$$2\mathbf{x} + 3\mathbf{y} = (4, -2, 8, 10) + (0, 9, -3, -12) = (4, 7, 5, -2). \quad \square$$

Physical interpretations are important. If **x** and **y** represent forces, then $\mathbf{x} + \mathbf{y}$ represents the resultant force produced by the two acting simultaneously. The vector $(-2)\mathbf{x}$ represents a force of magnitude twice that of **x** and applied in the opposite direction. Velocity vectors may be added to produce a resultant velocity, and acceleration vectors behave in a similar way. These operations are fundamental in applied studies such as physics, mechanics, and electronics.

A **zero vector** is any vector with all components zero. The symbol **0** is used for any zero vector, so **0** may represent $(0, 0)$ in one context and $(0, 0, 0)$ in another, for example. The **negative** of $\mathbf{x}$, denoted $-\mathbf{x}$, is defined as the unique vector $\mathbf{y}$ such that

$$\mathbf{x} + \mathbf{y} = \mathbf{y} + \mathbf{x} = \mathbf{0}.$$

From these definitions and the definitions of the vector operations above, we may obtain the following basic rules of vector algebra.

Associative laws:

$$(\mathbf{x} + \mathbf{y}) + \mathbf{z} = \mathbf{x} + (\mathbf{y} + \mathbf{z}),$$

$$(ab)\mathbf{x} = a(b\mathbf{x}).$$

Distributive laws:

$$a(\mathbf{x} + \mathbf{y}) = a\mathbf{x} + a\mathbf{y},$$

$$(a + b)\mathbf{x} = a\mathbf{x} + b\mathbf{x}.$$

Commutative law:

$$\mathbf{x} + \mathbf{y} = \mathbf{y} + \mathbf{x}.$$

The *identity vector* and the *identity scalar:*

$$\mathbf{0} + \mathbf{x} = \mathbf{x} + \mathbf{0} = \mathbf{x},$$

$$1\mathbf{x} = \mathbf{x}.$$

The *negative* of a vector:

$$-\mathbf{x} = (-1)\mathbf{x}.$$

More precisely, the "commutative law" is the "commutative law of vector addition." In the notation for a scalar multiple of a vector, the scalar may appear on either side of the vector, i.e., $a\mathbf{x} = \mathbf{x}a$. In one sense, this is not a commutative law since $a\mathbf{x}$ and $\mathbf{x}a$ are merely distinct notations for the same operation.

In proving these properties, we work with the individual components of the vectors. We establish the associative law for multiplication by scalars to illustrate a method of proof:

$$\begin{aligned} (ab)\mathbf{x} &= ((ab)x_1, \ldots, (ab)x_n) && \text{Definition of scalar multiple} \\ &= (a(bx_1), \ldots, a(bx_n)) && \text{Associative law for real numbers} \\ &= a(bx_1, \ldots, bx_n) && \text{Definition of scalar multiple} \end{aligned}$$

$$= a(b(x_1, \ldots, x_n)) \qquad \text{Definition of scalar multiple}$$

$$= a(b\mathbf{x}). \qquad \text{Definition of } \mathbf{x}$$

In these equations, $(ab)\mathbf{x}$ involves one ordinary multiplication and one multiplication of a vector by a scalar, while $a(b\mathbf{x})$ involves two multiplications of a vector by a scalar. Proofs for some of the other rules above are assigned in Exercises 41–47.

Using the rules of vector algebra, we can solve any vector equation of the form

$$a\mathbf{x} + b\mathbf{c} = \mathbf{0},$$

with $a \neq 0$, for $\mathbf{x}$ by subtracting $b\mathbf{c}$ from both sides, multiplying through by $1/a$, and simplifying the result.

Example 4

Solve and interpret the vector equation

$$2\mathbf{x} + 3(2, 4) = \mathbf{0}.$$

Solution

The solution is given by

$$\mathbf{x} = (1/2)[-3(2, 4)] = (-3, -6).$$

To interpret the given equation, substitute the components for the original unknown $\mathbf{x}$:

$$2(x_1, x_2) + 3(2, 4) = (0, 0)$$

$$(2x_1 + 6, 2x_2 + 12) = (0, 0).$$

Since two vectors are equal if and only if their corresponding components agree, the single vector equation is equivalent to the linear system

$$\begin{aligned} 2x_1 \qquad + 6 &= 0 \\ 2x_2 + 12 &= 0, \end{aligned}$$

whose solution is $(-3, -6)$. □

Inner Product

Besides vector addition and multiplication by scalars, we need one more operation for matrix algebra.

DEFINITION 2.3

If $\mathbf{x} = (x_1, \ldots, x_1)$ and $\mathbf{y} = (y_1, \ldots, y_n)$, the **inner product of x and y** is given by

$$\langle \mathbf{x}, \mathbf{y} \rangle = x_1 y_1 + \cdots + x_n y_n.$$

Example 5

If $\mathbf{x} = (2, 1, -4, 5)$ and $\mathbf{y} = (0, 5, 1, -1)$, then

$$\langle \mathbf{x}, \mathbf{y} \rangle = (2)(0) + (1)(5) + (-4)(1) + (5)(-1) = -4,$$

$$\langle \mathbf{x}, \mathbf{x} \rangle = (2)(2) + (1)(1) + (-4)(-4) + (5)(5) = 46,$$

$$\langle \mathbf{y}, \mathbf{y} \rangle = (0)(0) + (5)(5) + (1)(1) + (-1)(-1) = 27,$$

$$\langle 2\mathbf{x}, \mathbf{y} \rangle = (4)(0) + (2)(5) + (-8)(1) + (10)(-1) = -8 = 2\langle \mathbf{x}, \mathbf{y} \rangle. \quad \square$$

Example 6

Any linear form may be written as an inner product. For instance,

$$5x - y - 3z$$

may be expressed as $\langle \mathbf{a}, \mathbf{v} \rangle$, with $\mathbf{a} = (5, -1, -3)$ and $\mathbf{v} = (x, y, z)$. Similarly,

$$2x_1 - x_2 + 2x_3 + 4x_4$$

is expressible as $\langle \mathbf{b}, \mathbf{x} \rangle$, where $\mathbf{b} = (2, -1, 2, 4)$ and $\mathbf{x} = (x_1, x_2, x_3, x_4)$. $\square$

Since any linear form may be written as an inner product, any linear equation represents a condition requiring the value of an inner product to equal a given number. Equations of lines in the plane and planes in space may be expressed in terms of inner products. The inner product of **x** and **y** is also called the **dot product**, and is often denoted by $\mathbf{x} \cdot \mathbf{y}$. Since the inner product of two vectors is a scalar, it is also called the **scalar product**.

In the remainder of this section, we describe properties of the inner product to be used in matrix operations and in our study of *vector spaces* in Chapters 4 and 5. Note first that the norm of a vector can be expressed in terms of the inner product:

$$\|\mathbf{x}\| = \sqrt{\langle \mathbf{x}, \mathbf{x} \rangle}\,.$$

THEOREM 2.1

If $\mathbf{x}$ and $\mathbf{y}$ are vectors in the plane or in space, then

$$\langle \mathbf{x}, \mathbf{y} \rangle = \|\mathbf{x}\|\,\|\mathbf{y}\|\cos\theta,$$

where θ is the angle $(0 \leq \theta < 180°)$ between the lines determined by $\mathbf{x}$ and $\mathbf{y}$.

Proof We prove the result for plane vectors that determine distinct lines, leaving other cases as exercises. The directed segments determined by such vectors $\mathbf{x}$ and $\mathbf{y}$ form two sides of a triangle, as in Fig. 2.5. By the law of cosines,

$$\|\mathbf{y} - \mathbf{x}\|^2 = \|\mathbf{x}\|^2 + \|\mathbf{y}\|^2 - 2\|\mathbf{x}\|\,\|\mathbf{y}\|\cos\theta,$$

or

$$\left(\|\mathbf{x}\|^2 + \|\mathbf{y}\|^2 - \|\mathbf{y} - \mathbf{x}\|^2\right)/2 = \|\mathbf{x}\|\,\|\mathbf{y}\|\cos\theta.$$

Substitute the relations

$$\|\mathbf{x}\|^2 = x_1^2 + x_2^2, \qquad \|\mathbf{y}\|^2 = y^2 + y_2^2, \quad \text{and}$$

$$\|\mathbf{y} - \mathbf{x}\|^2 = (y_1 - x_1)^2 + (y_2 - x_2)^2$$

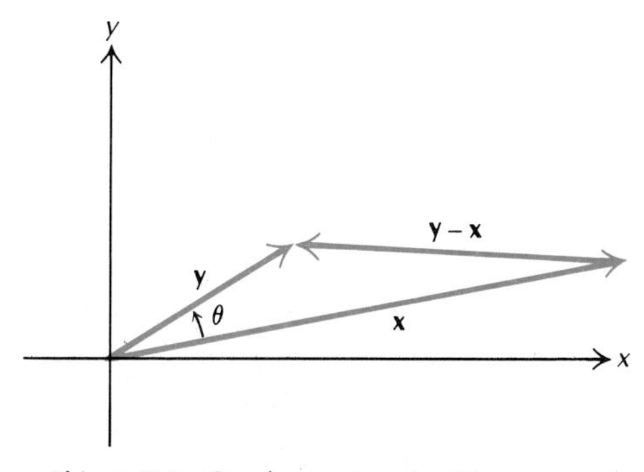

Figure 2.5 Configuration for Theorem 2.1.

into the second equation and simplify. The left side reduces to

$$x_1y_1 + x_2y_2,$$

and the desired result follows from Definition 2.3. □

In the plane or in space, the lines determined by the nonzero vectors **x** and **y** are perpendicular if they form a right angle. By Theorem 2.1, the lines determined by **x** and **y** are perpendicular if and only if $\langle \mathbf{x}, \mathbf{y} \rangle = 0$, since $\cos \pi/2 = 0$. This special case leads to the following general definition.

DEFINITION 2.4

If **x** and **y** are vectors, we say that **x is orthogonal to y** if $\langle \mathbf{x}, \mathbf{y} \rangle = 0$.

Example 7

Find all vectors in space that are orthogonal to $\mathbf{u} = (1,1,1)$ and $\mathbf{v} = (2,3,6)$.

Solution

Let $\mathbf{w} = (x, y, z)$ be any vector orthogonal to the given vectors. By Definition 2.4, we must find **w** such that

$$\langle \mathbf{w}, \mathbf{u} \rangle = x + y + z = 0$$

$$\langle \mathbf{w}, \mathbf{v} \rangle = 2x + 3y + 6z = 0.$$

The solution of this system consists of all vectors of the form $(3z, -4z, z)$, where z is a real number. □

Any vector of norm one is called a **unit vector**. If $\mathbf{x} = (x_1, \ldots, x_n)$ and $\mathbf{x} \neq \mathbf{0}$, then $x_i \neq 0$ for at least one i, so

$$\|\mathbf{x}\| = \sqrt{x_1^2 + \cdots + x_n^2} \geq |x_i| > 0,$$

hence $\|\mathbf{x}\| \neq 0$. Since

$$\|c\mathbf{x}\| = |c|\,\|\mathbf{x}\|$$

for any scalar c, we have that if $c = 1/\|\mathbf{x}\|$, then $\|c\mathbf{x}\| = 1$. Thus, for a nonzero vector **x**,

$$\mathbf{x}/\|\mathbf{x}\|$$

is a unit vector in the direction of $\mathbf{x}$. (We have used $\mathbf{x}/\|\mathbf{x}\|$ for the scalar multiple $(1/\|\mathbf{x}\|)\mathbf{x}$.)

In some problems, we need to find the effect of one vector in the direction of another. If $\mathbf{x}$ and $\mathbf{y}$ are plane or space vectors with $\mathbf{x} \neq \mathbf{0}$, we may construct a line through the terminal point of $\mathbf{y}$ perpendicular to the line through $\mathbf{x}$. This construction determines a vector $\mathbf{z}$ on the line through $\mathbf{x}$, as illustrated in Fig. 2.6. Let $\mathbf{z} = c\mathbf{x}$. To determine c, first suppose $\|\mathbf{x}\| = 1$. Since $c/\|\mathbf{y}\| = \cos\theta$ (Fig. 2.6), we have

$$c = \|\mathbf{x}\|\,\|\mathbf{y}\|\cos\theta = \langle \mathbf{x}, \mathbf{y} \rangle.$$

If $\|\mathbf{x}\| \neq 1$, we multiply $\mathbf{x}$ by $1/\|\mathbf{x}\|$ before finding the inner product. This idea is generalized in the following definition.

DEFINITION 2.5

If $\mathbf{x}$ and $\mathbf{y}$ are vectors with the same number of components, and if $\mathbf{x} \neq \mathbf{0}$, then the unit vector in the direction of $\mathbf{x}$ is $\mathbf{x}' = \mathbf{x}/\|\mathbf{x}\|$. The **orthogonal projection of y onto x** is the vector

$$\langle \mathbf{y}, \mathbf{x}' \rangle \mathbf{x}'.$$

In particular, if $\|\mathbf{x}\| = 1$, the projection is

$$\langle \mathbf{y}, \mathbf{x} \rangle \mathbf{x}.$$

Projections may also be made obliquely, i.e., parallel to a given line. Unless otherwise specified, however, "projection" will refer to an orthogonal projection.

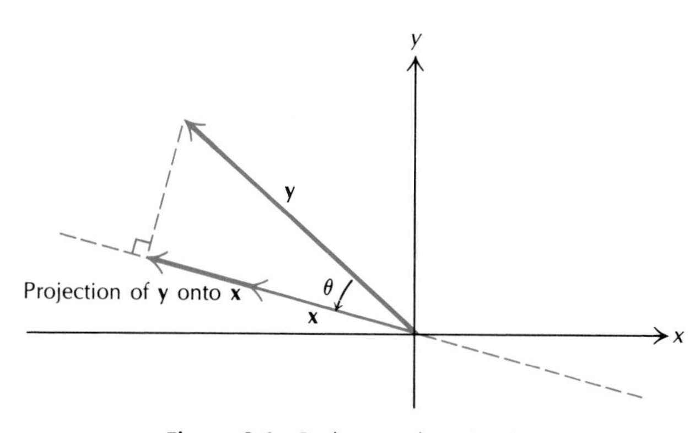

Figure 2.6 Orthogonal projection.

Example 8

If $\mathbf{x} = (2, 2, 1)$ and $\mathbf{y} = (1, 5, 4)$, the unit vector in the direction of $\mathbf{x}$ is

$$\mathbf{x}' = (1/\|\mathbf{x}\|)\mathbf{x} = (1/3)(2, 2, 1) = (2/3, 2/3, 1/3).$$

The component of $\mathbf{y}$ in the direction of $\mathbf{x}$ is

$$\langle \mathbf{y}, \mathbf{x}' \rangle = (1)(2/3) + (5)(2/3) + (4)(1/3) = 16/3,$$

and the orthogonal projection of $\mathbf{y}$ on $\mathbf{x}$ is

$$\langle \mathbf{y}, \mathbf{x}' \rangle \mathbf{x}' = \tfrac{16}{3} \cdot (2/3, 2/3, 1/3) = (32/9, 32/9, 16/9). \quad \square$$

REVIEW CHECKLIST

1. Formulate definitions of vector, component, norm, vector sum and scalar multiple, vector equality, inner product, unit vector, orthogonality, and the projection of one vector onto another.
2. Interpret vectors, vector sums, and scalar multiples geometrically and physically.
3. Find sums, scalar multiples, and inner products of given vectors.
4. Determine whether two given vectors are orthogonal.
5. Find the unit vector in a specified direction.
6. Find the orthogonal projection of one vector onto another.
7. Interpret linear systems, including equations of lines and planes, in terms of inner products.
8. Solve vector equations for unknown vectors or scalars.
9. Establish basic properties of algebraic vector operations.

EXERCISES

In Exercises 1–8, sketch the directed segment determined by the given vector and find the norm.

1. $(1, 0)$ **2.** $(0, 1)$ **3.** $(1, -2)$ **4.** $(-3, -4)$
5. $(1, 0, 0)$ **6.** $(1, 1, 1)$ **7.** $(1, -2, 2)$ **8.** $(6, 3, 2)$

In Exercises 9–13, sketch a graph for the indicated vector operation.

9. $(2, -1) + (1, 4)$ **10.** $(-1, 0) + (4, 2)$ **11.** $2(-1, 3)$
12. $(-1)(2, 2)$ **13.** $(1/2)(1, 3)$

In Exercises 14–17, perform the indicated operations, expressing the final result as a single vector.

14. $3(-1,4) + (1,2)$

15. $2(1,3) - 5(1,2)$

16. $(-4,2,2) + (1,3,2)$

17. $4(1,2,1,-1) + 3(-2,3,1,2)$

In Exercises 18–19, solve the given vector equation for **x**.

18. $3\mathbf{x} - 2(1,2,-1,3) = \mathbf{0}$

19. $2\mathbf{x} + 3(-1,2,2) = 4(0,1,1)$

In Exercises 20–23, find the inner product of the given pair of vectors.

20. $(2,1), (0,1)$

21. $(3,1,-2), (4,1,3)$

22. $(5,0,1,4), (3,2,4,1)$

23. $(-1,2,4,4,3), (2,3,-4,-1,2)$

In Exercises 24–25, determine whether the given vectors are orthogonal.

24. $(1,2,1), (-1,1,-1)$

25. $(3,4,-1,2), (2,-2,4,3)$

26. Find the value of a for which $(2,-1,a,2)$ and $(-1,1-2,4)$ are orthogonal.

In Exercises 27–29, find the orthogonal projection of $(2,3)$ onto the given vector, and sketch a graph for each.

27. $(1,0)$

28. $(-1/\sqrt{2},1/\sqrt{2})$

29. $(1,1)$

In Exercises 30–32, find the unit vector in the direction of the given vector.

30. $(3,4)$

31. $(-1,2,2)$

32. $(1,2,3,4)$

33. Find an equation of the line through $(0,0)$ and $(1,2)$ in terms of an inner product involving the vector $\mathbf{v} = (x, y)$.

34. Find the set of all vectors orthogonal to the vectors

$$(1,-1,2,2), \quad (1,2,3,2), \quad (2,1,5,4), \quad \text{and} \quad (1,-7,0,2).$$

35. Complete the proof of Theorem 2.1 by establishing the desired result for collinear plane vectors and for space vectors.

In Exercises 36–37, find all scalars c_1 and c_2 for which the given vector equation is satisfied.

36. $c_1(1,2) + c_2(-1,1) = \mathbf{0}$

37. $c_1(-1,3) + c_2(2,-6) = (1,1)$

In Exercises 38–39, find all scalars c_1, c_2, and c_3 for which the given vector equation is satisfied.

38. $c_1(1,1,2) + c_2(-1,-1,4) + c_3(4,4,2) = \mathbf{0}$
39. $c_1(1,2,3) + c_2(1,-1,-1) + c_3(1,-1,-2) = (4,2,3)$
40. For $\mathbf{x} = (2,1,-3)$, $\mathbf{y} = (-1,2,1)$, $\mathbf{z} = (3,4,7)$, verify that

$$\langle \mathbf{x} + \mathbf{y}, \mathbf{z} \rangle = \langle \mathbf{x}, \mathbf{z} \rangle + \langle \mathbf{y}, \mathbf{z} \rangle.$$

In Exercises 41–47, establish the indicated relation, assuming in each exercise that all vectors have the same number of components.

41. $\mathbf{x} + (\mathbf{y} + \mathbf{z}) = (\mathbf{x} + \mathbf{y}) + \mathbf{z}$
42. $a(\mathbf{x} + \mathbf{y}) = a\mathbf{x} + a\mathbf{y}$
43. $\mathbf{x} + \mathbf{y} = \mathbf{y} + \mathbf{x}$
44. $\langle \mathbf{x} + \mathbf{y}, \mathbf{z} \rangle = \langle \mathbf{x}, \mathbf{z} \rangle + \langle \mathbf{y}, \mathbf{z} \rangle$
45. $\langle \mathbf{x}, \mathbf{y} + \mathbf{z} \rangle = \langle \mathbf{x}, \mathbf{y} \rangle + \langle \mathbf{x}, \mathbf{z} \rangle$
46. $\langle c\mathbf{x}, \mathbf{y} \rangle = c\langle \mathbf{x}, \mathbf{y} \rangle$
47. $\langle a\mathbf{x} + b\mathbf{y}, \mathbf{z} \rangle = a\langle \mathbf{x}, \mathbf{z} \rangle + b\langle \mathbf{y}, \mathbf{z} \rangle$

2.2 MATRIX OPERATIONS

The operations studied in this section are extensions of vector operations to matrices. These operations will be used to solve linear systems in Sections 2.4 and 2.5 and to solve applied problems in Section 2.6.

Sums and Scalar Multiples

For matrix operations and subsequent work, we use

$$A = [a_{ij}]_{m \times n}$$

to denote the $m \times n$ matrix with entry a_{ij} at row i and column j.

DEFINITION 2.6

Let $A = [a_{ij}]_{m \times n}$ and $B = [b_{ij}]_{m \times n}$ be matrices, and c be a scalar. We say **A equals B** and write $A = B$ if

$$a_{ij} = b_{ij} \quad \text{for} \quad i = 1, \ldots, m \quad \text{and} \quad j = 1, \ldots, n.$$

The **sum of A and B** is given by

$$A + B = [a_{ij} + b_{ij}]_{m \times n}.$$

The **scalar multiple of A by c** is given by

$$cA = [ca_{ij}]_{m\times n}.$$

Example 1

The following illustrate the operations in Definition 2.6:

$$\begin{bmatrix} 3-2 & 2^2 \\ -1 & 5 \end{bmatrix} = \begin{bmatrix} 1 & 5-1 \\ -1 & 2+3 \end{bmatrix};$$

$$\begin{bmatrix} 3 & 1 \\ -1 & 4 \end{bmatrix} + \begin{bmatrix} 1 & 2 \\ 3 & -1 \end{bmatrix} = \begin{bmatrix} 3+1 & 1+2 \\ -1+3 & 4-1 \end{bmatrix} = \begin{bmatrix} 4 & 3 \\ 2 & 3 \end{bmatrix};$$

$$3\begin{bmatrix} 1 & 4 \\ -1 & -2 \end{bmatrix} = \begin{bmatrix} 3 & 12 \\ -3 & -6 \end{bmatrix}.$$ □

A matrix with all entries equal to zero is called a **zero matrix** and is denoted by 0. If A denotes an $m \times n$ matrix and 0 the $m \times n$ zero matrix, then

$$A + 0 = 0 + A = A,$$

so the $m \times n$ matrix 0 is the **additive identity** for $m \times n$ matrices. The **negative of A** is given by

$$-A = (-1)A$$

To see why, let $A = [a_{ij}]_{m\times n}$. By Definition 2.6, we have

$$(-1)A = [-a_{ij}]_{m\times n}.$$

Therefore

$$\begin{aligned} A + (-1)A &= [a_{ij}]_{m\times n} + [-a_{ij}]_{m\times n} \\ &= [0]_{m\times n} \\ &= 0. \end{aligned}$$

Since $-a_{ij} + a_{ij} = 0$, we have the second relation for $-A$:

$$(-1)A + A = 0.$$

In a similar way, the number 1 is the **identity for multiplication by a scalar**. That is, for any matrix A,

$$1A = A.$$

Multiplication by a scalar is associative, for if $A = [a_{ij}]_{m\times n}$, then

$$\begin{aligned}(bc)A &= \left[(bc)a_{ij}\right]_{m\times n}\\ &= \left[b(ca_{ij})\right]_{m\times n}\\ &= b[ca_{ij}]_{m\times n}\\ &= b(cA).\end{aligned}$$

We can solve certain matrix equations using associativity, the additive identity, and the negative of a matrix. For instance, if a, c, and B are known and $a \neq 0$, then the solution of

$$aA + cB = 0$$

for A is given by

$$A = -\frac{c}{a}B.$$

Example 2

Solve and interpret the matrix equation

$$3A - 4B = 0,$$

where A is unknown and

$$B = \begin{bmatrix} 3 & -6 \\ 2 & 1 \end{bmatrix}.$$

Solution

Add $4B$ to both sides and multiply through by 1/3:

$$A = \frac{4}{3}B = \begin{bmatrix} 4 & -8 \\ \frac{8}{3} & \frac{4}{3} \end{bmatrix}.$$

To interpret this result, let

$$A = \begin{bmatrix} a_{11} & a_{12} \\ a_{21} & a_{22} \end{bmatrix},$$

and substitute into the given equation. This gives

$$3\begin{bmatrix} a_{11} & a_{12} \\ a_{21} & a_{22} \end{bmatrix} - 4\begin{bmatrix} 3 & -6 \\ 2 & 1 \end{bmatrix} = \begin{bmatrix} 0 & 0 \\ 0 & 0 \end{bmatrix},$$

from which

$$\begin{bmatrix} 3a_{11} - 12 & 3a_{12} + 24 \\ 3a_{12} - \ 8 & 3a_{22} - \ 4 \end{bmatrix} = \begin{bmatrix} 0 & 0 \\ 0 & 0 \end{bmatrix}.$$

By Definition 2.6, these matrices are equal if and only if their corresponding entries are equal. Thus the given matrix equation is equivalent to the linear system

$$\begin{aligned} 3a_{11} \qquad\qquad\qquad\qquad - 12 &= 0 \\ 3a_{12} \qquad\qquad\quad + 24 &= 0 \\ 3a_{21} \qquad\quad - \ 8 &= 0 \\ 3a_{22} - \ 4 &= 0. \end{aligned}$$

The solution of this system is $(4, -8, 8/3, 4/3)$, in agreement with the direct solution of the matrix equation above. □

Example 3

Solve the matrix equation

$$\begin{bmatrix} x + 2y + \ z & 1 \\ x + 4y + 2z & 2x + y - z \end{bmatrix} = \begin{bmatrix} 3 & 1 \\ 4 & 0 \end{bmatrix}.$$

Solution

Since the entries at row 1 and column 2 agree, this matrix equation is equivalent to the system

$$\begin{aligned} x + 2y + \ z &= 3 \\ x + 4y + 2z &= 4 \\ 2x + \ y - \ z &= 0. \end{aligned}$$

The solution $(2, -1, 3)$ can be found by Gaussian elimination. □

Matrix Multiplication

We use the inner product to define *matrix multiplication*. For this operation, we regard each row and each column of any matrix a vector, as illustrated in the following example.

Example 4

For the 2×3 matrix

$$A = \begin{bmatrix} 2 & -7 & 4 \\ 3 & 1 & 5 \end{bmatrix},$$

the vectors formed from the rows are

$$\mathbf{row}_1 = (2, -7, 4), \qquad \mathbf{row}_2 = (3, 1, 5).$$

The vectors formed from the columns of A may be written more descriptively as *column vectors*, in which the components are arranged vertically (a formal definition will follow):

$$\mathbf{col}_1 = \begin{bmatrix} 2 \\ 3 \end{bmatrix}, \quad \mathbf{col}_2 = \begin{bmatrix} -7 \\ 1 \end{bmatrix}, \quad \mathbf{col}_3 = \begin{bmatrix} 4 \\ 5 \end{bmatrix}. \quad \square$$

If the number of columns of a matrix A equals the number of rows of B, the inner product of any row vector of A and any column vector of B is defined. Thus if

$$A = [a_{ij}]_{m \times n} \quad \text{and} \quad B = [b_{jk}]_{n \times p},$$

then for $i = 1, \ldots, m$ and $k = 1, \ldots, p$, we may form the inner product

$$\langle \mathbf{row}_i(\mathbf{A}), \mathbf{col}_k(\mathbf{B}) \rangle = a_{i1}b_{1k} + \cdots + a_{in}b_{nk}.$$

Example 5

If

$$A = \begin{bmatrix} 2 & 1 & -3 \\ 1 & 2 & 4 \end{bmatrix} \quad \text{and} \quad B = \begin{bmatrix} 4 & -1 & -7 & 5 \\ 3 & 2 & 0 & -2 \\ 1 & -1 & 4 & 3 \end{bmatrix},$$

then A has three columns and B has three rows, so the inner product of any row vector of A and any column vector of B is defined. For instance, from row 1 of A

and column 1 of B, we have

$$\langle \mathbf{row_1(A)}, \mathbf{col_1(B)} \rangle = (2)(4) + (1)(3) + (-3)(1) = 8.$$

From row 2 of A and column 3 of B,

$$\langle \mathbf{row_2(A)}, \mathbf{col_3(B)} \rangle = (1)(-7) + (2)(0) + (4)(4) = 9. \quad \square$$

In all, eight inner products can be found for the matrices in Example 5, since A has 2 rows and B has 4 columns. More generally, for an $m \times n$ matrix A and an $n \times p$ matrix B, the number of ways of pairing a row of A with a column of B is mp. Each inner product of a row of A and a column of B is an entry in the *matrix product* AB.

DEFINITION 2.7

If $A = [a_{ij}]_{m\times n}$ and $B = [b_{jk}]_{n\times p}$, then the **matrix product AB** is given by

$$AB = [c_{ik}]_{m\times p},$$

where

$$c_{ik} = \langle \mathbf{row_i(A)}, \mathbf{col_k(B)} \rangle = a_{i1}b_{1k} + \cdots + a_{in}b_{nk}.$$

Example 6

For the matrices

$$A = \begin{bmatrix} 3 & -2 \\ 2 & 3 \\ 4 & 0 \end{bmatrix}, \qquad B = \begin{bmatrix} 1 & -1 \\ 4 & 5 \end{bmatrix},$$

the product AB is given by

$$AB = \begin{bmatrix} 3 & -2 \\ 2 & 3 \\ 4 & 0 \end{bmatrix}\begin{bmatrix} 1 & -1 \\ 4 & 6 \end{bmatrix} = \begin{bmatrix} (3)(1) + (-2)(4) & (3)(-1) + (-2)(6) \\ (2)(1) + \quad (3)(4) & (2)(-1) + \quad (3)(6) \\ (4)(1) + \quad (0)(4) & (4)(-1) + \quad (0)(6) \end{bmatrix}$$

$$= \begin{bmatrix} -5 & -15 \\ 14 & 16 \\ 4 & -4 \end{bmatrix}. \quad \square$$

Figure 2.7 shows the matrix product with the entry at row i and column k in AB formed from row i of A and column k of B.

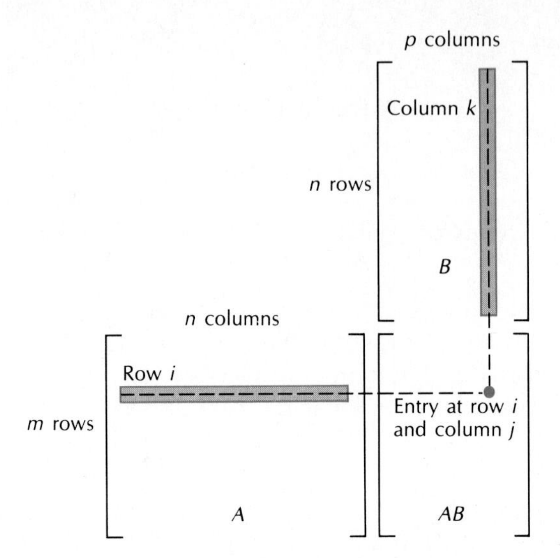

Figure 2.7 Forming the entry at row i and column k in the matrix product AB.

Any $m \times n$ linear system can be written conveniently as a single matrix equation. If

$$A = \begin{bmatrix} a_{11} & \cdots & a_{1n} \\ \vdots & \cdots & \vdots \\ a_{m1} & \cdots & a_{mn} \end{bmatrix}, \qquad X = \begin{bmatrix} x_1 \\ \vdots \\ x_n \end{bmatrix}, \qquad B = \begin{bmatrix} b_1 \\ \vdots \\ b_m \end{bmatrix},$$

then

$$AX = \begin{bmatrix} a_{11} & \cdots & a_{1n} \\ \vdots & & \vdots \\ a_{m1} & \cdots & a_{mn} \end{bmatrix} \begin{bmatrix} x_1 \\ \vdots \\ x_n \end{bmatrix} = \begin{bmatrix} a_{11}x_1 + \cdots + a_{1n}x_n \\ \vdots \qquad\qquad \vdots \\ a_{m1}x_1 + \cdots + a_{mn}x_n \end{bmatrix}.$$

Since two matrices are equal if and only if their corresponding entries agree, a single matrix equation denotes an entire linear system. That is,

$$AX = B \quad \text{is equivalent to} \quad \left\{ \begin{matrix} a_{11}x_1 + \cdots + a_{1n}x_n = b_1 \\ \vdots \qquad\qquad \vdots \quad \vdots \\ a_{m1}x_1 + \cdots + a_{mn}x_n = b_m \end{matrix} \right\}.$$

Example 7

Solve the matrix equation

$$AX = B,$$

where

$$A = \begin{bmatrix} 1 & 4 \\ -1 & 2 \end{bmatrix}, \quad X = \begin{bmatrix} x_1 \\ x_2 \end{bmatrix}, \quad B = \begin{bmatrix} -1 \\ -5 \end{bmatrix}.$$

Solution

The given matrix equation is equivalent to the linear system

$$\begin{aligned} x_1 + 4x_2 &= -1 \\ -x_1 + 2x_2 &= -5, \end{aligned}$$

and Gaussian elimination yields the solution $(3, -1)$. □

Matrices with one row or one column are especially important in matrix multiplication. A $1 \times n$ matrix

$$A = [a_1 \ \cdots \ a_n]$$

is called a **row vector**, since it represents the n-tuple

$$\mathbf{a} = (a_1, \ldots, a_n)$$

arranged in a row. Similarly, an $n \times 1$ matrix

$$B = \begin{bmatrix} b_1 \\ \vdots \\ b_n \end{bmatrix}$$

is called a **column vector**, since it represents the n-tuple

$$\mathbf{b} = (b_1, \ldots, b_n)$$

arranged in a column. By Definition 2.7,

$$AB = [a_1 \ \cdots \ a_n] \begin{bmatrix} b_1 \\ \vdots \\ b_n \end{bmatrix} = [a_1 b_1 + \cdots + a_n b_n].$$

When A is $n \times 1$ and B is $1 \times n$, we often write

$$AB = \langle \mathbf{a}, \mathbf{b} \rangle,$$

regarding a 1×1 matrix as a scalar.

Now we formulate a concept used briefly for row and column vectors here and extensively for general matrices later.

DEFINITION 2.8

If $A = [a_{ij}]_{m\times n}$, the **transpose of A** is given by

$$A^t = [a_{ji}]_{n\times m}.$$

Thus the rows of A^t are the columns of A, and the columns of A^t are the rows of A. Another common symbol for the transpose of A is A'. The symbol A^t will be used throughout this book.

Example 8

The transpose of

$$A = \begin{bmatrix} -4 & 1 & 2 & 1 \\ 3 & 0 & 1 & 5 \\ 4 & 7 & -1 & 8 \end{bmatrix}$$

is given by

$$A^t = \begin{bmatrix} -4 & 3 & 4 \\ 1 & 0 & 7 \\ 2 & 1 & -1 \\ 1 & 5 & 8 \end{bmatrix}.$$

Example 9

If $\mathbf{x} = (x_1, \ldots, x_n)$ and $\mathbf{y} = (y_1, \ldots, y_n)$, let X and Y be the corresponding column vectors. Then the inner product of $\mathbf{x}$ and $\mathbf{y}$ is given by

$$\langle \mathbf{x}, \mathbf{y} \rangle = X^t Y,$$

and the norm of $\mathbf{x}$ satisfies

$$\|\mathbf{x}\|^2 = X^t X. \quad \square$$

Properties of matrix products are very important, and are studied in the next section. We recommend a thorough grounding in the mechanics of matrix operations in preparation for this study.

REVIEW CHECKLIST

1. For matrices A and B and a scalar c, define matrix equality $A = B$, the matrix sum $A + B$, and the scalar multiple of A by c, cA.
2. Given $m \times n$ matrices A and B and a scalar c, find $A + B$ and cA.
3. Solve matrix equations using the definition of matrix equality.
4. Define the matrix product AB.
5. Find AB for any A and B for which the product is defined.
6. Define row vector, column vector, and the transpose of a matrix.
7. Find the transpose of a given matrix.

EXERCISES

In Exercises 1–8, find the indicated matrix if

$$A = \begin{bmatrix} 2 & 3 \\ -1 & 0 \end{bmatrix} \quad \text{and} \quad B = \begin{bmatrix} 1 & -1 \\ 4 & 2 \end{bmatrix}.$$

1. $A + B$
2. $3A$
3. $2A + 3B$
4. $A - B$
5. $4A - 2B$
6. $A + 0$
7. $2(3A)$
8. $6A$
9. Find the matrix A satisfying the equation

$$4A - 3\begin{bmatrix} 1 & 2 \\ -3 & 2 \end{bmatrix} = 0.$$

10. Find a and b such that

$$2\begin{bmatrix} a & -4 \\ 1 & b \end{bmatrix} - 5\begin{bmatrix} 1 & -3 \\ 2 & 1 \end{bmatrix} = \begin{bmatrix} 11 & 7 \\ -8 & 3 \end{bmatrix}.$$

In Exercises 11–16, find the indicated matrix if c_1 and c_2 are scalars, and

$$A = \begin{bmatrix} 1 & 3 \\ -1 & 2 \\ 3 & 4 \end{bmatrix}, \quad B = \begin{bmatrix} 0 & 1 \\ 3 & 2 \\ -2 & 3 \end{bmatrix}, \quad \text{and} \quad C = \begin{bmatrix} 2 & -3 \\ 1 & 2 \\ 4 & -1 \end{bmatrix}.$$

11. $(A + B) + C$
12. $A + (B + C)$
13. $c_1(A + B)$
14. $c_1A + c_1B$
15. $(c_1 + c_2)A$
16. $c_1A + c_2A$
17. Express the following sum as a single 2×2 matrix.

$$a_{11}\begin{bmatrix} 1 & 0 \\ 0 & 0 \end{bmatrix} + a_{12}\begin{bmatrix} 0 & 1 \\ 0 & 0 \end{bmatrix} + a_{21}\begin{bmatrix} 0 & 0 \\ 1 & 0 \end{bmatrix} + a_{22}\begin{bmatrix} 0 & 0 \\ 0 & 1 \end{bmatrix}$$

In Exercises 18–25, find the indicated matrix product.

18. $\begin{bmatrix} 2 & 1 \\ 3 & -1 \end{bmatrix}\begin{bmatrix} 2 & 0 \\ 1 & 4 \end{bmatrix}$
19. $\begin{bmatrix} 2 & 0 \\ 1 & 4 \end{bmatrix}\begin{bmatrix} 2 & 1 \\ 3 & -1 \end{bmatrix}$

20. $\begin{bmatrix} 1 & 3 \end{bmatrix}\begin{bmatrix} 2 \\ 5 \end{bmatrix}$

21. $\begin{bmatrix} 3 & 4 & 1 \\ 2 & 2 & 5 \end{bmatrix}\begin{bmatrix} 1 & 2 & 3 \\ -2 & 1 & 1 \\ 1 & 0 & 1 \end{bmatrix}$

22. $\begin{bmatrix} 1 & 4 & 1 \\ -2 & 1 & 3 \\ 0 & 1 & 5 \end{bmatrix}\begin{bmatrix} 2 & 2 & 1 \\ -1 & -1 & -3 \\ 2 & -4 & 2 \end{bmatrix}$

23. $\begin{bmatrix} 2 & 2 & 1 \\ -1 & -1 & -3 \\ 2 & -4 & 2 \end{bmatrix}\begin{bmatrix} 1 & 4 & 1 \\ -2 & 1 & 3 \\ 0 & 1 & 5 \end{bmatrix}$

24. $\begin{bmatrix} 2 & 1 \\ -1 & 2 \end{bmatrix}\begin{bmatrix} 3 & 4 \\ -4 & 3 \end{bmatrix}$

25. $\begin{bmatrix} 3 & 4 \\ -4 & 3 \end{bmatrix}\begin{bmatrix} 2 & 1 \\ -1 & 2 \end{bmatrix}$

In Exercises 26–45, find the indicated matrix if

$$A = \begin{bmatrix} 2 & -1 \\ 3 & 1 \end{bmatrix}, \quad B = \begin{bmatrix} -1 & 2 \\ 1 & 4 \end{bmatrix}, \quad \text{and} \quad C = \begin{bmatrix} 1 & 3 \\ 0 & 1 \end{bmatrix}.$$

26. AB **27.** BA **28.** $A(B + C)$ **29.** $AB + AC$
30. A^t **31.** B^t **32.** A^tB^t **33.** B^tA^t
34. $(AB)^t$ **35.** $(BA)^t$ **36.** AB^t **37.** A^tB
38. $(A^t)^t$ **39.** $(AB^t)^t$ **40.** $(A^tB)^t$ **41.** $2A^t$
42. $(2A)^t$ **43.** AA^t **44.** A^tA **45.** $(A + B)^t$

In Exercises 46–53, if possible, find the indicated matrix for

$$I = \begin{bmatrix} 1 & 0 \\ 0 & 1 \end{bmatrix} \quad \text{and} \quad A = \begin{bmatrix} 1 & 3 & 1 \\ -1 & 2 & 4 \end{bmatrix}.$$

46. IA **47.** AA **48.** $3I$ **49.** $3A$
50. $(3I)A$ **51.** $A - 3I$ **52.** $3I - A$ **53.** $0A$

54. Give an example of 2×2 matrices A and B for which $A \neq 0$, $B \neq 0$, and $AB = 0$.

55. Solve the linear system $AX = B$ where

$$A = \begin{bmatrix} 2 & -1 \\ 4 & -3 \end{bmatrix}, \quad X = \begin{bmatrix} x_1 \\ x_2 \end{bmatrix}, \quad \text{and} \quad B = \begin{bmatrix} 5 \\ 7 \end{bmatrix}.$$

56. Solve the linear system $AX = 0$ where

$$A = \begin{bmatrix} 1 & 4 & 3 \\ 2 & 7 & 7 \\ 1 & 6 & 1 \end{bmatrix} \quad \text{and} \quad X = \begin{bmatrix} x_1 \\ x_2 \\ x_3 \end{bmatrix}.$$

57. Find AA^t and A^tA if

$$A = \begin{bmatrix} \frac{2}{3} & -\frac{1}{3} & -\frac{2}{3} \\ \frac{2}{3} & \frac{2}{3} & \frac{1}{3} \\ \frac{1}{3} & -\frac{2}{3} & \frac{2}{3} \end{bmatrix}.$$

In Exercises 58–61, find AB and BA for the given A and B.

58. $A = \begin{bmatrix} 1 & 0 & 0 \\ 0 & 1 & 0 \\ 0 & 0 & 4 \end{bmatrix}, \quad B = \begin{bmatrix} -1 & 0 & 0 \\ 0 & 2 & 0 \\ 0 & 0 & -2 \end{bmatrix}$

59. $A = \begin{bmatrix} 1 & 3 & -1 \\ 0 & 2 & 1 \\ 0 & 0 & 2 \end{bmatrix}, \quad B = \begin{bmatrix} 2 & -1 & 2 \\ 0 & 4 & -2 \\ 0 & 0 & 3 \end{bmatrix}$

60. $A = \begin{bmatrix} 0 & 1 & 0 \\ 1 & 0 & 0 \\ 0 & 0 & 1 \end{bmatrix}, \quad B = \begin{bmatrix} 2 & 4 & -2 \\ 9 & -7 & 3 \\ 5 & -2 & -6 \end{bmatrix}$

61. $A = \begin{bmatrix} 1 & 0 & 0 \\ 2 & 1 & 0 \\ -1 & 0 & 1 \end{bmatrix}, \quad B = \begin{bmatrix} 2 & 0 & 0 \\ 4 & -1 & 3 \\ 3 & 5 & 2 \end{bmatrix}$

2.3 PROPERTIES OF MATRIX MULTIPLICATION

When matrix operations are used to solve linear systems or to find information such as a steady-state distribution associated with a physical system, the rules governing these operations must be followed. The rules, or properties, of matrix operations form an essential part of a rich mathematical structure. They give guidance and insight for advanced studies in modern algebra and other important areas of mathematics such as *operator theory*.

Associative and Distributive Laws

The operations used in solving scalar equations are applied in accordance with the rules of algebra. For instance, factoring is governed by the *left* and *right distributive* laws

$$x(y+z) = xy + xz \qquad \text{and} \qquad (x+y)z = xz + yz$$

where x, y, and z are real numbers. Matrix operations play a corresponding role in the solution of linear systems, and similar rules govern their application. Properties of matrix addition and multiplication of a matrix by a scalar were noted in Section 2.2. This section is devoted to matrix multiplication.

The following example illustrates the associative law.

Example 1

Verify that $(AB)C = A(BC)$ for the matrices

$$A = \begin{bmatrix} 2 & -1 \\ 3 & 4 \end{bmatrix}, \quad B = \begin{bmatrix} 0 & 2 \\ 5 & -3 \end{bmatrix}, \quad C = \begin{bmatrix} 4 & 1 \\ -2 & 2 \end{bmatrix}.$$

Solution

Apply Definition 2.7 to the four products in question:

$$AB = \begin{bmatrix} (2)(0) + (-1)(5) & (2)(2) + (-1)(-3) \\ (3)(0) + \quad (4)(5) & (3)(2) + \quad (4)(-3) \end{bmatrix} = \begin{bmatrix} -5 & 7 \\ 20 & -6 \end{bmatrix};$$

$$(AB)C = \begin{bmatrix} (-5)(4) + \quad (7)(-2) & (-5)(1) + \quad (7)(2) \\ (20)(4) + (-6)(-2) & (20)(1) + (-6)(2) \end{bmatrix} = \begin{bmatrix} -34 & 9 \\ 92 & 8 \end{bmatrix};$$

$$BC = \begin{bmatrix} (0)(4) + \quad (2)(-2) & (0)(1) + \quad (2)(2) \\ (5)(4) + (-3)(-2) & (5)(1) + (-3)(2) \end{bmatrix} = \begin{bmatrix} -4 & 4 \\ 26 & -1 \end{bmatrix};$$

$$A(BC) = \begin{bmatrix} (2)(-4) + (-1)(26) & (2)(4) + (-1)(-1) \\ (3)(-4) + \quad (4)(26) & (3)(4) + \quad (4)(-1) \end{bmatrix} = \begin{bmatrix} -34 & 9 \\ 92 & 8 \end{bmatrix}.$$

Since the corresponding entries in $(AB)C$ and $A(BC)$ agree, these matrices are equal by Definition 2.1. ☐

Example 1 illustrates a fundamental property of matrix multiplication. The general result is stated here and proved at the end of this section.

THEOREM 2.2 (Associative law)

If A, B, and C are $m \times n$, $n \times p$, and $p \times q$ matrices, respectively, then

$$A(BC) = (AB)C.$$

The calculations in Example 1 were straightforward, but not very revealing. Can the work be arranged in a way that suggests why the law holds more generally?

Example 2

Verify the associative law of matrix products for the matrices in Example 1, arranging the calculations as in the general proof of the associative law.

Solution

In Example 1, the entries in $(AB)C$ were ultimately obtained as:

$$[(2)(0) + (-1)(5)](4) + [(2)(2) + (-1)(-3)](-2) = -34$$

$$[(2)(0) + (-1)(5)](1) + [(2)(2) + (-1)(-3)] \quad (2) = \quad 9$$

$$[(3)(0) + (4)(5)](4) + [(3)(2) + (4)(-3)](-2) = 92$$

$$[(3)(0) + (4)(5)](1) + [(3)(2) + (4)(-3)](2) = 8.$$

In each lefthand side, remove the brackets, group the first and third terms, group the second and fourth terms, and remove the new common factors. The resulting expressions are

$$(2)[(0)(4) + (2)(-2)] + (-1)[(5)(4) + (-3)(-2)]$$

$$(2)[(0)(1) + (2)(2)] + (-1)[(5)(1) + (-3)(2)]$$

$$(3)[(0)(4) + (2)(-2)] + (4)[(5)(4) + (-3)(-2)]$$

$$(3)[(0)(1) + (2)(2)] + (4)[(5)(1) + (-3)(2)].$$

Combining within the brackets gives the expressions from which the entries in the product $A(BC)$ were formed in Example 1. □

Example 3

Verify that $A(B + C) = AB + AC$ for the matrices

$$A = \begin{bmatrix} 3 & 1 & -2 \\ 1 & 2 & 4 \\ 2 & 1 & 1 \end{bmatrix}, \quad B = \begin{bmatrix} 1 & 2 \\ -1 & -3 \\ 4 & 5 \end{bmatrix}, \quad \text{and} \quad C = \begin{bmatrix} 0 & -1 \\ 2 & 5 \\ -6 & -3 \end{bmatrix}.$$

Solution

Apply Definitions 2.1 and 2.7:

$$A(B + C) = \begin{bmatrix} 3 & 1 & -2 \\ 1 & 2 & 4 \\ 2 & 1 & 1 \end{bmatrix}\begin{bmatrix} 1 & 1 \\ 1 & 2 \\ -2 & 2 \end{bmatrix} = \begin{bmatrix} 8 & 1 \\ -5 & 13 \\ 1 & 6 \end{bmatrix};$$

$$AB + AC = \begin{bmatrix} -6 & -7 \\ 15 & 16 \\ 5 & 6 \end{bmatrix} + \begin{bmatrix} 14 & 8 \\ -20 & -3 \\ -4 & 0 \end{bmatrix} = \begin{bmatrix} 8 & 1 \\ -5 & 13 \\ 1 & 6 \end{bmatrix}.$$

Since the corresponding entries agree, $A(B + C) = AB + AC$. □

THEOREM 2.3 (Left and right distributive laws)

If A is an $m \times n$ matrix and B and C are $n \times p$ matrices, then

$$A(B + C) = AB + AC.$$

If A and B are $m \times n$ matrices and C is an $n \times p$ matrix, then

$$(A + B)C = AC + BC.$$

Proof We prove the left distributive law, leaving the right as an exercise. Let $A = [a_{ij}]_{m \times n}$, $B = [b_{jk}]_{n \times p}$, and $C = [c_{jk}]_{n \times p}$. Then

$$B + C = \left[b_{jk} + c_{jk}\right]_{n \times p},$$

hence

$$\begin{aligned} A(B + C) &= \left[a_{i1}(b_{1k} + c_{1k}) + \cdots + a_{in}(b_{nk} + c_{nk})\right]_{m \times p} \\ &= \left[(a_{i1}b_{1k} + \cdots + a_{in}b_{nk}) + (a_{i1}c_{1k} + \cdots + a_{in}c_{nk})\right]_{m \times p} \\ &= [a_{i1}b_{1k} + \cdots + a_{in}b_{nk}]_{m \times p} + [a_{i1}c_{1k} + \cdots + a_{in}c_{nk}]_{m \times p} \\ &= AB + AC. \quad \square \end{aligned}$$

Commutativity Questions

For matrices A and B, we say A commutes with B if $AB = BA$. If A is $m \times n$ and B is $n \times p$, then AB is defined. If also $p = m$, then BA is defined. Thus, both AB and BA are defined if and only if A is $m \times n$ and B is $n \times m$. In this case, AB is $m \times m$ and BA is $n \times n$. Thus, AB cannot equal BA unless both A and B are $n \times n$. When A and B are both $n \times n$, AB may or may not equal BA.

Example 4

If

$$A = \begin{bmatrix} 1 & 2 \\ -1 & 3 \end{bmatrix} \quad \text{and} \quad B = \begin{bmatrix} 2 & 2 \\ 3 & 1 \end{bmatrix},$$

then

$$AB = \begin{bmatrix} 8 & 4 \\ 7 & 1 \end{bmatrix} \quad \text{and} \quad BA = \begin{bmatrix} 0 & 10 \\ 2 & 9 \end{bmatrix}.$$

On the other hand, if

$$A = \begin{bmatrix} 1 & -1 \\ 1 & 1 \end{bmatrix} \quad \text{and} \quad B = \begin{bmatrix} 2 & -1 \\ 1 & 2 \end{bmatrix},$$

then

$$AB = \begin{bmatrix} 1 & -3 \\ 3 & 1 \end{bmatrix} \quad \text{and} \quad BA = \begin{bmatrix} 1 & -3 \\ 3 & 1 \end{bmatrix}.$$

For the first pair, $AB \neq BA$, but for the second, $AB = BA$. ☐

Example 4 establishes the following basic result.

If A and B are $n \times n$ matrices, then A may or may not commute with B.

Certain types of matrices do have special commutativity properties.

Example 5

For the matrices

$$A = \begin{bmatrix} 1 & 5 \\ 4 & -2 \end{bmatrix}, \quad I = \begin{bmatrix} 1 & 0 \\ 0 & 1 \end{bmatrix}, \quad \text{and} \quad 0 = \begin{bmatrix} 0 & 0 \\ 0 & 0 \end{bmatrix},$$

verify that $AI = IA = A$ and $A0 = 0A = 0$.

Solution

Carry out the indicated multiplications. For I,

$$AI = \begin{bmatrix} 1 & 5 \\ 4 & -2 \end{bmatrix}\begin{bmatrix} 1 & 0 \\ 0 & 1 \end{bmatrix} = \begin{bmatrix} 1 & 5 \\ 4 & -2 \end{bmatrix} = A$$

and

$$IA = \begin{bmatrix} 1 & 0 \\ 0 & 1 \end{bmatrix}\begin{bmatrix} 1 & 5 \\ 4 & -2 \end{bmatrix} = \begin{bmatrix} 1 & 5 \\ 4 & -2 \end{bmatrix} = A.$$

For 0,

$$A0 = \begin{bmatrix} 1 & 5 \\ 4 & -2 \end{bmatrix}\begin{bmatrix} 0 & 0 \\ 0 & 0 \end{bmatrix} = \begin{bmatrix} 0 & 0 \\ 0 & 0 \end{bmatrix} = 0$$

and

$$0A = \begin{bmatrix} 0 & 0 \\ 0 & 0 \end{bmatrix}\begin{bmatrix} 1 & 5 \\ 4 & -2 \end{bmatrix} = \begin{bmatrix} 0 & 0 \\ 0 & 0 \end{bmatrix} = 0.$$ ☐

The matrix I in Example 5 acts as a *multiplicative identity* for A, that is, it satisfies

$$AI = IA = A.$$

This example illustrates an important general concept.

DEFINITION 2.9

The $n \times n$ identity matrix $I = [d_{ij}]_{n \times n}$ is defined by

$$d_{ij} = \begin{Bmatrix} 1, & \text{if } i = j \\ 0, & \text{if } i \neq j \end{Bmatrix}.$$

Example 6

The identity matrices for $n = 2$, 3, and 4 are given by

$$I_2 = \begin{bmatrix} 1 & 0 \\ 0 & 1 \end{bmatrix}, \quad I_3 = \begin{bmatrix} 1 & 0 & 0 \\ 0 & 1 & 0 \\ 0 & 0 & 1 \end{bmatrix}, \quad I_4 = \begin{bmatrix} 1 & 0 & 0 & 0 \\ 0 & 1 & 0 & 0 \\ 0 & 0 & 1 & 0 \\ 0 & 0 & 0 & 1 \end{bmatrix},$$

respectively. □

Example 5 above also illustrates the characteristic property of an identity matrix in general.

IDENTITY PROPERTY

If A is an $n \times n$ matrix and I the $n \times n$ identity, then

$$AI = IA = A.$$

If A is any $n \times n$ matrix, the **powers of A**, $A^0, A^1, A^2, \ldots,$ are defined as follows:

$$A^0 = I,$$

and for $k = 1, 2, 3, \ldots,$

$$A^k = A(A^{k-1}).$$

Thus, for instance,

$$A^1 = A(A^0) = AI = A.$$

The **square of A** is given by

$$A^2 = A(A^1) = AA$$

and the **cube of A** by

$$A^3 = A(A^2) = A(AA) = (AA)A = (A^2)A.$$

We may establish by mathematical induction (Appendix A) that for every nonnegative integer k, the kth power of A is defined and satisfies

$$AA^k = A^kA = A^{k+1}.$$

With this result established, another mathematical induction argument proves that any two powers of A commute. Details of the proofs are assigned in Exercises 36–38.

COMMUTATIVITY OF POWERS OF A

If A is an $n \times n$ matrix, then for any nonnegative integers j and k,

$$A^jA^k = A^kA^j = A^{j+k}.$$

Commutativity affects certain rules of algebra. For instance, for the square of a binomial, we have

$$(A + B)^2 = A^2 + BA + AB + B^2.$$

To see why, apply the distributive laws:

$$(A + B)(A + B) = (A + B)A + (A + B)B = AA + BA + AB + BB.$$

The center two terms combine to $2AB$ if and only if $AB = BA$. Similarly,

$$(A - B)(A + B) = A^2 - BA + AB - B^2.$$

The right side reduces to $A^2 - B^2$ if A and B commute. The laws of exponents are also affected. For instance,

$$(AB)^3 = ABABAB.$$

If A and B commute, then the right side reduces to A^3B^3. (See Exercises 39–41.)

The following example illustrates another class of matrices with a useful commutativity property.

Example 7

Show that $AB = BA$ for the matrices

$$A = \begin{bmatrix} 2 & 0 & 0 \\ 0 & 5 & 0 \\ 0 & 0 & -2 \end{bmatrix} \quad \text{and} \quad B = \begin{bmatrix} 3 & 0 & 0 \\ 0 & 1 & 0 \\ 0 & 0 & 4 \end{bmatrix}.$$

Solution

Actually,

$$\begin{bmatrix} 2 & 0 & 0 \\ 0 & 5 & 0 \\ 0 & 0 & -2 \end{bmatrix}\begin{bmatrix} 3 & 0 & 0 \\ 0 & 1 & 0 \\ 0 & 0 & 4 \end{bmatrix} = \begin{bmatrix} 3 & 0 & 0 \\ 0 & 1 & 0 \\ 0 & 0 & 4 \end{bmatrix}\begin{bmatrix} 2 & 0 & 0 \\ 0 & 5 & 0 \\ 0 & 0 & -2 \end{bmatrix} = \begin{bmatrix} 6 & 0 & 0 \\ 0 & 5 & 0 \\ 0 & 0 & -8 \end{bmatrix}.$$

Matrices like those in Example 7 play an important role in linear algebra.

DEFINITION 2.10

If $A = [a_{ij}]_{n\times n}$, the **main diagonal** of A consists of the entries $a_{11}, a_{22}, \ldots, a_{nn}$. A matrix D is said to be a **diagonal matrix** if all its entries, except possibly those on the main diagonal, are zero.

Example 8

The main diagonal is shaded in the following matrix:

$$\begin{bmatrix} 4 & -1 & 2 \\ 3 & 1 & 5 \\ 2 & -3 & -4 \end{bmatrix}. \quad \square$$

A diagonal matrix D is often denoted

$$D = \text{diag}(d_1, \ldots, d_n),$$

where $d_i = d_{ii}$ for $i = 1, \ldots, n$. If $D = \text{diag}(d_1, \ldots, d_n)$ and $E = \text{diag}(e_1, \ldots, e_n)$, then

$$DE = \text{diag}(d_1 e_1, \ldots, d_n e_n) = \text{diag}(e_1 d_1, \ldots, e_n d_n) = ED.$$

That is, *diagonal matrices commute.*

We establish a useful formula for the transpose of a product, and note an immediate corollary.

THEOREM 2.4

If $A = [a_{ij}]_{m\times n}$ and $B = [b_{jk}]_{n\times p}$, then

$$(AB)^t = B^tA^t.$$

Proof Denote the entry at row i and column k in B^tA^t by d_{ik} and the entry at row k and column i in AB by c_{ki}. Then

$$d_{ik} = \mathbf{row_i(B^t)} \cdot \mathbf{col_k(A^t)} = \mathbf{col_i(B)} \cdot \mathbf{row_k(A)} = \mathbf{row_k(A)} \cdot \mathbf{col_i(B)} = c_{ki},$$

from which

$$B^tA^t = [d_{ik}]_{p\times m} = [c_{ki}]_{p\times m} = (AB)^t. \quad \square$$

Corollary If A and B are $n \times n$ matrices, then

$$AB = BA \qquad \text{if and only if} \qquad A^tB^t = B^tA^t.$$

Proof of the corollary is assigned in Exercise 42.

Divisors of Zero and Cancellation

The real number system has no elements that "divide zero," i.e., for real numbers a and b,

$$ab = 0 \qquad \text{if and only if} \qquad a = 0 \quad \text{or} \quad b = 0.$$

Is a similar statement true for matrices?

Example 8

If

$$A = \begin{bmatrix} 1 & 2 \\ 2 & 4 \end{bmatrix} \quad \text{and} \quad B = \begin{bmatrix} 2 & -6 \\ -1 & 3 \end{bmatrix},$$

then $A \neq 0$ and $B \neq 0$, but $AB = 0$. $\square$

An $n \times n$ matrix A is called a **left divisor of zero** if $A \neq 0$ and there is a $B \neq 0$ for which $AB = 0$. A similar definition holds for a **right divisor of zero**. If the homogeneous linear system with coefficient matrix A has nontrivial solutions, then A is a left divisor of zero, since we may form B as we did in Example 8 using such solutions as columns. If this system has only the trivial solution, then A is not a left divisor of zero.

Existence of zero divisors affects the ability to cancel common factors in equations. The real number system has left and right cancellation laws:

$$\text{if} \quad ab = ac \quad \text{and} \quad a \neq 0, \quad \text{then} \quad b = c;$$

$$\text{if} \quad ac = bc \quad \text{and} \quad c \neq 0, \quad \text{then} \quad a = b.$$

Cancellation is also lost in matrix algebra.

Example 9

If

$$A = \begin{bmatrix} 1 & 2 \\ 2 & 4 \end{bmatrix}, \quad B = \begin{bmatrix} 4 & -12 \\ -2 & 6 \end{bmatrix}, \quad \text{and} \quad C = \begin{bmatrix} 2 & -6 \\ -1 & 3 \end{bmatrix},$$

then $A(B - C) = 0$, so $AB = AC$. Since $A \neq 0$ and $B \neq C$, the left cancellation law fails for matrices. □

The condition $A \neq 0$ is not the real test for a divisor of zero or for cancellation. What is crucial is whether the homogeneous linear system with coefficient matrix A has nontrivial solutions. This question is especially important for square matrices and is studied extensively in the next two sections of this chapter and in the next two chapters. We close this section with a proof of the general associative law for matrix multiplication.

Proof of the Associative Law Let $A = [a_{ij}]_{m\times n}$, $B = [b_{jk}]_{n\times p}$, and $C = [c_{kl}]_{p\times q}$. Then

$$AB = [d_{ik}]_{m\times p},$$

where

$$d_{ik} = a_{i1}b_{1k} + \cdots + a_{in}b_{nk}.$$

Thus

$$(AB)C = [e_{il}]_{m\times q},$$

where

$$\begin{aligned} e_{il} &= d_{i1}c_{1l} + \cdots + d_{ip}c_{pl} \\ &= (a_{i1}b_{11} + \cdots + a_{in}b_{n1})c_{1l} + \cdots + (a_{i1}b_{1p} + \cdots + a_{in}b_{np})c_{pl}. \end{aligned}$$

Similarly,

$$BC = [f_{jl}]_{n \times q},$$

where

$$f_{jl} = b_{j1}c_{1l} + \cdots + b_{jp}c_{pl}.$$

Hence

$$A(BC) = [g_{il}]_{m \times q},$$

where

$$\begin{aligned} g_{il} &= a_{i1}f_{1l} + \cdots + a_{in}f_{nl} \\ &= a_{i1}(b_{11}c_{1l} + \cdots + b_{1p}c_{pl}) + \cdots + a_{in}(b_{n1}c_{1l} + \cdots + b_{np}c_{pl}). \end{aligned}$$

Removing the parentheses in this expression, rearranging the terms, and factoring $c_{1l}, \ldots, c_{pl}$ yields

$$g_{il} = (a_{i1}b_{11} + \cdots + a_{in}b_{n1})c_{1l} + \cdots + (a_{i1}b_{1p} + \cdots + a_{in}b_{np})c_{pl}.$$

Comparison with the expression for e_{il} above shows that

$$e_{il} = g_{il}$$

for $i = 1, \ldots, m$ and $l = 1, \ldots, p$, so $(AB)C = A(BC)$. ☐

REVIEW CHECKLIST

1. State the associative law for matrix products and the distributive laws for matrix multiplication over matrix addition. Apply these laws in carrying out matrix operations.
2. Give examples to show that the commutative law fails to hold for matrix multiplication.
3. Formulate definitions for the $n \times n$ identity matrix and for the powers of a square matrix.
4. Formulate definitions for the main diagonal of a matrix and for a diagonal matrix.
5. Give examples to illustrate commutativity properties for special types of matrices.
6. Formulate definitions for divisors of zero, and illustrate by example.
7. Give examples to illustrate that cancellation laws do not hold for matrix multiplication and identify conditions under which cancellation is possible.

EXERCISES

In Exercises 1–2, verify the left distributive law for the given matrices.

1. $A = \begin{bmatrix} 2 & 1 \\ -1 & 3 \end{bmatrix}$, $B = \begin{bmatrix} 1 & 4 \\ 1 & 2 \end{bmatrix}$, $C = \begin{bmatrix} -1 & 0 \\ 1 & 3 \end{bmatrix}$

2. $A = \begin{bmatrix} 3 & 1 & 2 \\ 1 & 2 & -1 \\ 2 & -1 & -2 \end{bmatrix}$, $B = \begin{bmatrix} 0 & 2 \\ -1 & 3 \\ 2 & -2 \end{bmatrix}$, $C = \begin{bmatrix} 1 & 1 \\ 2 & -1 \\ 1 & -2 \end{bmatrix}$

In Exercises 3–4, verify the right distributive law for the given matrices.

3. A, B, and C from Exercise 1

4. $A = \begin{bmatrix} 4 & -1 & 0 \\ 0 & 1 & 2 \end{bmatrix}$, $B = \begin{bmatrix} -2 & 1 & 3 \\ -1 & -2 & -2 \end{bmatrix}$, $C = \begin{bmatrix} 0 & 2 \\ 3 & -1 \\ 2 & 1 \end{bmatrix}$

In Exercises 5–6, verify the associative law of matrix multiplication for the given matrices.

5. A, B, and C from Exercise 1

6. $A = \begin{bmatrix} -1 & 1 & 2 \\ 2 & 0 & 3 \\ -2 & -1 & 1 \end{bmatrix}$, $B = \begin{bmatrix} 2 & 2 \\ 1 & -2 \\ 3 & 0 \end{bmatrix}$, $C = \begin{bmatrix} 1 \\ -1 \end{bmatrix}$

In Exercises 7–12, determine whether $AB = BA$ for the given matrices A and B.

7. $\begin{bmatrix} 3 & -1 \\ 1 & 2 \end{bmatrix}$, $\begin{bmatrix} -2 & 4 \\ 3 & -1 \end{bmatrix}$

8. $\begin{bmatrix} 5 & 0 \\ 3 & 1 \end{bmatrix}$, $\begin{bmatrix} -2 & 2 \\ 3 & -2 \end{bmatrix}$

9. $\begin{bmatrix} 1 & 4 \\ -1 & 1 \end{bmatrix}$, $\begin{bmatrix} -3 & 8 \\ -2 & -3 \end{bmatrix}$

10. $\begin{bmatrix} 2 & 1 \\ 3 & 2 \end{bmatrix}$, $\begin{bmatrix} 2 & -1 \\ -3 & 2 \end{bmatrix}$

11. $\begin{bmatrix} 1 & -1 & 2 \\ 2 & 0 & 1 \\ 3 & 1 & -1 \end{bmatrix}$, $\begin{bmatrix} 2 & -1 & -2 \\ 1 & -2 & 1 \\ 2 & 3 & -1 \end{bmatrix}$

12. $\begin{bmatrix} -1 & 0 & 1 \\ 2 & 1 & -1 \\ 1 & 2 & 2 \end{bmatrix}$, $\begin{bmatrix} 2 & 2 & 1 \\ -1 & -1 & -1 \\ 5 & 6 & 3 \end{bmatrix}$

13. Verify that $AB = BA = I$ for

$$A = \begin{bmatrix} -2 & -2 & -1 \\ 1 & 1 & 0 \\ 1 & 2 & 2 \end{bmatrix} \quad \text{and} \quad B = \begin{bmatrix} -2 & -2 & -1 \\ 2 & 3 & 1 \\ -1 & -2 & 0 \end{bmatrix}.$$

14. For the following matrices, find $ABC + ABD$ in two ways and compare the calculations for relative difficulty:

$$A = \begin{bmatrix} 1 & 2 \\ -1 & 0 \end{bmatrix}, \quad B = \begin{bmatrix} 2 & -2 \\ 3 & 1 \end{bmatrix}, \quad C = \begin{bmatrix} 3 & 2 \\ 1 & -2 \end{bmatrix}, \quad D = \begin{bmatrix} 1 & -2 \\ 1 & 3 \end{bmatrix}.$$

In Exercises 15–18, find the indicated matrix if $A = [a_{ij}]_{2\times 2}$, where $a_{ij} = ij$ for $i = 1,2$ and $j = 1,2$.

15. A **16.** $A^2 - 2A + I$ **17.** $(A - I)^2$ **18.** $A^2 - 5A$

19. If A is any square matrix, write $(A + I)^3$ as a sum of powers of A.

In Exercises 20–22, write the given expression as a matrix product, assuming that all the matrices involved are $n \times n$.

20. $(ABC)^t$ **21.** $(A_1A_2 \cdots A_k)^t$ **22.** $(A^tA)^t$

23. Write diag$(1, -1, 2)$ in standard matrix form, i.e., with nine entries enclosed in brackets.

24. If $D_1 = \text{diag}(3, -2, 0, 5)$ and $D_2 = \text{diag}(-2, 7, 1, 4)$, find D_1D_2.

25. Find a matrix A for which $A \neq 0$ and $A^2 = 0$.

26. Find a matrix A for which $A \neq 0$, $A^2 \neq 0$, and $A^3 = 0$.

27. Verify that $AB = 0$ for the matrices

$$A = \begin{bmatrix} 1 & -2 \\ -2 & 4 \end{bmatrix} \quad \text{and} \quad B = \begin{bmatrix} 6 & 8 \\ 3 & 4 \end{bmatrix}.$$

28. For the following matrix A, find a 3×3 matrix B having no zero columns and satisfying $AB = 0$:

$$A = \begin{bmatrix} 1 & -1 & 0 \\ 2 & 1 & 1 \\ 1 & -4 & -1 \end{bmatrix}.$$

29. Using properties of matrix operations, show that if $AB = AC$ and there is a matrix D such that $DA = I$, then $B = C$.

30. Show that if A and B are $n \times n$ matrices and $AB = BA$, then $(2A^2 - 3A + I)(B^3 + 2B) = (B^3 + 2B)(2A^2 - 3A + I)$.

31. Solve the matrix equation $A(A - I)X = B$ if

$$A = \begin{bmatrix} 2 & 1 \\ -1 & 3 \end{bmatrix}, \quad X = \begin{bmatrix} x \\ y \end{bmatrix}, \quad B = \begin{bmatrix} 5 \\ 1 \end{bmatrix}.$$

If $B^2 = A$, then B is called a **square root** of A. In Exercises 34–35, find one square root of the given matrix.

32. $\begin{bmatrix} 2 & 2 \\ 2 & 2 \end{bmatrix}$ **33.** $\begin{bmatrix} 1 & 0 \\ 0 & 1 \end{bmatrix}$

34. A real number must be nonnegative in order to have a real square root. Show by example that a real matrix with one or more negative entries may have a real square root.

35. Prove the right distributive law for matrices, i.e.,

$$(A + B)C = AC + BC.$$

Exercises 36–38 involve mathematical induction. Information on mathematical induction can be found in Appendix A.

36. Prove that A^k is defined for every positive integer k by the "recursive definition" given after the fundamental "Identity property."

37. Show that for every positive integer k,

$$A^kA = AA^k.$$

38. Use the result of Exercise 37 and an induction on j to establish that for all positive integers j and k,

$$A^jA^k = A^kA^j.$$

39. Prove that if A and B are $n \times n$ matrices, then

$$(A - B)(A + B) = A^2 - B^2 \text{ if and only if } AB = BA.$$

40. Show that for matrices A and B, and for any positive integer k,

$$\text{if } AB = BA, \text{ then } (AB)^k = A^kB^k.$$

41. If $(AB)^k = A^kB^k$ for $k = 2, 3, 4, \ldots,$ is it necessarily true that $AB = BA$? If true, prove it; if false, find an example to show why.

42. Prove the corollary to Theorem 2.2, i.e., if A and B are $n \times n$ matrices, then

$$AB = BA \qquad \text{if and only if} \qquad A^tB^t = B^tA^t.$$

43. Establish the uniqueness of the identity matrix. That is, show that if J is an $n \times n$ matrix such that $AJ = JA = A$ for every $n \times n$ matrix A, then $J = I$, the $n \times n$ identity.

2.4 MATRIX INVERSION AND LINEAR ALGEBRA

Matrix multiplication leads naturally to the idea of matrix inversion. The inverse of a matrix is analogous to the reciprocal of a number. The inverse of a matrix is used extensively in the study of linear systems.

Matrix Inversion

The *reciprocal*, or *multiplicative inverse*, of a nonzero real number a is the number b such that

$$ab = ba = 1.$$

The reciprocal of a is indicated by a^{-1}, or by $1/a$. Moreover, a^{-1} exists if and only if $a \neq 0$. When a^{-1} exists, it is unique by the cancellation law. Since $a1 = 1a = a$ for numbers and $AI = IA = A$ for matrices, the $n \times n$ identity I is the $n \times n$ matrix analogue of 1. Does the reciprocal have a corresponding matrix analogue?

Example 1

For the matrices

$$A = \begin{bmatrix} 2 & 1 \\ 3 & 2 \end{bmatrix} \quad \text{and} \quad B = \begin{bmatrix} 2 & -1 \\ -3 & 2 \end{bmatrix},$$

we may readily find AB and BA to verify that $AB = BA = I$, where I is the 2×2 identity. ☐

Since matrix algebra lacks commutative and cancellation laws, we must account for commutativity and uniqueness in framing a definition of the inverse of a matrix. Let A be an $n \times n$ matrix and I the $n \times n$ identity. If there exists a matrix B such that $AB = BA = I$, then B is unique. In fact,

$$\text{if} \quad AB = BA = I \quad \text{and} \quad AC = I, \quad \text{then} \quad C = B.$$

(A proof is assigned in Exercise 41.) We define the inverse of a matrix assuming this uniqueness property.

DEFINITION 2.11

Let A be an $n \times n$ matrix and I the $n \times n$ identity. If there exists an $n \times n$ matrix B such that

$$AB = BA = I,$$

then A is said to be **invertible**, or **nonsingular**, B is called the **inverse of A**, and we write

$$B = A^{-1}.$$

Invertibility is an easy question for diagonal matrices.

Example 2

We may verify the relation

$$\begin{bmatrix} 2 & 0 & 0 \\ 0 & \frac{1}{3} & 0 \\ 0 & 0 & 4 \end{bmatrix}^{-1} = \begin{bmatrix} \frac{1}{2} & 0 & 0 \\ 0 & 3 & 0 \\ 0 & 0 & \frac{1}{4} \end{bmatrix}$$

by finding the two products indicated in Definition 2.11. ☐

As Example 2 illustrates, if $D = \operatorname{diag}(d_1, \ldots, d_n)$, then D is invertible if and only if $d_i \neq 0$ for $i = 1, \ldots, n$. When D is invertible, $D^{-1} = \operatorname{diag}(1/d_1, \ldots, 1/d_n)$. Invertibility is also easy for 2×2 matrices.

Example 3

As in Examples 1 and 2, we have

$$\begin{bmatrix} 1 & 2 \\ -2 & 5 \end{bmatrix}^{-1} = \frac{1}{9}\begin{bmatrix} 5 & -2 \\ 2 & 1 \end{bmatrix} \quad \text{and} \quad \begin{bmatrix} 2 & 3 \\ 1 & 4 \end{bmatrix}^{-1} = \frac{1}{5}\begin{bmatrix} 4 & -3 \\ -1 & 2 \end{bmatrix}$$

by Definition 2.11. □

In general, for $A = [a_{ij}]_{2\times 2}$, we may form a matrix A_1 from A by interchanging a_{11} and a_{22} and changing the signs of a_{12} and a_{21} as in each pair in Example 3. Then

$$AA_1 = dI,$$

where

$$d = a_{11}a_{22} - a_{12}a_{21}$$

(Exercise 48). If $d \neq 0$, we may multiply this equation by $1/d$ to obtain I on the right. Thus if $d \neq 0$, then A is invertible and the inverse can be found. The following definition summarizes.

A 2 × 2 MATRIX INVERSION FORMULA

If $A = [a_{ij}]_{2\times 2}$, let $d = a_{11}a_{22} - a_{12}a_{21}$. Then A is invertible if and only if $d \neq 0$. In this case,

$$A^{-1} = \frac{1}{d}\begin{bmatrix} a_{22} & -a_{12} \\ -a_{21} & a_{11} \end{bmatrix}.$$

Example 4

Find the inverse of the matrix

$$A = \begin{bmatrix} 2 & -3 \\ 3 & 1 \end{bmatrix}.$$

Solution

Apply the inversion formula directly to obtain

$$A^{-1} = \frac{1}{11}\begin{bmatrix} 1 & 3 \\ -3 & 2 \end{bmatrix}. \quad \square$$

Solving Linear Systems Using the Inverse

A linear system with coefficient matrix A may be written in the matrix form $AX = B$. If A is invertible, then by the associative law and the definition of the inverse,

$$A^{-1}(AX) = (A^{-1}A)X = IX = X \quad \text{and} \quad A(A^{-1}B) = (AA^{-1})B = IB = B.$$

From these relations,

$$AX = B \quad \text{if and only if} \quad X = A^{-1}B$$

(Exercise 49). Thus once A^{-1} is known, we may solve the given system for X by calculating the product $A^{-1}B$.

Example 5

Solve the linear system

$$\begin{aligned} 3x - 2y &= 4 \\ -4x + 3y &= -5 \end{aligned}$$

by matrix inversion.

Solution

The matrix form of the given system is

$$AX = B,$$

where

$$A = \begin{bmatrix} 3 & -2 \\ -4 & 3 \end{bmatrix}, \quad X = \begin{bmatrix} x \\ y \end{bmatrix}, \quad \text{and} \quad B = \begin{bmatrix} 4 \\ -5 \end{bmatrix}.$$

The coefficient matrix A is invertible, and by the formula above,

$$A^{-1} = \begin{bmatrix} 3 & 2 \\ 4 & 3 \end{bmatrix}.$$

Therefore the required solution is given by

$$X = A^{-1}B = \begin{bmatrix} 3 & 2 \\ 4 & 3 \end{bmatrix}\begin{bmatrix} 4 \\ -5 \end{bmatrix} = \begin{bmatrix} 2 \\ 1 \end{bmatrix}. \quad \square$$

Example 6

Given that

$$\begin{bmatrix} 1 & 1 & -1 \\ 1 & 2 & 1 \\ 2 & 0 & -5 \end{bmatrix}^{-1} = \begin{bmatrix} -10 & 5 & 3 \\ 7 & -3 & -2 \\ -4 & 2 & 1 \end{bmatrix},$$

solve the linear system

$$\begin{aligned} x + y - z &= -1 \\ x + 2y + z &= -3 \\ 2x \qquad - 5z &= 2. \end{aligned}$$

Solution

The matrix form of the given system is $AX = B$ and A^{-1} is given, hence

$$X = A^{-1}B = \begin{bmatrix} -10 & 5 & 3 \\ 7 & -3 & -2 \\ -4 & 2 & 1 \end{bmatrix}\begin{bmatrix} -1 \\ -3 \\ 2 \end{bmatrix} = \begin{bmatrix} 1 \\ -2 \\ 0 \end{bmatrix}. \quad \square$$

Examples 5 and 6 are deceptively easy because little or no effort was needed to find the inverse of the coefficient matrix. In general, matrix inversion is very expensive, and more efficient methods, such as Gaussian elimination, are used. We use the inverse primarily in developing practical methods of solving linear systems and in communicating ideas concerning them.

A General Matrix Inversion Procedure

When the inverse of a given matrix is needed, it can be found by a procedure related to Gauss-Jordan elimination. We describe this procedure and illustrate its use by examples. A justification is given in Section 2.5.

MATRIX INVERSION BY GAUSS-JORDAN ELIMINATION

Let A be an invertible $n \times n$ matrix. If Gauss-Jordan elimination is applied to A augmented by the $n \times n$ identity, the final n columns of the resulting matrix constitute A^{-1}.

Example 7

Find the inverse of the matrix

$$A = \begin{bmatrix} 1 & -2 \\ 3 & -2 \end{bmatrix}$$

by Gauss-Jordan elimination.

Solution

Apply Gauss-Jordan elimination to A augmented by the 2×2 identity matrix:

$$\left[\begin{array}{cc|cc} 1 & -2 & 1 & 0 \\ 3 & -2 & 0 & 1 \end{array}\right] \rightarrow \left[\begin{array}{cc|cc} 1 & -2 & 1 & 0 \\ 0 & 4 & -3 & 1 \end{array}\right]$$

$$\rightarrow \left[\begin{array}{cc|cc} 1 & -2 & 1 & 0 \\ 0 & 1 & -\frac{3}{4} & \frac{1}{4} \end{array}\right]$$

$$\rightarrow \left[\begin{array}{cc|cc} 1 & 0 & -\frac{1}{2} & \frac{1}{2} \\ 0 & 1 & -\frac{3}{4} & \frac{1}{4} \end{array}\right].$$

The first two columns form the 2×2 identity matrix. Thus according to the description above, the final two columns contain A^{-1}:

$$A^{-1} = \begin{bmatrix} -\frac{1}{2} & \frac{1}{2} \\ -\frac{3}{4} & \frac{1}{4} \end{bmatrix}.$$

A check that this matrix satisfies the conditions of Definition 2.11 is assigned in Exercise 42. ☐

Example 8

Find the inverse of the matrix

$$A = \begin{bmatrix} 1 & 2 & -2 \\ 3 & 1 & -1 \\ 2 & -1 & 0 \end{bmatrix}$$

by Gauss-Jordan elimination.

Solution

Carry out Gauss-Jordan elimination on A augmented by I:

$$\left[\begin{array}{rrr|rrr} 1 & 2 & -2 & 1 & 0 & 0 \\ 3 & 1 & -1 & 0 & 1 & 0 \\ 2 & -1 & 0 & 0 & 0 & 1 \end{array}\right] \to \left[\begin{array}{rrr|rrr} 1 & 2 & -2 & 1 & 0 & 0 \\ 0 & -5 & 5 & -3 & 1 & 0 \\ 0 & -5 & 4 & -2 & 0 & 1 \end{array}\right]$$

$$\to \left[\begin{array}{rrr|rrr} 1 & 2 & -2 & 1 & 0 & 0 \\ 0 & -5 & 5 & -3 & 1 & 0 \\ 0 & 0 & -1 & 1 & -1 & 1 \end{array}\right]$$

$$\to \left[\begin{array}{rrr|rrr} 1 & 2 & -2 & 1 & 0 & 0 \\ 0 & 1 & -1 & \frac{3}{5} & -\frac{1}{5} & 0 \\ 0 & 0 & 1 & -1 & 1 & -1 \end{array}\right]$$

$$\to \left[\begin{array}{rrr|rrr} 1 & 2 & -2 & 1 & 0 & 0 \\ 0 & 1 & 0 & -\frac{2}{5} & \frac{4}{5} & -1 \\ 0 & 0 & 1 & -1 & 1 & -1 \end{array}\right]$$

$$\to \left[\begin{array}{rrr|rrr} 1 & 0 & 0 & -\frac{1}{5} & \frac{2}{5} & 0 \\ 0 & 1 & 0 & -\frac{2}{5} & \frac{4}{5} & -1 \\ 0 & 0 & 1 & -1 & 1 & -1 \end{array}\right].$$

The final three columns of this matrix constitute the inverse:

$$A^{-1} = \begin{bmatrix} -\frac{1}{5} & \frac{2}{5} & 0 \\ -\frac{2}{5} & \frac{4}{5} & -1 \\ -1 & 1 & -1 \end{bmatrix}.$$

A check of this result against Definition 2.11 is assigned in Exercise 43. ☐

In an application of the Gauss-Jordan method, even a single arithmetic error can be costly. We recommend especially careful work with frequent checks in using this procedure.

Matrix Equations and the Definition of the Inverse

By Definition 2.11, the inverse of a square matrix A, when it exists, is the solution of the equations $AB = I$ and $BA = I$ for B. As will be shown in Section 2.5, either of these equations yields A^{-1}.

Example 9

Set up and solve the equations for the entries in A^{-1}, where

$$A = \begin{bmatrix} 1 & 3 \\ 2 & 5 \end{bmatrix}.$$

Solution

Let

$$B = \begin{bmatrix} b_{11} & b_{12} \\ b_{21} & b_{22} \end{bmatrix}$$

be the unknown inverse of A. Substitute into the condition $AB = I$ of Definition 2.11 to obtain

$$\begin{bmatrix} 1 & 3 \\ 2 & 5 \end{bmatrix}\begin{bmatrix} b_{11} & b_{12} \\ b_{21} & b_{22} \end{bmatrix} = \begin{bmatrix} 1 & 0 \\ 0 & 1 \end{bmatrix}.$$

Multiplying the matrices on the left gives

$$\begin{bmatrix} b_{11} + 3b_{21} & b_{12} + 3b_{22} \\ 2b_{11} + 5b_{21} & 2b_{12} + 5b_{22} \end{bmatrix} = \begin{bmatrix} 1 & 0 \\ 0 & 1 \end{bmatrix}.$$

The resulting system of 4 equations in 4 unknowns can be separated into two systems, each with 2 equations in 2 unknowns:

$$\begin{aligned} b_{11} + 3b_{21} &= 1 \qquad & b_{12} + 3b_{22} &= 0 \\ 2b_{11} + 5b_{21} &= 0, \qquad & 2b_{12} + 5b_{22} &= 1. \end{aligned}$$

These systems have the same coefficient matrix. We augment this matrix by adjoining the two columns of constants on the right, and solve the two systems together by Gauss-Jordan elimination:

$$\left[\begin{array}{cc|cc} 1 & 3 & 1 & 0 \\ 2 & 5 & 0 & 1 \end{array}\right] \rightarrow \left[\begin{array}{cc|cc} 1 & 3 & 1 & 0 \\ 0 & -1 & -2 & 1 \end{array}\right] \rightarrow \left[\begin{array}{cc|cc} 1 & 0 & -5 & 3 \\ 0 & 1 & 2 & -1 \end{array}\right].$$

The solution is given by

$$b_{11} = -5, \qquad b_{21} = 2, \qquad b_{12} = 3, \qquad b_{22} = -1,$$

from which

$$A^{-1} = \begin{bmatrix} -5 & 3 \\ 2 & -1 \end{bmatrix}.$$

The second condition in Definition 2.11, $BA = I$, yields the systems

$$\begin{aligned} b_{11} + 2b_{12} &= 1 \\ 3b_{11} + 5b_{12} &= 0 \end{aligned} \quad \text{and} \quad \begin{aligned} b_{21} + 2b_{22} &= 0 \\ 3b_{21} + 5b_{22} &= 1 \end{aligned}$$

These systems are different from those determined by $AB = I$, but the same procedure yields the entries in B:

$$\left[\begin{array}{cc|cc} 1 & 2 & 1 & 0 \\ 3 & 5 & 0 & 1 \end{array}\right] \to \left[\begin{array}{cc|cc} 1 & 2 & 1 & 0 \\ 0 & -1 & -3 & 1 \end{array}\right] \to \left[\begin{array}{cc|cc} 1 & 0 & -5 & 2 \\ 0 & 1 & 3 & -1 \end{array}\right].$$

The solution

$$b_{11} = -5, \qquad b_{12} = 3, \qquad b_{21} = 2, \qquad b_{22} = -1,$$

agrees with that obtained from the condition $AB = I$. □

The procedure in Example 9 suggests that more general matrix equations may be solved by Gaussian elimination.

Example 10

Solve the matrix equation $AC = B$ for C, where

$$A = \begin{bmatrix} 1 & 2 & -1 \\ 2 & -1 & 3 \\ -2 & 1 & 2 \end{bmatrix} \quad \text{and} \quad B = \begin{bmatrix} 1 & -1 & 3 & -5 \\ 0 & 1 & 1 & 5 \\ 1 & 2 & 4 & 2 \end{bmatrix}.$$

Solution

To find C, apply Gaussian or Gauss-Jordan elimination to the augmented matrix

$$\left[\begin{array}{ccc|cccc} 1 & 2 & -1 & 1 & -1 & 3 & -5 \\ 2 & -1 & 3 & 0 & 1 & 1 & 5 \\ -2 & 1 & 2 & 1 & 2 & 4 & 2 \end{array}\right].$$

We omit the details, noting only the solution

$$C = \begin{bmatrix} 0 & -\frac{2}{5} & 0 & -\frac{2}{5} \\ \frac{3}{5} & 0 & 2 & -\frac{8}{5} \\ \frac{1}{5} & \frac{3}{5} & 1 & \frac{7}{5} \end{bmatrix}.$$ □

Of course, the matrix equation in Example 10 can be solved by a matrix inversion since $C = A^{-1}B$. However, finding the inverse is *not* recommended, even for solving matrix equations, since direct elimination is more efficient.

REVIEW CHECKLIST

1. Define the inverse of a matrix, and apply the defining condition as a check.
2. Find the inverse of an invertible diagonal matrix.
3. Find the inverse of an invertible 2×2 matrix by the special formula in this section.
4. Use the Gauss-Jordan method to determine whether a given matrix is invertible and, if so, find the inverse.
5. Set up and solve linear systems for the entries in the inverse of a given matrix.
6. Solve matrix equations of the form $AC = B$ for C, given A and B, by Gaussian or Gauss-Jordan elimination.

EXERCISES

In Exercises 1–6, find the inverse of the given matrix using the 2×2 inversion formula.

1. $\begin{bmatrix} 1 & 4 \\ 1 & 5 \end{bmatrix}$ **2.** $\begin{bmatrix} 2 & 1 \\ 4 & 3 \end{bmatrix}$ **3.** $\begin{bmatrix} 4 & 3 \\ 2 & 2 \end{bmatrix}$

4. $\begin{bmatrix} 1 & -1 \\ 2 & 3 \end{bmatrix}$ **5.** $\begin{bmatrix} 1 & 0 \\ 2 & 5 \end{bmatrix}$ **6.** $\begin{bmatrix} 3 & 4 \\ 2 & 1 \end{bmatrix}$

In Exercises 7–12, solve the given system by matrix inversion.

7. $\begin{aligned} 3x + 4y &= 2 \\ 4x + 5y &= 3 \end{aligned}$ **8.** $\begin{aligned} 3x - y &= 1 \\ 2x - 2y &= 2 \end{aligned}$ **9.** $\begin{aligned} 4x + y &= -1 \\ x + 2y &= 2 \end{aligned}$

10. $\begin{aligned} 2x + 5y &= 0 \\ x + 3y &= 2 \end{aligned}$ **11.** $\begin{aligned} -3x + 2y &= 1 \\ 2x + 4y &= -6 \end{aligned}$ **12.** $\begin{aligned} 5x - y &= 3 \\ 2x - y &= 1 \end{aligned}$

In Exercises 13–14, solve as indicated, given

$$A = \begin{bmatrix} 1 & 2 & 1 \\ 1 & 3 & 0 \\ 2 & 6 & 1 \end{bmatrix} \quad \text{and} \quad A^{-1} = \begin{bmatrix} 3 & 4 & -3 \\ -1 & -1 & 1 \\ 0 & -2 & 1 \end{bmatrix}.$$

13. Solve the linear system $AX = [1 \ \ 1 \ \ 1]^t$ for X.

14. Solve the matrix equation $AC = B$ for C, if

$$B = \begin{bmatrix} 1 & 2 & 0 & 1 \\ 1 & 2 & -1 & 0 \\ 1 & -1 & 3 & 2 \end{bmatrix}.$$

In Exercises 15–16, solve as directed, given

$$A = \begin{bmatrix} 1 & -1 & 2 & 1 \\ 2 & -1 & 2 & 3 \\ 2 & 0 & 1 & 6 \\ -1 & -1 & 1 & -4 \end{bmatrix} \quad \text{and} \quad A^{-1} = \begin{bmatrix} 1 & 1 & -2 & -2 \\ 7 & -3 & -3 & -5 \\ 4 & -2 & -1 & -2 \\ -1 & 0 & 1 & 1 \end{bmatrix}.$$

15. Solve the linear system $AX = [2 \quad -1 \quad 1 \quad 3]^t$ for X.

16. Solve the matrix equation $AC = B$ for C, if

$$B = \begin{bmatrix} 1 & 2 & -1 & 1 \\ -1 & 1 & 2 & -2 \\ 3 & 0 & -1 & 1 \\ 0 & 2 & 1 & -2 \end{bmatrix}.$$

17. Using Gauss-Jordan elimination, solve the matrix equation $AC = B$ for C, where

$$A = \begin{bmatrix} 1 & -1 & 1 \\ -1 & 2 & 1 \\ -2 & 3 & 1 \end{bmatrix} \quad \text{and} \quad B = \begin{bmatrix} 1 & 1 & 1 & -1 \\ 1 & 0 & 2 & 2 \\ 1 & -2 & 2 & 0 \end{bmatrix}.$$

In Exercises 18–23, use the method of Example 9 to find the inverse of the given matrix.

18. $\begin{bmatrix} 1 & -3 \\ 1 & -4 \end{bmatrix}$ **19.** $\begin{bmatrix} 2 & -1 \\ 3 & 1 \end{bmatrix}$ **20.** $\begin{bmatrix} 1 & 4 \\ 3 & 5 \end{bmatrix}$

21. $\begin{bmatrix} 1 & -1 & 2 \\ -1 & 2 & 1 \\ 2 & -3 & 2 \end{bmatrix}$ **22.** $\begin{bmatrix} 1 & 1 & -1 \\ -2 & 0 & 3 \\ -3 & -1 & 5 \end{bmatrix}$ **23.** $\begin{bmatrix} 1 & 1 & 2 \\ 2 & 3 & 2 \\ 1 & 3 & -1 \end{bmatrix}$

In Exercises 24–39, use the Gauss-Jordan method to decide whether the given matrix is invertible and, if so, find the inverse.

24. $\begin{bmatrix} 1 & 2 \\ -2 & 5 \end{bmatrix}$ **25.** $\begin{bmatrix} 1 & 2 \\ 2 & 4 \end{bmatrix}$ **26.** $\begin{bmatrix} 1 & 2 \\ 2 & 3 \end{bmatrix}$

27. $\begin{bmatrix} 2 & 3 \\ 1 & 4 \end{bmatrix}$ **28.** $\begin{bmatrix} -1 & 4 \\ 0 & 2 \end{bmatrix}$ **29.** $\begin{bmatrix} 2 & -3 \\ -6 & 9 \end{bmatrix}$

30. $\begin{bmatrix} 1 & 1 & 0 \\ 1 & 2 & 2 \\ -2 & -2 & -1 \end{bmatrix}$ **31.** $\begin{bmatrix} 1 & -1 & -1 \\ -1 & 1 & -1 \\ 2 & 2 & 0 \end{bmatrix}$ **32.** $\begin{bmatrix} 1 & 2 & -1 \\ -2 & 1 & 3 \\ -1 & 8 & 3 \end{bmatrix}$

33. $\begin{bmatrix} 2 & 1 & 1 \\ 1 & 2 & 1 \\ 1 & 1 & 2 \end{bmatrix}$ **34.** $\begin{bmatrix} 3 & -2 & 2 \\ 3 & -2 & 4 \\ 1 & -1 & -2 \end{bmatrix}$ **35.** $\begin{bmatrix} 3 & 2 & 0 \\ 2 & 1 & 3 \\ 4 & -2 & -1 \end{bmatrix}$

36. $\begin{bmatrix} 1 & -1 & 1 & -1 \\ 1 & 0 & -1 & 2 \\ -1 & 2 & -2 & 7 \\ 2 & -1 & -1 & 0 \end{bmatrix}$ **37.** $\begin{bmatrix} 1 & -2 & 1 & -1 \\ -1 & 4 & -2 & 3 \\ 2 & 0 & 1 & 3 \\ -2 & 6 & 0 & 5 \end{bmatrix}$

38. $\begin{bmatrix} 1 & -3 & 2 & 0 \\ -2 & -4 & 1 & 2 \\ 2 & 3 & 1 & 3 \\ -3 & 8 & -4 & 5 \end{bmatrix}$ **39.** $\begin{bmatrix} 2 & -1 & 0 & 3 \\ 1 & 1 & 2 & -1 \\ -1 & 2 & 3 & 1 \\ 0 & 1 & 2 & 1 \end{bmatrix}$

40. Show that if A and B are nonsingular, then so is AB, and

$$(AB)^{-1} = B^{-1}A^{-1}.$$

41. Let A and B be square matrices such that $AB = BA = I$. Show that if $AC = I$, then $B = C$. (Since we used this property in formulating Definition 2.11, use only results known up to this point.)

42. Carry out the matrix multiplications prescribed by Definition 2.11 to check that the matrix obtained in Example 7, page 103, is the inverse of the given A.

43. Perform the multiplications needed to verify that the matrix obtained in Example 8, page 103, is the inverse of the given A.

44. (*Mathematical induction required.*) Extend the result of Exercise 40 to the product of any n invertible matrices $A_1, A_2, \ldots, A_n$.

45. If A is invertible, show that A^{-1} is also invertible, and find $(A^{-1})^{-1}$.

46. Show that if A is invertible, then A^t is also invertible, and find $(A^t)^{-1}$.

47. Prove that if A is invertible, then A^k is invertible for every positive integer k, and find $(A^k)^{-1}$.

48. Show that if $A = [a_{ij}]_{2\times 2}$ and we form a matrix A_1 from A by interchanging a_{11} and a_{22} and changing the signs of a_{12} and a_{21}, then $AA_1 = (a_{11}a_{22} - a_{12}a_{21})I$.

49. Using the relations $A^{-1}(AX) = X$ and $A(A^{-1}B) = IB$, show that $AX = B$ if and only if $X = A^{-1}B$.

2.5 MATRIX INVERSION AND ELEMENTARY ROW OPERATIONS

Matrix multiplication can be used to perform the elementary row operations of Gaussian elimination. This idea is used to determine the structure of an invertible matrix and to determine conditions for the existence of solutions to linear systems in terms of invertibility.

Elementary Row Operations by Matrix Multiplication

The first example takes a fresh look at the elementary row operations of Gaussian elimination from the perspective of matrix algebra.

Example 1

Use matrix products to perform the elementary row operations

$$\text{R1} \leftrightarrow \text{R2}, \qquad \text{R2} \leftarrow 3\text{R2}, \qquad \text{and} \qquad \text{R3} \leftarrow \text{R3} + 2\text{R1}$$

on any matrix A with three rows.

Solution

Let

$$E_1 = \begin{bmatrix} 0 & 1 & 0 \\ 1 & 0 & 0 \\ 0 & 0 & 1 \end{bmatrix}, \quad E_2 = \begin{bmatrix} 1 & 0 & 0 \\ 0 & 3 & 0 \\ 0 & 0 & 1 \end{bmatrix}, \quad \text{and} \quad E_3 = \begin{bmatrix} 1 & 0 & 0 \\ 0 & 1 & 0 \\ 2 & 0 & 1 \end{bmatrix}.$$

Then the products E_1A, E_2A, and E_3A, respectively, are the matrices resulting from the given operations. For instance, let

$$A = \begin{bmatrix} 2 & -1 \\ 3 & 1 \\ -2 & 3 \end{bmatrix}.$$

Then check that

$$E_1A = \begin{bmatrix} 3 & 1 \\ 2 & -1 \\ -2 & 3 \end{bmatrix}, \quad E_2A = \begin{bmatrix} 2 & -1 \\ 9 & 3 \\ -2 & 3 \end{bmatrix}, \quad \text{and} \quad E_3A = \begin{bmatrix} 2 & -1 \\ 3 & 1 \\ 2 & 1 \end{bmatrix}$$

are the results of applying the given row operations to A. □

The matrix of an elementary row operation is called an **elementary matrix**. We generalize Example 1 to any elementary row operation.

FORMING AN ELEMENTARY MATRIX

The matrix E for which EA is the result of an elementary row operation on an $m \times n$ matrix A may be obtained from the $m \times m$ identity matrix I as follows.

Ri ↔ Rj Form E by exchanging rows i and j in I.

Ri ← cRi, c ≠ 0 Form E by substituting c for the 1 at row i and column i in I.

Ri ← Ri + cRj, i ≠ j Form E by substituting c for the 0 at row i and column j in I.

Example 2

Find the matrix E such that EA results from interchanging the second and fourth rows of any $4 \times n$ matrix A.

Solution

Form E by interchanging rows 2 and 4 in the 4×4 identity:

$$E = \begin{bmatrix} 1 & 0 & 0 & 0 \\ 0 & 0 & 0 & 1 \\ 0 & 0 & 1 & 0 \\ 0 & 1 & 0 & 0 \end{bmatrix}. \quad \square$$

Can we find the inverse of an elementary matrix?

Example 3

Let E_1, E_2, and E_3 be the elementary matrices given by

$$E_1 = \begin{bmatrix} 1 & 0 & 0 \\ 0 & 2 & 0 \\ 0 & 0 & 1 \end{bmatrix}, \quad E_2 = \begin{bmatrix} 0 & 0 & 1 \\ 0 & 1 & 0 \\ 1 & 0 & 0 \end{bmatrix}, \quad \text{and} \quad E_3 = \begin{bmatrix} 1 & 0 & 0 \\ 0 & 1 & 0 \\ 2 & 0 & 1 \end{bmatrix}.$$

Then

$$E_1^{-1} = \begin{bmatrix} 1 & 0 & 0 \\ 0 & \frac{1}{2} & 0 \\ 0 & 0 & 1 \end{bmatrix}, \quad E_2^{-1} = \begin{bmatrix} 0 & 0 & 1 \\ 0 & 1 & 0 \\ 1 & 0 & 0 \end{bmatrix}, \quad \text{and} \quad E_3^{-1} = \begin{bmatrix} 1 & 0 & 0 \\ 0 & 1 & 0 \\ -2 & 0 & 1 \end{bmatrix}. \quad \square$$

As Example 3 illustrates, any elementary matrix is invertible, and the inverse is an elementary matrix of the same type. If E multiplies a row by a nonzero constant c, then E^{-1} multiplies the same row by $1/c$. If E interchanges two rows, then $E^{-1} = E$. If E adds c times row i to row j, with $i \neq j$, then E^{-1} is the matrix that adds $-c$ times row i to row j. We summarize in the following result.

Proposition 1 If E is an elementary matrix, then E is invertible, and E^{-1} is an elementary matrix of the same type as E.

Invertible Matrices as Products of Elementary Matrices

Exercise 40 of Section 2.4 involved finding the inverse of the product of two invertible matrices. Exercise 46 involved finding the inverse of the transpose of an invertible matrix. These results are worthy of special note, and we cite them in the following statements.

Proposition 2 If A and B are invertible $n \times n$ matrices, then AB is invertible, and

$$(AB)^{-1} = B^{-1}A^{-1}.$$

Proposition 3 If A is invertible, then A^t is invertible, and

$$(A^t)^{-1} = (A^{-1})^t.$$

In the next example, the elementary matrices in a particular application of Gauss-Jordan elimination are examined carefully.

Example 4

The matrix

$$A = \begin{bmatrix} 1 & 2 \\ 2 & 3 \end{bmatrix}$$

may be reduced to the identity as follows:

$$\begin{bmatrix} 1 & 2 \\ 2 & 3 \end{bmatrix} \to \begin{bmatrix} 1 & 2 \\ 0 & -1 \end{bmatrix} \to \begin{bmatrix} 1 & 2 \\ 0 & 1 \end{bmatrix} \to \begin{bmatrix} 1 & 0 \\ 0 & 1 \end{bmatrix}.$$

The elementary matrices for these operations are, respectively,

$$E_1 = \begin{bmatrix} 1 & 0 \\ -2 & 1 \end{bmatrix}, \quad E_2 = \begin{bmatrix} 1 & 0 \\ 0 & -1 \end{bmatrix}, \quad \text{and} \quad E_3 = \begin{bmatrix} 1 & -2 \\ 0 & 1 \end{bmatrix}.$$

The reduction to the identity may be expressed in matrix form as

$$(E_3(E_2(E_1A))) = I.$$

Thus if $B = E_3E_2E_1$, then $BA = I$. Performing the two indicated multiplications gives

$$B = E_3E_2E_1 = \begin{bmatrix} 1 & -2 \\ 0 & 1 \end{bmatrix}\begin{bmatrix} 1 & 0 \\ 0 & -1 \end{bmatrix}\begin{bmatrix} 1 & 0 \\ -2 & 1 \end{bmatrix} = \begin{bmatrix} -3 & 2 \\ 2 & -1 \end{bmatrix}.$$

That $AB = I$ may also be verified. Hence, by Definition 2.11, A is invertible and $A^{-1} = B$. In addition,

$$E_1^{-1} = \begin{bmatrix} 1 & 0 \\ 2 & 1 \end{bmatrix}, \quad E_2^{-1} = \begin{bmatrix} 1 & 0 \\ 0 & -1 \end{bmatrix}, \quad \text{and} \quad E_3^{-1} = \begin{bmatrix} 1 & 2 \\ 0 & 1 \end{bmatrix}.$$

Performing the indicated multiplications gives

$$A = E_1^{-1}E_2^{-1}E_3^{-1}. \quad \square$$

The results of Example 4 illustrate an important general concept. In applying Gauss-Jordan elimination to an $n \times n$ matrix A of row rank n, we construct the identity matrix I by a sequence of elementary row operations. If these matrices, in order of application, are $E_1, \ldots, E_q$, then, as in Example 4,

$$E_q \cdots E_1 A = I.$$

Thus the matrix

$$B = E_q \cdots E_1$$

is a **left inverse of A**, that is,

$$BA = I.$$

The converse is also true. If A has a left inverse B, then A can be reduced to the identity I by elementary row operations. We can see why by recalling from Chapter 1 that A can be reduced to I in this way if and only if the equation

$$AX = 0$$

has only the trivial solution. But if $AX = 0$, then

$$B(AX) = B0 = 0.$$

By the associative law and the definition of left inverse, we also have

$$B(AX) = (BA)X = IX = X,$$

hence $X = 0$. Thus the system $AX = 0$ has only the trivial solution. Proposition 4 summarizes.

Proposition 4 A square matrix A has a left inverse if and only if there exist elementary matrices $E_1, \ldots, E_q$ such that

$$E_q \cdots E_1 A = I.$$

If A has a left inverse, let $E_1, \ldots, E_q$ be elementary matrices satisfying the equation in Proposition 4. Since each E_i is invertible by Proposition 1, we may multiply the equation in Proposition 4 on the *left* successively by the inverses to obtain

$$(*) \qquad A = E_1^{-1} \cdots E_q^{-1}.$$

This was also illustrated in Example 4. We may then multiply this identity on the *right* successively by $E_q, \ldots, E_1$ to find

$$AE_q \cdots E_1 = I.$$

Thus $E_q \cdots E_1$ is also a **right inverse** of A. Hence, for a square matrix A,

$$\text{if} \quad BA = I, \quad \text{then} \quad AB = I.$$

Thus if A has a left inverse B, then A is invertible and $B = A^{-1}$. Equation (*) shows that A can be expressed as a product of elementary matrices in this case, since the inverse of an elementary matrix is a matrix of the same type.

The converse of this last statement is also true. Suppose A is expressible as a product of elementary matrices,

$$A = E_q \cdots E_1.$$

By Proposition 1, each E_i is invertible. Hence, by Proposition 2 (extended to the product of any finite number of matrices as in Exercise 44, Section 2.4), A is invertible.

If A has a right inverse B, then B has a left inverse and hence is invertible. Multiplying the equation $AB = I$ on the right by B^{-1} gives $A = B^{-1}$. Thus A is invertible and $A^{-1} = B$. Theorem 2.5 summarizes.

THEOREM 2.5

The following are equivalent for $n \times n$ matrices A and B:

i. $AB = I$;

ii. $BA = I$;

iii. A is invertible and $B = A^{-1}$;

iv. A can be expressed as a product of elementary matrices.

When A is invertible, conditions (i)–(iv) of Theorem 2.5 hold, and A^{-1} is expressible as a product of elementary matrices. These conditions were illustrated in Example 4.

Conditions for the Existence of Solutions of Linear Systems

Theorem 2.5 justifies the Gauss-Jordan procedure for finding the inverse of a matrix in Section 2.4. Our previous study of the relationship between elementary matrices, Gaussian elimination, and matrix inversion also establishes the following result.

THEOREM 2.6

The following are equivalent for an $n \times n$ matrix A:

a. A is invertible;

b. A has row rank n;

c. the homogeneous linear system represented by the matrix equation $AX = 0$ has only the trivial solution;

d. any linear system of the form $AX = B$ has a unique solution.

The equivalence of conditions (a) and (b) in Theorem 2.6 follows from the discussion leading up to Theorem 2.5. The equivalence of (b), (c), and (d) follows from the solution concepts of Sections 1.3 and 1.4.

Example 5

The matrix

$$A = \begin{bmatrix} 1 & 3 & 2 \\ 1 & 4 & 1 \\ 1 & 2 & 2 \end{bmatrix}$$

is invertible, and its inverse is given by

$$A^{-1} = \begin{bmatrix} -6 & 2 & 5 \\ 1 & 0 & -1 \\ 2 & -1 & -1 \end{bmatrix}.$$

The elementary row operations

$$\begin{bmatrix} 1 & 3 & 2 \\ 1 & 4 & 1 \\ 1 & 2 & 2 \end{bmatrix} \rightarrow \begin{bmatrix} 1 & 3 & 2 \\ 0 & 1 & -1 \\ 0 & -1 & 0 \end{bmatrix} \rightarrow \begin{bmatrix} 1 & 3 & 2 \\ 0 & 1 & -1 \\ 0 & 0 & -1 \end{bmatrix}$$

show that A has row rank 3. This final matrix also shows that for any linear system with coefficient matrix A, including the homogeneous system, back substitution will yield a unique solution. □

In Chapters 3 and 4, we shall find additional conditions equivalent to the conditions in Theorem 2.6. In fact, in Section 4.6, we present a grand summary of various conditions for existence of solutions of linear systems developed in Chapters 1–4.

REVIEW CHECKLIST

1. Given an elementary row operation, find the corresponding matrix. Conversely, given an elementary matrix, identify the elementary row operation it represents.
2. Find the inverse of an elementary matrix.
3. Describe the relationship between the left inverse, the right inverse, and the inverse of a square matrix.
4. Identify an invertible matrix as a matrix that is expressible as a product of elementary matrices.

5. Identify a square linear system with a unique solution as a system with an invertible coefficient matrix.

EXERCISES

In Exercises 1–5, assuming that the given elementary row operation is applied to a $2 \times n$ matrix, find the matrix of the operation and its inverse.

1. $R1 \leftrightarrow R2$ **2.** $R1 \leftarrow 4R1$ **3.** $R2 \leftarrow -3R2$
4. $R2 \leftarrow R2 + 4R1$ **5.** $R1 \leftarrow R1 - R2$

In Exercises 6–10, find the elementary row operation represented by the given matrix, and find the inverse of the given matrix.

6. $\begin{bmatrix} 1 & 1 \\ 0 & 1 \end{bmatrix}$ **7.** $\begin{bmatrix} 1 & 0 \\ 2 & 1 \end{bmatrix}$ **8.** $\begin{bmatrix} 3 & 0 \\ 0 & 1 \end{bmatrix}$

9. $\begin{bmatrix} 0 & 1 \\ 1 & 0 \end{bmatrix}$ **10.** $\begin{bmatrix} 1 & 0 \\ 0 & 3 \end{bmatrix}$

In Exercises 10–16, assuming the given operation is applied to a $3 \times n$ matrix, find the matrix of the operation and its inverse.

11. $R1 \leftrightarrow R2$ **12.** $R2 \leftrightarrow R3$ **13.** $R2 \leftarrow R2 - 2R1$
14. $R1 \leftarrow -R1$ **15.** $R3 \leftarrow 4R3$ **16.** $R1 \leftarrow R1 + 2R3$

In Exercises 17–21, find the elementary row operation represented by the given matrix, and find the inverse of the given matrix.

17. $\begin{bmatrix} 1 & 0 & 0 \\ 0 & -3 & 0 \\ 0 & 0 & 1 \end{bmatrix}$ **18.** $\begin{bmatrix} 1 & 2 & 0 \\ 0 & 1 & 0 \\ 0 & 0 & 1 \end{bmatrix}$ **19.** $\begin{bmatrix} 0 & 0 & 1 \\ 0 & 1 & 0 \\ 1 & 0 & 0 \end{bmatrix}$

20. $\begin{bmatrix} 1 & 0 & 0 \\ 0 & 1 & 2 \\ 0 & 0 & 1 \end{bmatrix}$ **21.** $\begin{bmatrix} 1 & 0 & -3 \\ 0 & 1 & 0 \\ 0 & 0 & 1 \end{bmatrix}$

In Exercises 22–27, if possible, express the given matrix as a product of elementary matrices.

22. $\begin{bmatrix} 1 & 3 \\ 1 & 2 \end{bmatrix}$ **23.** $\begin{bmatrix} 2 & 1 \\ 3 & -2 \end{bmatrix}$ **24.** $\begin{bmatrix} 1 & -2 \\ -2 & 4 \end{bmatrix}$

25. $\begin{bmatrix} 1 & 0 & -1 \\ 1 & -1 & 2 \\ 2 & -1 & -1 \end{bmatrix}$ **26.** $\begin{bmatrix} 1 & 3 & 3 \\ 2 & 4 & 3 \\ 1 & -1 & -3 \end{bmatrix}$ **27.** $\begin{bmatrix} 1 & 2 & 1 \\ 2 & 5 & 4 \\ 1 & 1 & 0 \end{bmatrix}$

28. Let A be a square matrix. Show that if the homogeneous linear system $AX = 0$ has only the trivial solution, then so does the system $A^2X = 0$.

29. Generalize the result of Exercise 28 to include all powers of A, positive and negative.

30. Let A be an $n \times n$ matrix and p be a positive integer. Show that if the homogeneous system $AX = 0$ has only the trivial solution for an $n \times 1$ matrix, then the matrix equation $AC = 0$ has only the trivial solution for an $n \times p$ matrix.

31. Find the 3×3 matrix E that simultaneously adds 2 times row 1 to row 2 and substracts 4 times row 1 from row 3 in any $3 \times n$ matrix A when we multiply A on the left by E.

32. Describe the effects of multiplying a matrix A on the right by an elementary matrix.

(o) 2.6 COMPUTING AND APPLICATIONS

Matrix algebra can be used to develop numerical methods for solving linear systems. A typical illustration is an *LU decomposition*, which is widely used in solving linear systems by computer. The application of matrix algebra to physical problems is also illustrated.

LU Decomposition

In solving a square linear system

$$AX = B$$

on a computer, we often decompose A into a matrix product

$$A = LU,$$

where the factors L and U have an especially useful form.

Example 1

Let

$$A = \begin{bmatrix} 1 & 2 & 2 \\ 2 & 6 & 5 \\ 1 & 0 & 6 \end{bmatrix}.$$

We form an upper triangular matrix from A as in Gaussian elimination. Regard the elementary row operations as *subtractions*, and record the coefficients:

$$\begin{array}{c} \\ 2 \\ 1 \end{array} \begin{bmatrix} 1 & 2 & 2 \\ 0 & 2 & 1 \\ 0 & -2 & 4 \end{bmatrix} \qquad \begin{array}{l} \\ R2 \leftarrow R2 - (2)R1 \\ R3 \leftarrow R3 - (1)R1 \end{array}$$

$$\begin{array}{cc} & \\ 2 & \\ 1 & -1 \end{array} \begin{bmatrix} 1 & 2 & 2 \\ 0 & 2 & 1 \\ 0 & 0 & 5 \end{bmatrix} \qquad \begin{array}{l} \\ \\ R3 \leftarrow R3 - (-1)R2. \end{array}$$

Let U be the matrix obtained from the elimination. A second matrix, L, is formed from the identity matrix by entering the recorded coefficients below the main diagonal. Thus

$$L = \begin{bmatrix} 1 & 0 & 0 \\ 2 & 1 & 0 \\ 1 & -1 & 1 \end{bmatrix} \quad \text{and} \quad U = \begin{bmatrix} 1 & 2 & 2 \\ 0 & 2 & 1 \\ 0 & 0 & 5 \end{bmatrix}.$$

Performing the indicated multiplication shows that

$$LU = \begin{bmatrix} 1 & 0 & 0 \\ 2 & 1 & 0 \\ 1 & -1 & 1 \end{bmatrix} \begin{bmatrix} 1 & 2 & 2 \\ 0 & 2 & 1 \\ 0 & 0 & 5 \end{bmatrix} = \begin{bmatrix} 1 & 2 & 2 \\ 2 & 6 & 5 \\ 1 & 0 & 6 \end{bmatrix} = A. \quad \square$$

DEFINITION 2.12

A square matrix is said to be **upper [lower] triangular** if every entry below [above] the main diagonal equals 0.

Example 2

The matrices

$$\begin{bmatrix} 1 & 1 \\ 0 & 1 \end{bmatrix}, \quad \begin{bmatrix} 2 & 2 & 1 \\ 0 & 1 & 1 \\ 0 & 0 & 3 \end{bmatrix}, \quad \text{and} \quad \begin{bmatrix} 1 & -1 & 2 & 1 \\ 0 & 3 & 0 & 1 \\ 0 & 0 & 1 & 3 \\ 0 & 0 & 0 & 2 \end{bmatrix}$$

are upper triangular. The matrices

$$\begin{bmatrix} 2 & 0 \\ 1 & 1 \end{bmatrix}, \quad \begin{bmatrix} 1 & 0 & 0 \\ 1 & 1 & 0 \\ 1 & 1 & 1 \end{bmatrix}, \quad \text{and} \quad \begin{bmatrix} 1 & 0 & 0 & 0 \\ -1 & 2 & 0 & 0 \\ 2 & -1 & 3 & 0 \\ -3 & 2 & 3 & 4 \end{bmatrix}$$

are lower triangular. $\square$

In Example 1, L is a lower triangular matrix, and U is an upper triangular matrix. The factorization of A in this example illustrates the following more general concept.

DEFINITION 2.13

If a matrix A can be written in the form

$$A = LU,$$

where L is a lower and U is an upper triangular matrix, the expression on the right is called an **LU decomposition** of A.

In general, an LU decomposition is not unique.

Example 3

Carrying out the multiplication shows that the matrix A of Example 1 also has the LU decomposition

$$\begin{bmatrix} 1 & 0 & 0 \\ 2 & 2 & 0 \\ 1 & -2 & 5 \end{bmatrix} \begin{bmatrix} 1 & 2 & 2 \\ 0 & 1 & \frac{1}{2} \\ 0 & 0 & 1 \end{bmatrix} = \begin{bmatrix} 1 & 2 & 2 \\ 2 & 6 & 5 \\ 1 & 0 & 6 \end{bmatrix}.$$

To obtain this decomposition, divide row 2 of the original U by 2 and multiply column 2 of the original L by 2. Also, divide row 3 of the original U by 5 and multiply column 3 of the original L by 5. □

We may justify the method in Example 1 using concepts from Section 2.5. The reduction of A to upper triangular form requires three elementary row operations. Thus we have

$$E_3E_2E_1A = U,$$

where E_1, E_2, and E_3 are the corresponding elementary matrices. From this relation, we find

$$A = \left(E_1^{-1}E_2^{-1}E_3^{-1}\right)U = LU.$$

The recorded *subtraction* coefficients are used to form the product $E_1^{-1}E_2^{-1}E_3^{-1}$. Subtraction is used because the elementary row operation

$$\text{R}i \leftarrow \text{R}i - c\text{R}j$$

is the *inverse* of the corresponding additive operation.

Once an LU decomposition of A is known, the linear system

$$AX = B$$

can be solved as follows:

$$X = A^{-1}B = U^{-1}L^{-1}B.$$

Example 4

Solve the linear system

$$\begin{aligned} x + 2y - 2z &= 1 \\ 2x + 8y - 8z &= -2 \\ -x + 10y - 8z &= -3 \end{aligned}$$

using the LU decomposition with each diagonal entry in L equal to 1.

Solution

First obtain the required LU decomposition by the procedure described in Example 1:

$$\begin{matrix} \\ 2 \\ -1 \end{matrix} \begin{bmatrix} 1 & 2 & -2 \\ 0 & 4 & -4 \\ 0 & 12 & -10 \end{bmatrix} \quad \begin{matrix} \\ R2 \leftarrow R2 - 2R1 \\ R3 \leftarrow R3 - (-1)R1 \end{matrix}$$

$$\begin{matrix} & \\ 2 & \\ -1 & 3 \end{matrix} \begin{bmatrix} 1 & 2 & -2 \\ 0 & 4 & -4 \\ 0 & 0 & 2 \end{bmatrix} \quad \begin{matrix} \\ \\ R3 \leftarrow R3 - 3R2. \end{matrix}$$

Thus

$$L = \begin{bmatrix} 1 & 0 & 0 \\ 2 & 1 & 0 \\ -1 & 3 & 1 \end{bmatrix} \quad \text{and} \quad U = \begin{bmatrix} 1 & 2 & -2 \\ 0 & 4 & -4 \\ 0 & 0 & 2 \end{bmatrix}.$$

Then

$$L^{-1}B = \begin{bmatrix} 1 & 0 & 0 \\ -2 & 1 & 0 \\ 7 & -3 & 1 \end{bmatrix} \begin{bmatrix} 1 \\ -2 \\ -3 \end{bmatrix} = \begin{bmatrix} 1 \\ -4 \\ 10 \end{bmatrix}$$

and, in turn,

$$X = U^{-1}L^{-1}B = \begin{bmatrix} 1 & -\frac{1}{2} & 0 \\ 0 & \frac{1}{4} & \frac{1}{2} \\ 0 & 0 & \frac{1}{2} \end{bmatrix} \begin{bmatrix} 1 \\ -4 \\ 10 \end{bmatrix} = \begin{bmatrix} 3 \\ 4 \\ 5 \end{bmatrix}. \quad \square$$

In Example 4, we found L^{-1} essentially by inspection. Each entry immediately below the main diagonal required only a sign reversal. The entry 7 at row 3 and column 1 satisfied the condition $x(1) + (-3)(2) + 1(-1) = 0$ required by the definition of the inverse. In calculating U^{-1}, we also had to account for the effects of the diagonal entries.

In one popular variation of the LU concept, the diagonal entries are factored from both L and U and displayed in a diagonal matrix D. The result is the LDU decomposition of A.

Example 5

Express the matrix A of Example 1 in the form

$$A = LDU,$$

where L is lower triangular, D is diagonal, and U is upper triangular.

Solution

The LDU decomposition of A is given by

$$\begin{bmatrix} 1 & 0 & 0 \\ 2 & 1 & 0 \\ 1 & -1 & 1 \end{bmatrix}\begin{bmatrix} 1 & 0 & 0 \\ 0 & 2 & 0 \\ 0 & 0 & 5 \end{bmatrix}\begin{bmatrix} 1 & 2 & 2 \\ 0 & 1 & \frac{1}{2} \\ 0 & 0 & 1 \end{bmatrix} = \begin{bmatrix} 1 & 2 & 2 \\ 2 & 6 & 5 \\ 1 & 0 & 6 \end{bmatrix}.$$

To find this decomposition, we factored the diagonal entries from the rows of the original upper triangular matrix in Example 1 to form the matrix D. The matrix L is unchanged. □

This discussion was intended as a brief introduction to the LU concept. We leave additional study of this topic to numerical analysis, suggesting *Applied Numerical Analysis, Third Edition*, by C. F. Gerald and P. O. Wheatley, for further details. Algorithms based on an LU decomposition are often identified by the names Crout, Cholesky, and Doolittle.

Population Modeling

Matrix algebra provides for convenient modeling of distributions changing linearly in time or space. One illustration is the Leslie population model, in which a given female population is partitioned into age brackets of equal time length. The size of each bracket is noted at equally spaced time points, with the time step size equal to the bracket length.

Example 6

Suppose the set of all females under 60 in a particular country is partitioned into six age brackets: under 10 years, 10 through 19, . . . , 50 through 59. The size of each bracket is noted every 10 years when a census is taken. For instance, in this century, the census figures for 1900, 1910, ..., 1980 may be used. Let

$$X = [x_1 \quad x_2 \quad \cdots \quad x_6]^t$$

be the column vector whose ith component x_i denotes the number of females in the ith age bracket at the time of a particular census.

In the Leslie model, we assume a constant birthrate of b_i daughters per member of the ith bracket. Also assumed is a constant survival rate, s_i, the fraction of members of the ith bracket surviving to the next census. If all other factors such as immigration are ignored, the distribution vector X changes from one census to the next as follows:

$$\text{from} \quad \begin{bmatrix} x_1 \\ x_2 \\ x_3 \\ x_4 \\ x_5 \\ x_6 \end{bmatrix} \quad \text{to} \quad \begin{bmatrix} b_1x_1 + \cdots + b_6x_6 \\ s_1x_1 \\ s_2x_2 \\ s_3x_3 \\ s_4x_4 \\ s_5x_5 \end{bmatrix}.$$

Let

$$L = \begin{bmatrix} b_1 & b_2 & b_3 & b_4 & b_5 & b_6 \\ s_1 & 0 & 0 & 0 & 0 & 0 \\ 0 & s_2 & 0 & 0 & 0 & 0 \\ 0 & 0 & s_3 & 0 & 0 & 0 \\ 0 & 0 & 0 & s_4 & 0 & 0 \\ 0 & 0 & 0 & 0 & s_5 & 0 \end{bmatrix}.$$

If the distribution at one census is X, then the distribution at the next census is LX. The distribution at the next census is $L(LX) = L^2X$, and the pattern continues. Therefore powers of L, that is, $L^2, L^3, \ldots,$ can be used to describe and predict future population distributions based on this model. □

Exercises 16 and 17 illustrate the Leslie model with specific numbers. The model will be explored further in Chapter 7. In some applications, the *relative* distribution of the population at the nth census for large n is given (approximately) by a constant vector X. For large n, there is also a scalar c such that if X gives the distribution at one census, then cX is the distribution for the next.

Genetics

Another type of distribution changing linearly in time describes genetic traits and inheritance. These traits are many and varied, ranging from human disease to livestock food yield to floral color and size.

Under *autosomal* inheritance, the trait in question is determined by a pair of genes, a *dominant* gene A and a *recessive* gene a. Any individual in the population has exactly one of three *genotypes*: AA and aa are the *purebreds* relative to this trait, Aa is the *hybrid*. One parent is *fertilized* by a second to produce an offspring with one gene received from each parent. We are interested in the relative fractions of each genotype in any generation.

Example 7

Suppose the color of a certain variety of flower grown at a particular nursery is determined genetically by:

$$AA \leftrightarrow \text{red}; \qquad aa \leftrightarrow \text{white}; \qquad Aa \leftrightarrow \text{coral}.$$

In the initial generation, 1/3 of the flowers are red, 1/3 white, and 1/3 coral. Let

$$X^{(n)} = [r^{(n)} \quad w^{(n)} \quad c^{(n)}]^t$$

be the column vector whose entries are the fractions of red, white, and coral flowers in the nth generation, respectively. The initial distribution is

$$X^{(0)} = [\tfrac{1}{3} \quad \tfrac{1}{3} \quad \tfrac{1}{3}]^t.$$

Suppose a breeding program is carried out in such a way that the distribution at the $(n+1)$st generation is obtained from the distribution at the nth generation by the following equations:

$$r^{(n+1)} = \tfrac{2}{3}r^{(n)} + \tfrac{1}{3}c^{(n)},$$

$$w^{(n+1)} = \tfrac{1}{3}w^{(n)} + \tfrac{1}{6}c^{(n)},$$

$$c^{(n+1)} = \tfrac{1}{3}r^{(n)} + \tfrac{2}{3}w^{(n)} + \tfrac{1}{2}c^{(n)}.$$

Then the distribution for the next generation may be described by a matrix product:

$$X^{(n+1)} = AX^{(n)},$$

where

$$A = \begin{bmatrix} \frac{2}{3} & 0 & \frac{1}{3} \\ 0 & \frac{1}{3} & \frac{1}{6} \\ \frac{1}{3} & \frac{2}{3} & \frac{1}{2} \end{bmatrix}.$$

Thus

$$X^{(1)} = AX^{(0)},$$

$$X^{(2)} = AX^{(1)} = A^2X^{(0)},$$

and the pattern continues by the associative law. The powers of A applied to the initial vector yield the distributions for subsequent generations. Matrix algebra provides a natural and convenient way of modeling these quantitative results of the breeding program. □

Example 7 illustrates a *finite Markov chain* modeling a system with m states. The system may be in any state i from 1 to m at time $t = n$. The ith component of the vector

$$X^{(n)} = \begin{bmatrix} p_1^{(n)} & \cdots & p_m^{(n)} \end{bmatrix}^t$$

is a number between 0 and 1 showing the relative frequency with which the system is in state i at time n. In the time step from n to $n + 1$, the system may move from one state to another. The distribution at time $n + 1$ is given by

$$X^{(n+1)} = AX^{(n)},$$

where A is an $m \times m$ matrix. The matrix A is called the **transition matrix** of the chain. In any such A, every entry is nonnegative, and the sum of the entries in any column is 1.

One important question about a Markov chain concerns the behavior for large n—does the system settle down to a steady-state distribution? The following theorem, which is cited without proof, provides the answer.

A Theorem from Markov Chain Theory

Let A be the transition matrix of a finite Markov chain. If for some n every entry in A^n is positive, then:

1. there is a unique vector X such that

$$AX = X;$$

2. for any vector $X^{(0)}$, each component of $A^n X^{(0)}$ is approximately equal to the corresponding component of X when n is large;
3. each column vector of A^n is approximately equal to X when n is large.

The vector X in this theorem is called the **limiting distribution** of the chain.

Example 8

Find the limiting distribution for the transition matrix of Example 7.

Solution

Let X be the unknown distribution vector. By the theorem on Markov chains above, X must satisfy

$$AX = X.$$

Since $X = IX$, where I is the identity matrix, this equation may be rewritten in the form

$$(I - A)X = 0.$$

First find the coefficient matrix:

$$I - A = \begin{bmatrix} 1 & 0 & 0 \\ 0 & 1 & 0 \\ 0 & 0 & 1 \end{bmatrix} - \begin{bmatrix} \frac{2}{3} & 0 & \frac{1}{3} \\ 0 & \frac{1}{3} & \frac{1}{6} \\ \frac{1}{3} & \frac{2}{3} & \frac{1}{2} \end{bmatrix} = \begin{bmatrix} \frac{1}{3} & 0 & -\frac{1}{3} \\ 0 & \frac{2}{3} & -\frac{1}{6} \\ -\frac{1}{3} & -\frac{2}{3} & \frac{1}{2} \end{bmatrix}.$$

After clearing the fractions on the right, carry out a Gaussian elimination:

$$\begin{bmatrix} 1 & 0 & -1 \\ 0 & 4 & -1 \\ -2 & -4 & 3 \end{bmatrix} \to \begin{bmatrix} 1 & 0 & -1 \\ 0 & 4 & -1 \\ 0 & -4 & 1 \end{bmatrix} \to \begin{bmatrix} 1 & 0 & -1 \\ 0 & 4 & -1 \\ 0 & 0 & 0 \end{bmatrix}.$$

From the back substitution, the general solution is $(4z, z, 4z)$. When $z = 1/9$,

$$X = \begin{bmatrix} \frac{4}{9} & \frac{1}{9} & \frac{4}{9} \end{bmatrix}^t.$$

We may readily verify, for instance, that

$$A^4 = \begin{bmatrix} \frac{68}{144} & \frac{56}{144} & \frac{62}{144} \\ \frac{14}{144} & \frac{20}{144} & \frac{17}{144} \\ \frac{62}{144} & \frac{68}{144} & \frac{65}{144} \end{bmatrix},$$

which is already approaching the form of the limiting matrix predicted by the theorem. The limiting distribution X indicates an ultimate distribution for the flowers in Example 6 of 4/9 red, 1/9 white, and 4/9 coral. □

REVIEW CHECKLIST

1. Find LU decompositions of a given square matrix.
2. Solve linear systems using LU decomposition.
3. Find the Leslie matrix for a particular population.
4. Use the Leslie matrix for predicting future population distributions.
5. Use the transition matrix to determine distribution of genotypes in succeeding generations.
6. Given appropriate information, formulate and use matrix models for more general finite Markov chains.

EXERCISES

In Exercises 1–5, find an LU decomposition of the given matrix.

1. $\begin{bmatrix} 1 & 2 \\ 2 & 5 \end{bmatrix}$ **2.** $\begin{bmatrix} 1 & 3 \\ -2 & 2 \end{bmatrix}$ **3.** $\begin{bmatrix} 1 & -2 & 1 \\ 2 & -3 & 5 \\ -1 & 4 & 6 \end{bmatrix}$

4. $\begin{bmatrix} 1 & 1 & 1 \\ -1 & 1 & 1 \\ 3 & -1 & 1 \end{bmatrix}$ **5.** $\begin{bmatrix} 1 & 1 & 0 & -2 \\ 2 & 3 & 2 & -2 \\ -1 & 1 & 5 & 9 \\ -2 & -4 & -1 & 10 \end{bmatrix}$

In Exercises 6–7, find the LU decomposition satisfying the given condition for the matrix

$$\begin{bmatrix} 1 & -1 & 3 \\ -1 & 4 & 3 \\ -2 & 5 & 5 \end{bmatrix}.$$

6. Each main diagonal entry in U equals 1.
7. Each main diagonal entry in L equals 1.

In Exercises 8–12, given A and B, solve the linear system $AX = B$ for X using any LU decomposition of A.

8. $A = \begin{bmatrix} 1 & 4 \\ 2 & 3 \end{bmatrix}, \quad B = \begin{bmatrix} 1 \\ 1 \end{bmatrix}$ **9.** $A = \begin{bmatrix} 3 & 1 \\ 2 & 2 \end{bmatrix}, \quad B = \begin{bmatrix} 1 \\ 1 \end{bmatrix}$

10. $A = \begin{bmatrix} 1 & 2 & 0 \\ 1 & 4 & -2 \\ 2 & 0 & 7 \end{bmatrix}, \quad B = \begin{bmatrix} 1 \\ 2 \\ 2 \end{bmatrix}$ **11.** $A = \begin{bmatrix} 4 & 0 & 1 \\ -8 & 1 & -4 \\ 8 & 1 & 2 \end{bmatrix}, \quad B = \begin{bmatrix} 2 \\ -1 \\ 1 \end{bmatrix}$

12. $A = \begin{bmatrix} 1 & 2 & 1 & 0 \\ -2 & -2 & 1 & -2 \\ 1 & -4 & -9 & 7 \\ 2 & 0 & -3 & 5 \end{bmatrix}, \quad B = \begin{bmatrix} 3 \\ 0 \\ -2 \\ -1 \end{bmatrix}$

In Exercises 13–15, find the LDU decomposition of the given matrix.

13. $\begin{bmatrix} 1 & 2 \\ 2 & 5 \end{bmatrix}$ **14.** $\begin{bmatrix} 1 & -1 & 2 \\ 2 & 5 & 2 \\ 1 & 5 & 12 \end{bmatrix}$ **15.** $\begin{bmatrix} 1 & 2 & 0 \\ 1 & 3 & 1 \\ 2 & 4 & -3 \end{bmatrix}$

In Exercises 16–17, if the given matrix is the transition matrix of a finite Markov chain, find the limiting distribution X.

16. $\begin{bmatrix} \frac{1}{2} & \frac{1}{4} \\ \frac{1}{2} & \frac{3}{4} \end{bmatrix}$ **17.** $\begin{bmatrix} \frac{1}{3} & \frac{1}{3} & \frac{3}{4} \\ \frac{1}{3} & \frac{1}{2} & \frac{1}{8} \\ \frac{1}{3} & \frac{1}{6} & \frac{1}{8} \end{bmatrix}$

18. For a certain female population divided into three age brackets, suppose the corresponding Leslie matrix is

$$L = \begin{bmatrix} 1 & 1 & 1 \\ .5 & 0 & 0 \\ 0 & .5 & 0 \end{bmatrix}.$$

a. If the initial population of 1200 females is divided equally into the three age brackets, find the distributions predicted by the model for the next five censuses.

b. Find L^2, L^3, L^4, and L^5.

c. In Chapter 7 we shall find a steady-state constant growth rate of approximately 1.460 in each bracket, and *relative* distribution $[.685 \quad .235 \quad .080]^t$. Compare these figures with the final results in parts (a) and (b).

19. Suppose the data for the Leslie matrix for the population in Example 6 on page 121 are: $b_2 = 0.58$, $b_3 = 1.06$, $b_4 = 0.08$; $s_1 = 0.96$, $s_2 = 0.98$, $s_3 = 0.98$, $s_4 = 0.98$, $s_5 = 0.96$. Each remaining entry equals 0.

a. If the population distribution in the 1970 census is

$$[26.5 \quad 21.2 \quad 17.3 \quad 14.2 \quad 11.6 \quad 9.2]^t,$$

with components representing the number of million females in the age brackets, predict the 1980 distribution.

b. Compare the result of part (a) with the result of a constant growth rate of 1.2 in each age bracket.

20. In Example 7 on page 123, suppose we change the breeding scheme so that the change in the distribution from the nth generation to the $(n + 1)$st is given by:

$$r^{(n+1)} = \tfrac{1}{2}r^{(n)} + \tfrac{1}{2}w^{(n)} + \tfrac{1}{4}c^{(n)},$$

$$w^{(n+1)} = \tfrac{1}{4}r^{(n)} + \tfrac{1}{3}w^{(n)} + \tfrac{1}{2}c^{(n)},$$

$$c^{(n+1)} = \tfrac{1}{4}r^{(n)} + \tfrac{1}{6}w^{(n)} + \tfrac{1}{4}c^{(n)}.$$

a. Find the transition matrix A.

b. Determine the distribution of colors in the second generation (still assuming an equal distribution in the first generation).
c. Find the vector X such that $AX = X$.
d. Find the matrix that approximates A^n for large values of n.

21. Suppose a certain business telephone is able to hold a call while the line is in use. The following three states may be identified: in state 1 the phone is idle; in state 2 the phone is busy with no calls waiting; in state 3 the phone is busy and a call is holding. Suppose also that informal observation indicates the matrix

$$A = \begin{bmatrix} \frac{1}{2} & \frac{1}{4} & \frac{1}{8} \\ \frac{3}{8} & \frac{1}{2} & \frac{3}{8} \\ \frac{1}{8} & \frac{1}{4} & \frac{1}{2} \end{bmatrix}$$

approximates the transition matrix for change of state in a unit of time.
a. If we assume the initial distribution

$$X^{(0)} = \begin{bmatrix} \frac{1}{3} & \frac{1}{3} & \frac{1}{3} \end{bmatrix}^t,$$

find the distribution for one, two, and three units of time later.
b. Find the limiting distribution X.
c. Find the matrix that approximates A^n for large values of n.

REVIEW EXERCISES

In Exercises 1–4, find the indicated matrix product.

1. $\begin{bmatrix} 2 & 3 \\ -1 & 1 \end{bmatrix}\begin{bmatrix} 1 & 2 \\ 4 & 3 \end{bmatrix}$

2. $\begin{bmatrix} 1 & 4 & 1 \end{bmatrix}\begin{bmatrix} 2 \\ 1 \\ -2 \end{bmatrix}$

3. $\begin{bmatrix} 2 \\ 1 \\ -2 \end{bmatrix}\begin{bmatrix} 1 & 4 & 1 \end{bmatrix}$

4. $\begin{bmatrix} 2 & 0 & -3 \\ 1 & -2 & 2 \\ 3 & 3 & -1 \end{bmatrix}\begin{bmatrix} 1 & 5 & 2 \\ 3 & -1 & 2 \\ 2 & 0 & -3 \end{bmatrix}$

In Exercises 5–10, if possible, find the inverse of the given matrix.

5. $\begin{bmatrix} 1 & 2 \\ 5 & 3 \end{bmatrix}$

6. $\begin{bmatrix} 2 & 1 \\ 3 & 2 \end{bmatrix}$

7. $\begin{bmatrix} 1 & -4 \\ -2 & 8 \end{bmatrix}$

8. $\begin{bmatrix} 1 & 2 & -1 \\ 1 & -1 & 1 \\ 2 & -5 & -3 \end{bmatrix}$

9. $\begin{bmatrix} 1 & 1 & 1 \\ 1 & 0 & 2 \\ 2 & 1 & 5 \end{bmatrix}$

In Exercises 10–11, given A and B, solve the matrix equation $AC = B$ for C by Gaussian or Gauss-Jordan elimination.

10. $A = \begin{bmatrix} 1 & 2 \\ 1 & 3 \end{bmatrix}, \quad B = \begin{bmatrix} 1 & 2 & 1 \\ 1 & 0 & 4 \end{bmatrix}$

11. $A = \begin{bmatrix} 1 & -2 & 2 \\ -2 & 5 & -1 \\ 2 & -6 & -1 \end{bmatrix}, \quad B = \begin{bmatrix} 2 & 1 \\ 3 & 0 \\ 1 & 1 \end{bmatrix}$

In Exercises 12–15, find the indicated matrix, where

$$A = \begin{bmatrix} 1 & 2 & 3 \\ 1 & 3 & 3 \\ -1 & -1 & -2 \end{bmatrix}.$$

12. A^t **13.** A^2 **14.** A^{-1} **15.** A^{-2}

16. Express the matrix

$$\begin{bmatrix} 1 & 4 \\ 2 & 3 \end{bmatrix}$$

as a product of elementary matrices.

Determinants

The determinant *provides a condition used to determine whether a given square linear system has a* unique *solution. This condition is a key factor in the eigenvalue problem in Chapter 7. The determinant also gives a matrix inversion technique that is especially useful for finding the inverse of a 3 × 3 matrix "by hand."*

3.1 EVALUATION OF DETERMINANTS

Except for 2×2 and 3×3 matrices, we usually evaluate the determinant of a matrix using elementary row operations. In this section, we define the determinant function, describe its basic properties, and apply the definition and these properties in evaluating determinants.

Determinants by Cofactor Expansions

The determinant is a function that assigns a number to a given $n \times n$ matrix A. The determinant of A is denoted by $\det(A)$ or by the rectangular array of entries in A surrounded by vertical bars instead of brackets. Before the general definition, let's look at some specific examples.

Example 1

By definition, the determinant of a 1×1 matrix $A = [a]$ is:

$$\det(A) = a.$$

The vertical bars are not used here, since they would denote the absolute value of a. This case is included primarily for completeness. Most of our work will be with the determinant of an $n \times n$ matrix where $n > 1$. □

Example 2

For $A = [a_{ij}]_{2\times 2}$, the definition of the determinant function will yield

$$\det(A) = \begin{vmatrix} a_{11} & a_{12} \\ a_{21} & a_{22} \end{vmatrix} = a_{11}a_{22} - a_{12}a_{21}.$$

As a specific illustration,

$$\begin{vmatrix} 2 & 1 \\ -1 & 3 \end{vmatrix} = 2 \cdot 3 - 1(-1) = 7. \quad \square$$

The first term in the expansion of $\det(A)$ in Example 2 is the product of a_{11} by the determinant of the 1×1 matrix $[a_{22}]$ left after we delete row 1 and column 1 from A. The second term is the product of a_{12} by the determinant of the matrix $[a_{21}]$ left when we delete row 1 and column 2 from A. The following display shows the formation of these terms:

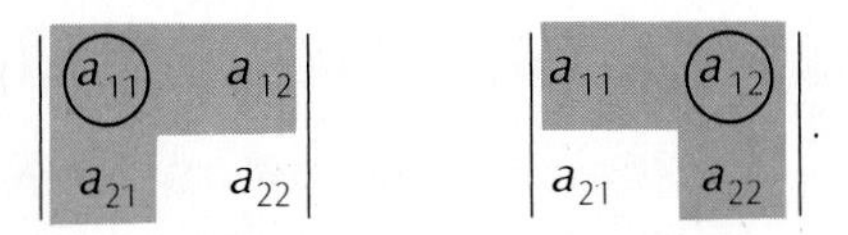

In addition, a plus sign is attached to the first product and a minus to the second.

Example 3

For $A = [a_{ij}]_{3\times 3}$, the definition of the determinant function will give

$$\det(A) = a_{11}\begin{vmatrix} a_{22} & a_{23} \\ a_{32} & a_{33} \end{vmatrix} - a_{12}\begin{vmatrix} a_{21} & a_{23} \\ a_{31} & a_{33} \end{vmatrix} + a_{13}\begin{vmatrix} a_{21} & a_{22} \\ a_{31} & a_{32} \end{vmatrix}.$$

To illustrate in more detail,

$$\begin{vmatrix} 1 & 2 & -3 \\ -1 & 3 & 5 \\ -4 & -2 & 0 \end{vmatrix} = 1\begin{vmatrix} 3 & 5 \\ -2 & 0 \end{vmatrix} - 2\begin{vmatrix} -1 & 5 \\ -4 & 0 \end{vmatrix} + (-3)\begin{vmatrix} -1 & 3 \\ -4 & -2 \end{vmatrix}$$

$$= 1(3 \cdot 0 - 5(-2)) - 2((-1)(0) - 5(-4)$$

$$+ (-3)((-1)(-2) - 3(-4))$$

$$= 10 - 40 - 42 = -72. \quad \square$$

For the first term in the expansion of $\det(A)$ in Example 3, a_{11} is multiplied by the determinant of the matrix left after row 1 and column 1 are deleted from A.

For the second, a_{12} is multiplied by the determinant of the matrix left when row 1 and column 2 are deleted from A. For the third, a_{13} is multiplied by the determinant of the matrix left after row 1 and column 3 are deleted from A. The formation of these terms is shown by:

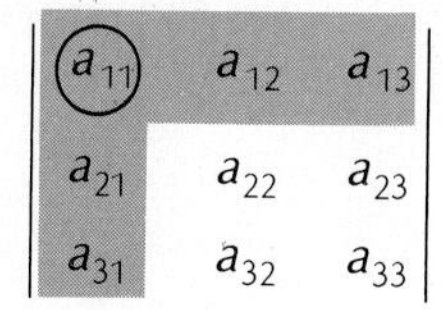

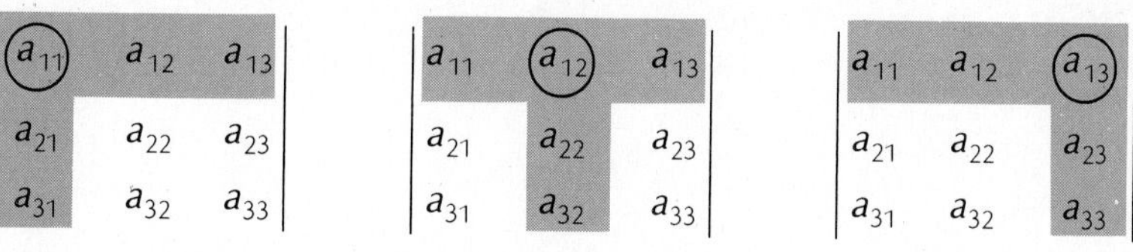

The attached signs alternate from plus to minus to plus.

In defining the determinant function, we use a *recursive definition*. We begin with an explicit definition of $\det(A)$ for a 1×1 matrix A. For $n \geq 2$, the determinant of any $n \times n$ matrix is defined in terms of determinants of $(n-1) \times (n-1)$ matrices. Examples 2 and 3 illustrate the recursive nature of this definition.

DEFINITION 3.1

The **determinant of a 1 × 1 matrix [a]** is the entry a. Now suppose the determinant has been defined for $k \times k$ matrices for all k from 2 through $n-1$, where $n \geq 2$. Let $A = [a_{ij}]_{n \times n}$ be an $n \times n$ matrix. For $i, j = 1, \ldots, n$, let M_{ij} denote the determinant of the matrix left when row i and column j are deleted from A, and let $C_{ij} = (-1)^{i+j} M_{ij}$. Then the **determinant of A** is defined as

$$\det(A) = a_{11}C_{11} + \cdots + a_{1n}C_{1n}.$$

Example 4

Evaluate $\det(A)$ for

$$A = \begin{vmatrix} 1 & 2 & 0 & 3 \\ 1 & 1 & 2 & 4 \\ 5 & 2 & 1 & 0 \\ 3 & 4 & 5 & 2 \end{vmatrix}.$$

Solution

By Definition 3.1,

$$\det(A) = 1\begin{vmatrix} 1 & 2 & 4 \\ 2 & 1 & 0 \\ 4 & 5 & 2 \end{vmatrix} - 2\begin{vmatrix} 1 & 2 & 4 \\ 5 & 1 & 0 \\ 3 & 5 & 2 \end{vmatrix} + 0\begin{vmatrix} 1 & 1 & 4 \\ 5 & 2 & 0 \\ 3 & 4 & 2 \end{vmatrix} - 3\begin{vmatrix} 1 & 1 & 2 \\ 5 & 2 & 1 \\ 3 & 4 & 5 \end{vmatrix}.$$

Each of these 3 × 3 determinants may, in turn, be expanded along its first row as in Example 2. For instance,

$$\begin{vmatrix} 1 & 2 & 4 \\ 2 & 1 & 0 \\ 4 & 5 & 2 \end{vmatrix} = 1\begin{vmatrix} 1 & 0 \\ 5 & 2 \end{vmatrix} - 2\begin{vmatrix} 2 & 0 \\ 4 & 2 \end{vmatrix} + 4\begin{vmatrix} 2 & 1 \\ 4 & 5 \end{vmatrix}$$

$$= 1 \cdot 2 - 2 \cdot 4 + 4 \cdot 6 = 18.$$

We leave the evaluation of the three remaining 3 × 3 determinants as an exercise, citing only the final results:

$$\det(A) = 1(18) - 2(70) + 0(50) - 3(12) = -158. \quad \square$$

The determinants M_{ij} in Definition 3.1 are called the **minors of A**, and the signed minors C_{ij} in this definition are called the **cofactors of A**.

Example 5

Find the minor M_{22} and the cofactor C_{34} for $A = [a_{ij}]_{4\times 4}$.

Solution

By definition,

$$M_{22} = \begin{vmatrix} a_{11} & a_{12} & a_{13} & a_{14} \\ a_{21} & a_{22} & a_{23} & a_{24} \\ a_{31} & a_{32} & a_{33} & a_{34} \\ a_{41} & a_{42} & a_{43} & a_{44} \end{vmatrix} = \begin{vmatrix} a_{11} & a_{13} & a_{14} \\ a_{31} & a_{33} & a_{34} \\ a_{41} & a_{43} & a_{44} \end{vmatrix},$$

and

$$C_{34} = (-1)^{3+4}\begin{vmatrix} a_{11} & a_{12} & a_{13} & a_{14} \\ a_{21} & a_{22} & a_{23} & a_{24} \\ a_{31} & a_{32} & a_{33} & a_{34} \\ a_{41} & a_{42} & a_{43} & a_{44} \end{vmatrix} = -\begin{vmatrix} a_{11} & a_{12} & a_{13} \\ a_{21} & a_{22} & a_{23} \\ a_{41} & a_{42} & a_{43} \end{vmatrix}. \quad \square$$

The next example illustrates that the choice of row 1 for the expansion of a determinant in Definition 3.1 was arbitrary.

Example 6

For the matrix

$$\begin{bmatrix} 1 & 4 & 0 \\ 2 & -1 & 2 \\ 5 & -2 & 0 \end{bmatrix},$$

find the cofactor expansions along row 1 and down column 3.

Solution

Along row 1,

$$\begin{vmatrix} 1 & 4 & 0 \\ 2 & -1 & 2 \\ 5 & -2 & 0 \end{vmatrix} = 1\begin{vmatrix} -1 & 2 \\ -2 & 0 \end{vmatrix} - 4\begin{vmatrix} 2 & 2 \\ 5 & 0 \end{vmatrix} + 0\begin{vmatrix} 2 & -1 \\ 5 & -2 \end{vmatrix} = 4 + 40 = 44.$$

Expanding down column 3, we capitalize on the two zeros:

$$\begin{vmatrix} 1 & 4 & 0 \\ 2 & -1 & 2 \\ 5 & -2 & 0 \end{vmatrix} = 0\begin{vmatrix} 2 & -1 \\ 5 & -2 \end{vmatrix} - 2\begin{vmatrix} 1 & 4 \\ 5 & -2 \end{vmatrix} + 0\begin{vmatrix} 1 & 4 \\ 2 & -1 \end{vmatrix} = 44.$$

Any row or column for the expansion yields the value 44. □

Example 6 illustrates the following general theorem. A proof is given in Section 3.3.

THEOREM 3.1

For $A = [a_{ij}]_{n \times n}$, let C_{ij} denote the cofactors of A for $i, j = 1, \ldots, n$. Then for each i,

$$\det(A) = a_{i1}C_{i1} + \cdots + a_{in}C_{in},$$

and for each j,

$$\det(A) = a_{1j}C_{1j} + \cdots + a_{nj}C_{nj}.$$

Sometimes the determinant of a 2×2 matrix is indicated by:

$$\begin{vmatrix} a_{11} & a_{12} \\ a_{21} & a_{22} \end{vmatrix} = a_{11}a_{22} - a_{12}a_{21}.$$

A similar notation may be used for a 3×3 determinant:

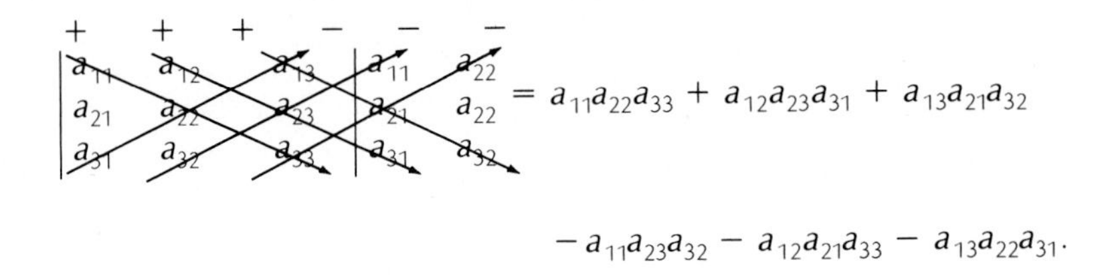

$$= a_{11}a_{22}a_{33} + a_{12}a_{23}a_{31} + a_{13}a_{21}a_{32}$$

$$- a_{11}a_{23}a_{32} - a_{12}a_{21}a_{33} - a_{13}a_{22}a_{31}.$$

Caution. The expansion scheme indicated by this special notation does *not* apply to the determinant of an $n \times n$ matrix for $n > 3$.

We comment also on the use of $(-1)^{i+j}$ in the definition of the cofactor C_{ij}. This notation simply means that a plus $(+)$ is assigned if $i + j$ is even and a minus $(-)$ is assigned if it is odd. This scheme may also be represented by the following sign diagram:

$$\begin{bmatrix} + & - & + & - & \cdots \\ - & + & - & + & \cdots \\ + & - & + & - & \cdots \\ - & + & - & + & \cdots \\ \vdots & \vdots & \vdots & \vdots & \end{bmatrix}.$$

Properties of Determinants

The determinant function has a number of useful basic properties. We list nine of them as a group for convenient reference, following with examples. Throughout this list, A denotes an $n \times n$ matrix.

PROPERTIES OF THE DETERMINANT FUNCTION

D1. If A has a zero row [column], then $\det(A) = 0$.

D2. If A is triangular, then $\det(A) = a_{11} \cdots a_{nn}$, the product of the main diagonal entries.

D3. If B is obtained from A by a row [column] interchange, then $\det(B) = -\det(A)$.

D4. If we obtain B from A by multiplying a row [column] by c, then $\det(B) = c \cdot \det(A)$.

D5. If A has two rows [columns] with equal entries or two with proportional entries, then $\det(A) = 0$.

D6. If B and C are also $n \times n$ matrices, and if the corresponding entries of A, B, and C are equal in all rows except row i, where

$$a_{ij} + b_{ij} = c_{ij} \quad \text{for} \quad j = 1, \ldots, n,$$

then

$$\det(A) + \det(B) = \det(C).$$

D7. The corresponding result holds for columns. If B is obtained from A by the addition of c times row i to row j, then $\det(B) = \det(A)$. The corresponding result holds for columns.

D8. For matrix products, $\det(AB) = \det(A) \cdot \det(B)$.

D9. For the transpose, $\det(A^t) = \det(A)$.

Properties D1–D9 will be established in Section 3.3. For now, we show how to use them in evaluating determinants. Each of the following illustrations may be verified by Definition 3.1.

Example 7

Property D1 gives

$$\begin{vmatrix} 3 & 4 & -2 \\ 0 & 0 & 0 \\ 2 & -1 & 5 \end{vmatrix} = 0. \quad \square$$

Example 8

By Property D3,

$$\begin{vmatrix} 4 & 0 & 5 \\ -2 & 5 & -1 \\ 3 & -4 & 2 \end{vmatrix} = -\begin{vmatrix} 3 & -4 & 2 \\ -2 & 5 & -1 \\ 4 & 0 & 5 \end{vmatrix}$$

since the determinant on the right is obtained from that on the left by the interchange of rows 1 and 3. $\square$

Example 9

Property D4 yields

$$\begin{vmatrix} 0 & 1 & -6 \\ 3 & -5 & 4 \\ 2 & 7 & 2 \end{vmatrix} = 2\begin{vmatrix} 0 & 1 & -3 \\ 3 & -5 & 2 \\ 2 & 7 & 1 \end{vmatrix}$$

since the factor 2 on the right may be distributed to the entries in column 3. □

Example 10

Property D8 shows that $\det(AB) = \det(A)\det(B)$ for

$$A = \begin{vmatrix} 2 & -3 \\ 4 & 1 \end{vmatrix} \quad \text{and} \quad B = \begin{vmatrix} 3 & 1 \\ -2 & -1 \end{vmatrix}.$$

This relation may also be verified by direct calculation and Definition 3.1:

$$AB = \begin{vmatrix} 12 & 5 \\ 10 & 3 \end{vmatrix}, \qquad \det(A) = 14, \qquad \det(B) = -1, \qquad \det(AB) = -14.$$ □

Example 11

Property D6 shows that

$$\begin{vmatrix} 2 & 1 & -2 \\ 3 & -1 & 0 \\ 1 & 3 & -2 \end{vmatrix} + \begin{vmatrix} 2 & 1 & -2 \\ 3 & -1 & 0 \\ 2 & -2 & 1 \end{vmatrix} = \begin{vmatrix} 2 & 1 & -2 \\ 3 & -1 & 0 \\ 3 & 1 & -1 \end{vmatrix}.$$

The corresponding entries in row 1 agree, the corresponding entries in row 2 agree, and the entries in row 3 on the right are obtained by summing the corresponding entries on the left. □

Evaluation of Determinants by Elementary Row Operations

Property D3 (illustrated in Example 8) shows the effect of a row interchange on the value of the determinant. Property D4 (illustrated in Example 9) accounts for multiplication of a row by a constant. Property D7 accounts for addition of a multiple of one row to another, as the next example illustrates.

Example 12

A particular illustration of the elementary row operation R2 ← R2 − 3R1 is:

$$\begin{bmatrix} 1 & 2 & 1 \\ 3 & 1 & 4 \\ -1 & 2 & 2 \end{bmatrix} \rightarrow \begin{bmatrix} 1 & 2 & 1 \\ 0 & -5 & 1 \\ -1 & 2 & 2 \end{bmatrix}.$$

By Property D6,

$$\begin{vmatrix} 1 & 2 & 1 \\ 3 & 1 & 4 \\ -1 & 2 & 2 \end{vmatrix} + \begin{vmatrix} 1 & 2 & 1 \\ -3 & -6 & -3 \\ -1 & 2 & 2 \end{vmatrix} = \begin{vmatrix} 1 & 2 & 1 \\ 0 & -5 & 1 \\ -1 & 2 & 2 \end{vmatrix}$$

since corresponding entries in row 1 agree, as do those in row 3. The middle determinant equals zero by Property D5, since the entries in rows 1 and 2 are proportional. Hence if the original matrix is denoted by A and the matrix resulting from the elementary row operation by B, then $\det(A) = \det(B)$. □

In practice, we usually evaluate determinants using Properties D3, D4, and D7. To evaluate $\det(A)$ for $A = [a_{ij}]_{n \times n}$, eliminate the nonzero entries below a_{11} and expand down column 1. The product of a_{11} by the resulting cofactor equals $\det(A)$ by Theorem 3.1 and Property D7. We continue this procedure until we reach a 2×2 determinant.

Example 13

Evaluate the determinant

$$\begin{vmatrix} 1 & 0 & 2 \\ 2 & 2 & -1 \\ 3 & 4 & 3 \end{vmatrix}.$$

Solution

$$\begin{vmatrix} 1 & 0 & 2 \\ 2 & 2 & -1 \\ 3 & 4 & 3 \end{vmatrix} = \begin{vmatrix} 1 & 0 & 2 \\ 0 & 2 & -5 \\ 0 & 4 & -3 \end{vmatrix} = 1\begin{vmatrix} 2 & -5 \\ 4 & -3 \end{vmatrix} = 14. \quad \square$$

Example 14

Evaluate the determinant

$$\begin{vmatrix} 2 & 6 & -2 & 4 \\ 3 & 9 & 0 & 4 \\ 2 & 5 & 1 & 1 \\ -1 & 2 & 1 & 4 \end{vmatrix}.$$

Solution

$$\begin{vmatrix} 2 & 6 & -2 & 4 \\ 3 & 9 & 0 & 4 \\ 2 & 5 & 1 & 1 \\ -1 & 2 & 1 & 4 \end{vmatrix} = 2\begin{vmatrix} 1 & 3 & -1 & 2 \\ 3 & 9 & 0 & 4 \\ 2 & 5 & 1 & 1 \\ -1 & 2 & 1 & 4 \end{vmatrix} = 2\begin{vmatrix} 1 & 3 & -1 & 2 \\ 0 & 0 & 3 & -2 \\ 0 & -1 & 3 & -3 \\ 0 & 5 & 0 & 6 \end{vmatrix}$$

$$= 2\begin{vmatrix} 0 & 3 & -2 \\ -1 & 3 & -3 \\ 5 & 0 & 6 \end{vmatrix} = -2\begin{vmatrix} -1 & 3 & -3 \\ 0 & 3 & -2 \\ 5 & 0 & 6 \end{vmatrix}$$

$$= -2\begin{vmatrix} -1 & 3 & -3 \\ 0 & 3 & -2 \\ 0 & 15 & -9 \end{vmatrix}$$

$$= -2(-1)\begin{vmatrix} 3 & -2 \\ 15 & -9 \end{vmatrix} = 6. \quad \square$$

If a zero column appears at any stage, the value of the given determinant is zero by Property D1, and the procedure terminates.

Expanding down a column other than the first or along some row may be easier than using column 1, for instance, if one or more zeros appear in a row or a column at any stage. We used column 1 for harmony with the Gaussian elimination procedure.

If the first pass of a Gaussian elimination is carried out on a given matrix A, $\det(A)$ can be obtained from the resulting upper triangular matrix. The determinant is the product of the diagonal entries in the upper triangular matrix, multiplied by -1 (if the number of row interchanges is odd) and by the reciprocal of each constant by which a row is multiplied. Hand implementation of this method may require special care to avoid errors. On a computer, $\det(A)$ can usually be found conveniently in conjunction with Gaussian elimination.

REVIEW CHECKLIST

1. Formulate a recursive definition of the determinant function.
2. Formulate definitions of minor and cofactor.
3. Evaluate determinants by cofactor expansions along any row or down any column.

4. Apply Properties D1–D9 in evaluating determinants.
5. Evaluate determinants by the technique illustrated in Examples 13 and 14 on pages 139–140.

EXERCISES

In Exercises 1–18, evaluate the given determinant.

1. $\begin{vmatrix} 1 & 2 \\ 3 & 2 \end{vmatrix}$ **2.** $\begin{vmatrix} 1 & 1 \\ 3 & 3 \end{vmatrix}$ **3.** $\begin{vmatrix} 1 & 1 \\ 0 & 1 \end{vmatrix}$

4. $\begin{vmatrix} 3 & -1 \\ 2 & 5 \end{vmatrix}$ **5.** $\begin{vmatrix} 4 & 5 \\ 1 & 0 \end{vmatrix}$ **6.** $\begin{vmatrix} 1 & 5 \\ 2 & 4 \end{vmatrix}$

7. $\begin{vmatrix} 1 & 1 & 2 \\ 1 & 1 & 2 \\ 3 & 1 & 4 \end{vmatrix}$ **8.** $\begin{vmatrix} 1 & 2 & -1 \\ 2 & 4 & 3 \\ 0 & 0 & 0 \end{vmatrix}$ **9.** $\begin{vmatrix} 1 & -1 & 1 \\ 3 & 2 & 3 \\ 4 & 2 & 4 \end{vmatrix}$

10. $\begin{vmatrix} 2 & 4 & 3 \\ -4 & -8 & -6 \\ 3 & 1 & 1 \end{vmatrix}$ **11.** $\begin{vmatrix} 4 & 5 & 2 \\ 2 & 8 & 1 \\ 6 & 0 & 3 \end{vmatrix}$ **12.** $\begin{vmatrix} 1 & 0 & -2 \\ 2 & 0 & -4 \\ 3 & 0 & -1 \end{vmatrix}$

13. $\begin{vmatrix} 1 & 0 & 0 \\ 0 & 1 & 0 \\ 0 & 0 & 2 \end{vmatrix}$ **14.** $\begin{vmatrix} 3 & 1 & 5 \\ 0 & 2 & 8 \\ 0 & 0 & -3 \end{vmatrix}$

15. $\begin{vmatrix} 1 & 3 & 1 & 3 \\ 0 & 1 & 0 & 2 \\ 0 & 1 & 1 & 2 \\ 0 & -2 & 1 & 0 \end{vmatrix}$ **16.** $\begin{vmatrix} 1 & -1 & 2 & 1 \\ 3 & 1 & 0 & 4 \\ 2 & -2 & 4 & 2 \\ 3 & 1 & 0 & 2 \end{vmatrix}$

17. $\begin{vmatrix} 2 & 1 & -1 & 3 \\ 0 & 2 & 5 & 4 \\ 0 & 0 & -1 & 3 \\ 0 & 0 & 2 & 5 \end{vmatrix}$ **18.** $\begin{vmatrix} 3 & 3 & 1 & -1 \\ 1 & -6 & 2 & 2 \\ 5 & -3 & 4 & 1 \\ 4 & 9 & 5 & -3 \end{vmatrix}$

19. Write out and evaluate all nine minors for the matrix of Exercise 11, and find the corresponding cofactors.

20. Construct a "sign pattern matrix" for the cofactors of a 4×4 matrix. That is, form a 4×4 array of *signs*, entering at row i and column j a plus sign $(+)$ if $M_{ij} = C_{ij}$ and a negative sign $(-)$ if $M_{ij} = -C_{ij}$.

In Exercises 21–26, find the value of the determinant, given that

$$\begin{vmatrix} a & b & c \\ 3 & 1 & -1 \\ 2 & 1 & 2 \end{vmatrix} = 12.$$

21. $\begin{vmatrix} 2a & 2b & 2c \\ 3 & 1 & -1 \\ 2 & 1 & 2 \end{vmatrix}$ **22.** $\begin{vmatrix} 3 & 1 & -1 \\ a & b & c \\ 2 & 1 & 2 \end{vmatrix}$ **23.** $\begin{vmatrix} c & b & a \\ -1 & 1 & 3 \\ 2 & 1 & 2 \end{vmatrix}$

24. $\begin{vmatrix} 2 & 1 & 2 \\ 3 & 1 & -1 \\ a & b & c \end{vmatrix}$ **25.** $\begin{vmatrix} b & c & a \\ 1 & -1 & 3 \\ 1 & 2 & 2 \end{vmatrix}$ **26.** $\begin{vmatrix} a & 3 & 2 \\ b & 1 & 1 \\ c & -1 & 2 \end{vmatrix}$

27. Give an example of 2×2 matrices A and B for which

$$\det(A) + \det(B) \neq \det(A + B).$$

In Exercises 28–29, evaluate $\det(AB)$ and $\det(A)\det(B)$ for the given matrices A and B.

28. $A = \begin{bmatrix} 2 & -1 \\ 1 & 2 \end{bmatrix}, \quad B = \begin{bmatrix} 3 & 3 \\ 1 & -1 \end{bmatrix}$

29. $A = \begin{bmatrix} 1 & -1 & 1 \\ 2 & 0 & 1 \\ 3 & 3 & -2 \end{bmatrix}, \quad B = \begin{bmatrix} 2 & 0 & 1 \\ 1 & -2 & 2 \\ 2 & 1 & -1 \end{bmatrix}$

In Exercises 30–34, find the value(s) of a for which the determinant is zero.

30. $\begin{vmatrix} a & 1 \\ 2 & 3 \end{vmatrix}$ **31.** $\begin{vmatrix} a & 1 \\ 4 & a \end{vmatrix}$ **32.** $\begin{vmatrix} a-2 & 2 \\ 4 & a+2 \end{vmatrix}$

33. $\begin{vmatrix} a-1 & 0 & 1 \\ 0 & a-1 & 0 \\ 1 & 0 & a-1 \end{vmatrix}$ **34.** $\begin{vmatrix} a-2 & -2 & -2 \\ 1 & a-2 & -1 \\ -1 & 2 & a+1 \end{vmatrix}$

35. If $\det(A) = 3$, find $\det(A^2)$ and $\det(A^3)$.

36. If A is invertible and $\det(A) = 4$, find $\det(A^{-1})$.

37. If $\det(A) = 5$ and $B^2 = A$, find $\det(B)$.

38. If $\det(A) = 10$, find $\det(A^tA)$.

39. Find the relation between $\det(AB)$ and $\det(BA)$ for A and B both $n \times n$ matrices.

Suggested Project

An alternate approach to determinants is based on properties of *permutations*. A **permutation** of a finite set is an arrangement, or ordering, of the elements of the set. For instance, the permutations of the set $\{1, 2\}$ may be written as

$$12 \quad \text{and} \quad 21$$

(read one-two and two-one), and the permutations of $\{1, 2, 3\}$ as

$$123, \quad 132, \quad 213, \quad 231, \quad 312, \quad \text{and} \quad 321.$$

A permutation of $\{1, \ldots, n\}$ is called **even** or **odd** according to whether the number of pairwise interchanges required to restore the natural order is even or odd. For instance, 12 is even, 21 is odd, 123, 231, and 312 are even, and 132, 213, and 321 are odd. For $n > 1$, the set $\{1, \ldots, n\}$ has $n!$ permutations, half of which are even, the other half odd.

For a given matrix $A = [a_{ij}]_{n \times n}$, we may form products of n factors, taking one entry from each row and column. For instance, for $n = 3$, one such product is

$$a_{12}a_{23}a_{31}.$$

If the row subscripts appear in natural order, then each such product corresponds uniquely to a permutation of the column subscripts, hence there are $n!$ such products. An alternate definition of $\det(A)$ is the algebraic sum of all such products, in which the products corresponding to even permutations are added and those corresponding to the odd ones are subtracted. For $n = 3$, for instance, this definition yields

$$\det(A) = a_{11}a_{22}a_{33} + a_{12}a_{23}a_{31} + a_{13}a_{21}a_{32} - a_{11}a_{23}a_{32} - a_{12}a_{21}a_{33} - a_{13}a_{22}a_{31}.$$

Explore the permutation approach. In the previous expression for a 3×3 determinant, for example, factor a_{11} from the first and fourth terms to obtain the cofactor expansion along row 1. Starting from this definition, derive Properties D1–D9 and Theorem 3.1 (not necessarily in order).

3.2 DETERMINANTS AND LINEAR SYSTEMS

The determinant provides a condition for the existence of a unique solution to a square linear system. It is also used to test for invertibility of a square matrix and to find the inverse when it exists. The determinant also gives *Cramer's rule*, a method for solving linear systems that is historically significant in linear algebra and other areas of study, including integral equations.

Square Linear Systems with a Unique Solution

We begin with an example illustrating a condition for existence and uniqueness of a solution to a square linear system.

Example 1

For the linear system

$$\begin{aligned} x + 2y + 4z &= 3 \\ 2x + 2y + z &= -3 \\ 3x + 4y + 3z &= -2, \end{aligned}$$

elementary row operations may be applied to the augmented matrix:

$$\left[\begin{array}{ccc|c} 1 & 2 & 4 & 3 \\ 2 & 2 & 1 & -3 \\ 3 & 4 & 3 & -2 \end{array}\right] \to \left[\begin{array}{ccc|c} 1 & 2 & 4 & 3 \\ 0 & -2 & -7 & -9 \\ 0 & -2 & -9 & -11 \end{array}\right] \to \left[\begin{array}{ccc|c} 1 & 2 & 4 & 3 \\ 0 & 2 & 7 & 9 \\ 0 & 0 & -2 & -2 \end{array}\right].$$

Suppose the coefficient matrix is denoted by A and let

$$U = \begin{bmatrix} 1 & 2 & 4 \\ 0 & 2 & 7 \\ 0 & 0 & -2 \end{bmatrix}.$$

Each main diagonal entry in U is nonzero, so back substitution yields a unique solution to the given system. By Property D2 of Section 3.1, $\det(U)$ equals the product of the main diagonal entries in U. Since each of these entries is nonzero, $\det(U) \neq 0$. Since each elementary row operation used to obtain U from A leaves the determinant unchanged, $\det(A) = \det(U)$ by Property D7. That is, $\det(A) \neq 0$ and the system $AX = B$ has a unique solution. □

Example 1 illustrates the following general theorem.

THEOREM 3.2

A linear system with a square coefficient matrix A has a unique solution if and only if $\det(A) \neq 0$.

Proof Suppose we start with A and perform a sequence of elementary row operations to reach an upper triangular matrix U. By Properties D3, D4, and D7 from Section 3.1, the effect of an elementary row operation on the determinant of a matrix is a multiplication by a nonzero scalar (which is -1 if Property D3 applies and 1 if Property D7 applies). Thus, $\det(A) \neq 0$ if and only if $\det(U) \neq 0$. By Property D2, $\det(U)$ equals the product of its main diagonal entries. Hence, $\det(U) \neq 0$ if and only if every main diagonal entry in U is nonzero. In turn, every main diagonal entry in U is nonzero if and only if every linear system with coefficient matrix A has a unique solution, by properties of back substitution. Thus we obtain the statement of the theorem.

Caution. 1. If $\det(A) = 0$, then a linear system with coefficient matrix A *may or may not have a solution*. When it does, the solution is *not* unique.

2. Properties D2, D3, D4, and D7 were used in the proof of Theorem 3.2. We shall make special note in Section 3.3 that the proofs of these properties are independent of this theorem.

Corollary to Theorem 3.2. A homogeneous linear system with a square coefficient matrix A has nontrivial solutions if and only if $\det(A) = 0$.

Proof If $\det(A) = 0$, then a linear system with coefficient matrix A must have nontrivial solutions; otherwise, the trivial solution would be unique, hence $\det(A) \neq 0$ by Theorem 3.2. Conversely, if any homogeneous system with coefficient matrix A has a nontrivial solution, then it has at least two solutions, hence by Theorem 3.2, $\det(A) = 0$. ☐

Example 2

Find the value(s) of a for which the system $(aI - A)X = 0$ has nontrivial solutions, where

$$A = \begin{bmatrix} 1 & 4 \\ 1 & -2 \end{bmatrix}.$$

Solution

Apply the Corollary to Theorem 3.2 with $aI - A$ in the role of A. The system $(aI - A)X = 0$ has a nontrivial solution if and only if $\det(aI - A) = 0$. Since

$$aI - A = a\begin{bmatrix} 1 & 0 \\ 0 & 1 \end{bmatrix} - \begin{bmatrix} 1 & 4 \\ 1 & -2 \end{bmatrix} = \begin{bmatrix} a-1 & -4 \\ -1 & a+2 \end{bmatrix}$$

and

$$\begin{vmatrix} a-1 & -4 \\ -1 & a+2 \end{vmatrix} = (a-1)(a+2) - 4 = a^2 + a - 6,$$

the required condition is

$$a^2 + a - 6 = 0.$$

The roots of this equation, -3 and 2, are the required values of a. ☐

Example 2 is closely related to Exercises 31–34 of Section 3.1, and illustrates the way Theorem 3.2 is used to solve eigenvalue problems in Chapter 7. Theorem 3.2 is also used in other areas of mathematics, such as differential equations. We shall indicate some of these areas in Section 3.4.

Determinants and Matrix Inversion

The following example illustrates a relationship between the invertibility of a matrix A and the value of $\det(A)$.

Example 3

If

$$A = \begin{vmatrix} 2 & 5 \\ -3 & -4 \end{vmatrix},$$

then $\det(A) = 7$. In addition, A is invertible, and

$$A^{-1} = \frac{1}{7}\begin{vmatrix} -4 & -5 \\ 3 & 2 \end{vmatrix}.$$

That is, $\det(A) \neq 0$ and A is invertible. ☐

Example 3 illustrates the following general principle.

THEOREM 3.3

If A is a square matrix, then A is invertible if and only if $\det(A) \neq 0$.

Proof By Theorem 2.6, A is invertible if and only if every linear system with coefficient matrix A has a unique solution. By Theorem 3.2, every such system has a unique solution if and only if $\det(A) \neq 0$. The desired result follows. ☐

We shall need a result related to the cofactor expansions described in Definition 3.1 and Theorem 3.1. If a dot product is formed with coefficients from one row [column] and cofactors from a different row [column], the resulting sum equals zero.

Example 4

Let

$$A = \begin{bmatrix} 1 & 2 & 3 \\ -1 & 1 & 2 \\ 3 & 2 & -2 \end{bmatrix}.$$

Coefficients from row 1 and cofactors from row 3 give

$$a_{11}C_{31} + a_{12}C_{32} + a_{13}C_{33} = \begin{vmatrix} 2 & 3 \\ 1 & 2 \end{vmatrix} - 2\begin{vmatrix} 1 & 3 \\ -1 & 2 \end{vmatrix} + 3\begin{vmatrix} 1 & 2 \\ -1 & 1 \end{vmatrix}$$

$$= 1 - 2 \cdot 5 + 3 \cdot 3 = 0.$$ ☐

We formalize the general result illustrated by Example 4.

Lemma. Let A be an $n \times n$ matrix, and let i, j, k, l be integers between 1 and n, inclusive. If $i \neq k$, then

$$a_{i1}C_{k1} + \cdots + a_{in}C_{kn} = 0,$$

and if $j \neq l$, then

$$a_{1j}C_{1l} + \cdots + a_{nj}C_{nl} = 0.$$

Proof If we form a matrix B by replacing row k of A with row i, then $\det(B) = 0$ since B has two equal rows. The cofactors of B along row k are the same as the corresponding cofactors of A, that is, C_{kj} for $j = 1, \ldots, n$. We also have $b_{kj} = a_{ij}$ for $j = 1, \ldots, n$. Thus expanding $\det(B)$ along row k according to Theorem 3.1 yields

$$\begin{aligned} 0 = \det(B) &= b_{k1}C_{k1} + \cdots + b_{kn}C_{kn} \\ &= a_{i1}C_{k1} + \cdots + a_{in}C_{kn}. \end{aligned}$$

The result for columns may be proved in a similar way. (See Exercise 20.) □

We give a method for inverting a matrix using the lemma and Theorem 3.1. The **adjoint of an n × n matrix A** is defined by

$$\operatorname{Adj}(A) = \begin{bmatrix} C_{11} & \cdots & C_{n1} \\ \vdots & & \vdots \\ C_{1n} & \cdots & C_{nn} \end{bmatrix},$$

where the C_{ij} are the cofactors of A. Note carefully: in this definition the row and column subscripts are transposed relative to the usual notation for a general $n \times n$ matrix A.

THEOREM 3.4

Let A be an $n \times n$ matrix. If $\det(A) \neq 0$, then

$$A^{-1} = \frac{1}{\det(A)} \operatorname{Adj}(A).$$

Proof If $\det(A) \neq 0$, then A is invertible by Theorem 3.2. By definition of the adjoint,

$$A(\mathrm{Adj}(A)) = \begin{bmatrix} a_{11} & \cdots & a_{1n} \\ \vdots & & \vdots \\ a_{n1} & \cdots & a_{nn} \end{bmatrix} \begin{bmatrix} C_{11} & \cdots & C_{n1} \\ \vdots & & \vdots \\ C_{1n} & \cdots & C_{nn} \end{bmatrix}.$$

For $i = 1, \ldots, n$, the main diagonal entries in the product are

$$a_{i1}C_{i1} + \cdots + a_{in}C_{in},$$

all of which equal $\det(A)$ by Theorem 3.1. The remaining entries are of the form

$$a_{i1}C_{k1} + \cdots + a_{in}C_{kn}, \qquad k \neq i.$$

Each of these entries equals zero by the lemma. Hence

$$A(\mathrm{Adj}(A)) = (\det(A))I.$$

Dividing through by $\det(A)$, we obtain the desired result. ☐

Example 5

Find the inverse of the matrix

$$A = \begin{bmatrix} 1 & -1 & 3 \\ 2 & 1 & 0 \\ 1 & 2 & -2 \end{bmatrix}$$

using the adjoint method.

Solution

Calculate the cofactors of A to be entered in the matrix $\mathrm{Adj}(A)$. For instance, along *row* 1

$$C_{11} = \begin{vmatrix} 1 & 0 \\ 2 & -2 \end{vmatrix} = -2, \qquad C_{12} = -\begin{vmatrix} 2 & 0 \\ 1 & -2 \end{vmatrix} = 4, \qquad C_{13} = \begin{vmatrix} 2 & 1 \\ 1 & 2 \end{vmatrix} = 3.$$

These values are entered in *column* 1 of the adjoint. Similar calculations yield the cofactors along row 2 [3] to be entered in column 2 [3] of $\mathrm{Adj}(A)$. The resulting adjoint is given by

$$\text{Adj}(A) = \begin{bmatrix} -2 & 4 & -3 \\ 4 & -5 & 6 \\ 3 & -3 & 3 \end{bmatrix}.$$

The determinant of A may be found by Definition 3.1 since the cofactors of A along row 1 are known:

$$\det(A) = a_{11}C_{11} + a_{12}C_{12} + a_{13}C_{13} = 1(-2) + (-1)(4) + 3(3) = 3.$$

Therefore by Theorem 3.4,

$$A^{-1} = \frac{1}{3}\begin{bmatrix} -2 & 4 & -3 \\ 4 & -5 & 6 \\ 3 & -3 & 3 \end{bmatrix}. \quad \square$$

For a 2×2 matrix, the method of adjoints should be familiar: it is the special method given in Section 2.4. The adjoint method is reasonably appropriate for inverting 2×2 and 3×3 matrices, especially with relatively small integral entries.

If a given matrix A is not invertible, calculation of all the adjoint entries may be unnecessary. One way to save work is to find $\det(A)$ by a cofactor expansion first—if the value is zero, then A is singular. If not, we have one row or column of $\text{Adj}(A)$ already calculated.

Cramer's Rule

*Cramer's rule** is a method for solving a square solution with a unique solution. Once popular, this method is seldom used today. We include this result for its historical significance and because of its presence in the mathematical literature.

THEOREM 3.5 (CRAMER'S RULE)

Let A be an $n \times n$ matrix and B be an $n \times 1$ matrix. For $j = 1, \ldots, n$, the matrix formed by replacing column j of A with B is denoted A_j. If $\det(A) \neq 0$, then the unique solution of $AX = B$ is given by

$$x_i = \frac{\det(A_i)}{\det(A)} \quad \text{for} \quad i = 1, \ldots, n.$$

* Gabriel Cramer (1704–1752) was a Swiss mathematician who is primarily remembered for this once popular rule for solving linear systems.

Proof Let j be a fixed integer between 1 and n, inclusive. If

$$\begin{array}{ccccc} a_{11}x_1 & + \cdots + & a_{1n}x_n & = & b_1 \\ \vdots & & \vdots & & \vdots \\ a_{n1}x_1 & + \cdots + & a_{nn}x_n & = & b_n, \end{array}$$

then for $i = 1, \ldots, n$, multiply the ith equation by C_{ij}:

$$\begin{array}{ccccccc} x_1 a_{11} C_{1j} & + \cdots + & x_j a_{1j} C_{1j} & + \cdots + & x_n a_{1n} C_{1j} & = & b_1 C_{1j} \\ \vdots & & \vdots & & \vdots & & \vdots \\ x_1 a_{n1} C_{nj} & + \cdots + & x_j a_{nj} C_{nj} & + \cdots + & x_n a_{nn} C_{nj} & = & b_n C_{nj}. \end{array}$$

If we add these equalities, regroup the lefthand side by columns, and factor the x_j's, the jth term on the left is

$$x_j(a_{1j}C_{1j} + \cdots + a_{nj}C_{nj}).$$

By Theorem 3.1, this term equals

$$x_j \cdot \det(A).$$

The remaining $n - 1$ terms have the form

$$x_k(a_{1k}C_{1j} + \cdots + a_{nk}C_{nj}) \qquad \text{with} \quad k \neq j.$$

By the lemma of this section, each of these terms equals zero. Hence the left side of the regrouped expression above reduces to

$$x_j \cdot \det(A).$$

On the right side,

$$b_1C_{1j} + \cdots + b_nC_{nj} = \det(A_j)$$

by definition of A_j and Theorem 3.1. Since j was chosen arbitrarily between 1 and n, the proof is complete. ☐

Example 6

Solve the linear system

$$\begin{aligned} 2x - 3y &= 1 \\ 3x + 5y &= 11 \end{aligned}$$

by Cramer's rule.

Solution

In the notation of Theorem 3.5,

$$A = \begin{vmatrix} 2 & -3 \\ 3 & 5 \end{vmatrix}, \quad A_1 = \begin{vmatrix} 1 & -3 \\ 11 & 5 \end{vmatrix}, \quad \text{and} \quad A_2 = \begin{vmatrix} 2 & 1 \\ 3 & 11 \end{vmatrix}.$$

Hence

$$\begin{aligned} x &= \det(A_1)/\det(A) = 38/19 = 2, \\ y &= \det(A_2)/\det(A) = 19/19 = 1. \quad \square \end{aligned}$$

Example 7

Solve the linear system

$$\begin{aligned} x - 2y - 2z &= -1 \\ 2x + y + 4z &= 2 \\ 3x - y + 4z &= -3 \end{aligned}$$

by Cramer's rule.

Solution

In the notation of Theorem 3.5,

$$A = \begin{bmatrix} 1 & -2 & -2 \\ 2 & 1 & 4 \\ 3 & -1 & 4 \end{bmatrix}, \quad A_1 = \begin{bmatrix} -1 & -2 & -2 \\ 2 & 1 & 4 \\ -3 & -1 & 4 \end{bmatrix},$$

$$A_2 = \begin{bmatrix} 1 & -1 & -2 \\ 2 & 2 & 4 \\ 3 & -3 & 4 \end{bmatrix}, \quad \text{and} \quad A_3 = \begin{bmatrix} 1 & -2 & -1 \\ 2 & 1 & 2 \\ 3 & -1 & -3 \end{bmatrix}.$$

Hence

$$x = 30/10 = 3, \qquad y = 40/10 = 4, \qquad \text{and} \qquad z = -20/10 = -2. \quad \square$$

REVIEW CHECKLIST

1. Characterize square linear systems with unique solutions in terms of the determinant of the coefficient matrix.
2. Characterize square homogeneous systems with nontrivial solutions in terms of the determinant of the coefficient matrix.
3. For a square matrix A, recognize the value of dot products formed with coefficients from one row [column] and cofactors from another as zero.
4. Characterize invertible matrices in terms of determinants.
5. Formulate a definition of the adjoint of a square matrix.
6. Find the inverse of an invertible square matrix by the method of adjoints.
7. Apply Cramer's rule in solving small square linear systems with unique solutions.

EXERCISES

In Exercises 1–9, use the method of adjoints to determine whether the given matrix is invertible and, if so, to find the inverse.

1. $\begin{bmatrix} 2 & -1 \\ 3 & 3 \end{bmatrix}$ **2.** $\begin{bmatrix} 1 & 4 \\ 2 & 5 \end{bmatrix}$ **3.** $\begin{bmatrix} -1 & 2 \\ 3 & -6 \end{bmatrix}$

4. $\begin{bmatrix} 1 & 2 & 0 \\ 1 & -1 & 2 \\ 3 & 1 & 1 \end{bmatrix}$ **5.** $\begin{bmatrix} 2 & 1 & -1 \\ 0 & 1 & 3 \\ 2 & 1 & 1 \end{bmatrix}$ **6.** $\begin{bmatrix} 1 & 2 & 1 \\ 1 & -1 & 2 \\ 3 & 1 & -2 \end{bmatrix}$

7. $\begin{bmatrix} 1 & -1 & 2 \\ 0 & 1 & 2 \\ 0 & 0 & 1 \end{bmatrix}$ **8.** $\begin{bmatrix} 3 & 1 & -1 \\ 2 & 4 & 1 \\ 7 & -5 & 5 \end{bmatrix}$ **9.** $\begin{bmatrix} 1 & 4 & 2 \\ -1 & 0 & 1 \\ 2 & 2 & 3 \end{bmatrix}$

In Exercises 10–17, solve the given system by Cramer's rule.

10. $\begin{aligned} x - 2y &= 2 \\ x - 4y &= -2 \end{aligned}$

11. $\begin{aligned} 2x + 5y &= 1 \\ 4x + 5y &= -5 \end{aligned}$

12. $\begin{aligned} 3x + 4y &= 5 \\ 2x - 3y &= 9 \end{aligned}$

13. $\begin{aligned} x + 2y &= 4 \\ 2x + y &= 3 \end{aligned}$

14. $\begin{aligned} 3x + y + z &= 3 \\ -2x + y + 2z &= 4 \\ x + y &= 3 \end{aligned}$

15. $\begin{aligned} 2x - 2y - z &= -1 \\ y + z &= 1 \\ -x + y + z &= -1 \end{aligned}$

16. $\begin{aligned} -2x + z &= 0 \\ x + 2y - 2z &= 4 \\ 3x + y - 4z &= 2 \end{aligned}$

17. $\begin{aligned} x - y + z &= 1 \\ 2x + y + z &= 2 \\ 3x + y + 2z &= 0 \end{aligned}$

Exercises 18–19 are True-False: answer T if the statement is true and F if it is false.

18. By Cramer's rule, if $\det(A) = 0$, then a nonhomogeneous linear system with coefficient matrix A has no solution.

19. By Cramer's rule, if A is a square matrix and the linear system $AX = 0$ has nontrivial solutions, then $\det(A) = 0$.

20. Let A be an $n \times n$ matrix, and let j and l be integers between 1 and n, inclusive. Show that if $j \neq l$, then

$$a_{1j}C_{1l} + \cdots + a_{nj}C_{nl} = 0.$$

21. Show that for any square matrix A, $\text{Adj}(A^t) = (\text{Adj}(A))^t$.

22. Show that $\det(\text{Adj}(A)) = (\det(A))^{n-1}$ for any $n \times n$ matrix with $n \geq 2$.

23. Let A be an $n \times n$ matrix. Show that if $\det(A) \neq 0$ or $n \geq 3$, then

$$\text{Adj}(\text{Adj}(A)) = (\det(A))^{n-2}A.$$

(o) 3.3 PROOF OF PROPERTIES OF DETERMINANTS

In Section 3.1, we defined the determinant function for a square matrix and, without proof, stated Properties D1–D9 and Theorem 3.1. These results have useful implications for linear systems, as we found in Section 3.2. In this section, we give proofs of Properties D1–D9 and Theorem 3.1.

We take a closer look at Definition 3.1. Expanding along the first row of an $n \times n$ matrix A yields

$$\det(A) = a_{11}M_{11} - a_{12}M_{12} + \cdots + (-1)^{1+n}a_{1n}M_{1n},$$

where M_{ij} denotes the minor corresponding to row i and column j. Each minor is an $(n-1) \times (n-1)$ determinant, which may be expanded along its first row if $n > 2$. Thus in two steps $\det(A)$ is expressed as a sum of $n(n-1)$ terms of the form

$$\pm a_{1j}a_{2k}d_{jk},$$

where $j \neq k$ and d_{jk} is the determinant of the $(n-2) \times (n-2)$ matrix left after rows 1 and 2 and columns j and k are deleted from A.

Example 1

Let

$$A = \begin{bmatrix} 4 & 8 & 3 & 5 \\ 2 & 6 & -1 & 7 \\ 0 & 9 & 1 & -2 \\ 1 & 3 & 0 & 4 \end{bmatrix}.$$

By Definition 3.1,

$$\det(A) = 4M_{11} - 8M_{12} + 3M_{13} - 5M_{14}.$$

Expanding M_{11} gives

$$4M_{11} = 4\begin{vmatrix} 6 & -1 & 7 \\ 9 & 1 & -2 \\ 3 & 0 & 4 \end{vmatrix} = 4\left(6\begin{vmatrix} 1 & -2 \\ 0 & 4 \end{vmatrix} - (-1)\begin{vmatrix} 9 & -2 \\ 3 & 4 \end{vmatrix} + 7\begin{vmatrix} 9 & 1 \\ 3 & 0 \end{vmatrix}\right).$$

The signed coefficients appearing in this expansion are

$$(4)(6), \qquad -(4)(-1), \qquad \text{and} \qquad (4)(7).$$

The remaining nine terms from this step may be found by expanding M_{12}, M_{13}, and M_{14}. The signed coefficients in these terms, which we note for comparison with Example 2, are

$$-(8)(2), \quad (8)(-1), \quad -(8)(7), \quad (3)(2), \quad -(3)(6), \quad (3)(7),$$

$$-(5)(2), \quad (5)(6), \quad \text{and} \quad -(5)(-1).$$

These are the coefficients of the determinants of the 2×2 matrices formed from the bottom two rows of A in the second step expansion of $\det(A)$. □

For a 3×3 matrix, only the first row expansion is needed since each term is the product of a signed coefficient by the determinant of a 2×2 matrix. For an $n \times n$ matrix with $n > 3$, we need only $n - 2$ expansion steps to reach the desired form, as illustrated in Example 1 with a 4×4 matrix.

PROPERTY D1

If an $n \times n$ matrix A has a row or a column of zeros, then $\det(A) = 0$.

Proof for a Row of Zeros Suppose each entry in row k of A equals zero. If $k \leq n - 2$, then each coefficient in the expansion of $\det(A)$ at the kth step has a zero factor. If row k is one of the bottom two rows of A, then one factor of each term in the final expansion is the determinant of a 2×2 matrix with a zero row and hence equals zero. In either case, $\det(A) = 0$. □

For a proof of Property D1 for columns, see Exercise 1. For a proof of Property D2, that the determinant of a triangular matrix equals the product of its main diagonal entries, see Exercise 2.

Property D3 is more difficult to prove than are Properties D1 and D2. An illustration will help prepare for the argument.

Example 2

Let B be the matrix obtained when rows 1 and 2 in the matrix A of Example 1 are interchanged:

$$B = \begin{bmatrix} 2 & 6 & -1 & 7 \\ 4 & 8 & 3 & 5 \\ 0 & 9 & 1 & -2 \\ 1 & 3 & 0 & 4 \end{bmatrix}.$$

The first term in the expansion of $\det(B)$ along row 1 is

$$4M_{11} = 2\begin{vmatrix} 8 & 3 & 5 \\ 9 & 1 & -2 \\ 3 & 0 & 4 \end{vmatrix} = 2\left(8\begin{vmatrix} 1 & -2 \\ 0 & 4 \end{vmatrix} - 3\begin{vmatrix} 9 & -2 \\ 3 & 4 \end{vmatrix} + 5\begin{vmatrix} 9 & 1 \\ 3 & 0 \end{vmatrix}\right).$$

The signed coefficients appearing in this expansion are

$$(2)(8), \qquad -(2)(3), \qquad \text{and} \qquad (2)(5).$$

These expressions appear among the coefficients in the second step expansion of $\det(A)$ in Example 1, each with the sign reversed. Expanding the determinants in the remaining terms in the expansion of $\det(B)$ along row 1 yields the remaining nine coefficients found in Example 1, each with the sign reversed. (See Exercise 3.) Hence for this A and B, $\det(B) = -\det(A)$. □

PROPERTY D3

If B is obtained from an $n \times n$ matrix A by the interchange of two rows or columns, then $\det(B) = -\det(A)$.

Proof for a Row Interchange Suppose the interchanged rows are rows 1 and 2. Then the expansions of $\det(A)$ and $\det(B)$ at step 2 are identical except for a sign reversal of each term. (The idea underlying this statement is illustrated in Examples 1 and 2 and a proof for the general case is assigned in Exercise 4.) Thus in this case, $\det(B) = -\det(A)$.

Next, suppose $i \geq 2$ and adjacent rows i and $i + 1$ are interchanged. Then the corresponding signed coefficients for $\det(A)$ and $\det(B)$ are identical through the $(i - 1)$st step. Each of these coefficients multiplies the determinant of an $(n - (i - 1)) \times (n - (i - 1))$ matrix, and the corresponding matrices for $\det(A)$ and $\det(B)$ agree except that their rows 1 and 2 are interchanged. By the proof of

Property D3 for rows 1 and 2 above, corresponding terms in the expansions of $\det(A)$ and $\det(B)$ at the $(i+1)$st step are equal in magnitude and opposite in sign. Thus in this case also, $\det(B) = -\det(A)$.

Finally, if $i \leq j-2$ and B is obtained from A by the interchange of rows i and j, we can construct A from B by

$$(j-i) + (j-i-1) = 2(j-i) - 1$$

interchanges of adjacent rows (Exercise 5). Since each of these interchanges reverses the sign of the determinant and the number of adjacent row interchanges is odd, the final sign is reversed. Thus in any case, $\det(B) = -\det(A)$. ☐

Note This proof of Property D3 is independent of Theorem 3.2 and consequent results in Section 3.2.

A proof of Property D3 for a column exchange is assigned in Exercise 6. For a proof of Property D4, which asserts that if we obtain B from A by multiplying a row or a column of A by a scalar c, then $\det(B) = c \cdot \det(A)$, see Exercise 7.

PROPERTY D5

If an $n \times n$ matrix A has two rows [columns] with equal or proportional entries, then $\det(A) = 0$.

Proof for Equal Rows Suppose the entries in row i of A equal the corresponding entries of row j, where $i \neq j$. If we form B from A by interchanging rows i and j, then $A = B$, so $\det(A) = \det(B)$. Since also $\det(A) = -\det(B)$ by Property D3, $\det(A) = 0$. ☐

Property D5 for a matrix with entries in one row proportional to those in another row follows from Property D4 and the result just established for equal rows. Details are assigned in Exercise 8, as is a proof of Property D5 for columns.

PROPERTY D6

Let A, B, and C be $n \times n$ matrices with corresponding entries equal except in row i. If

$$a_{ij} + b_{ij} = c_{ij}$$

for $j = 1, \ldots, n$, then $\det(A) + \det(B) = \det(C)$. The corresponding result holds for columns.

Proof for Rows Suppose A, B, and C satisfy the hypotheses of this property. If $i = 1$, then

$$\begin{aligned}\det(A) + \det(B) &= a_{11}C_{11} + \cdots + a_{1n}C_{1n} + b_{11}C_{11} + \cdots + b_{1n}C_{1n} \\ &= (a_{11} + b_{11})C_{11} + \cdots + (a_{1n} + b_{1n})C_{1n} \\ &= c_{11}C_{11} + \cdots + c_{1n}C_{1n}.\end{aligned}$$

If $i > 1$, we may obtain the desired conclusion by interchanging rows 1 and i, and applying Property D3 and the result just established for row 1. See Exercise 9 for details. □

A proof of Property D6 for columns is assigned in Exercise 10.

PROPERTY D7

If B is obtained by the addition of c times one row [column] of an $n \times n$ matrix A to another row [column], then $\det(B) = \det(A)$.

Proof for Rows Suppose $i \neq j$ and we add c times row i to row j of A to obtain B. If A_j is the matrix formed from A with c times row i replacing row j, the corresponding entries of A, A_j, and B agree except in row j, where

$$a_{jk} + ca_{ik} = b_{jk} \qquad \text{for } j = 1, \ldots, n.$$

By Property D6,

$$\det(A) + \det(A_j) = \det(B).$$

Since $\det(A_j) = 0$ by Property D5, the desired result follows. □

Note This proof of Property D7 is independent of Theorem 3.2 and consequent results in Section 3.2.

A proof of Property D7 for columns is assigned in Exercise 11. In preparation for proving Property D8 in general, we first establish a special case.

Lemma If E is an $n \times n$ elementary matrix and B any $n \times n$ matrix, then

$$\det(EB) = \det(E)\det(B).$$

Proof Consider the three types of elementary row operations. (1) If E interchanges two rows, we may obtain E from the identity I by a row exchange. Since

$\det(I) = 1$ by Property D2, $\det(E) = -1$ by Property D3. Since $\det(EB) = -\det(B)$, also by Property D3, we have $\det(EB) = \det(E)\det(B)$. (2) If E multiplies a row by c, the desired result follows from Properties D2 and D4 (Exercise 12). (3) If E adds a multiple of one row to another, the desired result follows from Properties D2, D6, and D7 (Exercise 13). □

PROPERTY D8

If A and B are $n \times n$ matrices, then

$$\det(AB) = \det(A)\det(B).$$

Proof If A is invertible, then by Theorem 2.5, E may be expressed as a product of elementary matrices:

$$A = E_1 \cdots E_p.$$

By the lemma above (applied $p - 1$ times),

$$\det(AB) = \det(E_1) \cdots \det(E_p)\det(B).$$

By the same lemma, the product of the first p factors on the right equals $\det(A)$. Hence in this case,

$$\det(AB) = \det(A)\det(B).$$

If A is not invertible, we may start with A and construct an upper triangular matrix U with at least one zero row by a sequence of elementary row operations. (Otherwise, we could carry out a Gauss-Jordan elimination and construct an inverse, as in Section 2.4.) Thus for some q,

$$E_1 \cdots E_q A = U.$$

By the previous lemma and Property D1,

$$\det(E_1) \cdots \det(E_q)\det(A) = \det(U) = 0.$$

Since each $\det(E_i) \neq 0$, we must have $\det(A) = 0$. Since

$$E_1 \cdots E_q AB = UB$$

and UB inherits at least one zero row from U, $\det(UB) = 0$, and the same argument shows $\det(AB) = 0$. Hence in either case, $\det(AB) = \det(A)\det(B)$. □

PROPERTY D9

If A is an $n \times n$ matrix, then $\det(A^t) = \det(A)$.

Proof Construct an upper triangular matrix U from A by a sequence of elementary row operations:

$$E_1 \cdots E_p A = U.$$

Since U is upper triangular, U^t is lower triangular. Since U and U^t have the same main diagonal entries, we have

$$\det(U) = \det(U^t)$$

by Property D2. Thus if

$$B = E_1 \cdots E_p,$$

then $BA = U$, $\det(B) \neq 0$, and

$$\det(B)\det(A) = \det(BA) = \det(U) = \det(U^t) = \det\big((BA)^t\big).$$

By Theorem 2.4,

$$(BA)^t = A^t B^t,$$

hence we may continue to obtain

$$(*) \qquad \det(B)\det(A) = \det(A^t B^t) = \det(A^t)\det(B^t).$$

Since $\det(E^t) = \det(E)$ for any elementary matrix E (Exercise 14), and since

$$B^t = E_p^t \cdots E_1^t$$

by Theorem 2.4, we have

$$\det(B^t) = \det(B).$$

Since $\det(B) \neq 0$, canceling this common value from the left and right sides of $(*)$ yields

$$\det(A^t) = \det(A). \quad \square$$

Theorem 3.1 follows from Definition 3.1 and Properties D3 and D9. Details are assigned in Exercise 15.

REVIEW CHECKLIST

1. Describe the recursive nature of Definition 3.1 as a sequence of expansion steps, identifying the number of terms generated at each step and their general form.
2. Outline a proof of Properties D1–D9 and Theorem 3.1 of Section 3.1.

EXERCISES

1. Show that if a square matrix A has a zero column, then $\det(A) = 0$.
2. Show that if a matrix A is upper triangular or lower triangular, then $\det(A)$ equals the product of the main diagonal entries.
3. Complete the expansion of the first row minors of the determinant in Example 2, and compare the twelve terms in the second step expansion in Example 2 with the twelve terms obtained in Example 1.
4. Prove that if a matrix B is obtained from a square matrix A by the interchange of rows 1 and 2, then the expansions of $\det(A)$ and $\det(B)$ at step 2 are identical except for a sign reversal in each term.
5. Let A be an $n \times n$ matrix, and let i and j be integers between 1 and n, inclusive. Show that if $i \leq j - 2$ and we obtain B from A by interchanging rows i and j, we can construct A from B by

$$(j - i) + (j - i - 1) = 2(j - i) - 1$$

interchanges of adjacent rows.
6. Prove that if A is a square matrix and B is obtained from A by an interchange of two columns, then $\det(B) = -\det(A)$. (If Property D9 is used, check carefully to ensure that the proof of Property D9 is independent of the results in Section 3.2.)
7. Let A be a square matrix, and c a real number. Show that if we obtain a matrix B by multiplying a row or a column of A by c, then $\det(B) = c \cdot \det(A)$. (Be careful that the proof is independent of the results in Section 3.2.)
8. Let A be a square matrix with corresponding entries in two rows proportional, i.e., if these two rows are regarded as vectors, then one is a scalar multiple of the other. Prove that $\det(A) = 0$, and prove the corresponding result for columns.
9. Complete the proof of Property D6 for rows.
10. Let A, B, and C be $n \times n$ matrices with corresponding entries equal except in column j. Prove that if

$$a_{ij} + b_{ij} = c_{ij}$$

for $i = 1, \ldots, n$, then $\det(A) + \det(B) = \det(C)$.
11. Show that if B is obtained by the addition of c times one column of an $n \times n$ matrix A to another column, then $\det(B) = \det(A)$.
12. Show that if B is an $n \times n$ matrix and E is an $n \times n$ elementary matrix that multiplies a row of B by a constant c, then $\det(EB) = \det(E)\det(B)$.
13. Show that if B is an $n \times n$ matrix and E is an $n \times n$ elementary matrix that adds c times one row of B to another, then $\det(EB) = \det(E)\det(B)$.
14. Show that $\det(E^t) = \det(E)$ for any elementary matrix E.
15. Prove Theorem 3.1 using Definition 3.1 and Properties D3 and D9.

(o) 3.4 APPLICATIONS

Determinants are used in several areas of mathematics and its application fields. The examples in this section are primarily from geometry and analysis.

Equations of Curves

In Section 1.5, we used linear algebra to find equations of curves and surfaces through given data points. We found the equation of an interpolating curve or surface by solving a linear system. Since the determinant provides a condition for the existence of solutions of linear systems, perhaps determinants can be used in finding equations of curves and surfaces.

Example 1

Find an equation in determinant form for the line through two distinct points (x_1, y_1) and (x_2, y_2) in the plane.

Solution

Any line in the plane is determined by an equation of the form

$$Ax + By + C = 0,$$

where A and B are not both zero. Thus for the distinct points (x_1, y_1) and (x_2, y_2), there exist constants A, B, and C, not all zero, such that for any point (x, y) on the line,

$$\begin{aligned} Ax + By + C &= 0 \\ Ax_1 + By_1 + C &= 0 \\ Ax_2 + By_2 + C &= 0. \end{aligned}$$

Hence if A, B, and C are regarded as unknowns, then these three relations constitute a homogeneous linear system in A, B, and C with a nontrivial solution. By the corollary to Theorem 3.2, the determinant of the coefficient matrix must equal zero, i.e.,

$$\begin{vmatrix} x & y & 1 \\ x_1 & y_1 & 1 \\ x_2 & y_2 & 1 \end{vmatrix} = 0.$$

This relation may be used as an equation of the interpolating line since the lefthand side is the sum of a linear form in x and y plus a constant and is zero when x_1 $[x_2]$ is substituted for x and y_1 $[y_2]$ is substituted for y. The linear form is explicit in the cofactor expansion along row 1:

$$\begin{vmatrix} y_1 & 1 \\ y_2 & 1 \end{vmatrix} x - \begin{vmatrix} x_1 & 1 \\ x_2 & 1 \end{vmatrix} y + \begin{vmatrix} x_1 & y_1 \\ x_2 & y_2 \end{vmatrix}. \quad \square$$

Example 2

Find an equation for the line through the points (2, 5) and (4, 1) in the plane.

Solution

The form in Example 1 yields

$$\begin{vmatrix} x & y & 1 \\ 2 & 5 & 1 \\ 4 & 1 & 1 \end{vmatrix} = 0.$$

Expanding by cofactors along row 1 gives

$$4x + 2y - 18 = 0. \quad \square$$

A similar form can be found for the equation of a plane through three noncollinear points in space. (See Exercises 13–17.)

Example 3

Find an equation in determinant form for the parabola through three noncollinear points (x_1, y_1), (x_2, y_2), and (x_3, y_3) in the plane if the axis of symmetry is vertical.

Solution

A parabola through three noncollinear points and having vertical axis of symmetry is determined by an equation of the form

$$Ax^2 + Bx + Cy + D = 0,$$

where $A \neq 0$ and $C \neq 0$. Thus for the given points there exist constants A, B, C, and D, not all zero, such that for any point (x, y) on the parabola,

$$\begin{aligned} Ax^2 + Bx + Cy + D &= 0 \\ Ax_1^2 + Bx_1 + Cy_1 + D &= 0 \\ Ax_2^2 + Bx_2 + Cy_2 + D &= 0 \\ Ax_3^2 + Bx_3 + Cy_3 + D &= 0. \end{aligned}$$

Hence these relations constitute a homogeneous linear system in A, B, C, and D with a nontrivial solution. By the corollary to Theorem 3.2, the determinant of the coefficient matrix must equal zero, i.e.,

$$\begin{vmatrix} x^2 & x & y & 1 \\ x_1^2 & x_1 & y_1 & 1 \\ x_2^2 & x_2 & y_2 & 1 \\ x_3^2 & x_3 & y_3 & 1 \end{vmatrix} = 0.$$

This relation serves as an equation of the interpolating parabola. The value of the determinant equals zero when the coordinates of any of the given points are substituted for x and y. The cofactor expansion of this determinant along row 1 shows that this equation has the desired form:

$$\begin{vmatrix} x_1 & y_1 & 1 \\ x_2 & y_2 & 1 \\ x_3 & y_3 & 1 \end{vmatrix} x^2 - \begin{vmatrix} x_1^2 & y_1 & 1 \\ x_2^2 & y_2 & 1 \\ x_3^2 & y_3 & 1 \end{vmatrix} x + \begin{vmatrix} x_1^2 & x_1 & 1 \\ x_2^2 & x_2 & 1 \\ x_3^2 & x_3 & 1 \end{vmatrix} y - \begin{vmatrix} x_1^2 & x_1 & y_1 \\ x_2^2 & x_2 & y_2 \\ x_3^2 & x_3 & y_3 \end{vmatrix} = 0.$$

The coefficient of x^2 is nonzero since the given points are noncollinear. (See Example 1.) The coefficient of y equals

$$-(x_1 - x_2)(x_2 - x_3)(x_3 - x_1)$$

(Exercise 42), which is nonzero because any two of the given points have the distinct x-coordinates. (See Exercise 43.) ☐

Example 4

Find the equation of the parabola through the points $(-1, 7)$, $(1, -1)$, and $(2, -2)$, if the axis of symmetry is vertical.

Solution

From Example 3, use the equation

$$\begin{vmatrix} x^2 & x & y & 1 \\ 1 & -1 & 7 & 1 \\ 1 & 1 & -1 & 1 \\ 4 & 2 & -2 & 1 \end{vmatrix} = 0.$$

Expand along row 1 to get

$$6x^2 - 24x - 6y + 12 = 0.$$

This equation simplifies to

$$y = x^2 - 4x + 2,$$

which was also obtained in Example 1 of Section 1.5. $\square$

Area and Volume by Determinants

We may use determinants to find the sine of the angle between two vectors in the plane or in space, and hence to find areas and volumes. For a proof of our first result, see Exercise 41.

Proposition 1 If the vectors $\mathbf{x} = (x_1, x_2, x_3)$ and $\mathbf{y} = (y_1, y_2, y_3)$ form an angle θ in space, then

$$\|\mathbf{x}\|^2\|\mathbf{y}\|^2\sin^2\theta = \begin{vmatrix} x_2 & x_3 \\ y_2 & y_3 \end{vmatrix}^2 + \begin{vmatrix} x_3 & x_1 \\ y_3 & y_1 \end{vmatrix}^2 + \begin{vmatrix} x_1 & x_2 \\ y_1 & y_2 \end{vmatrix}^2.$$

The area of the parallelogram formed by $\mathbf{x}$ and $\mathbf{y}$ is given by

$$A = \|\mathbf{x}\|\,\|\mathbf{y}\|\sin\theta,$$

if $0° < \theta < 180°$ (see Fig. 3.1). Thus we may use the formula in Proposition 1 to find this area. For a given $\mathbf{x}$ and $\mathbf{y}$, we evaluate the righthand side of this formula and take the square root. This produces the following result.

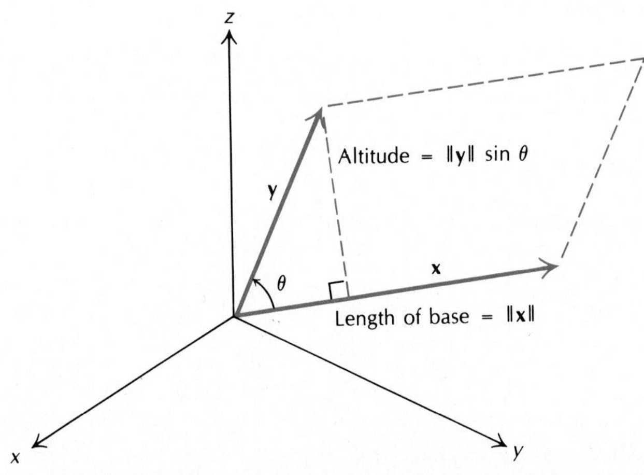

Figure 3.1 Area of the parallelogram formed by $\mathbf{x}$ and $\mathbf{y}$ given by $A = \|\mathbf{x}\|\,\|\mathbf{y}\|\sin\theta$.

Proposition 2 The area A of the parallelogram in space formed by the vectors $\mathbf{x} = (x_1, x_2, x_3)$ and $\mathbf{y} = (y_1, y_2, y_3)$ is given by

$$A = \sqrt{\begin{vmatrix} x_2 & x_3 \\ y_2 & y_3 \end{vmatrix}^2 + \begin{vmatrix} x_3 & x_1 \\ y_3 & y_1 \end{vmatrix}^2 + \begin{vmatrix} x_1 & x_2 \\ y_1 & y_2 \end{vmatrix}^2}.$$

Example 5

Find the area of the (plane) parallelogram formed by the vectors $\mathbf{x} = (1, -2)$ and $\mathbf{y} = (2, -1)$ as illustrated in Fig. 3.2.

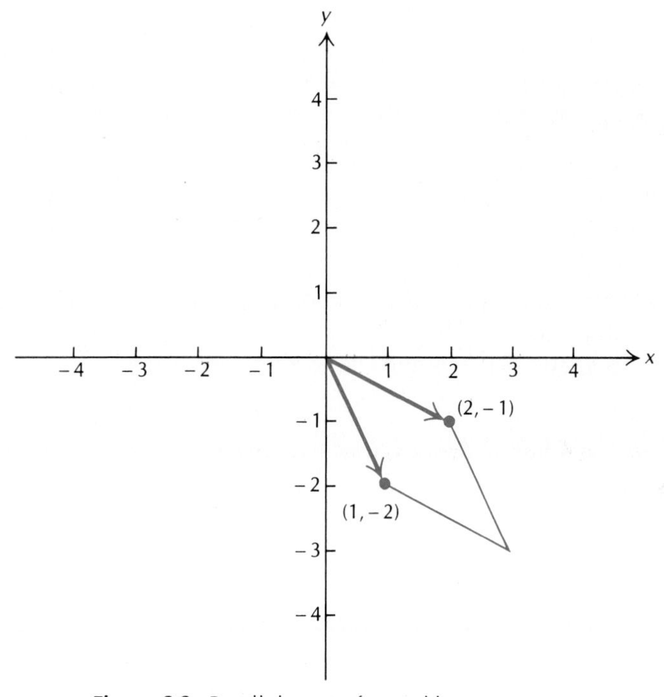

Figure 3.2 Parallelogram formed by vectors.

Solution

In xyz-space, the given vectors are represented as the ordered triples $(1, -2, 0)$ and $(2, -1, 0)$. Hence,

$$A^2 = \begin{vmatrix} -2 & 0 \\ -1 & 0 \end{vmatrix}^2 + \begin{vmatrix} 0 & 1 \\ 0 & 2 \end{vmatrix}^2 + \begin{vmatrix} 1 & -2 \\ 2 & -1 \end{vmatrix}^2 = 9,$$

from which $A = 3$. □

As Example 5 illustrates, the area of the parallelogram formed by two vectors in the plane can be obtained as a special case of the formula in Proposition 2.

Proposition 3 The area A of the parallelogram formed by the vectors $\mathbf{x} = (x_1, x_2)$ and $\mathbf{y} = (y_1, y_2)$ may be obtained from the relation

$$A^2 = \begin{vmatrix} x_1 & x_2 \\ y_1 & y_2 \end{vmatrix}^2 .$$

The next result gives the volume of a parallelepiped formed by three vectors in space.

Proposition 4 The volume V of the parallelepiped in space determined by the vectors $\mathbf{x} = (x_1, x_2, x_3)$, $\mathbf{y} = (y_1, y_2, y_3)$, and $\mathbf{z} = (z_1, z_2, z_3)$ (Fig. 3.3) can be obtained from the relation

$$V^2 = \begin{vmatrix} x_1 & x_2 & x_3 \\ y_1 & y_2 & y_3 \\ z_1 & z_2 & z_3 \end{vmatrix}^2 .$$

Proof The volume of a parallelepiped is given by the area of any base times the altitude to that base. Thus if A is the area of the base parallelogram formed by $\mathbf{x}$ and $\mathbf{y}$, and h is the altitude to this base, then

$$V = Ah.$$

The formula in Proposition 2 yields A, and the problem is to find h.

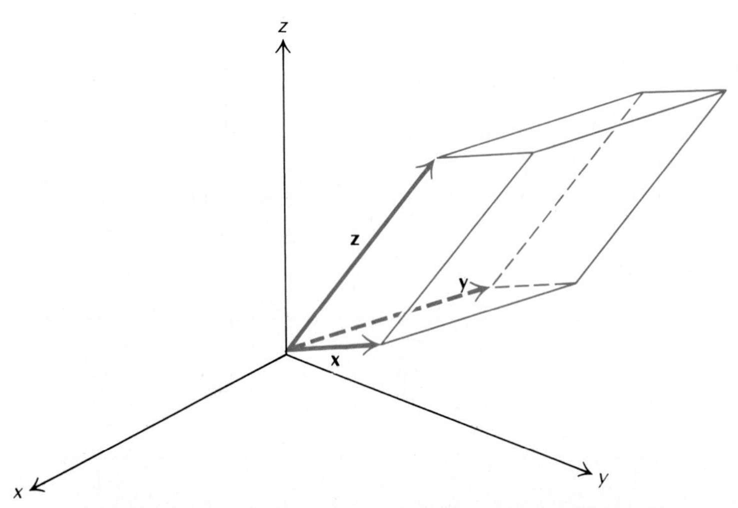

Figure 3.3 Parallelepiped formed by three vectors in space.

The altitude h is the length of the orthogonal projection (Definition 2.5) of **z** onto any vector **v** that is perpendicular to the plane of **x** and **y**. We may find such a vector **v** using properties of determinants. By Property D1,

$$\begin{vmatrix} x_1 & x_2 & x_3 \\ x_1 & x_2 & x_3 \\ y_1 & y_2 & y_3 \end{vmatrix} = 0.$$

Expanding this determinant along row 1 yields

$$x_1 = \begin{vmatrix} x_2 & x_3 \\ y_2 & y_3 \end{vmatrix} - x_2 \begin{vmatrix} x_1 & x_3 \\ y_1 & y_3 \end{vmatrix} + x_3 \begin{vmatrix} x_1 & x_2 \\ y_1 & y_2 \end{vmatrix} = 0.$$

Thus if

$$\mathbf{v} = \left(\begin{vmatrix} x_2 & x_3 \\ y_2 & y_3 \end{vmatrix}, - \begin{vmatrix} x_1 & x_3 \\ y_1 & y_3 \end{vmatrix}, \begin{vmatrix} x_1 & x_2 \\ y_1 & y_2 \end{vmatrix} \right),$$

then **v** is orthogonal to **x** (Definition 2.4) since $\langle \mathbf{x}, \mathbf{v} \rangle = 0$. By a similar argument starting from the relation

$$\begin{vmatrix} y_1 & y_2 & y_3 \\ x_1 & x_2 & x_3 \\ y_1 & y_2 & y_3 \end{vmatrix} = 0,$$

we have also $\langle \mathbf{y}, \mathbf{v} \rangle = 0$. By Definition 2.5, the orthogonal projection of **z** onto **v** is

$$\langle \mathbf{z}, \mathbf{v}' \rangle \mathbf{v}',$$

where $\mathbf{v}'$ is a unit vector in the direction of **v**. Since $\|\mathbf{v}\| = A$ by Proposition 1,

$$\mathbf{v}' = \mathbf{v}/\|\mathbf{v}\| = \mathbf{v}/A.$$

From the discussion immediately preceding Definition 2.5, we have

$$\langle \mathbf{z}, \mathbf{v}' \rangle \mathbf{v}' = |\langle \mathbf{z}, \mathbf{v}' \rangle| \|\mathbf{v}'\| = |\langle \mathbf{z}, \mathbf{v}' \rangle|,$$

since $\mathbf{v}'$ is a unit vector. Thus the length of the orthogonal projection of **z** onto **v** is given by

$$h = |\langle \mathbf{z}, \mathbf{v}' \rangle| = |\langle \mathbf{z}, \mathbf{v}/A \rangle| = |\langle \mathbf{z}, \mathbf{v} \rangle|/A.$$

Hence for the volume V, we have

$$V = Ah = |\langle \mathbf{z}, \mathbf{v} \rangle|,$$

from which

$$V^2 = \langle \mathbf{z}, \mathbf{v} \rangle^2.$$

Expanding the determinant in the asserted formula for V^2 along the third row gives

$$\begin{vmatrix} x_1 & x_2 & x_3 \\ y_1 & y_2 & y_3 \\ z_1 & z_2 & z_3 \end{vmatrix} = \langle \mathbf{z}, \mathbf{v} \rangle.$$

Squaring both sides of this equation produces the desired formula, since $V^2 = \langle \mathbf{z}, \mathbf{v} \rangle^2$. □

Example 6

Find the volume V of the parallelepiped formed by the vectors $\mathbf{x} = (1, 4, 2)$, $\mathbf{y} = (3, 1, -1)$, and $\mathbf{z} = (2, 1, 3)$.

Solution

Use the formula in Proposition 4. Since

$$\begin{vmatrix} 1 & 3 & 2 \\ 3 & 1 & -1 \\ 2 & 1 & 3 \end{vmatrix} = -38,$$

the required volume is given by $V = |-38| = 38$. □

Other Applications (Calculus Based)

The vector $\mathbf{v}$ defined in proof of Proposition 4 is called the **cross product of x and y**. The cross product is used heavily in multivariable calculus, vector analysis, mechanics, electrostatic fields, and fluid dynamics. An excellent discussion of the cross product can be found in *Calculus and Analytic Geometry*, 7th ed. by G. B. Thomas and R. L. Finney.

Determinants are used in solving ordinary differential equations. For second and higher order linear equations, we use a special determinant, called the *Wronskian*, to determine whether the general solution of an equation has been found and in a method of finding solutions known as *variation of parameters*. For information, see *Elementary Differential Equations and Boundary Value Problems*, 4th ed., by W. E. Boyce and R. C. DiPrima.

In multivariable calculus, we use a special determinant, the *Jacobian*, to calculate derivatives and differentials with respect to the new variables when a change of variables is made. The Jacobian is used in physical applications of calculus to simplify a mathematical model and/or to facilitate calculations based on a model. A definition of the Jacobian can be found in the Thomas/Finney calculus book.

REVIEW CHECKLIST

1. Use determinants to interpolate data with curves and surfaces.
2. Use determinants to find the area of the parallelogram formed by two noncollinear vectors.
3. Use determinants to find the volume of the parallelepiped formed by three vectors.

EXERCISES

In Exercises 1–3, find a determinant equation for the line through the given points.

1. $(4,1), (3,2)$ **2.** $(5,0), (1,4)$ **3.** $(1,-2), (2,1)$

In Exercises 4–7, find a determinant equation for the parabola with a vertical axis of symmetry through the following points.

4. $(1,-2), (2,-1), (-1,2)$ **5.** $(1,-2), (2,2), (-2,-2)$
6. $(1,-2), (2,1), (-2,1)$ **7.** $(-1,2), (2,4), (3,4)$
8. Find a determinant equation for the parabola with horizontal axis of symmetry through three noncollinear points (x_1, y_1), (x_2, z_2), and (x_3, y_3).

In Exercises 9–12, use the form found in Exercise 8 to find the equation of the parabola with a horizontal axis of symmetry through the given points.

9. $(-1,1), (4,2), (7,-1)$ **10.** $(-1,1), (4,2), (-4,-2)$
11. $(1,1), (0,2), (1,3)$ **12.** $(-2,2), (-2,3), (4,5)$
13. Find a determinant equation for the plane through three noncollinear points (x_1, y_1, z_1), (x_2, y_2, z_2), and (x_3, y_3, z_3).

In Exercises 14–17, use the form found in Exercise 13 to find the equation of the plane through the given points.

14. $(1,1,1), (1,1,0), (1,0,0)$ **15.** $(1,1,0), (1,0,1), (0,1,1)$
16. $(1,1,1), (2,1,0), (-1,0,2)$ **17.** $(1,1,1), (4,1,1), (3,1,0)$
18. Find a determinant equation for the cubic

$$Ax^3 + Bx^2 + Cx + Dy + E = 0, \qquad A \neq 0, \qquad D \neq 0,$$

interpolating the points (x_1, y_1), (x_2, y_2), (x_3, y_3), (x_4, y_4) in the plane.

In Exercises 19–20, use the form found in Exercise 18 to find the equation of the cubic through the given points.

19. $(1,0), (-1,-4), (2,8), (-2,-8)$

20. $(-1,1), (1,1), (2,-5), (3,-3)$

21. Find the determinant equation for the circle

$$A(x^2+y^2)+Bx+Cy+D=0, \qquad A \neq 0,$$

interpolating noncollinear points $(x_1,y_1), (x_2,y_2), (x_3,y_3)$ in the plane.

In Exercises 22–23, use the form found in Exercise 21 to find the equation of the circle through the given points.

22. $(-3,2), (-2,3), (1,4)$

23. $(1,1), (-2,2), (-2,-2)$

24. Find an equation of the form

$$Ax^2+Bxy+Cy^2+D=0,$$

through the points $(1,4), (2,2), (4,1)$ in the plane.

In Exercises 25–28, use determinants to find the area of the parallelogram formed by the given pair of vectors.

25. $(2,1,0), (-1,1,0)$

26. $(2,0,4), (2,0,3)$

27. $(1,1,1), (1,1,0)$

28. $(2,1,3), (3,1,2)$

In Exercises 29–32, find the volume of the parallelepiped formed by the given vectors.

29. $(1,1,0), (1,0,1), (0,1,1)$

30. $(1,1,1), (1,1,0), (1,0,0)$

31. $(2,1,0), (-1,3,1), (4,1,2)$

32. $(3,1,1), (1,-2,3), (2,2,3)$

33. Use the formula for the area of a parallelogram in the plane to express the area of the triangle with vertices at three noncollinear points $(x_1,x_2), (y_1,y_2), (z_1,z_2)$ as a determinant.

In Exercises 34–37, use the formula found in Exercise 33 to find the area of the triangle formed by the given points.

34. $(0,0), (1,0), (1,1)$

35. $(1,1), (2,0), (3,3)$

36. $(3,2), (-2,1), (-1,-1)$

37. $(4,1), (4,3), (2,1)$

38. Express the volume of the tetrahedron with vertices at four noncollinear points $(x_1,x_2,x_3), (y_1,y_2,y_3), (z_1,z_2,z_3)$, and (w_1,w_2,w_3) as a determinant. *Hint.* The volume of a tetrahedron is one-sixth of the volume of the parallelepiped formed by the vectors along the three edges from any vertex.

In Exercises 39–40, use the formula found in Exercise 38 to find the volume of the tetrahedron formed by the given points.

39. (1,0,0), (0,1,0), (0,0,1), (0,0,0)

40. (2,1,1), (1,0,1), (0,1,1), (1,0,0)

41. Using Definitions 2.1 and 2.3, and Theorem 2.1, establish Proposition 1.

42. Show that for any numbers x_1, x_2, and x_3,

$$\begin{vmatrix} x_1^2 & x_1 & 1 \\ x_2^2 & x_2 & 1 \\ x_3^2 & x_3 & 1 \end{vmatrix} = -(x_1 - x_2)(x_2 - x_3)(x_3 - x_1).$$

43. Show that if (x_1, y_1) and (x_2, y_2) are distinct points and both satisfy the same equation

$$Ax^2 + Bx + Cy + D = 0, \qquad \text{with } C \neq 0$$

then $x_1 \neq x_2$.

REVIEW EXERCISES

In Exercises 1–8, find the determinant of the given matrix by applying elementary row operations to reach an upper triangular form.

1. $\begin{bmatrix} 1 & 2 & 4 \\ 3 & 1 & -1 \\ 2 & 4 & 8 \end{bmatrix}$ **2.** $\begin{bmatrix} 2 & 5 & 3 \\ 2 & 1 & -1 \\ -2 & 1 & 3 \end{bmatrix}$ **3.** $\begin{bmatrix} 1 & 1 & 1 \\ 1 & 2 & 3 \\ 1 & 4 & 9 \end{bmatrix}$

4. $\begin{bmatrix} -1 & 1 & 3 \\ 4 & 1 & 0 \\ 1 & 0 & 1 \end{bmatrix}$ **5.** $\begin{bmatrix} 1 & 3 & 3 \\ 1 & 1 & 1 \\ 1 & 2 & 2 \end{bmatrix}$ **6.** $\begin{bmatrix} 1 & 2 & 0 \\ 3 & 1 & -4 \\ 2 & 1 & -3 \end{bmatrix}$

7. $\begin{bmatrix} 1 & 1 & 0 & 2 \\ 1 & 2 & -1 & 0 \\ 2 & 1 & 3 & 7 \\ 3 & 5 & 0 & -1 \end{bmatrix}$ **8.** $\begin{bmatrix} 1 & 2 & 0 & 2 \\ 3 & 7 & 1 & 4 \\ 2 & 5 & 0 & 4 \\ -1 & 0 & 3 & -5 \end{bmatrix}$

In Exercises 9–14, use the adjoint method to find the inverse of the given matrix when it exists.

9. $\begin{bmatrix} 1 & 2 \\ 5 & 3 \end{bmatrix}$ **10.** $\begin{bmatrix} 2 & 1 \\ 3 & 2 \end{bmatrix}$ **11.** $\begin{bmatrix} 1 & -4 \\ -2 & 8 \end{bmatrix}$

12. $\begin{bmatrix} 1 & 2 & -1 \\ 1 & -1 & 1 \\ 2 & -5 & -3 \end{bmatrix}$ **13.** $\begin{bmatrix} 1 & 3 & 1 \\ 2 & 1 & 3 \\ 1 & -7 & 3 \end{bmatrix}$ **14.** $\begin{bmatrix} 1 & 1 & 1 \\ 1 & 0 & 2 \\ 2 & 1 & 5 \end{bmatrix}$

In Exercises 15–16, evaluate the given determinant using Properties D1–D9 from Section 3.1.

15. $\begin{vmatrix} 3 & -5 & 2 \\ 0 & 0 & 0 \\ 4 & -3 & -2 \end{vmatrix}$

16. $\begin{vmatrix} 3 & 5 & 3 \\ 1 & 0 & 1 \\ 2 & 4 & 2 \end{vmatrix}$

In Exercises 17–18, solve the given linear system by Cramer's rule.

17. $\begin{aligned} 2x - 5y &= 3 \\ 3x - 4y &= 4 \end{aligned}$

18. $\begin{aligned} 3x + 2y &= 5 \\ 2x - 3y &= 1 \end{aligned}$

Vector Spaces

Vector spaces provide a mathematical structure for linear systems and other concepts used widely in theory and practice. Although the study of vector spaces may at first seem abstract, this work opens the door to more advanced studies in mathematics and its applications, including modern algebra, functional analysis, calculus of variations, mathematical physics, environmental sciences, engineering, and operations research. Vector space concepts constitute the core of the body of knowledge known as "linear algebra."

4.1 VECTOR SPACES

Vectors, vector addition, and multiplication by a scalar (Section 2.1) have many special properties, some of which we saw in Chapter 2 on matrix algebra. In this section, we identify ten basic vector properties from which a large and powerful body of knowledge may be derived. Since other important and useful mathematical systems have the same basic properties, the knowledge derived from these properties applies to all such systems as well.

Vector Space Concept

According to Definition 2.1 (page 63), a vector is an ordered set of real numbers called components:

$$\mathbf{x} = (x_1, \ldots, x_n).$$

By Definition 2.2 (page 65), if vectors $\mathbf{x}$ and $\mathbf{y}$ have the same number of components, we may add them to obtain

$$\mathbf{x} + \mathbf{y} = (x_1, \ldots, x_n) + (y_1, \ldots, y_n) = (x_1 + y_1, \ldots, x_n + y_n).$$

Any vector $\mathbf{x}$ may also be multiplied by a scalar c:

$$c\mathbf{x} = c(x_1, \ldots, x_n) = (cx_1, \ldots, cx_n).$$

Example 1

If $\mathbf{x} = (1, -5, 3, 4)$, $\mathbf{y} = (2, 0, -3, 2)$, and $c = 2$, then

$$\mathbf{x} + \mathbf{y} = (3, -5, 0, 6) \quad \text{and} \quad c\mathbf{x} = (2, -10, 6, 8). \quad \square$$

For any positive integer n, let $\mathbb{R}^n$ denote the set of all vectors with n components. Geometrically, $\mathbb{R}^1$ may be represented as the real line, $\mathbb{R}^2$ as the xy-plane, and $\mathbb{R}^3$ as xyz-space. Although we have not formally identified the basic algebraic properties of $\mathbb{R}^n$, many of them appeared in Sections 2.2 and 2.3 on matrix algebra, since a vector in $\mathbb{R}^n$ may be regarded as a special matrix. Example 2 contains a statement of the basic properties of $\mathbb{R}^n$.

Since one goal is to derive information from properties of vectors, we shift our point of view and regard the properties as *axioms*. A *vector space* is defined in terms of these axioms, and then a *vector* is any element in a vector space. This gives new meaning to the term "vector."

DEFINITION 4.1

A **real vector space** consists of:

1. an underlying set V whose elements are called **vectors**;
2. a pair of operations, denoted by $\mathbf{x} + \mathbf{y}$ and $c\mathbf{x}$ for vectors $\mathbf{x}$ and $\mathbf{y}$ and a real scalar c, satisfying the following ten axioms.

The axioms for a vector space are organized into three groups, the first of which pertains to addition only.

AXIOMS OF VECTOR ADDITION

A1. Closure. If $\mathbf{x}$ and $\mathbf{y}$ are in V, then $\mathbf{x} + \mathbf{y}$ is also in V.

A2. Associativity. If $\mathbf{x}$, $\mathbf{y}$, and $\mathbf{z}$ are in V, then

$$\mathbf{x} + (\mathbf{y} + \mathbf{z}) = (\mathbf{x} + \mathbf{y}) + \mathbf{z}.$$

A3. Identity. There is an element $\mathbf{0}$ in V satisfying $\mathbf{x} + \mathbf{0} = \mathbf{0} + \mathbf{x} = \mathbf{x}$ for every $\mathbf{x}$ in V.

A4. Inverse. If $\mathbf{x}$ is in V, then there is an element $-\mathbf{x}$ in V satisfying

$$-\mathbf{x} + \mathbf{x} = \mathbf{x} + -\mathbf{x} = \mathbf{0}.$$

A5. Commutativity. If $\mathbf{x}$ and $\mathbf{y}$ are in V, then

$$\mathbf{x} + \mathbf{y} = \mathbf{y} + \mathbf{x}.$$

The next three vector space axioms pertain to multiplication by a scalar only.

AXIOMS FOR MULTIPLICATION BY A SCALAR

A6. Closure. If $\mathbf{x}$ is in V and c is a scalar, then $c\mathbf{x}$ is also in V.

A7. Associativity. If $\mathbf{x}$ is in V and b and c are scalars, then

$$(bc)\mathbf{x} = b(c\mathbf{x}).$$

A8. Identity. If $\mathbf{x}$ is in V, then

$$1\mathbf{x} = \mathbf{x}.$$

The final two vector space axioms relate the vector operations.

AXIOMS CONNECTING THE VECTOR OPERATIONS

A9. Left distributivity. If $\mathbf{x}$ is in V and b and c are scalars, then

$$(b + c)\mathbf{x} = b\mathbf{x} + c\mathbf{x}.$$

A10. Right distributivity. If $\mathbf{x}$ and $\mathbf{y}$ are in V and c is a scalar, then

$$c(\mathbf{x} + \mathbf{y}) = c\mathbf{x} + c\mathbf{y}.$$

You may recognize the forms and the names of most of these axioms from matrix algebra. In the remainder of this section, we illustrate these axioms and derive some immediate consequences.

Examples of Vector Spaces

In the examples below, V and W denote sets in which vector operations are defined. The phrase, "V is a vector space," means that V is a set and the vector operations in V satisfy axioms A1–A10. The first example includes a whole collection of vector spaces.

Example 2

For every positive integer n, $\mathbb{R}^n$ is a vector space. We show that any $\mathbb{R}^n$ satisfies axioms A1 and A5, leaving proofs for the remaining axioms as exercises (see Exercise 1 for axioms A2–A4, Exercise 2 for A6–A8, and Exercise 3 for A9–A10). For axiom A1:

$$\mathbf{x} + \mathbf{y} = (x_1 + y_1, \ldots, x_n + y_n) \qquad \text{(Definition 2.2)}$$

$$x_i + y_i \text{ is in } \mathbb{R} \text{ for } i = 1, \ldots, n \qquad (\mathbb{R} \text{ is closed under addition})$$

$$\mathbf{x} + \mathbf{y} \text{ is in } \mathbb{R}^n \qquad \text{(Definition 2.1 and definition of } \mathbb{R}^n\text{)}$$

For axiom A5:

$$\mathbf{x} + \mathbf{y} = (x_1 + y_1, \ldots, x_n + y_n) \qquad \text{(Definition 2.2)}$$

$$\mathbf{y} + \mathbf{x} = (y_1 + x_1, \ldots, y_n + x_n) \qquad \text{(Definition 2.2)}$$

$$x_i + y_i = y_i + x_i \quad \text{for } i = 1, \ldots, n \qquad \text{(addition in } \mathbb{R} \text{ is commutative)}$$

$$\mathbf{x} + \mathbf{y} = \mathbf{y} + \mathbf{x} \qquad \text{(Definition 2.2, vector equality)} \quad \square$$

Example 3

Let V consist of all vectors $\mathbf{x} = (x_1, x_2)$ in $\mathbb{R}^2$ satisfying

$$x_2 = x_1/2$$

with the vector operations of Definition 2.2. For axiom A1, let $\mathbf{x}$ and $\mathbf{y}$ be in V. By definition of V,

$$x_2 = x_1/2 \qquad \text{and} \qquad y_2 = y_1/2.$$

Adding these equations gives

$$x_2 + y_2 = (x_1 + y_1)/2,$$

so $\mathbf{x} + \mathbf{y}$ is also in V. For axiom A6, let $\mathbf{x}$ be in V and let c be a scalar. Since $\mathbf{x}$ is in V,

$$x_2 = x_1/2.$$

Multiplying both sides of this equation by c yields

$$cx_2 = (cx_1)/2.$$

Hence $c\mathbf{x}$ is also in V.

Axiom A3 is satisfied since $0 = 0/2$. If $\mathbf{x}$ is in V, then by axiom A6, $(-1)\mathbf{x}$ is also in V. Hence axiom A4 is satisfied. (See Exercise 1.) The remaining six axioms for a vector space are satisfied since each vector in V is also in $\mathbb{R}^2$. (See Example 2 and Exercises 1–3.) Thus V satisfies axioms A1–A10 and hence is a vector space. □

When does a set of vectors *not* form a vector space?

Example 4

Let V consist of all vectors $\mathbf{x} = (x_1, x_2)$ in $\mathbb{R}^2$ satisfying

$$x_2 = x_1/2 + 1$$

with the vector operations of Definition 2.2. Then V is *not* a vector space. For instance, let $\mathbf{x} = (2, 2)$ and $\mathbf{y} = (4, 3)$. Then $\mathbf{x}$ and $\mathbf{y}$ are in V. However,

$$\mathbf{x} + \mathbf{y} = (6, 5),$$

which is *not* in V because

$$5 \neq 6/2 + 1.$$

Since V is not closed under vector addition, it is not a vector space. (See also Exercise 6.) □

Example 5

The set V of all (real) polynomials of degree 2 or less is a vector space under the following vector operations. For

$$p = a_0 + a_1x + a_2x^2 \quad \text{and} \quad q = b_0 + b_1x + b_2x^2$$

in V and any scalar c, we let

$$p + q = (a_0 + b_0) + (a_1 + b_1)x + (a_2 + b_2)x^2$$

and

$$cp = ca_0 + (ca_1)x + (ca_2)x^2.$$

From these definitions, it follows immediately that if p and q are in V and c is a scalar, then $p + q$ and cp are also in V, since each is a polynomial of degree 2 or less. Thus V satisfies axioms A1 and A6. For axiom A10, we represent p and q in V as above. Then for any scalar c, we have

$$\begin{aligned} c(p+q) &= c\left[(a_0 + b_0) + (a_1 + b_1)x + (a_2 + b_2)x^2\right] \\ &= c(a_0 + b_0) + c(a_1 + b_1)x + c(a_2 + b_2)x^2 \\ &= ca_0 + cb_0 + ca_1x + ca_1x + ca_2x^2 + cb_2x^2 \\ &= c(a_0 + a_1x + a_2)x^2 + c(b_0 + b_1x + b_2x^2) \\ &= cp + cq. \end{aligned}$$

The first step is justified by the definition of addition in V and the remaining steps by properties of addition and multiplication in $\mathbb{R}$. For the remaining axioms, see Exercise 15. $\square$

Example 6

Let W consist of all polynomials of degree 1 or less, with the vector operations of Example 5. For axioms A1 and A6, let

$$p = a_0 + a_1x \qquad \text{and} \qquad q = b_0 + b_1x$$

be in W, and let c be a scalar. Then

$$p + q = (a_0 + a_1x) + (b_0 + b_1x) = (a_0 + b_0) + (a_1 + b_1)x$$

and

$$cp = c(a_0 + a_1x) = (ca_0) + (ca_1)x.$$

Thus $p + q$ and cp are also in W. Since 0 is a polynomial of degree 1 or less, axiom A3 is satisfied. For A4, if $a_0 + a_1x$ is in W, then

$$-a_0 - a_1x$$

is also in W. Since each element in W is in the vector space V of Example 5, and since the vector operations in W coincide with those in V, W also satisfies the remaining axioms for a vector space. (See Exercise 15.) $\square$

Example 7

The set W of all polynomials of degree 2 is *not* a vector space since it is not closed under addition. For instance,

$$(x^2 + x) + (-x^2 + x) = 2x,$$

and $2x$ is not a polynomial of degree 2. (See also Exercise 7.) $\square$

The next three examples illustrate *sets of functions* satisfying axioms A1–A10. The ability to regard a function as an element of a set is one key to the application of linear algebra in important theoretical and applied studies, including functional analysis, numerical analysis, and mathematical physics.

The functions in these examples have domain and codomain both equal to $\mathbb{R}$. For instance, the functions f and g defined by

$$f(x) = x^3 \qquad \text{and} \qquad g(x) = \cos x \qquad \text{for all real } x$$

are included, but the function h defined by

$$h(x) = \sqrt{x} \qquad \text{for all } x \geq 0$$

is not, since the domain does not equal $\mathbb{R}$. We also use the familiar definition of equality. Two functions (with the same codomain) are **equal** if their domains coincide and their values at every domain point agree.

Example 8

Let V be the set of all functions f with domain and codomain both equal to $\mathbb{R}$. For vector addition, if f and g are in V, then $f + g$ is defined for any real x by

$$(f + g)(x) = f(x) + g(x).$$

For multiplication by a scalar c, cf is the function defined for any real x by

$$(cf)(x) = cf(x).$$

Closure of V under both operations follows immediately from these definitions. For associativity of addition (axiom A2), we prove that

$$(f + g) + h = f + (g + h)$$

using the definition of equality noted above. The domain of each function in V is $\mathbb{R}$, and for any x in the domain, we have

$$\begin{aligned}
[(f + g) + h](x) &= (f + g)(x) + h(x) && \text{(Definition of addition in } V\text{)} \\
&= [f(x) + g(x)] + h(x) && \text{(Definition of addition in } V\text{)} \\
&= f(x) + [g(x) + h(x)] && \text{(Associative law in } \mathbb{R}\text{)} \\
&= f(x) + (g + h)(x) && \text{(Definition of addition in } V\text{)} \\
&= [f + (g + h)](x) && \text{(Definition of addition in } V\text{)}
\end{aligned}$$

The zero function defined by

$$f_0(x) = 0 \qquad \text{for all } x,$$

serves as an additive identity (the 0 in axiom A3). If f is any element of V, then for any real x,

$$\begin{aligned}
(f + f_0)(x) &= f(x) + f_0(x) && \text{(Definition of addition in } V\text{)} \\
&= f(x) + 0 && \text{(Definition of } f_0\text{)} \\
&= f(x) && \text{(Addition in } \mathbb{R}\text{)} \\
f + f_0 &= f && \text{(Definition of equality of functions)}
\end{aligned}$$

For the same reasons,

$$(f_0 + f)(x) = f_0(x) + f(x) = 0 + f(x) = f(x),$$

hence $f_0 + f = f$. Thus f_0 satisfies the conditions for an additive identity (axiom A3).

For an additive inverse of a vector (axiom A4), if f is in V, then $(-1)f$ serves as an additive inverse of f. For any real x,

$$\begin{aligned}
[f + (-1)f](x) &= f(x) + [(-1)f](x) && \text{(Definition of addition in } V\text{)} \\
&= f(x) + (-1)f(x) && \text{(Definition of multiplication by a scalar in } V\text{)} \\
&= 0 && \text{(Addition in } \mathbb{R}\text{)}
\end{aligned}$$

Similarly,

$$[(-1)f + f](x) = [(-1)f](x) + f(x) = (-1)f(x) + f(x) = 0.$$

Thus $f + (-1)f$ and $(-1)f + f$ are both identically zero on $\mathbb{R}$. By the definitions of f_0 and equality of functions,

$$f + (-1)f = (-1)f + f = f_0.$$

Since f_0 is an additive identity, $(-1)f$ satisfies the conditions in axiom A4. Proofs that V satisfies axioms A5 and A7–A10 are assigned in Exercise 8. With these proofs it will be established that V is a vector space. $\square$

Example 9

Let V be the vector space of Example 8. Let W consist of all f in V such that $f(1) = 0$. If f and g are in W, then

$$\begin{aligned} (f+g)(1) &= f(1) + g(1) && \text{(Definition of addition in } W\text{)} \\ &= 0 + 0 && \text{(Definition of } W\text{)} \\ &= 0 && \text{(Addition in } \mathbb{R}\text{)} \end{aligned}$$

Thus $f + g$ is in W. Similarly for any scalar c,

$$\begin{aligned} (cf)(1) &= cf(1) && \text{(Definition of multiplication by a scalar in } W\text{)} \\ &= c \cdot 0 && \text{(Definition of } W\text{)} \\ &= 0 && \text{(Multiplication in } \mathbb{R}\text{)} \end{aligned}$$

Hence cf is in W. Thus W satisfies A1 and A5. Since $f_0(1) = 0$, f_0 is in W and A3 is satisfied. Since $(-1)f$ serves as an additive inverse of f, A4 is satisfied by the closure of W under multiplication by a scalar just established. That W satisfies the remaining axioms for a vector space was established in Example 9. Hence W is a vector space. ☐

Example 10

Let V be the vector space of Example 8. Let W consist of all f in V such that $f(0) = 1$. If

$$f(x) = x + 1 \quad \text{and} \quad g(x) = x^2 + 1 \quad \text{for all real } x,$$

then f and g are in W. (Why?) Adding f and g yields

$$(f+g)(x) = x^2 + x + 2.$$

(Again, why?) Then W is not closed under addition, since

$$(f+g)(0) = 2.$$

Hence W is not a vector space. ☐

Example 11

Let V consist of all 2×2 matrices with real entries. For vector addition and multiplication by a scalar, we use the operations given in Definition 2.6. That is, if $A = [a_{ij}]_{2\times 2}$, $B = [b_{ij}]_{2\times 2}$, and c is a scalar, then

$$A + B = \left[a_{ij} + b_{ij}\right]_{2\times 2} \quad \text{and} \quad cA = \left[ca_{ij}\right]_{2\times 2}.$$

From these definitions, it follows that if A and B are in V and c is a scalar, then $A + B$ and cA are also in V, since each is a 2×2 matrix with real entries. Thus V satisfies axioms A1 and A6. For the remaining axioms, see Exercise 16. ☐

Some Immediate Consequences of the Vector Space Axioms

Several properties of any vector space V can be obtained directly from axioms A1–A10 in a few short steps.

THEOREM 4.1

The following statements hold in any vector space V.

a. The identity vector $\mathbf{0}$ in V is unique.
b. The inverse of any vector $\mathbf{x}$ in V is unique.
c. For every $\mathbf{x}$ in V, $0\mathbf{x} = \mathbf{0}$.
d. For every $\mathbf{x}$ in V, $-\mathbf{x} = (-1)\mathbf{x}$.
e. For every scalar a, $a\mathbf{0} = \mathbf{0}$.
f. For every $\mathbf{x}$ in V and every scalar a, if $a\mathbf{x} = \mathbf{0}$ then $a = 0$ or $\mathbf{x} = \mathbf{0}$.

Proof We prove statements a–d, assigning e and f in Exercise 18.

a. Let $\mathbf{0}$ and $\mathbf{y}$ be additive identities in V, i.e.,

$$\mathbf{0} + \mathbf{x} = \mathbf{x} + \mathbf{0} = \mathbf{x} \quad \text{and} \quad \mathbf{x} + \mathbf{y} = \mathbf{y} + \mathbf{x} = \mathbf{x}$$

for all $\mathbf{x}$ in V. Then

$$\mathbf{0} = \mathbf{0} + \mathbf{y} = \mathbf{y}.$$

The first equality holds because $\mathbf{y}$ is an additive identity and the second equality holds because $\mathbf{0}$ is an additive identity.

b. If $\mathbf{x}$ is in V, let $\mathbf{y}$ and $\mathbf{z}$ be inverses of $\mathbf{x}$. Then

$$\begin{aligned}
\mathbf{y} &= \mathbf{0} + \mathbf{y} && \text{(Axiom A3)}\\
&= (\mathbf{z} + \mathbf{x}) + \mathbf{y} && \text{(Axiom A4)}\\
&= \mathbf{z} + (\mathbf{x} + \mathbf{y}) && \text{(Axiom A2)}\\
&= \mathbf{z} + \mathbf{0} && \text{(Axiom A4)}\\
&= \mathbf{z} && \text{(Axiom A3)}
\end{aligned}$$

With the uniqueness of the additive inverse established, we shall use $-\mathbf{x}$ for the additive inverse of a vector $\mathbf{x}$.

c. Let $\mathbf{x}$ be in V. Then

$$\begin{aligned}
0\mathbf{x} &= 0\mathbf{x} + \mathbf{0} && \text{(Axiom A3)}\\
&= 0\mathbf{x} + [0\mathbf{x} + (-0\mathbf{x})] && \text{(Axiom A4)}\\
&= (0\mathbf{x} + 0\mathbf{x}) + (-0\mathbf{x}) && \text{(Axiom A2)}\\
&= (0 + 0)\mathbf{x} + (-0\mathbf{x}) && \text{(Axiom A9)}\\
&= 0\mathbf{x} + (-0\mathbf{x}) && \text{(Addition in } \mathbb{R})\\
&= \mathbf{0} && \text{(Axiom A4)}
\end{aligned}$$

d. If $\mathbf{x}$ is in V, then

$$\begin{aligned}
\mathbf{x} + (-1)\mathbf{x} &= 1\mathbf{x} + (-1)\mathbf{x} && \text{(Axiom A8)}\\
&= [1 + (-1)]\mathbf{x} && \text{(Axiom A9)}\\
&= 0\mathbf{x} && \text{(Addition in } \mathbb{R})\\
&= \mathbf{0} && \text{(Proved in part c)}
\end{aligned}$$

By axiom A5, we have also $(-1)\mathbf{x} + \mathbf{x} = \mathbf{0}$. Hence $(-1)\mathbf{x}$ is an additive inverse of $\mathbf{x}$ by A4. By the uniqueness of the additive inverse proved in part b above, $(-1)\mathbf{x} = -\mathbf{x}$. ☐

Theorem 4.1 illustrates the utility of the axiomatic point of view: the six properties apply to all vector spaces at once, and they need not be verified individually for the spaces $\mathbb{R}^n$ (Example 2), the spaces of polynomials (Example 5), and the more general spaces of real functions (Example 8), or the spaces of matrices (Example 11). The remaining sections of this chapter contain deeper results that will also apply to vector spaces in general.

REVIEW CHECKLIST

1. Formulate a definition of a vector space, identifying the two main components and the ten basic axioms they satisfy.
2. Given a set V and a pair of operations, determine whether V forms a (real) vector space relative to these operations. (This includes testing whether V satisfies axioms A1–A10, complete with justification of each decision.)

3. Cite a variety of examples of vector spaces.
4. Cite and prove the basic properties in Theorem 4.1 and apply them to general vector spaces.

EXERCISES

These exercises differ from those in Chapters 1–3, due to their increased emphasis on proof. These arguments involve reasoning from hypothesis to conclusion in a series of steps, each requiring justification. Valid reasons may involve a law you can cite by name, or perhaps a theorem or an axiom you can identify by number in this book. Try to avoid "clearly" and "obviously"—if a statement is truly obvious, these terms are redundant. A little more thinking usually will produce a cogent reason. Proofs serving as models appear throughout this section.

1. Establish the following properties of vector addition in $\mathbb{R}^n$, where n is any positive integer. Assume the corresponding properties in $\mathbb{R}$ in justifying steps.
(A2) For any $\mathbf{x}$, $\mathbf{y}$, and $\mathbf{z}$ in $\mathbb{R}^n$,

$$(\mathbf{x} + \mathbf{y}) + \mathbf{z} = \mathbf{x} + (\mathbf{y} + \mathbf{z}).$$

(A3) If $\mathbf{0} = (0, \ldots, 0)$ in $\mathbb{R}^n$, then for any $\mathbf{x}$ in $\mathbb{R}^n$,

$$\mathbf{0} + \mathbf{x} = \mathbf{x} + \mathbf{0} = \mathbf{x}.$$

(A4) If $\mathbf{x}$ is in $\mathbb{R}^n$ and $-\mathbf{x}$ denotes $(-1)\mathbf{x}$, then

$$-\mathbf{x} + \mathbf{x} = \mathbf{x} + (-\mathbf{x}) = \mathbf{0}.$$

2. Establish the following properties of multiplication by a scalar in $\mathbf{R}^n$, assuming the corresponding properties in $\mathbb{R}$.
(A6) For any $\mathbf{x}$ in $\mathbb{R}^n$ and c in $\mathbb{R}$, $c\mathbf{x}$ is also in $\mathbb{R}^n$.
(A7) For any $\mathbf{x}$ in $\mathbb{R}^n$ and any b and c in $\mathbb{R}$,

$$(bc)\mathbf{x} = b(c\mathbf{x}).$$

(The unjustified steps can be found in Section 2.1.)
(A8) For any $\mathbf{x}$ in $\mathbb{R}^n$,

$$1\mathbf{x} = \mathbf{x}.$$

3. Establish the following relations between vector addition and multiplication by scalars in $\mathbb{R}^n$, assuming the corresponding properties in $\mathbb{R}$.
(A9) For any $\mathbf{x}$ in $\mathbb{R}^n$ and any scalars b and c,

$$(b + c)\mathbf{x} = b\mathbf{x} + c\mathbf{x}.$$

(A10) For any **x** and **y** in $\mathbb{R}^n$ and any scalar c,

$$c(\mathbf{x} + \mathbf{y}) = c\mathbf{x} + c\mathbf{y}.$$

4. Let W consist of all vectors $\mathbf{x} = (x_1, x_2)$ in $\mathbb{R}^2$ with both components positive (the first quadrant). Show that W is closed under vector addition, but not under multiplication by a scalar. Is W a vector space?

5. Let W consist of all vectors $\mathbf{x} = (x_1, x_2)$ in $\mathbb{R}^2$ with at least one component zero (the x- and y-axes). Show that W is closed under multiplication by a scalar, but not under vector addition. Is W a vector space?

6. Show that the set V of Example 4 (page 177) is not closed under multiplication by a scalar.

7. Show that the set W of Example 7 (page 178) is not closed under multiplication by a scalar.

8. Show that the set V of Example 8 (page 179) satisfies axioms A5 and A7–A10.

In Exercises 9–14, if V consists of all vectors $\mathbf{x} = (x_1, x_2)$ in $\mathbb{R}^2$ of the given form, determine whether V is a vector space.

9. $(x_1, 2x_1)$ **10.** $(x_1, 0)$ **11.** $(x_1, 1)$

12. $(0, x_2)$ **13.** (x_1, x_1) **14.** $(2x_2 - 1, x_2)$

15. Verify that the set V of Example 5 (page 177) satisfies axioms A2–A5 and A7–A9 for the given vector operations.

16. Verify that the set V of Example 11 (page 182) satisfies axioms A2–A5 and A7–A10 for the given vector operations.

In Exercises 17–22, let V be the vector space of Example 11 (page 182), and let W consist of all $A = [a_{ij}]_{2\times 2}$ in V satisfying the given condition. Determine whether W is a vector space.

17. $a_{12} = 1$ **18.** $a_{11} = 0$ **19.** $a_{11} + a_{12} = 0$

20. $a_{21} + a_{22} = 1$ **21.** $a_{11} - 2a_{12} + 4a_{21} + 5a_{22} = 0$

22. $a_{11}^2 + a_{22}^2 = 1$

In Exercises 23–28, let V be the vector space of Example 5 (page 177), and let W consist of all polynomials of the form $p = a_0 + a_1x + a_2x^2$ satisfying the given condition. Determine whether W is a vector space.

23. $a_0 = 0$ **24.** $a_0 - a_1 + a_2 = 0$ **25.** $a_1 + 2a_2 = 1$

26. $a_2 = 0$ **27.** $a_1a_2 = 0$ **28.** $a_1^2 = 0$

29. For the set V of all polynomials of degree 3 or less, formulate definitions of addition and multiplication by scalars analogous to the operations in Example 5 (page 177), and determine whether V is a vector space.

In Exercises 30–31, if V is the solution set of the given linear system, determine whether V forms a vector space relative to the vector operations of Definition 2.2.

30. $\begin{aligned} x + 2y + z &= 0 \\ 2x + y + z &= 0 \end{aligned}$

31. $\begin{aligned} x + y - z &= 1 \\ 3x - 3y + z &= 1 \end{aligned}$

32. Prove statements (e) and (f) of Theorem 4.1.

33. Prove the following cancellation laws in any vector space V.
  a. If $\mathbf{x} + \mathbf{z} = \mathbf{y} + \mathbf{z}$, then $\mathbf{x} = \mathbf{y}$.
  b. If $c\mathbf{x} = c\mathbf{y}$ and $c \neq 0$, then $\mathbf{x} = \mathbf{y}$.
  c. If $b\mathbf{x} = c\mathbf{x}$ and $\mathbf{x} \neq \mathbf{0}$, then $b = c$.

In Exercises 34–38, let V be the vector space of Example 8 (page 179) and let W consist of all functions f in V satisfying the given condition. Determine whether W is a vector space.

34. $f(10) = 0$ **35.** $f(1) + 2f(0) = 0$ **36.** $f(-1) = f(1)$

37. $f(-1) + f(1) = 1$ **38.** $f(0) + 2f(1) - 3f(2) + 4f(3) = 0$

39. Let V denote the set of all real functions on the closed interval $[0, 1]$, so the domain of any f in V consists of all real x such that $0 \leq x \leq 1$. Define addition and multiplication by a scalar pointwise, as in Example 8 (page 179). Determine whether V is a vector space.

4.2 SUBSPACES AND SPANS

The examples and exercises of Section 4.1 illustrate an important point. If V is a vector space and W a set of vectors in V, then W *may or may not form a vector space relative to the operations in* V. A careful study of this point will enhance our ability to relate vector spaces to linear systems.

Subspaces

The following concept was illustrated in Examples 3, 4, 6, 7, 9, and 10 of Section 4.1.

DEFINITION 4.2

Let V be a vector space relative to a pair of operations. If W is a subset of V, then W is called a **subspace of V** if W forms a vector space relative to the same operations.

The examples listed above also illustrate the following result.

THEOREM 4.2

Let V be a vector space and W be a nonempty subset of V. If W is closed under both vector operations in V, then W is a subspace of V.

Proof If $\mathbf{x}$ is in W, then $\mathbf{x}$ is in V, so by Theorem 4.1, part (d), $-\mathbf{x} = (-1)\mathbf{x}$. Since W is closed under multiplication by a scalar, $(-1)\mathbf{x}$ is in W, so axiom A4 is satisfied. Since W is nonempty, let $\mathbf{x}$ be an element of W. Then $-\mathbf{x}$ is in W, and since W is closed under vector addition, $-\mathbf{x} + \mathbf{x} = \mathbf{0}$ is in W. Axioms A2, A5, and A7–A10 are satisfied for W since they are satisfied for V. Hence W is a vector space. □

Example 1

Subspaces of $\mathbb{R}^2$ are easy to find. For instance,

$$\mathbf{0} + \mathbf{0} = \mathbf{0},$$

and

$$c\mathbf{0} = \mathbf{0}$$

for any scalar c. By Theorem 4.2, the set consisting of the vector $\mathbf{0}$ alone, that is, $\{\mathbf{0}\}$, *forms a subspace of* $\mathbb{R}^2$.

If W is a line through $\mathbf{0}$, then the equation of W is of the form

$$Ax + By = 0,$$

where A and B are not both zero. Closure may be established as follows. If $\mathbf{x} = (x_1, x_2)$ and $\mathbf{y} = (y_1, y_2)$ are in W and c is a scalar, then

$$Ax_1 + Bx_2 = 0 \quad \text{and} \quad Ay_1 + By_2 = 0,$$

from which

$$A(x_1 + y_1) + B(x_2 + y_2) = 0 \quad \text{and} \quad A(cx_1) + B(cx_2) = 0,$$

so $\mathbf{x} + \mathbf{y}$ and $c\mathbf{x}$ are in W. Hence *any line W through the origin is a subspace of* $\mathbb{R}^2$ *by Theorem* 4.2.

Finally, $\mathbb{R}^2$ itself satisfies Definition 4.2 in the role of W, so $\mathbb{R}^2$ *is a subspace of itself*. These subspaces of $\mathbb{R}^2$ are shown in Fig. 4.1. □

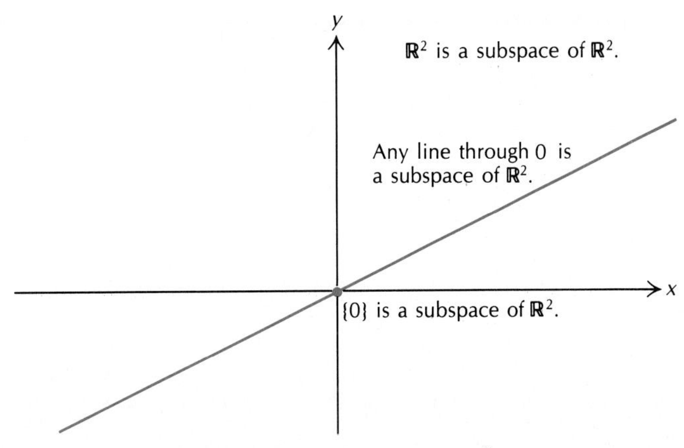

Figure 4.1 Subspaces of $\mathbb{R}^2$.

Example 2

Subspaces are also easy to find in $\mathbb{R}^3$. By the argument given in Example 1 for the identity in $\mathbb{R}^2$, the singleton set $\{\mathbf{0}\}$, *where* $\mathbf{0} = (0,0,0)$, *is a subspace of* $\mathbb{R}^3$.

If W is a line through the origin, then W consists of all vectors of the form

$$(at, bt, ct),$$

where t is any real number and (a, b, c) is a fixed nonzero point on the line. From this form, closure of W under the vector operations follows from the equalities

$$(at_1, bt_1, ct_1) + (at_2, bt_2, ct_2) = (a(t_1 + t_2), b(t_1 + t_2), c(t_1 + t_2))$$

and

$$k(at, bt, ct) = (a(kt), b(kt), c(kt)).$$

By Theorem 4.2, *any line through the origin is a subspace of* $\mathbb{R}^3$.

Since any plane through the origin has an equation of the form

$$Ax + By + Cz = 0 \qquad \text{where } A, B, \text{ and } C \text{ are not all zero,}$$

the argument in Example 1 for lines, modified to include the third component, shows that *any plane through the origin is a subspace of* $\mathbb{R}^3$.

Finally, $\mathbb{R}^3$ is a subspace of itself. These subspaces of $\mathbb{R}^3$ are shown in Fig. 4.2. □

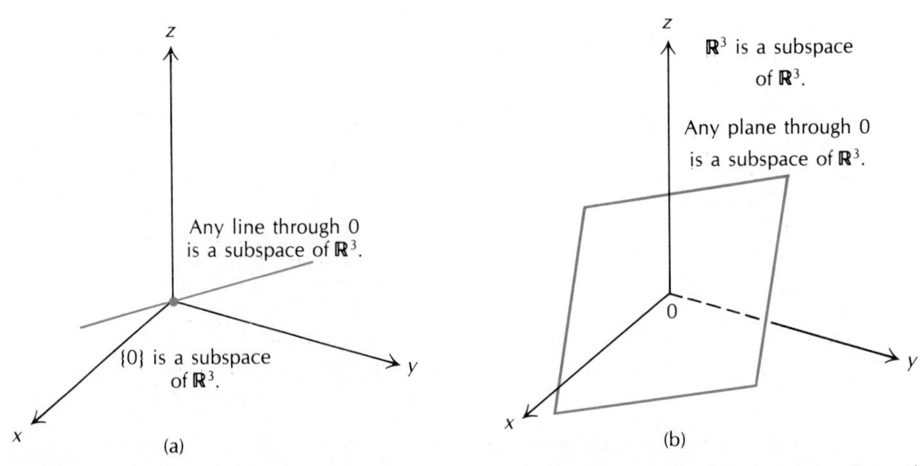

Figure 4.2 Subspaces of $\mathbb{R}^3$: (a) the origin and any line through it; (b) any plane through the origin and $\mathbb{R}^3$ itself.

Because of Theorem 4.2, the usual way to show that a nonempty set W in a vector space V is a subspace of V is to establish closure of W under the vector operations. In showing that W is *not* a subspace of V, we may show that W is not closed under one of the vector operations. However, any subspace of V must contain $\mathbf{0}$, so if a set W in V does not contain $\mathbf{0}$, we may conclude immediately that W is not a subspace of V. This point is worth remembering for the exercises and subsequent work.

Examples 1 and 2 illustrate the following important result.

THEOREM 4.3

The set of all solutions of a homogeneous linear system with an $m \times n$ coefficient matrix A is a subspace of $\mathbb{R}^n$. This space is called the **nullspace of A**.

Proof Any such system may be represented in matrix form as

$$A\mathbf{x} = \mathbf{0},$$

where $\mathbf{x}$ is the column vector of unknowns and $\mathbf{0}$ is the column vector of zeros. We denote the solution set of this linear system by W. Let $\mathbf{u}$ and $\mathbf{v}$ be in W, and let c be a scalar. Since $\mathbf{u}$ and $\mathbf{v}$ are solutions of the given system,

$$A\mathbf{u} = \mathbf{0} \quad \text{and} \quad A\mathbf{v} = \mathbf{0}.$$

By properties of matrix multiplication,

$$A(\mathbf{u} + \mathbf{v}) = A\mathbf{u} + A\mathbf{v} = \mathbf{0} + \mathbf{0} = \mathbf{0} \quad \text{and} \quad A(c\mathbf{u}) = cA\mathbf{u} = c\mathbf{0} = \mathbf{0}.$$

Thus, W is closed under the vector operations in $\mathbb{R}^n$. By Theorem 4.2, W is a subspace of $\mathbb{R}^n$. □

Example 3

Let W be the solution set of the homogeneous linear system

$$\begin{aligned} x - 2y + 2z &= 0 \\ 3x - 5y + 4z &= 0 \\ x - y &= 0 \end{aligned}$$

in $\mathbb{R}^3$. Then W is a subspace of $\mathbb{R}^3$ by Theorem 4.3. Solving this system shows that W consists of all ordered triples of the form $(2z, 2z, z)$. Thus W is the line through the origin and the point $(2, 2, 1)$. The space W illustrates one type of subspace of $\mathbb{R}^3$ described in Example 2. □

Example 4

Let W be the solution set of the nonhomogeneous linear system

$$x + 2y = 1$$
$$2x - y = 7.$$

Then W is *not* a subspace of $\mathbb{R}^2$, since $\mathbf{0}$ is not in W. □

Linear Combinations

The following concept plays a crucial role in vector spaces and their application to linear systems.

DEFINITION 4.3

In a vector space, an expression of the form

$$c_1\mathbf{x}_1 + \cdots + c_n\mathbf{x}_n$$

is called a **linear combination** of the vectors $\mathbf{x}_1, \ldots, \mathbf{x}_n$.

We may also refer to the expression in Definition 4.3 as "the linear combination of $\mathbf{x}_1, \ldots, \mathbf{x}_n$ by the scalar coefficients $c_1, \ldots, c_n$."

Example 5

In $\mathbb{R}^2$, the vector $(7, -1)$ may be expressed as a linear combination of the vectors $\mathbf{x}_1 = (2,1)$ and $\mathbf{x}_2 = (1, -1)$ since

$$2x_1 + 3x_2 = 2(2,1) + 3(1, -1) = (4, 2) + (3, -3) = (7, -1).$$

We may also express any vector (x, y) in $\mathbb{R}^2$ as a linear combination of $\mathbf{i} = (1,0)$ and $\mathbf{j} = (0,1)$ since

$$x\mathbf{i} + y\mathbf{j} = x(1,0) + y(0,1) = (x, y).$$ □

Example 6

In $\mathbb{R}^3$, $\mathbf{x} = (0, -1, 3)$ may be expressed as a linear combination of the vectors $\mathbf{x}_1 = (1,1,1)$, $\mathbf{x}_2 = (1,1,0)$, and $\mathbf{x}_3 = (1,0,0)$ since

$$\begin{aligned} 3\mathbf{x}_1 + (-4)\mathbf{x}_2 + \mathbf{x}_3 &= 3(1,1,1) - 4(1,1,0) + (1,0,0) \\ &= (3,3,3) - (4,4,0) + (1,0,0) \\ &= (0, -1, 3). \end{aligned}$$

We may also express any vector (x, y, z) in $\mathbb{R}^3$ as a linear combination of $\mathbf{i} = (1,0,0)$, $\mathbf{j} = (0,1,0)$, $\mathbf{k} = (0,0,1)$ since

$$x\mathbf{i} + y\mathbf{j} + z\mathbf{k} = x(1,0,0) + y(0,1,0) + z(0,0,1) = (x, y, z). \quad \square$$

In Examples 5 and 6, the scalar coefficients were either given or easy to find. The next examples address the problem of how to find a set of coefficients for a desired linear combination in $\mathbb{R}^n$ more generally.

Example 7

Express the vector $(7, -3)$ as a linear combination of $(1,1)$ and $(1, -1)$ in $\mathbb{R}^2$.

Solution

The unknown coefficients c_1 and c_2 must satisfy the vector equation

$$c_1(1,1) + c_2(1, -1) = (7, -3).$$

Performing the indicated operations yields

$$\begin{aligned}(7, -3) &= c_1(1,1) + c_2(1, -1)\\ &= (c_1, c_1) + (c_2, -c_2)\\ &= (c_1 + c_2, c_1 - c_2).\end{aligned}$$

Equating components gives the linear system

$$\begin{aligned}c_1 + c_2 &= 7\\ c_1 - c_2 &= -3.\end{aligned}$$

The solution $c_1 = 2$, $c_2 = 5$ may readily be checked:

$$2(1,1) + 5(1, -1) = (2,2) + (5, -5) = (7, -3),$$

as required. $\square$

Example 8

If possible, express the vector $(1,1,1)$ as a linear combination of $(1,2,-2)$, $(2,1,-3)$, $(2,-5,-1)$ in $\mathbb{R}^3$.

Solution

The problem is to find scalars c_1, c_2, and c_3 satisfying

$$c_1(1,2,-2) + c_2(2,1,-3) + c_3(2,-5,-1) = (1,1,1),$$

if they exist. As in Example 5, combine and equate components:

$$(c_1, 2c_1, -2c_1) + (2c_2, c_2, -3c_2) + (2c_3, -5c_3, -c_3) = (1,1,1)$$

$$(c_1 + 2c_2 + 2c_3, 2c_1 + c_2 - 5c_3, -2c_1 - 3c_2 - c_3) = (1,1,1).$$

The resulting system for the coefficients is

$$\begin{aligned} c_1 + 2c_2 + 2c_3 &= 1 \\ 2c_1 + c_2 - 5c_3 &= 1 \\ -2c_1 - 3c_2 - c_3 &= 1. \end{aligned}$$

Solve this system by Gaussian elimination:

$$\left[\begin{array}{rrr|r} 1 & 2 & 2 & 1 \\ 2 & 1 & -5 & 1 \\ -2 & -3 & -1 & 1 \end{array}\right] \to \left[\begin{array}{rrr|r} 1 & 2 & 2 & 1 \\ 0 & -3 & -9 & -1 \\ 0 & 1 & 3 & 3 \end{array}\right].$$

This system is inconsistent, as rows 2 and 3 of the final matrix show, so the system for c_1, c_2, and c_3 has no solution. Thus $(1,1,1)$ *cannot* be expressed as a linear combination of $(1,2,-2)$, $(2,1,-3)$, and $(2,-5,-1)$. □

Example 9

The same sum may result when a given set of vectors is combined with different sets of scalar coefficients. For instance,

$$5(1,1,0) + 3(1,2,1) + (-2)(2,3,1) = (4,5,1),$$

$$4(1,1,0) + 2(1,2,1) + (-1)(2,3,2) = (4,5,1).$$

These two sets of coefficients were found by solving the linear system

$$\begin{aligned} c_1 + c_2 + 2c_3 &= 4 \\ c_1 + 2c_2 + 3c_3 &= 5 \\ c_2 + c_3 &= 1. \end{aligned}$$

The general solution of this system is $(3 - z, 1 - z, z)$. One set of coefficients was found using $z = -2$ and the other was found using $z = -1$. □

Example 10

In the vector space V of all real functions on $\mathbb{R}$ (Section 4.1, Example 8, page 179) let f, g, h, j be defined by

$$f(x) = \cos^2 x, \qquad g(x) = \sin^2 x, \qquad h(x) = \cos 2x, \qquad j(x) = 1$$

for all real x. Then the trigonometric identities

$$\cos^2 x + \sin^2 x = 1 \qquad \text{and} \qquad \cos^2 x - \sin^2 x = \cos 2x$$

take the form

$$f + g = j \qquad \text{and} \qquad f - g = h$$

in V. These identities show that j and h may be expressed as *linear combinations* of f and g in V. □

The Span of a Set of Vectors

The next result will lead to an important concept in the study of vector spaces and linear systems.

THEOREM 4.4

If V is a vector space and W is the set of *all* linear combinations of the vectors $\mathbf{x}_1, \ldots, \mathbf{x}_n$ in V, then W is a subspace of V.

Proof Since $\mathbf{x}_1 = 1\mathbf{x}_1$, $\mathbf{x}_1$ is in W, so W is nonempty. To see why W is closed under the vector operations in V, let

$$\mathbf{x} = c_1\mathbf{x}_1 + \cdots + c_n\mathbf{x}_n \qquad \text{and} \qquad \mathbf{y} = k_1\mathbf{x}_1 + \cdots + k_n\mathbf{x}_n$$

be vectors in W and let c be a scalar. Then

$$\mathbf{x} + \mathbf{y} = (c_1 + k_1)\mathbf{x}_1 + \cdots + (c_n + k_n)\mathbf{x}_n,$$

so $\mathbf{x} + \mathbf{y}$ is in W. In addition, $c\mathbf{x}$ is in W since

$$c\mathbf{x} = (cc_1)\mathbf{x}_1 + \cdots + (cc_n)\mathbf{x}_n.$$

By Theorem 4.2, W is a subspace of V. □

DEFINITION 4.4

Let V be a vector space, let $\mathbf{x}_1, \ldots, \mathbf{x}_n$ be vectors in V, and let W be the set of all linear combinations of $\mathbf{x}_1, \ldots, \mathbf{x}_n$. We call W the **span of $\mathbf{x}_1, \ldots, \mathbf{x}_n$**, and we say that the vectors **$\mathbf{x}_1, \ldots, \mathbf{x}_n$ span W**. We also call the set consisting of these vectors, that is, $\{\mathbf{x}_1, \ldots, \mathbf{x}_n\}$, a **spanning set** for W.

Example 11

If $\mathbf{x} = (1, 2)$ in $\mathbb{R}^2$, then any linear combination of $\mathbf{x}$ is of the form $c\mathbf{x}$, that is, a scalar multiple of $\mathbf{x}$. The span of $\mathbf{x}$ is the line through $\mathbf{x}$ and the origin, i.e., the line determined by the equation $y = 2x$. (See Fig. 4.3.) ☐

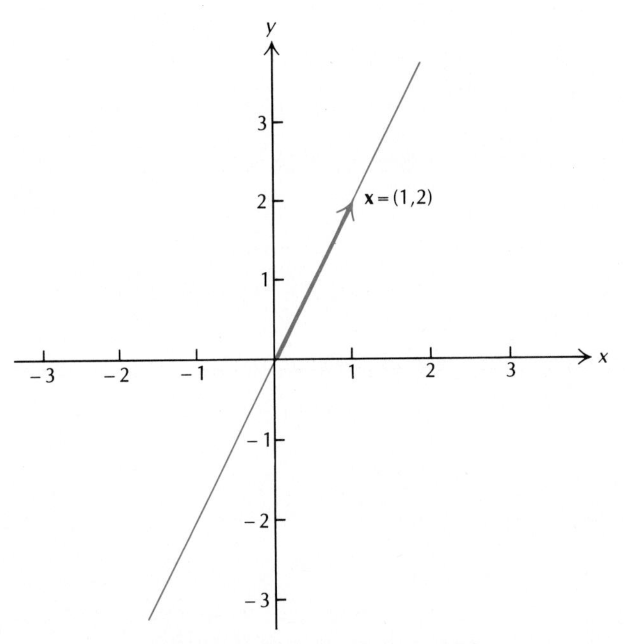

Figure 4.3 Span of $\mathbf{x} = (1, 2)$ in $\mathbb{R}^2$.

Example 12

For $\mathbf{x} = (-1, 2, 1)$ in $\mathbb{R}^3$, any linear combination of $\mathbf{x}$ is of the form $c\mathbf{x}$, so the span of $\mathbf{x}$ coincides with the line through $\mathbf{x}$ and the origin. This line consists of all triples of the form $(-t, 2t, t)$ for t a real number. (See Fig. 4.4.) ☐

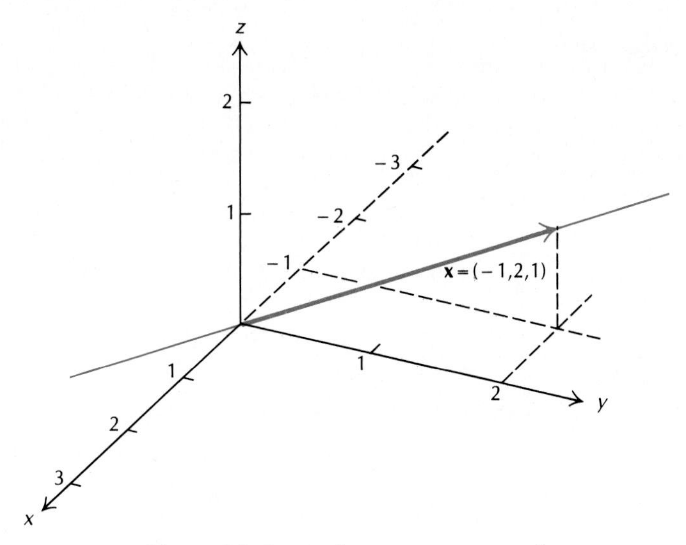

Figure 4.4 Span of $\mathbf{x} = (-1, 2, 1)$ in $\mathbb{R}^3$.

Example 13

For $\mathbf{x}_1 = (1, 0)$ and $\mathbf{x}_2 = (1, 1)$ in $\mathbb{R}^2$, the span of $\mathbf{x}_1, \mathbf{x}_2$ in $\mathbb{R}^2$ consists of all vectors in $\mathbb{R}^2$ of the form

$$\mathbf{x} = c_1(1, 0) + c_2(1, 1) = (c_1 + c_2, c_2).$$

If $\mathbf{b} = (b_1, b_2)$ is any vector in $\mathbb{R}^2$, then $\mathbf{b}$ may be written as a linear combination of $\mathbf{x}_1$ and $\mathbf{x}_2$ since the linear system

$$c_1 + c_2 = b_1$$

$$c_2 = b_2$$

has a solution. This system is consistent since both the row rank of the coefficient matrix and the row rank of the augmented matrix equal 2. Thus the span of $\mathbf{x}_1, \mathbf{x}_2$ is $\mathbb{R}^2$. In other words, the set $S = \{\mathbf{x}_1, \mathbf{x}_2\}$ is a spanning set for $\mathbb{R}^2$. ▭

Example 14

Find the span of $\mathbf{x}_1, \mathbf{x}_2$ in $\mathbb{R}^3$, if

$$\mathbf{x}_1 = (1, 2, 2) \qquad \text{and} \qquad \mathbf{x}_2 = (3, 2, 0).$$

Solution

By Definitions 4.3 and 4.4, the required span consists of all vectors in $\mathbb{R}^3$ of the form $c_1\mathbf{x}_1 + c_2\mathbf{x}_2$. The vectors $\mathbf{x}_1$ and $c_1\mathbf{x}_1$ lie on one line through the origin, and $\mathbf{x}_2$ and $c_2\mathbf{x}_2$ lie on another. The vector sum

$$c_1\mathbf{x}_1 + c_2\mathbf{x}_2$$

is found by completing the parallelogram with $c_1\mathbf{x}_1$ and $c_2\mathbf{x}_2$ as sides. (See Fig. 4.5.) Thus the geometry suggests that the required span is the plane determined by $\mathbf{x}_1$ and $\mathbf{x}_2$.

We test this conjecture using Definitions 4.3 and 4.4. A vector $\mathbf{v} = (x, y, z)$ is in the span of the given vectors if and only if there exist c_1 and c_2 such that

$$(x, y, z) = \mathbf{v} = c_1\mathbf{x}_1 + c_2\mathbf{x}_2 = c_1(1, 2, 2) + c_2(3, 2, 0).$$

In turn, this condition is true if and only if the linear system

$$\begin{aligned} c_1 + 3c_2 &= x \\ 2c_1 + 2c_2 &= y \\ 2c_1 \qquad &= z \end{aligned}$$

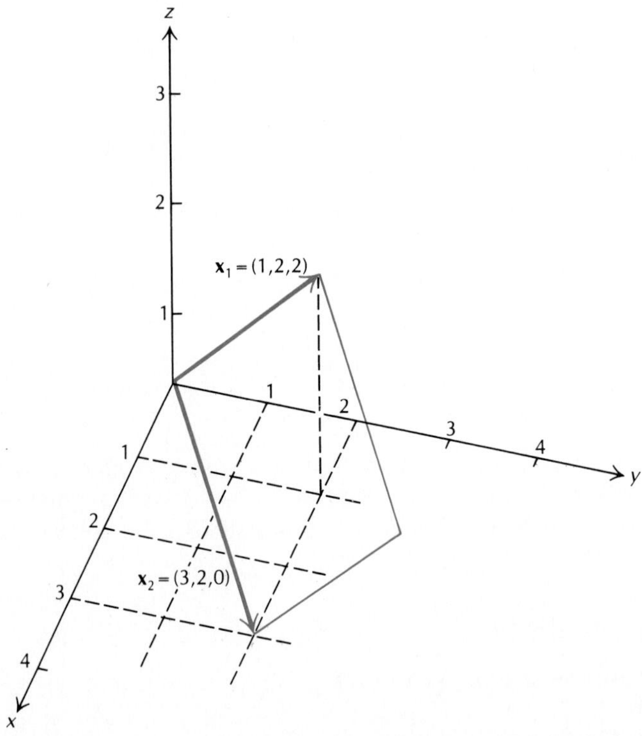

Figure 4.5 Parallelogram in space formed by the vectors $\mathbf{x}_1 = (1, 2, 2)$ and $\mathbf{x}_2 = (3, 2, 0)$.

is consistent. The first pass of Gaussian elimination yields

$$\left[\begin{array}{cc|c} 1 & 3 & x \\ 2 & 2 & y \\ 2 & 0 & z \end{array}\right] \rightarrow \left[\begin{array}{cc|c} 1 & 3 & x \\ 0 & -4 & y-2x \\ 0 & 0 & 2z-3y+2x \end{array}\right]$$

(here the intermediate steps are omitted). Thus the span of $\mathbf{x}_1$ and $\mathbf{x}_2$ is the plane determined by the equation

$$2x - 3y + 2z = 0. \quad \square$$

Examples 11 and 12 illustrate that *the span of a single nonzero* vector in $\mathbb{R}^2$ or $\mathbb{R}^3$ *is a line through the origin*. Examples 13 and 14 illustrate that *the span of two noncollinear vectors* in $\mathbb{R}^2$ or $\mathbb{R}^3$ *is a plane through the origin* (the former equal to $\mathbb{R}^2$ itself). What happens with two collinear vectors?

Example 15

Find the span of $\mathbf{x}_1, \mathbf{x}_2$ in $\mathbb{R}^3$ if

$$\mathbf{x}_1 = (1, 3, -2) \qquad \text{and} \qquad \mathbf{x}_2 = (2, 6, -4).$$

Solution

By Definitions 4.3 and 4.4, the required span consists of all vectors in $\mathbb{R}^3$ of the form

$$c_1\mathbf{x}_1 + c_2\mathbf{x}_2.$$

Since

$$\mathbf{x}_2 = 2\mathbf{x}_1,$$

it follows that

$$\begin{aligned} c_1\mathbf{x}_1 + c_2\mathbf{x}_2 &= c_1\mathbf{x}_1 + c_2(2\mathbf{x}_1) \\ &= c_1\mathbf{x}_1 + (2c_2)\mathbf{x}_1 \\ &= (c_1 + 2c_2)\mathbf{x}_1, \end{aligned}$$

so the span of $\mathbf{x}_1$ and $\mathbf{x}_2$ coincides with the line through $\mathbf{x}_1$ and $\mathbf{0}$. Hence the span of $\mathbf{x}_1$ and $\mathbf{x}_2$ coincides with the span of $\mathbf{x}_1$ (and with the span of $\mathbf{x}_2$). $\square$

Two observations account for our immediate interest in subspaces and spans.

1. The solution set of a homogeneous $m \times n$ linear system is a subspace of $\mathbb{R}^n$.
2. Two basic ways to describe these solution sets are: (a) parametrically, and (b) as the span of a set of solutions.

The choice of which way to represent a solution set may depend on the mathematical or physical context of the linear system, and we need to be able to use both methods.

These observations suggest only part of what is needed in practice. For instance, in some mathematical or physical contexts, subspaces of a given vector space V arise independently, i.e., *without regard to any linear system*. Thus we must be able to take a subspace as the point of departure, i.e., as given, and describe it in the following ways:

i. as the solution set of a linear system;
ii. parametrically;
iii. as the span of a set chosen from W.

Example 16

The plane through the origin orthogonal to $(1, -3, 4)$ in $\mathbb{R}^3$ may be described:

i. as the plane determined by the equation

$$x - 3y - 4z = 0;$$

ii. parametrically, as the set of all points of the form

$$(3s + 4t, s, t);$$

iii. as the span of $\mathbf{x}_1, \mathbf{x}_2$, where

$$\mathbf{x}_1 = (3,1,0) \qquad \text{and} \qquad \mathbf{x}_2 = (4,0,1).$$

(We chose these particular vectors by assigning first $s = 1$ and $t = 0$, then $s = 0$ and $t = 1$, and evaluating $3s + 4t$ to obtain the first component.) Each of these representations serves useful purposes—some you may recall from previous studies, others are yet to come. □

Our introduction to subspaces and spans leaves some issues open. For instance in Example 13, since the span of $\mathbf{x}_1$ and $\mathbf{x}_2$ coincides with the span of $\mathbf{x}_1$ alone, a sharper notion of a spanning set is needed for identification of a span as a "line," a "plane," or another type of subspace. These issues will be addressed in the remaining sections of this chapter.

REVIEW CHECKLIST

1. Formulate a definition for a subspace of a vector space.
2. Given a vector space V and a nonempty subset W of V, determine whether W is a subspace of V.
3. Give examples to illustrate the subspace concept.

4. Identify the nullspace of an $m \times n$ matrix as a subspace of $\mathbb{R}^n$.
5. Formulate a definition for a linear combination of a set of vectors.
6. Find linear combinations of given vectors and scalars.
7. Determine whether a given vector may be expressed as a linear combination of a given set of vectors and, if so, find an appropriate set of coefficients for the expression.
8. Define the span of a set of vectors and a spanning set of a subspace.
9. Express the solution set of a homogeneous linear system as the span of a set of solutions.
10. Find a parametric representation of the span of a given set of vectors.

EXERCISES

1. Express $\mathbf{0}$ as a linear combination of the vectors $(1,4)$ and $(3,7)$ in $\mathbb{R}^2$.

In Exercises 2–5, determine whether the set W of all vectors $\mathbf{x} = (x_1, x_2)$ in $\mathbb{R}^2$ satisfying the given condition(s) is a subspace of $\mathbb{R}^2$.

2. $x_2 = x_1$
3. $x_2 = x_1$ and $0 \le x_1 \le 1$
4. $x_2 = x_1 + 1$
5. $x_2 = x_1^2$

In Exercises 6–11, determine whether the set W of all vectors $\mathbf{x} = (x_1, x_2, x_3)$ in $\mathbb{R}^3$ satisfying the given condition(s) is a subspace of $\mathbb{R}^3$.

6. $x_1 = x_2 = x_3$
7. $x_1 + 2x_2 - x_3 = 0$
8. $x_1 + x_2 - 3x_3 = 1$
9. $x_1 = x_2$ and $x_3 = x_2^2$
10. $-1 \le x_2 \le 1$
11. $x_3 = 5x_1 + 2x_2$

In Exercises 12–14, find the indicated linear combination in $\mathbb{R}^2$ for $\mathbf{x}_1 = (1,1)$, $\mathbf{x}_2 = (0,1)$, $\mathbf{x}_3 = (-1,1)$, $c_1 = 2$, $c_2 = 3$, and $c_3 = -1$. Sketch a graph to show how each is formed.

12. $c_1\mathbf{x}_1 + c_2\mathbf{x}_2$
13. $c_1\mathbf{x}_1 + c_3\mathbf{x}_3$
14. $c_1\mathbf{x}_1 + c_2\mathbf{x}_2 + c_3\mathbf{x}_3$

In Exercises 15–18, find the indicated linear combination in $\mathbb{R}^3$ for $\mathbf{x}_1 = (1,-1,2)$, $\mathbf{x}_2 = (1,0,1)$, $\mathbf{x}_3 = (2,4,-3)$, $c_1 = 4$, $c_2 = -2$, and $c_3 = 2$.

15. $c_1\mathbf{x}_1 + c_2\mathbf{x}_2$
16. $c_1\mathbf{x}_1 + c_3\mathbf{x}_3$
17. $c_2\mathbf{x}_2 + c_3\mathbf{x}_3$
18. $c_1\mathbf{x}_1 + c_2\mathbf{x}_2 + c_3\mathbf{x}_3$

In Exercises 19–24, if possible, express the vector $\mathbf{x} = (2,3)$ as a linear combination of the given vectors in $\mathbb{R}^2$.

19. $(1,0), (0,1)$
20. $(1,0), (1,1)$
21. $(1,2), (2,4)$
22. $(4,6), (0,0)$
23. $(-2,-3), (1,1)$
24. $(1,2), (1,3)$

In Exercises 25–30, if possible, express the vector $\mathbf{x} = (-1,1,2)$ as a linear combination of the given vectors in $\mathbb{R}^3$.

25. $(1,0,0), (0,1,0), (0,0,1)$
26. $(1,0,0), (0,1,0)$
27. $(1,2,-1), (3,3,-4)$
28. $(1,0,0), (1,1,0), (1,1,1)$
29. $(4,-4,-8)$
30. $(1,1,3), (1,2,1), (1,0,5)$

In Exercises 31–36, if possible, express the vector $\mathbf{x} = (1,2,-3,-2)$ as a linear combination of the given vectors in $\mathbb{R}^4$.

31. $(1,0,0,0), (0,1,0,0), (0,0,1,0), (0,0,0,1)$
32. $(1,0,0,0), (1,1,0,0), (1,1,1,0), (1,1,1,1)$
33. $(1,4,2,2), (3,6,-9,-6)$
34. $(1,1,0,0), (1,0,0,1), (0,0,1,1)$
35. $(1,2,-1,-2), (1,2,0,-2)$
36. $(1,2,1,-1), (2,-1,-3,3), (1,5,4,-8), (-2,3,1,-2)$

In Exercises 37–40, determine whether the given vector can be expressed as a linear combination of the vectors $\mathbf{x}_1 = (1,2,-1)$, $\mathbf{x}_2 = (1,-1,2)$, and $\mathbf{x}_3 = (2,0,1)$ and if so, find an appropriate set of coefficients.

37. $(3,2,3)$
38. $(0,0,0)$
39. $(1,1,1)$
40. $(6,6,-3)$

In Exercises 41–45, determine whether the given vectors span $\mathbb{R}^2$.

41. $(1,0), (0,1)$
42. $(1,0), (0,0)$
43. $(1,0), (1,1)$
44. $(1,1), (1,-1)$
45. $(1,-2), (4,-8)$

In Exercises 46–51, determine whether the given vectors span $\mathbb{R}^3$.

46. $(1,0,0), (0,1,0), (0,0,1)$
47. $(1,0,0), (0,1,0)$
48. $(1,2,1), (0,0,0), (1,1,2)$
49. $(1,0,0), (1,1,0), (1,1,1)$
50. $(1,1,0), (1,0,1), (0,1,1)$
51. $(1,1,-2), (1,-1,8), (2,1,1)$

In Exercises 52–54, let W consist of all vectors $\mathbf{v} = (x, y)$ in $\mathbb{R}^2$ for which

$$2x - 3y = 0.$$

52. Show that W is a subspace of $\mathbb{R}^2$.
53. Find a parametric representation of W.
54. Find a vector $\mathbf{x}$ whose span coincides with W.

In Exercises 55–57, W is the nullspace of the matrix

$$\begin{bmatrix} 1 & 3 & -4 \\ 1 & 4 & -3 \\ 1 & 1 & -6 \end{bmatrix}.$$

55. Show that W is a subspace of $\mathbb{R}^3$.
56. Find a parametric representation of W.
57. Find a vector $\mathbf{x}$ whose span coincides with W.

In Exercises 58–60, W consists of all vectors $\mathbf{w} = (x, y, z)$ in $\mathbb{R}^3$ for which

$$2x - 4y + z = 0.$$

58. Prove that W is a subspace of $\mathbb{R}^3$.
59. Find a parametric representation of W.
60. Find vectors $\mathbf{u}$ and $\mathbf{v}$ such that the span of $\{\mathbf{u}, \mathbf{v}\}$ coincides with W.

In Exercises 61–62, let $\mathbf{x} = (3, -4)$ in $\mathbb{R}^2$, and let W be the span of $\mathbf{x}$.

61. Find an equation for W.
62. Find a parametric representation of W.

In Exercises 63–66, let $\mathbf{x} = (1, 2, 3)$ in $\mathbb{R}^3$, and let W be the span of $\mathbf{x}$.

63. Find another nonzero vector in W.
64. Find a parametric representation of W.
65. Find a 2×3 linear system whose solution set is W.
66. Find a 3×3 linear system of three distinct equations whose solution set is W.

In Exercises 67–69, let $\mathbf{x}_1 = (1, -1, 2)$ and $\mathbf{x}_2 = (1, 0, 1)$, and let W be the span of $\mathbf{x}_1, \mathbf{x}_2$ in $\mathbb{R}^3$.

67. Show that W is a subspace of $\mathbb{R}^3$.
68. Find an equation for W.
69. Find a parametric representation of W.

In Exercises 70–71, V is the vector space of 2×2 matrices (see Example 11, Section 4.1, page 182). Express

$$\begin{bmatrix} 4 & 2 \\ 5 & 1 \end{bmatrix}$$

as a linear combination of the given vectors.

70. $\begin{bmatrix} 1 & 0 \\ 0 & 0 \end{bmatrix}, \begin{bmatrix} 0 & 1 \\ 0 & 0 \end{bmatrix}, \begin{bmatrix} 0 & 0 \\ 1 & 0 \end{bmatrix}, \begin{bmatrix} 0 & 0 \\ 0 & 1 \end{bmatrix}$

71. $\begin{bmatrix} 1 & 0 \\ 0 & 0 \end{bmatrix}, \begin{bmatrix} 1 & 1 \\ 0 & 0 \end{bmatrix}, \begin{bmatrix} 1 & 1 \\ 1 & 0 \end{bmatrix}, \begin{bmatrix} 1 & 1 \\ 1 & 1 \end{bmatrix}$

In Exercises 72–74, V is the vector space of Exercises 70–71. Determine whether the given vector can be expressed as a linear combination of

$$\begin{bmatrix} 1 & 0 \\ 1 & 1 \end{bmatrix}, \begin{bmatrix} 1 & 2 \\ -1 & 0 \end{bmatrix}, \begin{bmatrix} 1 & -4 \\ 5 & 3 \end{bmatrix}, \begin{bmatrix} 0 & 2 \\ -2 & -1 \end{bmatrix}.$$

72. $\begin{bmatrix} 1 & 0 \\ 0 & 0 \end{bmatrix}$ **73.** $\begin{bmatrix} 0 & 0 \\ 0 & 0 \end{bmatrix}$ **74.** $\begin{bmatrix} 3 & -4 \\ 7 & 5 \end{bmatrix}$

75. If $\mathbf{x} = (1,1,2)$, $\mathbf{y} = (1,0,1)$, and $\mathbf{z} = (3,-2,1)$ in $\mathbb{R}^3$, show that the span of $\mathbf{x}$ is a subspace of the span of $\mathbf{y}$ and $\mathbf{z}$.

76. Show that the solution set of the system

$$x - 3y - z = 0$$

$$x + y - 5z = 0$$

is a subspace of the vector space in $\mathbb{R}^3$ determined by

$$x - 7y + 3z = 0.$$

In Exercises 77–79, V is the space of all real functions on $\mathbb{R}$ (Example 8, page 179). Express the indicated function as a linear combination of f_0, f_1, f_2, and f_3, where

$$f_0(x) = 1, \quad f_1(x) = x, \quad f_2(x) = x^2, \quad f_3(x) = x^3$$

for all real x.

77. $f(x) = (x-2)(x+2)$ **78.** $g(x) = (x+1)^3$

79. $h(x) = (x-2)^3$

80. Show that the set W of all vectors of the form $(x_1, x_2, 0)$ is a subspace of $\mathbb{R}^3$.

81. (*Mathematical induction needed*) Prove that if V is a vector space, W a subspace of V, and $\mathbf{x}_1, \ldots, \mathbf{x}_n$ are in W, then W must contain the span of $\mathbf{x}_1, \ldots, \mathbf{x}_n$.

82. Show that if V is a vector space, W a subspace of V, and U a subspace of W, then U is a subspace of V.

In Exercises 83–84, V is a vector space and W and U are subspaces. Determine whether the indicated set is necessarily a subspace of V. (Either prove that it is or show by example that it need not be.)

83. $W \cap U$ (The *intersection* consists of all vectors belonging to both spaces.)
84. $W \cup U$ (The *union* consists of all vectors belonging to at least one of the spaces.)

4.3 LINEAR DEPENDENCE AND INDEPENDENCE

Although spanning sets are useful in describing vector spaces and their subspaces they may contain more vectors than necessary. In this section, we identify redundancies in spanning sets and give a test to determine when a spanning set in $\mathbb{R}^n$ has unnecessary vectors. We also find all subspaces of $\mathbb{R}^2$ and $\mathbb{R}^3$.

The Concept and Some Examples

In identifying a subspace of a vector space as the span of a particular set, we need to know whether certain "algebraic relations" exist among the vectors in the spanning set.

DEFINITION 4.5

Let $S = \{\mathbf{x}_1, \ldots, \mathbf{x}_n\}$ be a set of vectors in a vector space. We say that S is **linearly dependent** if there exist scalars $c_1, \ldots, c_n$, *not all zero*, such that

$$c_1\mathbf{x}_1 + \cdots + c_n\mathbf{x}_n = \mathbf{0}.$$

We say that S is **linearly independent** if the condition

$$c_1\mathbf{x}_1 + \cdots + c_n\mathbf{x}_n = \mathbf{0}$$

implies that $c_1 = \cdots = c_n = 0$.

Example 1

The set $\{(1, -4), (-2, 8)\}$ is linearly dependent in $\mathbb{R}^2$, since

$$2 \cdot (1, -4) + 1 \cdot (-2, 8) = (0, 0)$$

and the coefficients are not *both* zero. □

Example 2

The singleton set $\{(1, 0, -2)\}$ is linearly independent in $\mathbb{R}^3$, since the condition

$$c(1, 0, -2) = (0, 0, 0)$$

implies $c = 0$. □

Example 2 illustrates the following general principle. For a proof, see Exercise 64.

Proposition 1 In a vector space, any set consisting of a single nonzero vector is linearly independent.

Example 3

The set $\{(2,3),(0,0)\}$ is linearly dependent in $\mathbb{R}^2$, since

$$0 \cdot (2,3) + 1 \cdot (0,0) = (0,0),$$

and the coefficients are not both zero. □

Example 3 illustrates the following principle. For a proof, see Exercise 65.

Proposition 2 In a vector space, any set containing the additive identity $\mathbf{0}$ is linearly dependent.

In view of Propositions 1 and 2, we concentrate on sets consisting of two or more nonzero vectors.

Example 4

Determine whether the set $S = \{(3,-6),(-2,4)\}$ is linearly independent in $\mathbb{R}^2$.

Solution

By Definition 4.5, S is linearly dependent if there exist scalars c_1 and c_2, not both zero, such that

$$c_1(3,-6) + c_2(-2,4) = (0,0).$$

This equation is equivalent to the equation

$$(3c_1 - 2c_2, -6c_1 + 4c_2) = (0,0).$$

Equating components gives the homogeneous linear system

$$\begin{aligned} 3c_1 - 2c_2 &= 0 \\ -6c_1 + 4c_2 &= 0 \end{aligned}$$

for c_1 and c_2. Since the row rank of the coefficient matrix is 1, this system has nontrivial solutions, for example, $c_1 = 2$ and $c_2 = 3$. Hence S is linearly *dependent*. □

In Example 4, linear dependence of S means that the vectors in S are collinear and the span of S consists of all vectors on this line. Here linear dependence also means that at least one of the vectors in S is expressible as a multiple of the other.

Example 5

Determine whether the set $S = \{(1, -1), (2, 3)\}$ is linearly independent in $\mathbb{R}^2$.

Solution

By Definition 4.5, S is linearly independent if the condition

$$c_1(1, -1) + c_2(2, 3) = (c_1 + 2c_2, -c_1 + 3c_2) = (0, 0)$$

implies $c_1 = c_2 = 0$. This vector equation is equivalent to the homogeneous linear system

$$\begin{aligned} c_1 + 2c_2 &= 0 \\ -c_1 + 3c_2 &= 0. \end{aligned}$$

Since the coefficient matrix has row rank 2, this system has only the trivial solution. Thus S is linearly *independent*. □

In Example 5, linear independence of S means that the vectors in S are noncollinear and that the span of S coincides with $\mathbb{R}^2$. To see why, let $\mathbf{b} = (b_1, b_2)$ be any vector in $\mathbb{R}^2$. Since the coefficient matrix has row rank 2, we may solve the system

$$\begin{aligned} c_1 + 2c_2 &= b_1 \\ -c_1 + 3c_2 &= b_2 \end{aligned}$$

for c_1 and c_2. Thus there exist scalars such that

$$c_1(1, -1) + c_2(2, 3) = (b_1, b_2),$$

so $\mathbf{b}$ is in the span of S. Since $\mathbf{b}$ was arbitrary in $\mathbb{R}^2$, the span of S contains and hence coincides with $\mathbb{R}^2$.

Next, we establish an important condition for linear dependence in $\mathbb{R}^n$.

Proposition 3 In $\mathbb{R}^n$, any set containing more than n vectors is linearly dependent.

Proof If $S = \{\mathbf{x}_1, \ldots, \mathbf{x}_m\}$ in $\mathbb{R}^n$ and $m > n$, let

$$\mathbf{x}_i = (a_{i1}, \ldots, a_{in})$$

for $i = 1, \ldots, m$. Then the equation

$$c_1 x_1 + \cdots + c_m x_m = 0$$

is equivalent to the homogeneous linear system

$$\begin{aligned} a_{11} c_1 + \cdots + a_{1m} c_m &= 0 \\ \vdots \qquad\qquad \vdots \quad &\quad \vdots \\ a_{n1} c_1 + \cdots + a_{nm} c_m &= 0. \end{aligned}$$

Since $n < m$, the row rank of the coefficient matrix must also be less than m. Thus this system has a nontrivial solution for the coefficients $c_1, \ldots, c_m$ (according to the summary for solution sets of homogeneous linear systems on page 30). ☐

Example 6

The set $S = \{(1,0), (1,2), (3,-1)\}$ is linearly dependent in $\mathbb{R}^2$. In fact, the linear system

$$\begin{aligned} c_1 + c_2 + 3c_3 &= 0 \\ 2c_2 - c_3 &= 0 \end{aligned}$$

has the nontrivial solution $c_1 = 7$, $c_2 = -1$, $c_3 = -2$, and

$$7\mathbf{x}_1 - \mathbf{x}_2 - 2\mathbf{x}_3 = \mathbf{0}.$$ ☐

A Test for Linear Independence in $\mathbb{R}^n$

By Proposition 3, only sets containing n or fewer vectors need be tested. For such sets, we use a different procedure than we did in proving Proposition 3. We describe the test and illustrate it with examples, characterize linearly dependent and linearly independent sets in $\mathbb{R}^2$ and $\mathbb{R}^3$, and then indicate why the test is valid.

Linear Independence Test in $\mathbb{R}^n$

For any set $S = \{\mathbf{x}_1, \ldots, \mathbf{x}_m\}$ in $\mathbb{R}^n$, let A be the matrix whose ith row vector coincides with $\mathbf{x}_i$. If r is the row rank of A, then S is linearly independent if and only if $r = m$.

Example 7

Determine whether the set $S = \{(1, 2, -1), (1, 0, 1), (2, 4, -3)\}$ is linearly independent in $\mathbb{R}^3$.

Solution

Form matrix A with the given vectors as rows and reduce A to row echelon form:

$$A = \begin{bmatrix} 1 & 2 & -1 \\ 1 & 0 & 1 \\ 2 & 6 & -3 \end{bmatrix} \to \begin{bmatrix} 1 & 2 & -1 \\ 0 & -2 & 2 \\ 0 & 2 & -1 \end{bmatrix} \to \begin{bmatrix} 1 & 2 & -1 \\ 0 & 1 & -1 \\ 0 & 0 & 1 \end{bmatrix}.$$

Since the row rank of A equals the number of vectors in S, this set is linearly independent. □

In Example 7, note also that $\det(A) \neq 0$ and that S spans $\mathbb{R}^3$.

Example 8

Determine whether the set $S = \{(1, 2, 3), (1, 3, 1), (-1, -4, 1)\}$ is linearly independent in $\mathbb{R}^3$.

Solution

Proceed as in Example 7:

$$A = \begin{bmatrix} 1 & 2 & 3 \\ 1 & 3 & 1 \\ -1 & -4 & 1 \end{bmatrix} \to \begin{bmatrix} 1 & 2 & 3 \\ 0 & 1 & -2 \\ 0 & -2 & 4 \end{bmatrix} \to \begin{bmatrix} 1 & 2 & 3 \\ 0 & 1 & -2 \\ 0 & 0 & 0 \end{bmatrix}.$$

Since the row rank of A is less than the number of vectors in S, this set is linearly dependent. □

In Example 8, note that $\det(A) = 0$ and that S spans a plane in $\mathbb{R}^3$. In fact, the equation of this plane is

$$7x - 2y - z = 0.$$

We may find the coefficients in this equation by solving the system $AX = 0$.

Example 9

Let V be the vector space of all real functions on $\mathbb{R}$. (See Example 8, Section 4.1, page 179.) Let $S = \{f, g, h\}$, where

$$f(x) = \cos^2 x, \qquad g(x) = \sin^2 x, \qquad \text{and} \qquad h(x) = 1$$

for all x. Show that S is linearly dependent.

Solution

By the Pythagorean theorem,

$$\sin^2 x + \cos^2 x = 1.$$

By definition of equality of functions, this identity is equivalent to the relation

$$1f + 1g + (-1)h = 0,$$

where 0 is the function identically zero. By Definition 4.5, S is linearly dependent. ☐

We summarize various results above to characterize linearly dependent sets and linearly independent sets in $\mathbb{R}^2$ and $\mathbb{R}^3$. In view of Propositions 1–3, the descriptions are limited to sets with two vectors in $\mathbb{R}^2$ and sets with two or three vectors in $\mathbb{R}^3$.

A Set S of Two Vectors in $\mathbb{R}^2$ or $\mathbb{R}^3$ The following statements are equivalent.

a. S is linearly dependent.
b. At least one vector in S is a multiple of the other.
c. S spans a line through the origin.

For $\mathbb{R}^2$, these conditions are equivalent to:

d. The determinant of the matrix formed by S equals zero.

The following are equivalent in the contrary case:

a′. S is linearly independent.
b′. Neither vector in S is a multiple of the other.
c′. S spans a plane ($\mathbb{R}^2$, if both vectors are in $\mathbb{R}^2$).

For $\mathbb{R}^2$, these conditions are equivalent to:

d′. The determinant of the matrix formed by S is nonzero.

A Set S of Three Vectors in $\mathbb{R}^3$ The following statements are equivalent.

a. S is linearly dependent.
b. At least one of the vectors in S is expressible as a linear combination of the other two.
c. S spans a line or a plane.
d. The determinant of the matrix formed by S equals zero.

The following are equivalent in the contrary case:

a′. S is linearly independent.
b′. The vectors in S are not coplanar.
c′. S spans $\mathbb{R}^3$.
d′. The determinant of the matrix formed by S is nonzero.

In the remainder of this section, we outline the idea behind the linear independence test for $\mathbb{R}^n$. For the set S in Example 8, we may find scalars c_1, c_2, and c_3 satisfying the condition for linear dependence in Definition 4.5 from the elementary row operations used in this example. If the vectors in S are denoted by $\mathbf{x}_1$, $\mathbf{x}_2$, and $\mathbf{x}_3$, then the bottom two row vectors in the matrix resulting from the first operations

$$\text{R2} \leftarrow \text{R2} - \text{R1} \qquad \text{and} \qquad \text{R3} \leftarrow \text{R3} + \text{R1}$$

are

$$\mathbf{x}_2 - \mathbf{x}_1 \qquad \text{and} \qquad \mathbf{x}_3 + \mathbf{x}_1,$$

in that order. Then from the next operation

$$\text{R3} \leftarrow \text{R3} + 2\text{R2}$$

we obtain a matrix with bottom row vector

$$(\mathbf{x}_3 + \mathbf{x}_1) + 2(\mathbf{x}_2 - \mathbf{x}_1) = -\mathbf{x}_1 + 2\mathbf{x}_2 + \mathbf{x}_3.$$

Since this is a zero row, we may take

$$c_1 = -1, \qquad c_2 = 2, \qquad \text{and} \qquad c_3 = 1$$

for the constants required in Definition 4.5.

This observation suggests that, in general, if C is a matrix obtained from a matrix A by a sequence of elementary row operations, then each row vector of C may be expressed as a linear combination of the row vectors of A. The converse is also true: any linear combination of the row vectors of a matrix can be achieved by a sequence of elementary row operations. Instead of attempting a general proof, we simply illustrate the idea with an example.

Example 10

If $\mathbf{x}_1$, $\mathbf{x}_2$, and $\mathbf{x}_3$ are the row vectors of a matrix A in order, express the linear combination

$$2\mathbf{x}_1 - 3\mathbf{x}_2 + \mathbf{x}_3$$

as a row vector obtained from a sequence of elementary row operations.

Solution

Simply write down a sequence of elementary row operations that produces the desired result. Application of

$$\text{R2} \leftarrow -3\text{R2}, \qquad \text{R2} \leftarrow \text{R2} + 2\text{R1}, \qquad \text{R3} \leftarrow \text{R3} + \text{R2}$$

to A yields the desired vector in the third row. □

In view of the relation between linear combinations of row vectors and elementary row operations, the following lemma justifies the test for linear independence in $\mathbb{R}^n$ (see Exercise 66).

Lemma A set of vectors S is linearly dependent if and only if S contains at least one vector that may be expressed as a linear combination of the others.

Proof Let $S = \{\mathbf{x}_1, \ldots, \mathbf{x}_n\}$. If S is linearly dependent, then by Definition 4.5 there exist $c_1, \ldots, c_n$, not all zero, such that

$$c_1\mathbf{x}_1 + \cdots + c_n\mathbf{x}_n = \mathbf{0}.$$

If $c_1 \neq 0$, then this equation can be solved for $\mathbf{x}_1$:

$$\mathbf{x}_1 = (-c_2/c_1)\mathbf{x}_2 + \cdots + (-c_n/c_1)\mathbf{x}_n.$$

If $c_1 = 0$, then since $c_j \neq 0$ for at least one j, the equation above can be solved in a similar way for $\mathbf{x}_j$.

Conversely, suppose one of the given vectors is expressible as a linear combination of the others. If $\mathbf{x}_1$ is one such vector, then there exist constants $c_2, \ldots, c_n$ such that

$$\mathbf{x}_1 = c_2\mathbf{x}_2 + \cdots + c_n\mathbf{x}_n.$$

Then

$$1\mathbf{x}_1 + (-c_2)\mathbf{x}_2 + \cdots + (-c_n)\mathbf{x}_n = \mathbf{0}.$$

Since the coefficients are not all zero, S is linearly dependent by Definition 4.5. A similar argument applies for any $\mathbf{x}_j$ that is expressible as a linear combination of the other vectors. ☐

Example 11

Let f, g, h, j be defined for all real x by

$$f(x) = x^2, \qquad g(x) = x, \qquad h(x) = 1, \qquad j(x) = (x+1)^2.$$

We may express j as a linear combination of f, g, h, since

$$x^2 + 2x + 1 = (x+1)^2$$

for all x. Hence by the lemma, the set $S = \{f, g, h, j\}$ is linearly dependent in the space of all real functions on $\mathbb{R}$. ☐

REVIEW CHECKLIST

1. Define linear dependence and linear independence for a set of vectors.
2. Test a given set of vectors for linear independence using the definition.
3. Identify any set containing $\mathbf{0}$ as linearly dependent by inspection.
4. Identify any set consisting of a single nonzero vector as linearly independent.
5. Identify any set with more than n vectors in $\mathbb{R}^n$ as linearly dependent.
6. Use matrix methods to test a set of m vectors in $\mathbb{R}^n$ for linear independence if $m \leq n$.
7. Use a determinant method to test a set of n vectors in $\mathbb{R}^n$ for linear independence.
8. Describe the relationship between linear combinations and linear dependence.
9. For sets in $\mathbb{R}^2$ and $\mathbb{R}^3$, describe the meaning of linear independence and linear dependence of a set in terms of its span.

EXERCISES

In Exercises 1–12, determine by inspection whether the given set is linearly independent.

1. $\{(0,0)\}$

2. $\{(1,0)\}$

3. $\{(1,0),(0,1)\}$

4. $\{(1,1),(0,0)\}$

5. $\{(-2,3),(4,-6)\}$

6. $\{(1,2),(3,1),(4,2)\}$

7. $\{(1,2,-1),(2,4,-2)\}$

8. $\{(1,2,4)\}$

9. $\{(0,0,0)\}$

10. $\{(0,0,0),(1,1,1)\}$

11. $\{(1,2,3),(2,1,1)\}$

12. $\{(0,1,0),(2,3,1),(4,1,1),(5,5,2)\}$

In Exercises 13–15, use Definition 4.5 to set up a linear system for the scalars c_1, c_2, and c_3 to determine whether the given set is linearly dependent in $\mathbb{R}^3$. When nontrivial solutions exist, find one and check it.

13. $\{(1,2,-2),(2,1,1),(3,0,1)\}$

14. $\{(1,-1,4),(1,0,6),(3,-5,8)\}$

15. $\{(2,1,0),(1,-2,2),(4,7,-4)\}$

In Exercises 16–18, determine whether the given set is linearly independent in $\mathbb{R}^3$ by forming a matrix A with the given vectors as rows and reducing A to row echelon form.

16. $\{(1,1,0),(1,0,1),(0,1,1)\}$

17. $\{(1,3,-2),(1,5,3),(2,4,-9)\}$

18. $\{(3,2,1),(1,1,1),(-2,-3,1)\}$

In Exercises 19–21, determine whether the given set is linearly independent in $\mathbb{R}^3$ by forming a matrix A with the given vectors as rows, and evaluating the determinant of A.

19. $\{(1,2,-3),(2,3,-1),(2,1,9)\}$

20. $\{(1,3,1),(2,2,3),(4,2,-1)\}$

21. $\{(3,1,-1),(2,1,0),(2,2,-1)\}$

In Exercises 22–28, find the value(s) of a for which the given set is linearly dependent in $\mathbb{R}^2$ or $\mathbb{R}^3$.

22. $\{(1,2),(a,4)\}$

23. $\{(a,9),(4,a)\}$

24. $\{(a-1,2),(3,a-2)\}$

25. $\{(6,a,-4),(-3,-1,2)\}$

26. $\{(1,2,1),(1,1,a),(2,-1,7)\}$

27. $\{(1,-1,0),(0,1,a),(0,4-a,4)\}$

28. $\{(a-2,3,1),(0,a+1,1),(-3,0,a-3)\}$

In Exercises 29–31, S is a set of three vectors in $\mathbb{R}^3$.

29. If S spans $\mathbb{R}^3$, what may we conclude about linear dependence or independence of S?

30. If S is linearly independent, find the span of S.

31. If the vectors in S are coplanar, what may we conclude about linear dependence or independence of S?

In Exercises 32–34, determine whether the given set is linearly independent in $\mathbb{R}^4$.

32. $\{(2,1,3,1),(1,2,-1,0),(3,0,1,2)\}$

33. $\{(1,2,-1,0),(1,3,-3,1),(3,7,-4,4),(2,3,2,5)\}$

34. $\{(1,0,1,1),(1,1,3,1),(1,-1,0,0),(2,-1,2,1)\}$

In Exercises 35–41, determine whether the given set is linearly independent in the vector space V of all polynomials of degree two or less.

35. $\{x^2\}$ **36.** $\{1,x\}$ **37.** $\{1,x,x^2\}$ **38.** $\{0\}$

39. $\{x,2x-1,1\}$ **40.** $\{2x^2,3x^2\}$

41. $\{1,2x+3,x^2+2x+1,3x^2-2x\}$

42. Let $S=\{\mathbf{x}_1,\mathbf{x}_2\}$ be a linearly dependent set in a vector space V. Show that for any $\mathbf{x}_3$ in V, the set $T=\{\mathbf{x}_1,\mathbf{x}_2,\mathbf{x}_3\}$ is also linearly dependent.

43. Generalize the result of Exercise 42 in terms of the number of elements in S and in terms of the relationship between S and T.

44. Show that if $S=\{\mathbf{x}_1,\mathbf{x}_2,\mathbf{x}_3\}$ is linearly independent in a vector space V, then $T=\{\mathbf{x}_1,\mathbf{x}_2\}$ is also linearly independent.

45. Generalize the result of Exercise 44, in terms of the number of elements in S and in terms of the relationship between S and T.

In Exercises 46–52, if

$$f(x)=\cos^2 x, \quad g(x)=\sin^2 x, \quad h(x)=1, \quad j(x)=\cos 2x, \quad \text{and} \quad k(x)=0,$$

determine whether the given set is linearly independent in the vector space V of all real functions on $\mathbb{R}$.

46. $\{f,g\}$ **47.** $\{f,h\}$ **48.** $\{f,g,h\}$

49. $\{f,k\}$ **50.** $\{g\}$ **51.** $\{f,g,j\}$

In Exercises 52–55, $S=\{(1,4,0),(2,-1,1),(1,-5,1)\}$ in $\mathbb{R}^3$.

52. Show that S is linearly dependent.

53. Find the equation of the plane spanned by S.

54. Show that $(5,2,2)$ lies in the span of S.

55. Find two distinct expressions of $(5,2,2)$ as a linear combination of the vectors in S.

56. Prove that if a set $\{\mathbf{x}_1,\mathbf{x}_2\}$ is linearly independent in a vector space V, then the set $\{\mathbf{x}_1+\mathbf{x}_2,\mathbf{x}_1-\mathbf{x}_2\}$ is also linearly independent.

57. Show that if $\{\mathbf{x}_1, \mathbf{x}_2, \mathbf{x}_3\}$ is linearly independent in a vector space V, then $\{\mathbf{x}_1 + \mathbf{x}_2 + \mathbf{x}_3, 2\mathbf{x}_1 + 2\mathbf{x}_3, \mathbf{x}_1 + 3\mathbf{x}_2\}$ is also linearly independent.

58. Let $\mathbf{x}_1 = (1, 1, -1, 0)$, $\mathbf{x}_2 = (1, 2, 0, 2)$, $\mathbf{x}_3 = (2, 4, 1, 2)$, and $\mathbf{x}_4 = (3, 2, -6, 2)$ in $\mathbb{R}^4$, let $S = \{\mathbf{x}_1, \mathbf{x}_2, \mathbf{x}_3, \mathbf{x}_4\}$, and let A be the matrix with rows $\mathbf{x}_1$, $\mathbf{x}_2$, $\mathbf{x}_3$, and $\mathbf{x}_4$, in that order. Reduce A to row echelon form to show that S is linearly dependent, recording the elementary row operations. Then use this record to find scalars c_1, c_2, c_3, and c_4, not all zero, such that

$$c_1\mathbf{x}_1 + c_2\mathbf{x}_2 + c_3\mathbf{x}_3 + c_4\mathbf{x}_4 = \mathbf{0}.$$

In Exercises 59–61, determine whether the given set is linearly independent in the vector space of all 2 × 2 real matrices.

59. $\begin{bmatrix} 1 & 0 \\ 0 & 0 \end{bmatrix}, \begin{bmatrix} 1 & 1 \\ 0 & 0 \end{bmatrix}, \begin{bmatrix} 1 & 1 \\ 1 & 0 \end{bmatrix}, \begin{bmatrix} 1 & 1 \\ 1 & 1 \end{bmatrix}$

60. $\begin{bmatrix} 1 & -1 \\ 0 & 1 \end{bmatrix}, \begin{bmatrix} 1 & 0 \\ 2 & 2 \end{bmatrix}, \begin{bmatrix} 3 & -1 \\ 5 & 4 \end{bmatrix}, \begin{bmatrix} 2 & -4 \\ -7 & 3 \end{bmatrix}$

61. $\begin{bmatrix} 1 & 2 \\ 0 & 1 \end{bmatrix}, \begin{bmatrix} 1 & 3 \\ 0 & 0 \end{bmatrix}, \begin{bmatrix} 0 & 1 \\ 1 & 0 \end{bmatrix}, \begin{bmatrix} 1 & 1 \\ -1 & 2 \end{bmatrix}, \begin{bmatrix} 0 & 0 \\ 1 & 4 \end{bmatrix}$

62. For S, a set of n vectors in $\mathbb{R}^n$, let A be a matrix with the vectors in S as rows, in any order. Show that S is linearly independent if and only if A is invertible.

63. For S, a set of m vectors in $\mathbb{R}^n$, let A be the matrix with the vectors in S as rows, in the given order. Let B be the coefficient matrix of the linear system for the scalars $c_1, \ldots, c_m$ in Definition 4.5. How are A and B related?

64. Prove that any set containing the identity $\mathbf{0}$ in a vector space is linearly dependent.

65. Prove that any set consisting of a single nonzero vector in a vector space is linearly independent.

66. Use the lemma of this section and the relation between linear combinations of row vectors and elementary row operations to justify the linear independence test in $\mathbb{R}^n$.

4.4 BASIS AND DIMENSION

In describing a vector space as the span of a set, we avoid algebraic redundancies by using a linearly independent spanning set. In this section, we shall find that for any given vector space, all linearly independent spanning sets have the same number of elements. This property leads to the important concept of the *dimension* of a vector space.

A Basis for a Vector Space

We have a name for spanning sets with no algebraic redundancies.

DEFINITION 4.6

Let V be a vector space. A set of vectors B is called a **basis for V** if:

i. B spans V; and
ii. B is linearly independent.

We emphasize that if B is a basis for V, then every vector in B must lie in V, since B is contained in its own span.

Example 1

Since the 2×2 identity matrix has row rank two, the set

$$B = \{(1,0),(0,1)\}$$

is linearly independent in $\mathbb{R}^2$ by the test on page 207. In addition, B spans $\mathbb{R}^2$ since for any (b_1, b_2) in $\mathbb{R}^2$,

$$(b_1, b_2) = b_1(1,0) + b_2(0,1).$$

This basis is called the **standard basis for** $\mathbb{R}^2$ and is shown in Fig. 4.6a. The standard basis vectors are often denoted by **i** and **j** or by $\mathbf{e}_1$ and $\mathbf{e}_2$.

Similarly, each of the sets

$$\{(1,1),(0,1)\}, \qquad \{(1,1),(1,-1)\}, \qquad \text{and} \qquad \{(4,1),(2,3)\}$$

is a basis for $\mathbb{R}^2$, since each is linearly independent and spans $\mathbb{R}^2$. For instance, the matrix with row vectors $(1,1)$ and $(0,1)$ has row rank 2, and for any (b_1, b_2) in

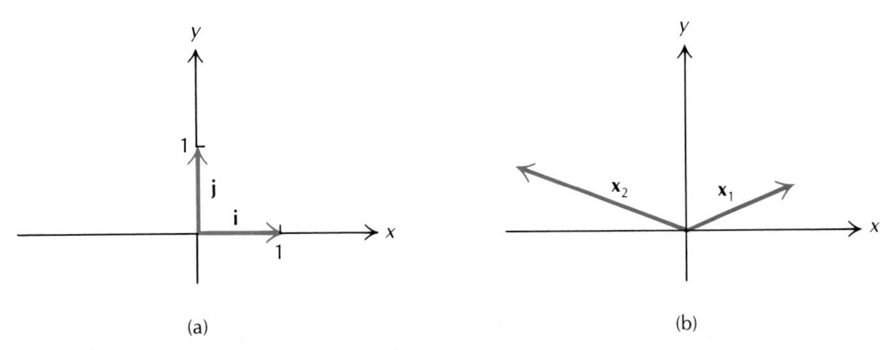

Figure 4.6 (a) The standard basis for $\mathbb{R}^2$. (b) Any linearly independent set of two vectors is a basis for $\mathbb{R}^2$.

$\mathbb{R}^2$, we have

$$(b_1, b_2) = b_1(1,1) + (b_2 - b_1)(0,1).$$

Figure 4.6b shows an arbitrary basis in $\mathbb{R}^2$. ☐

Example 2

We may readily verify (as in Example 1) that the set

$$B = \{(1,0,0), (0,1,0), (0,0,1)\}$$

is a basis for $\mathbb{R}^3$. This basis is called the **standard basis for** $\mathbb{R}^3$, and is shown in Fig. 4.7a. The vectors in B are often denoted by **i**, **j** and **k**, or by $\mathbf{e}_1$, $\mathbf{e}_2$, and $\mathbf{e}_3$.
Similarly, each of the sets

$$\{(1,1,1), (0,1,1), (0,0,1)\} \qquad \text{and} \qquad \{(1,2,-1), (2,2,1), (1,1,3)\}$$

is a basis for $\mathbb{R}^3$, since each is linearly independent and spans $\mathbb{R}^3$. For instance, the matrix with row vectors $(1,1,1)$, $(0,1,1)$, and $(0,0,1)$ has row rank three, and any **b** in $\mathbb{R}^3$ may be expressed as a linear combination of these vectors. Figure 4.7b shows an arbitrary basis of $\mathbb{R}^3$. ☐

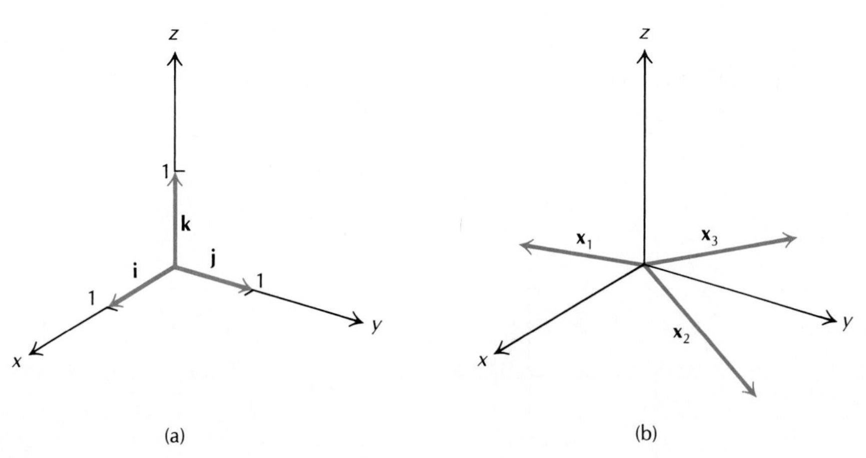

Figure 4.7 (a) The standard basis for $\mathbb{R}^3$. (b) Any linearly independent set of three vectors forms a basis for $\mathbb{R}^3$.

Example 3

Find a basis for the solution space of the homogeneous system

$$\begin{aligned} x_1 + x_2 - x_3 &= 0 \\ x_1 + 2x_2 + x_3 + x_4 &= 0 \\ 3x_1 + 5x_2 + x_3 + 2x_4 &= 0 \\ 2x_1 + x_2 - 4x_3 - x_4 &= 0. \end{aligned}$$

Solution

In other words, if A is the coefficient matrix of the given system, the problem is to find the nullspace of A. This system can be solved by Gaussian elimination:

$$\begin{bmatrix} 1 & 1 & -1 & 0 \\ 1 & 2 & 1 & 1 \\ 3 & 5 & 1 & 2 \\ 2 & 1 & -4 & -1 \end{bmatrix} \rightarrow \begin{bmatrix} 1 & 1 & -1 & 0 \\ 0 & 1 & 2 & 1 \\ 0 & 2 & 4 & 2 \\ 0 & -1 & -2 & -1 \end{bmatrix} \rightarrow \begin{bmatrix} 1 & 1 & -1 & 0 \\ 0 & 1 & 2 & 1 \\ 0 & 0 & 0 & 0 \\ 0 & 0 & 0 & 0 \end{bmatrix}.$$

Assign x_3 and x_4 arbitrarily and solve the system represented by the final matrix to obtain

$$x_2 = -2x_3 - x_4, \qquad x_1 = -x_2 + x_3 = 3x_3 + x_4.$$

Thus the solution consists of all vectors in $\mathbb{R}^4$ of the form

$$(3x_3 + x_4, -2x_3 - x_4, x_3, x_4) = x_3(3, -2, 1, 0) + x_4(1, -1, 0, 1).$$

(We may find the vectors on the right by inspection or by evaluating the general solution with $x_3 = 1$ and $x_4 = 0$ and with $x_3 = 0$ and $x_4 = 1$.) This equality shows that the set

$$B = \{(3, -2, 1, 0), (1, -1, 0, 1)\}$$

spans the nullspace of A, since any solution of the given linear system may be expressed as a linear combination of the vectors in B. Since B is also linearly independent (why?), B is a basis for the solution set of the given system. □

Example 4

Find a basis for the subspace of $\mathbb{R}^4$ spanned by the set

$$S = \{(1, -1, 0, 2), (1, 0, 1, 1), (2, -1, 2, 5), (2, 1, 5, 5), (3, -5, -5, 2)\}.$$

Solution

Form a matrix A with the vectors from S as rows, and bring it to row echelon form:

$$\begin{bmatrix} 1 & -1 & 0 & 2 \\ 1 & 0 & 1 & 1 \\ 2 & -1 & 2 & 5 \\ 2 & 1 & 5 & 5 \\ 3 & -5 & -5 & 2 \end{bmatrix} \rightarrow \begin{bmatrix} 1 & -1 & 0 & 2 \\ 0 & 1 & 1 & -1 \\ 0 & 1 & 2 & 1 \\ 0 & 3 & 5 & 1 \\ 0 & -2 & -5 & -4 \end{bmatrix} \rightarrow \begin{bmatrix} 1 & -1 & 0 & 2 \\ 0 & 1 & 1 & -1 \\ 0 & 0 & 1 & 2 \\ 0 & 0 & 2 & 4 \\ 0 & 0 & -3 & -6 \end{bmatrix}$$

$$\rightarrow \begin{bmatrix} 1 & -1 & 0 & 2 \\ 0 & 1 & 1 & -1 \\ 0 & 0 & 1 & 2 \\ 0 & 0 & 0 & 0 \\ 0 & 0 & 0 & 0 \end{bmatrix}.$$

Since each row vectors of the final matrix U is expressible as a linear combination of the row vectors of A (according to the discussion on page 210) the span of the set

$$B = \{(1, -1, 0, 2), (0, 1, 1, -1), (0, 0, 1, 2)\}$$

is contained in the span of S. But since we may recover A from U by elementary row operations, the span of S is contained in the span of B. Since B is linearly independent (why?), B is a basis for the span of S. □

We may also see that S and B span the same space in Example 4 as follows. If

$$S = \{\mathbf{x}_1, \mathbf{x}_2, \mathbf{x}_3, \mathbf{x}_4, \mathbf{x}_5\} \qquad \text{and} \qquad B = \{\mathbf{y}_1, \mathbf{y}_2, \mathbf{y}_3\},$$

then (Exercises 46–47)

$$\mathbf{x}_1 = \mathbf{y}_1, \qquad \mathbf{x}_2 = \mathbf{y}_1 + \mathbf{y}_2, \qquad \mathbf{x}_3 = 2\mathbf{y}_1 + \mathbf{y}_2 + \mathbf{y}_3,$$

$$\mathbf{x}_4 = 2\mathbf{y}_1 + 3\mathbf{y}_2 + 2\mathbf{y}_3, \qquad \text{and} \qquad \mathbf{x}_5 = 3\mathbf{y}_1 - 2\mathbf{y}_2 - 3\mathbf{y}_3,$$

and

$$\mathbf{y}_1 = \mathbf{x}_1, \qquad \mathbf{y}_2 = -\mathbf{x}_1 + \mathbf{x}_2, \qquad \text{and} \qquad \mathbf{y}_3 = -\mathbf{x}_1 - \mathbf{x}_2 + \mathbf{x}_3.$$

The first of these relations shows that the span of B contains the span of S, and the second shows that S contains the span of B.

Example 5

Let V denote the space of all polynomials of degree 2 or less (Example 5, page 177), and define p_0, p_1, and p_2 by

$$p_0 = 1, \qquad p_1 = x, \qquad \text{and} \qquad p_2 = x^2.$$

If

$$B = \{p_0, p_1, p_2\},$$

then the span of B coincides with V since every p in V is expressible in the form

$$p = a_0p_0 + a_1p_1 + a_2p_2.$$

In addition, B is linearly independent since the relation

$$c_0p_0 + c_1p_1 + c_2p_2 = 0$$

is equivalent to

$$c_0 + c_1x + c_2x^2 = 0 \cdot 1 + 0x + 0x^2.$$

Equating coefficients, we have $c_0 = c_1 = c_2 = 0$. By Definition 4.6, B is a basis for V. ☐

Dimension

As Examples 1 and 2 illustrate, a basis for $\mathbb{R}^n$ is a linearly independent set of n vectors. In fact, in every basis for a given vector space, V has the same number of elements. Before proving this theorem, we show that a linearly independent set in V cannot be larger than a set that spans V.

Lemma Let V be a vector space, let T be a linearly independent set in V, and suppose S spans V. If S contains m vectors and T contains n vectors, then $n \leq m$.

Proof Let $S = \{\mathbf{x}_1, \ldots, \mathbf{x}_m\}$ and $T = \{\mathbf{y}_1, \ldots, \mathbf{y}_n\}$. Since S spans V, each $\mathbf{y}_j$ in T is expressible as a linear combination of the vectors in S. That is, for $j = 1, \ldots, n$,

$$\mathbf{y}_j = a_{1j}\mathbf{x}_1 + a_{2j}\mathbf{x}_2 + \cdots + a_{mj}\mathbf{x}_m$$

for some set of scalars. We show that the assumption $n > m$ leads to a contradiction. If $n > m$, then there exist constants $c_1, \ldots, c_n$, not all zero, such that

$$\begin{array}{ccc} a_{11}c_1 + \cdots + a_{1n}c_n & = & 0 \\ \vdots \quad\quad \vdots & & \vdots \\ a_{m1}c_1 + \cdots + a_{mn}c_n & = & 0. \end{array}$$

This is because a homogeneous linear system with fewer equations than unknowns must have a nontrivial solution. Thus

$$(a_{11}c_1 + \cdots + a_{1n}c_n)\mathbf{x}_1 + \cdots + (a_{m1}c_1 + \cdots + a_{mn}c_n)\mathbf{x}_m = \mathbf{0}.$$

Expand the lefthand side and rearrange the terms to obtain

$$c_1(a_{11}\mathbf{x}_1 + \cdots + a_{m1}\mathbf{x}_m) + \cdots + c_n(a_{1n}\mathbf{x}_1 + \cdots + a_{mn}\mathbf{x}_m) = \mathbf{0}$$

or

$$c_1\mathbf{y}_1 + \cdots + c_n\mathbf{y}_n = \mathbf{0}.$$

This relation contradicts linear independence of T. Since the assumption $n > m$ leads to a contradiction, $n \leq m$. □

THEOREM 4.5

If one basis for a vector space V consists of n vectors, then so does every basis for V.

Proof Let B and B' be bases for V with m and n vectors, respectively. By the previous lemma with B as S and B' as T, $n \leq m$. By the same lemma with B' as S and B as T, we also have $m \leq n$. Hence $n = m$. □

Theorem 4.5 allows the formulation of an especially important concept for vector spaces.

DEFINITION 4.7

The **dimension** of a vector space V is given by

$$\dim(V) = \begin{cases} 0 & \text{if } V = \{\mathbf{0}\} \\ n & \text{if every basis of } V \text{ contains } n \text{ vectors.} \end{cases}$$

Example 6

For every positive integer n, $\dim(\mathbb{R}^n) = n$ since the row vectors of the $n \times n$ identity matrix, $\mathbf{e}_1, \ldots, \mathbf{e}_n$, form a basis for $\mathbb{R}^n$. □

Example 7

The solution space of the linear system in Example 3 has dimension 2 since it has a basis consisting of 2 vectors in $\mathbb{R}^4$. □

Example 8

The dimension of the space V of polynomials of degree n or less is $n + 1$, since the polynomials $p_0, p_1, \ldots, p_n$ defined by

$$p_0 = 1, \qquad \text{and} \qquad p_k = x^k \qquad \text{for } k = 1, \ldots, n$$

form a basis for V. (Cf. Example 5.) □

Properties of Bases and Dimension

We note several results concerning bases and the dimension of a vector space, assigning their proofs in Exercises 42–45.

THEOREM 4.6

A set $S = \{\mathbf{x}_1, \ldots, \mathbf{x}_n\}$ in a vector space V is a basis for V if and only if every $\mathbf{x}$ in V has a unique expression of the form

$$\mathbf{x} = c_1\mathbf{x}_1 + \cdots + c_n\mathbf{x}_n.$$

Example 9

The set

$$B = \{(1,1), (1,2)\}$$

is a basis for $\mathbb{R}^2$. For the particular vector $(1, -1)$,

$$3(1,1) - 2(1,2) = (1, -1).$$

The coefficients 3 and 2 are unique, for if

$$c_1(1,1) + c_2(1,2) = (1, -1)$$

in $\mathbb{R}^2$, then equating the expressions for $(1, -1)$ yields

$$c_1(1,1) + c_2(1,2) = 3(1,1) - 2(1,2).$$

This equation can be rearranged to

$$(c_1 - 3)(1,1) + (c_2 + 3)(1,2) = (0,0),$$

from which $c_1 = 3$ and $c_2 = -2$ by linear independence of B. □

THEOREM 4.7

If B is a basis for a vector space V, then:

a. B is not properly contained in any linearly independent set in V;
b. no proper subset of B spans V.

Example 10

We may readily verify that the set

$$B = \{(2,1,3), (-4,2,1), (3,0,2)\}$$

is a basis for $\mathbb{R}^3$. If another vector is adjoined to B, for instance if we form the set

$$\{(2,1,3), (-4,2,1), (3,0,2), (3,-2,4)\},$$

the result is a linearly dependent set of four vectors in $\mathbb{R}^3$. If a vector is removed from B, the result is a set that does not span $\mathbb{R}^3$. For instance, the span of

$$\{(-4,2,1), (3,0,2)\}$$

does not contain $(1,0,0)$. □

THEOREM 4.8

Let V be a vector space with a finite basis and let S be a set of vectors in V.

a. If S spans V, then S contains a basis for V.
b. If S is linearly independent in V, then S is contained in a basis B for V.

THEOREM 4.9

Let V be a vector space of dimension n and let S be a set of n vectors in V. If S spans V or if S is linearly independent in V, then S is a basis for V.

Corollary Any linearly independent set containing n vectors in $\mathbb{R}^n$ is a basis for $\mathbb{R}^n$.

Example 11

Since the set $S = \{(1,2,-4),(1,-2,8)\}$ is linearly independent, $\mathbb{R}^3$ has a basis B containing S by Theorem 4.8(b). One way to find such a B is to adjoin $(1,0,0)$ to S to form the set

$$S' = \{(1,2,-4),(1,-2,8),(1,0,0)\}.$$

If S' is linearly independent, then S' is a basis for $\mathbb{R}^3$ by Theorem 4.9. Otherwise, the process may be repeated with $(0,1,0)$ and, if necessary, again with $(0,0,1)$. The procedure must yield a basis for $\mathbb{R}^3$ for otherwise the span of S would include the standard basis, hence all of $\mathbb{R}^3$. In this illustration, the set

$$S' = \{(1,2,-4),(1,-2,8),(1,0,0)\}$$

is linearly dependent, but the set

$$B = \{(1,2,-4),(1,-2,8),(0,1,0)\}$$

is a basis for $\mathbb{R}^3$, as required. □

Part (a) of Theorem 4.8 is related to Example 4, in which a basis for the span of a linearly dependent set S was found. The basis contained vectors not in S. By Theorem 4.7, there is also a basis among the vectors in S.

Example 12

In $\mathbb{R}^4$, find a basis for V, the span of the set

$$S = \{(1,-1,0,2),(1,0,1,1),(2,-1,2,5),(2,1,5,5),(3,-5,-5,2)\}$$

consisting of vectors in S.

Solution

Denote the vectors in S by $\mathbf{x}_1,\ldots,\mathbf{x}_5$ and solve the vector equation

$$(*) \qquad c_1\mathbf{x}_1 + \cdots + c_5\mathbf{x}_5 = \mathbf{0}$$

for the unknown coefficients $c_1,\ldots,c_5$. This equation is equivalent to the homogeneous linear system

$$\begin{aligned} c_1 + c_2 + 2c_3 + 2c_4 + 3c_4 &= 0 \\ -c_1 \qquad - c_3 + c_4 - 5c_5 &= 0 \\ c_2 + 2c_3 + 5c_4 - 5c_5 &= 0 \\ 2c_1 + c_2 + 5c_3 + 5c_4 + 2c_5 &= 0. \end{aligned}$$

If Gaussian elimination is carried out, the resulting row echelon matrix is

$$\begin{bmatrix} 1 & 1 & 2 & 2 & 3 \\ 0 & 1 & 1 & 3 & -2 \\ 0 & 0 & 1 & 2 & -3 \\ 0 & 0 & 0 & 0 & 0 \end{bmatrix}.$$

If $c_4 = 1$ and $c_5 = 0$, then solving c_1, c_2, and c_3 gives $(3, -1, -2, 1, 0)$. Letting $c_4 = 0$ and $c_5 = 1$ gives the solution $(-8, -1, 3, 0, 1)$. Substituting these solutions into Eq. $(*)$ yields

$$3\mathbf{x}_1 - \mathbf{x}_2 - 2\mathbf{x}_3 + \mathbf{x}_4 = \mathbf{0} \quad \text{and} \quad -8\mathbf{x}_1 - \mathbf{x}_2 + 3\mathbf{x}_3 + \mathbf{x}_5 = \mathbf{0},$$

from which

$$\mathbf{x}_4 = -3\mathbf{x}_1 + \mathbf{x}_2 + 2\mathbf{x}_3 \quad \text{and} \quad \mathbf{x}_5 = 8\mathbf{x}_1 + \mathbf{x}_2 - 3\mathbf{x}_3.$$

Since $\mathbf{x}_1$, $\mathbf{x}_2$, and $\mathbf{x}_3$ are in V, these equations show that the set $B = \{\mathbf{x}_1, \mathbf{x}_2, \mathbf{x}_3\}$ spans V. Since V has a basis containing three vectors (this was established in Example 4), B is a basis for V by Theorem 4.9. □

Example 13

Let V be the space of all real polynomials of any degree. For each positive integer k, define a polynomial p_k by:

$$p_k = x^k.$$

Then for each positive integer n, the set $S_n = \{p_1, \ldots, p_n\}$ is linearly independent. (See Example 5.) Thus V *does not have finite dimension,* for suppose $\dim(V) = n$ for some n. Then by Theorem 4.9, S_n would be a basis for V since S_n is linearly independent and has n vectors. But by Theorem 4.7, S_n cannot be a basis for V since S_n is properly contained in S_{n+1} and S_{n+1} is also linearly independent. □

Infinite-dimensional spaces, as illustrated in Example 13, are very important and useful in mathematics. Further study of these spaces is beyond the scope of this book.

REVIEW CHECKLIST

1. Formulate a definition of a basis of a vector space.
2. Identify the standard basis in $\mathbb{R}^n$ for any n.
3. Characterize a basis for $\mathbb{R}^n$ as any linearly independent set of n vectors.
4. Find a basis for the solution set of a homogeneous linear system.
5. Find a basis for the span of a given set in $\mathbb{R}^n$ using matrix methods.
6. Formulate a statement of Theorem 4.5, which provides for the concept of dimension.
7. Define the dimension of a vector space.
8. Characterize a basis B of a vector space V in terms of unique representation of any vector in V as a linear combination of vectors in B (Theorem 4.6).
9. State the maximum and minimum properties of a basis of a vector space (Theorem 4.7).
10. Describe in your own words the principles embodied in Theorems 4.8 and 4.9.
11. Given a linearly independent set S in $\mathbb{R}^n$, find a basis for $\mathbb{R}^n$ containing S.

EXERCISES

In Exercises 1–3, determine whether the given set forms a basis for $\mathbb{R}^2$.

1. $\{(1,1),(1,2)\}$ **2.** $\{(2,4),(3,6)\}$ **3.** $\{(1,1),(1,0),(0,1)\}$

In Exercises 4–6, determine whether the given set forms a basis for $\mathbb{R}^3$.

4. $\{(2,6,-2),(1,2,3),(-3,9,3)\}$ **5.** $\{(1,1,1),(0,1,1),(0,0,1)\}$
6. $\{(1,1,1),(1,1,2),(1,2,2),(0,0,0)\}$

In Exercises 7–11, V is a vector space with a finite basis B. Answer T if the statement is true, F if false.

7. Every basis for a subspace of V is linearly independent.
8. Every linearly independent set in V is a basis for its span.
9. Every subset of B is a basis for a subspace of V.
10. Every subspace of V has a basis of vectors from B.
11. Every linearly independent set in V is contained in a basis for V.
12. Find a basis and the dimension of the plane in $\mathbb{R}^3$ determined by the equation

$$x + 2y - 3z = 0.$$

13. Find a basis and the dimension of the subspace in $\mathbb{R}^4$ determined by the equation

$$x_1 - x_2 + 4x_3 + 2x_4 = 0.$$

In Exercises 14–18, find the nullspace of the given matrix.

14. $\begin{bmatrix} 3 & -6 & 3 \\ -4 & 8 & -4 \\ 1 & -2 & 1 \end{bmatrix}$ **15.** $\begin{bmatrix} 1 & 2 & -2 \\ 3 & 7 & -4 \\ 2 & 1 & -10 \end{bmatrix}$ **16.** $\begin{bmatrix} 1 & -2 & 4 \\ 2 & -1 & 2 \\ 3 & 1 & 3 \end{bmatrix}$

17. $\begin{bmatrix} 1 & 2 & 1 & -1 \\ 1 & 3 & 1 & -3 \\ -1 & 0 & 1 & -3 \\ 2 & 3 & 1 & 0 \end{bmatrix}$ **18.** $\begin{bmatrix} 1 & 2 & 1 & -1 \\ 1 & 3 & 1 & -3 \\ 3 & 4 & 3 & 1 \\ 2 & 3 & 2 & 0 \end{bmatrix}$

In Exercises 19–24, find a basis and the dimension of the span of the given set in $\mathbb{R}^3$ or $\mathbb{R}^4$, as appropriate.

19. $\{(1,-1,2),(2,1,3),(-1,5,0)\}$
20. $\{(2,1,3),(1,2,-1),(1,0,2)\}$
21. $\{(2,4,-1),(3,6,-2),(-1,2,1/2)\}$
22. $\{(3,1,-1),(1,4,2),(5,-2,-4)\}$
23. $\{(1,0,1,-2),(1,1,3,-2),(2,1,5,-1),(1,-1,1,4)\}$
24. $\{(1,2,0,1),(1,3,1,1),(2,5,2,-1),(3,5,1,-3)\}$
25. Find a "standard basis" for the vector space of all real 2×2 matrices. What is the dimension of this space?

In Exercises 26–27, determine whether the given set forms a basis for the vector space of Exercise 25.

26. $\begin{bmatrix} 1 & 0 \\ 1 & 0 \end{bmatrix}, \begin{bmatrix} 2 & 1 \\ 1 & 0 \end{bmatrix}, \begin{bmatrix} 1 & 2 \\ 1 & -1 \end{bmatrix}, \begin{bmatrix} 3 & 1 \\ 1 & 1 \end{bmatrix}$

27. $\begin{bmatrix} 1 & 0 \\ 1 & 0 \end{bmatrix}, \begin{bmatrix} 2 & 1 \\ 1 & 0 \end{bmatrix}, \begin{bmatrix} 1 & 0 \\ 1 & -1 \end{bmatrix}, \begin{bmatrix} 3 & 1 \\ 0 & 1 \end{bmatrix}$

In Exercises 28–30, determine whether the given set is a basis for the vector space of real polynomials of degree 2 or less.

28. $\{1, 1+x, 1+x+x^2\}$ **29.** $\{1-x, 1+x-x^2, 2-x^2\}$
30. $\{1+x+x^2, 1-2x-2x^2, 2-x+x^2\}$
31. Find a basis and the dimension of the solution space of any homogeneous 4×6 linear system whose coefficient matrix is equivalent to the row echelon matrix

$$\begin{bmatrix} 1 & 2 & 1 & 3 & -4 & -2 \\ 0 & 0 & 1 & 0 & 1 & 1 \\ 0 & 0 & 0 & 0 & 1 & 3 \\ 0 & 0 & 0 & 0 & 0 & 0 \end{bmatrix}.$$

In Exercises 32–34, let

$$S = \{f_1, f_2, f_3, f_4\},$$

where

$$f_1(x) = 1, \quad f_2(x) = \cos^2 x, \quad f_3(x) = \sin^2 x, \quad \text{and} \quad f_4(x) = \cos 2x,$$

in the space of all real functions on $\mathbb{R}$. Determine whether the given set is a basis for the span of S.

32. $\{f_1, f_2, f_3\}$ **33.** $\{f_1, f_2\}$ **34.** $\{f_1, f_4\}$

35. In $\mathbb{R}^5$, find a basis for the span of the set

$$S = \{(1,2,0,-1,1), (2,5,2,-2,0), (1,1,-2,0,5), (-2,-3,2,3,-1), (1,5,6,1,-1)\}$$

consisting of vectors in S. What is the dimension of span (S)?

In Exercises 36–39, determine whether there exists a basis for $\mathbb{R}^4$ containing the given set and, if so, find one.

36. $\{(1,1,1,0), (1,0,1,0)\}$

37. $\{(1,-2,4,-1), (2,-4,8,-2)\}$

38. $\{(1,2,0,3), (1,1,1,2), (1,4,-2,5)\}$

39. $\{(1,3,1,3), (2,2,0,1), (3,1,-1,2)\}$

In Exercises 40–41, let

$$S = \{(2,1,-1), (1,2,0), (5,1,-3)\}$$

in $\mathbb{R}^3$.

40. Determine whether S is a basis for span (S).

41. Express the vector $(1,5,1)$ as a linear combination of the vectors in S in two ways (i.e., with distinct sets of coefficients).

42. Prove Theorem 4.6. In a vector space V, a set S is a basis for V if and only if every vector in V is uniquely expressible as a linear combination of the vectors in S.

43. Prove Theorem 4.7. A basis for a vector space V is maximal relative to linear independence and minimal relative to spanning V.

44. Prove Theorem 4.8. In a vector space V, every spanning set contains a basis, and every linearly independent set is contained in a basis.

45. Prove Theorem 4.9. In a vector space V of dimension n, every spanning set with n vectors must be linearly independent in V, and every linearly independent set with n vectors must span V.

46. Verify the relations showing that the span of S equals the span of B in the remarks immediately following Example 4, page 218.

47. Derive the relations cited in Exercise 46.

48. Let V be the vector space of all real valued functions with domain $\mathbb{R}$. Show that V does not have a finite basis.

4.5 ROW SPACE AND COLUMN SPACE

Studying the space spanned by the row vectors of a matrix and the space spanned by the column vectors enhances our understanding of linear systems. We find an important relation between the dimensions of these spaces and explore their connection with linear systems.

Definition and Basic Properties

We identify two important vector spaces associated with a given matrix.

DEFINITION 4.8

The **row space** of an $m \times n$ matrix A is the subspace of $\mathbb{R}^n$ spanned by the row vectors of A, and the **column space** is the subspace of $\mathbb{R}^m$ spanned by the column vectors.

Example 1

Describe the row space and the column space of the matrix

$$A = \begin{bmatrix} 1 & 2 & 0 \\ -1 & 1 & 2 \end{bmatrix}$$

geometrically.

Solution

By Definition 4.8, the row space of A is the subspace of $\mathbb{R}^3$ spanned by the set

$$S = \{(1,2,0), (-1,1,2)\}.$$

Apply the technique used in Example 14 on page 195. A vector $\mathbf{v} = (x, y, z)$ in $\mathbb{R}^3$ lies in the span of S if and only if there exist scalars c_1 and c_2 such that

$$c_1(1,2,0) + c_2(-1,1,2) = (x, y, z).$$

Thus $\mathbf{v}$ is in the span of S if and only if the system

$$\begin{aligned} c_1 - c_2 &= x \\ 2c_1 + c_2 &= y \\ 2c_2 &= z \end{aligned}$$

is consistent. The first pass of Gaussian elimination gives

$$\left[\begin{array}{rr|l} 1 & -1 & x \\ 2 & 1 & y \\ 0 & 2 & z \end{array}\right] \rightarrow \left[\begin{array}{rr|l} 1 & -1 & x \\ 0 & 3 & y-2x \\ 0 & 6 & 3z \end{array}\right] \rightarrow \left[\begin{array}{rr|l} 1 & -1 & x \\ 0 & 3 & y-2x \\ 0 & 0 & 3z-2y+4x \end{array}\right].$$

Thus the row space of A coincides with the plane

$$4x - 2y + 3z = 0.$$

By Definition 4.8, the column space of A is the subspace of $\mathbb{R}^2$ spanned by the set

$$\left\{\begin{bmatrix} 1 \\ -1 \end{bmatrix}, \begin{bmatrix} 2 \\ 1 \end{bmatrix}, \begin{bmatrix} 0 \\ 2 \end{bmatrix}\right\}.$$

Any two of these vectors form a linearly independent set, and the column space coincides with $\mathbb{R}^2$. □

By the linear independence test in Section 4.3, we may determine the dimension of the row space of A by counting the number of nonzero rows in any equivalent row echelon matrix U. This test is valid because each row vector in U is expressible as a linear combination of the row vectors in A and, conversely, each row vector in A may be expressed as a linear combination of the row vectors in U. The following result indicates why the test is so useful.

Lemma In a row echelon matrix, the nonzero row vectors form a linearly independent set.

We shall not prove this result in general, but instead illustrate the underlying idea by example.

Example 2

Use the definition of linear independence to show that the nonzero vectors in the row echelon matrix

$$A = \begin{bmatrix} 1 & -1 & 4 & 3 \\ 0 & 0 & 1 & 2 \\ 0 & 0 & 0 & 1 \\ 0 & 0 & 0 & 0 \end{bmatrix}$$

form a linearly independent set.

Solution

Suppose that a linear combination of the nonzero row vectors from A vanishes, i.e.,

$$c_1(1,-1,4,3) + c_2(0,0,1,2) + c_3(0,0,0,1) = (0,0,0,0).$$

Then c_1 must equal zero, since all entries below the first leading 1 are zero. Hence

$$c_2(0,0,1,2) + c_3(0,0,0,1) = (0,0,0,0).$$

Repeating the argument for c_1 gives $c_2 = 0$. Then

$$c_3(0,0,0,1) = (0,0,0,0),$$

from which $c_3 = 0$. Thus only the trivial linear combination of the nonzero row vectors of A vanishes, so these vectors form a linearly independent set. □

THEOREM 4.10

Let A be a nonzero matrix. Then the nonzero row vectors in any row echelon matrix U equivalent to A form a basis for the row space V of A. In addition, if r denotes the row rank of A, then $\dim(V) = r$.

Proof Let S denote the set of nonzero row vectors in U. Since the span of the set of all row vectors of A coincides with the span of the set of nonzero row vectors, S spans V. By the lemma, S is linearly independent, hence a basis for V by Definitions 4.6 and 4.8. Since r equals the number of vectors in S by Definition 1.7, we have $\dim(V) = r$ by Definition 4.7. □

Theorem 4.10 is closely related to the following result.

THEOREM 4.11

Let A be an $n \times n$ matrix, S the set of row vectors of A. Then S is linearly independent if and only if the row space of A coincides with $\mathbb{R}^n$.

Proof By Definition 4.8, the row space of A coincides with $\mathbb{R}^n$ if and only if the span of S coincides with $\mathbb{R}^n$. Since S contains n vectors, the desired result follows immediately from the corollary to Theorem 4.9. □

Row and Column Spaces and Matrix Multiplication

Let's have a fresh look at matrix multiplication from the perspective of row and column spaces.

Example 3

Interpret the matrix product

$$\begin{bmatrix} 1 & 2 & 1 \\ 2 & 1 & -2 \end{bmatrix} \begin{bmatrix} 3 & 1 & 0 & 1 \\ 1 & 1 & -1 & 1 \\ 2 & 1 & 4 & -2 \end{bmatrix} = \begin{bmatrix} 7 & 4 & 2 & 1 \\ 3 & 1 & -9 & 7 \end{bmatrix}$$

in terms of row space and column space.

Solution

Denote the given matrices from left to right by A, C, and B. Each row vector of B is expressible as a linear combination of the row vectors of C with coefficients constituting the corresponding row of A as follows:

$$\overbrace{(7,4,2,1)}^{\text{Rows of B}} = 1\overbrace{(3,1,0,1)} + 2\overbrace{(1,1,-1,1)}^{\text{Rows of C}} + 1\overbrace{(2,1,4,-2)},$$

$$(3,1,-9,7) = 2(3,1,0,1) + 1(1,1,-1,1) - 2(2,1,4,-2).$$

(Rows of A: the coefficients $2, 1, -2$.)

Similarly, each column vector of B is expressible as a linear combination of the column vectors of A with coefficients constituting the corresponding column of C:

(Columns of B; Columns of A)

$$\begin{bmatrix} 7 \\ 3 \end{bmatrix} = 3\begin{bmatrix} 1 \\ 2 \end{bmatrix} + 1\begin{bmatrix} 2 \\ 1 \end{bmatrix} + 2\begin{bmatrix} 1 \\ -2 \end{bmatrix}, \quad \begin{bmatrix} 4 \\ 1 \end{bmatrix} = 1\begin{bmatrix} 1 \\ 2 \end{bmatrix} + 1\begin{bmatrix} 2 \\ 1 \end{bmatrix} + 1\begin{bmatrix} 1 \\ -2 \end{bmatrix},$$

$$\begin{bmatrix} 2 \\ -9 \end{bmatrix} = 0\begin{bmatrix} 1 \\ 2 \end{bmatrix} - 1\begin{bmatrix} 2 \\ 1 \end{bmatrix} + 4\begin{bmatrix} 1 \\ -2 \end{bmatrix}, \quad \begin{bmatrix} 1 \\ 7 \end{bmatrix} = 1\begin{bmatrix} 1 \\ 2 \end{bmatrix} + 1\begin{bmatrix} 2 \\ 1 \end{bmatrix} - 2\begin{bmatrix} 1 \\ -2 \end{bmatrix}. \quad \square$$

(Columns of C: the coefficients.)

We generalize the concept illustrated in Example 3 without formal proof.

Row space interpretation of matrix multiplication If $B = AC$, then for each i, the ith row vector of B may be expressed as a linear combination of the row vectors of C with coefficients constituting the ith row of A.

Column space interpretation of matrix multiplication If $B = AC$, then for each j, the jth column vector of B may be expressed as a linear combination of the column vectors of A with coefficients constituting the jth column of C.

The next examples illustrate a converse for the row space interpretation and for the column space interpretation.

Example 4

Represent the relations

$$(-1,-1) = 2(1,2) + 3(-1,1) - 4(0,2)$$

$$(5,14) = 5(1,2) + 0(-1,1) + 2(0,2)$$

$$(4,11) = 3(1,2) - 1(-1,1) + 3(0,2)$$

by a matrix product whose row space interpretation consists of the given linear combinations.

Solution

Form B from the row vectors on the left, form C from those on the right, and form A with the given coefficients as rows. The required matrix product $B = AC$ is given by

$$\begin{bmatrix} -1 & -1 \\ 5 & 14 \\ 4 & 11 \end{bmatrix} = \begin{bmatrix} 2 & 3 & -4 \\ 5 & 0 & 2 \\ 3 & -1 & 3 \end{bmatrix} \begin{bmatrix} 1 & 2 \\ -1 & 1 \\ 0 & 2 \end{bmatrix}.$$

The row space interpretation of this product consists of the given linear combinations, as required. □

Example 5

Represent the relations

$$\begin{bmatrix} 3 \\ 2 \\ 5 \end{bmatrix} = 3\begin{bmatrix} 1 \\ 2 \\ -1 \end{bmatrix} - 4\begin{bmatrix} 0 \\ 1 \\ -2 \end{bmatrix}, \quad \begin{bmatrix} 2 \\ 5 \\ -4 \end{bmatrix} = 2\begin{bmatrix} 1 \\ 2 \\ -1 \end{bmatrix} + 1\begin{bmatrix} 0 \\ 1 \\ -2 \end{bmatrix}$$

by a matrix product whose column space interpretation consists of the given linear combinations.

Solution

Form B from the column vectors on the left, A from those on the right, and C with the coefficients as columns. The required matrix product $B = AC$ is given by

$$\begin{bmatrix} 3 & 2 \\ 2 & 5 \\ 5 & -4 \end{bmatrix} = \begin{bmatrix} 1 & 0 \\ 2 & 1 \\ -1 & -2 \end{bmatrix} \begin{bmatrix} 3 & 2 \\ -4 & 1 \end{bmatrix}.$$

The column space interpretation of this product consists of the given linear combinations, as required. □

We specialize the general relations illustrated by Examples 3–5 to a matrix product AC with $C = X$, a column vector, and indicate the relevance of this concept to linear systems.

Proposition Let $A = [a_{ij}]_{m \times n}$ and $X = [x_1, \ldots, x_n]^t$. Then the equation $AX = B$ is equivalent to the vector equation

$$x_1 \begin{bmatrix} a_{11} \\ \vdots \\ a_{m1} \end{bmatrix} + \cdots + x_n \begin{bmatrix} a_{1n} \\ \vdots \\ a_{mn} \end{bmatrix} = \begin{bmatrix} b_1 \\ \vdots \\ b_m \end{bmatrix}.$$

Proof Combine the expression on the left to obtain

$$\begin{bmatrix} a_{11}x_1 + \cdots + a_{1n}x_1 \\ \vdots \qquad\qquad \vdots \\ a_{11}x_1 + \cdots + a_{1n}x_1 \end{bmatrix},$$

which equals the product AX. □

Equality of Row and Column Space Dimensions

We next show that for any matrix A, the dimension of the row space equals the dimension of the column space.

DEFINITION 4.9

The **column rank** of a matrix A is the dimension of the column space of A.

Example 6

For the matrix

$$A = \begin{bmatrix} 1 & -2 & -2 & 0 \\ 2 & -3 & 0 & -1 \\ 1 & 2 & 3 & 1 \\ 0 & 2 & -3 & 3 \end{bmatrix},$$

find the row rank and the column rank.

Solution

The row rank of A is 3:

$$\begin{bmatrix} 1 & -2 & -2 & 0 \\ 2 & -3 & 0 & -1 \\ 1 & 2 & 3 & 1 \\ 0 & 2 & -3 & 3 \end{bmatrix} \to \begin{bmatrix} 1 & -2 & -2 & 0 \\ 0 & 1 & 4 & -1 \\ 0 & 4 & 5 & 1 \\ 0 & 2 & -3 & 3 \end{bmatrix} \to \begin{bmatrix} 1 & -2 & -2 & 0 \\ 0 & 1 & 4 & -1 \\ 0 & 0 & -11 & 5 \\ 0 & 0 & -11 & 5 \end{bmatrix}.$$

The column rank of A is also 3 and is found by the same procedure applied to A^t:

$$\begin{bmatrix} 1 & 2 & 1 & 0 \\ -2 & -3 & 2 & 2 \\ -2 & 0 & 3 & -3 \\ 0 & -1 & 1 & 3 \end{bmatrix} \to \begin{bmatrix} 1 & 2 & 1 & 0 \\ 0 & 1 & 4 & 2 \\ 0 & 4 & 5 & -3 \\ 0 & -1 & 1 & 3 \end{bmatrix} \to \begin{bmatrix} 1 & 2 & 1 & 0 \\ 0 & 1 & 4 & 2 \\ 0 & 0 & -11 & -11 \\ 0 & 0 & 5 & 5 \end{bmatrix}.$$

□

Example 6 illustrates the following general result.

THEOREM 4.12

For any matrix A, the row rank equals the column rank.

Proof Let E be a product of elementary matrices such that EA is in *reduced* row echelon form. Then the row rank of E equals the row rank of EA since elementary matrices preserve row rank. Moreover, equality of the row rank and the column rank of EA follows from the special form of EA, so we need only show that the column rank of A equals the column rank of EA. By Theorem 2.5, E is invertible, hence for any column vector X,

$$AX = 0 \quad \text{if and only if} \quad EAX = 0.$$

By the previous proposition, a linear combination of the column vectors of A equals zero if and only if the linear combination of the column vectors of EA with the same coefficients equals zero. Thus a set of column vectors from A is linearly independent if and only if the corresponding set from EA is, hence the column spaces of A and EA have the same dimension. □

In view of Theorem 4.12, we may formulate the following concept.

DEFINITION 4.10

The **rank** of a matrix is the common value of its row rank and its column rank.

We study the relationship of row and column spaces to linear systems further in Section 4.6, where we summarize conditions for existence and uniqueness of solutions to square linear systems.

REVIEW CHECKLIST

1. Define the row space and the column space of a matrix.
2. Find the row space and the column space of a given matrix.
3. Find a basis and the dimension of the row [column] space of a given matrix.
4. For a given matrix A, find a basis for the row [column] space consisting of row [column] vectors from A.
5. Identify the nonzero row vectors in a row echelon matrix as constituting a linearly independent set.
6. Formulate statements of the "row space interpretation" and the "column space interpretation" of a matrix product.
7. Apply the row space interpretation to express the rows of a given product matrix in terms of the rows of the factors. Do this also for columns.
8. Given a set of linear combinations of vectors from any $\mathbb{R}^n$, represent the set as a matrix product whose row (or column) space interpretation includes the given linear combinations.
9. Define the rank of a matrix.

EXERCISES

In Exercises 1–6, describe the row space of the given matrix geometrically, and find its dimension.

1. $\begin{bmatrix} 0 & 0 \\ 0 & 0 \end{bmatrix}$ **2.** $\begin{bmatrix} 1 & 1 \\ 1 & 1 \end{bmatrix}$ **3.** $\begin{bmatrix} 1 & 2 \\ 3 & 4 \end{bmatrix}$

4. $\begin{bmatrix} 1 & -3 & 3 \\ 2 & -1 & 2 \\ 3 & 0 & 5 \end{bmatrix}$ **5.** $\begin{bmatrix} 1 & -3 & 3 \\ 2 & -1 & 2 \\ -2 & -9 & 6 \end{bmatrix}$ **6.** $\begin{bmatrix} 3 & -6 & 9 \\ 2 & -4 & 6 \\ 1 & -2 & 3 \end{bmatrix}$

Exercises 7–12. For the matrices in Exercises 1–6, describe the column space geometrically, and find its dimension.

In Exercises 13–16, find a basis for the row space of the given matrix.

13. $\begin{bmatrix} 1 & 2 & 1 & -2 \\ 1 & 3 & 0 & -2 \\ 2 & 6 & 1 & -2 \\ 1 & -1 & 3 & -3 \end{bmatrix}$ **14.** $\begin{bmatrix} 3 & -4 & 1 & 2 \\ 6 & -8 & 2 & 4 \\ -3 & 4 & -1 & -2 \\ 12 & -16 & 4 & 8 \end{bmatrix}$

15. $\begin{bmatrix} 1 & 2 & 1 & 3 \\ 2 & 3 & 0 & 1 \\ 2 & 1 & -4 & -9 \\ 5 & 5 & 0 & 2 \end{bmatrix}$ **16.** $\begin{bmatrix} 1 & 0 & 1 & -1 \\ 3 & 1 & 4 & 1 \\ 2 & 1 & 3 & 2 \\ 1 & 1 & 2 & 3 \end{bmatrix}$

Exercises 17–20. For the matrices in Exercises 13–16, find a basis for the column space.

Exercises 21–24. For the matrices in Exercises 5, 6, 15, and 16, find a basis for the row space consisting of row vectors in the given matrix.

In Exercises 25–29, we are given a product of the form $AC = B$. Express each row vector of B as a linear combination of the row vectors of C.

25. $\begin{bmatrix} 2 & 1 \\ 3 & 2 \end{bmatrix}\begin{bmatrix} 2 \\ -3 \end{bmatrix} = \begin{bmatrix} 1 \\ 0 \end{bmatrix}$ **26.** $\begin{bmatrix} 2 & -1 \\ 4 & -3 \end{bmatrix}\begin{bmatrix} 1 & 2 \\ 0 & 2 \end{bmatrix} = \begin{bmatrix} 2 & 2 \\ 4 & 2 \end{bmatrix}$

27. $\begin{bmatrix} 1 & -1 \\ 2 & 4 \end{bmatrix}\begin{bmatrix} 3 & 2 & 2 \\ 1 & -1 & 0 \end{bmatrix} = \begin{bmatrix} 2 & 3 & 2 \\ 10 & 0 & 4 \end{bmatrix}$

28. $\begin{bmatrix} 1 & 2 & 0 \\ -1 & 2 & 1 \end{bmatrix}\begin{bmatrix} 1 \\ 2 \\ 1 \end{bmatrix} = \begin{bmatrix} 5 \\ 4 \end{bmatrix}$

29. $\begin{bmatrix} -3 & 1 & 2 \\ 2 & 2 & -1 \end{bmatrix}\begin{bmatrix} 2 & 1 \\ 1 & 2 \\ 1 & 2 \end{bmatrix} = \begin{bmatrix} -3 & 3 \\ 5 & 4 \end{bmatrix}$

30. $\begin{bmatrix} 0 & 1 & -3 \\ 2 & 4 & -5 \end{bmatrix}\begin{bmatrix} 1 & 2 & 3 \\ 1 & 1 & 2 \\ 1 & 2 & 3 \end{bmatrix} = \begin{bmatrix} -2 & -5 & -7 \\ 1 & -2 & -1 \end{bmatrix}$

Exercises 31–35. For the products in Exercises 26–30, express each column vector of B as a linear combination of the column vectors of A.

In Exercises 36–38, find a matrix product $AC = B$ whose row space interpretation consists of the given linear combination(s).

36. $(5,1,0) = 2(4,-1,3) + 3(-1,1,-2)$

37. $(-4,-5,2) = 3(0,1,2) - 4(1,2,1)$
$(3,8,7) = 2(0,1,2) + 3(1,2,1)$

38. $(-1,3,3) = 3(1,1,1) - 4(1,0,0)$
$(0,-1,1) = (1,1,1) - 2(1,1,0) + 3(1,0,0)$
$(9,7,2) = 2(1,1,1) + 5(1,1,0) + 2(1,0,0)$

In Exercises 39–41, find a matrix product $AC = B$ whose column space interpretation consists of the given linear combination(s).

39. $\begin{bmatrix} 9 \\ 2 \\ 0 \end{bmatrix} = 2\begin{bmatrix} 4 \\ 2 \\ 1 \end{bmatrix} - 1\begin{bmatrix} -1 \\ 2 \\ 2 \end{bmatrix}$

40. $\begin{bmatrix} 5 \\ 0 \\ -1 \end{bmatrix} = 3\begin{bmatrix} 1 \\ 0 \\ -1 \end{bmatrix} + 2\begin{bmatrix} 1 \\ 0 \\ 1 \end{bmatrix}, \quad \begin{bmatrix} 5 \\ 0 \\ -3 \end{bmatrix} = 4\begin{bmatrix} 1 \\ 0 \\ -1 \end{bmatrix} + 1\begin{bmatrix} 1 \\ 0 \\ 1 \end{bmatrix}$

41. $\begin{bmatrix} 2 \\ 0 \\ 2 \end{bmatrix} = 1\begin{bmatrix} -1 \\ 1 \\ 2 \end{bmatrix} + 1\begin{bmatrix} 2 \\ 1 \\ -1 \end{bmatrix} + 1\begin{bmatrix} 1 \\ -2 \\ 1 \end{bmatrix}, \quad \begin{bmatrix} 1 \\ 0 \\ 7 \end{bmatrix} = 3\begin{bmatrix} -1 \\ 1 \\ 2 \end{bmatrix} + 1\begin{bmatrix} 2 \\ 1 \\ -1 \end{bmatrix} + 2\begin{bmatrix} 1 \\ -2 \\ 1 \end{bmatrix}$

42. Let A be an $n \times n$ matrix. If the row space of A coincides with $\mathbb{R}^n$, how is the column space related to $\mathbb{R}^n$?

43. Let A be an $m \times n$ matrix. Show that if X is a column vector with n components, then AX lies in the column space of A, and if Y is a row vector with m components, then YA lies in the row space of A.

44. If A is a 2×4 matrix, find the possible values for the dimension of the column space.

45. If A is an $m \times n$ matrix, let p be the lesser of m and n. If $k \leq p$, we may form a $k \times k$ **submatrix** of A by deleting $m - k$ rows and $n - k$ columns. For instance, if

$$A = \begin{bmatrix} 1 & 2 & 1 & 3 \\ 2 & 1 & 0 & 1 \\ 3 & 1 & 1 & 2 \end{bmatrix},$$

then

$$\begin{bmatrix} 1 & 1 \\ 3 & 1 \end{bmatrix}$$

is the 2×2 submatrix formed by deleting row 2 and columns 2 and 4. Find the determinants of all 2×2 submatrices of A formed by deleting row 2.

46. Show that if a matrix A has a $k \times k$ submatrix with a nonzero determinant, and if B is obtained from A by an elementary row operation, then B has a $k \times k$ submatrix with a nonzero determinant.

47. Using the result of Exercise 46, show that if a matrix A has a $k \times k$ submatrix with a nonzero determinant and every $(k+1) \times (k+1)$ submatrix of A has determinant zero, then the rank of A equals k. (That is, the rank of a matrix is the dimension of the largest nonzero determinant contained in A.)

4.6 REVIEW OF LINEAR SYSTEMS THEORY

In this section, we organize much of the information about linear systems learned from Gaussian elimination, matrix algebra, determinants, and vector spaces. We give a summary of conditions for the existence and uniqueness of a solution to a linear system and a summary of conditions for the consistency of a linear system.

Unique Solution of a Linear System

Throughout this subsection, *A denotes an $n \times n$ matrix*. In Chapter 1 we solved linear systems of the form

$$AX = B$$

by Gaussian elimination. In particular,

> $AX = B$ has a unique solution for every B if and only if A is equivalent to a row echelon matrix with every main diagonal entry equal to 1.

Since this condition depends only on A,

> $AX = B$ has a unique solution for every B if and only if it has a unique solution for at least one B.

In particular,

> $AX = B$ has a unique solution for every B if and only if the associated *homogeneous* system $AX = 0$ has only the trivial solution.

We defined the row rank of A as the number of nonzero rows in any row echelon matrix equivalent to A. Since the row rank of A equals n if and only if each main diagonal entry of every equivalent row echelon matrix equals 1,

> $AX = B$ has a unique solution for every B if and only if the row rank of A equals n.

Finally from Chapter 1, any row echelon matrix equivalent to A has every main diagonal entry equal to 1 if and only if Gauss-Jordan elimination produces the $n \times n$ identity matrix:

> $AX = B$ has a unique solution for every B if and only if A is equivalent to the identity.

The study of matrix algebra in Chapter 2 showed that:

> $AX = B$ has a unique solution for every B if and only if A is invertible.

We introduced the transpose of a matrix, and found that:

> A is invertible if and only if A^t is invertible.

In Chapter 2, we saw also that:

> $AX = B$ has a unique solution for every B if and only if A is expressible as a product of elementary matrices.

From Chapter 3 (on determinants), A is invertible if and only if $\det(A) \neq 0$. Thus,

> $AX = B$ has a unique solution for every B if and only if $\det(A) \neq 0$.

Our studies so far in Chapter 4 have produced several more conditions, which are included with those reviewed above from Chapters 1–3 in the following summary.

THEOREM 4.13

The following statements are equivalent for an $n \times n$ matrix A.

a. $AX = B$ has a unique solution for every B.
b. $AX = B$ has a unique solution for at least one B.
c. $AX = 0$ has only the trivial solution.
d. In every row echelon matrix equivalent to A, each main diagonal entry equals 1.
e. A is equivalent to the $n \times n$ identity matrix.
f. A is invertible.
g. A may be expressed as a product of elementary matrices.
h. $\det(A) \neq 0$.
i. The rank of A equals n.
j. The row [column] space of A has dimension n.
k. The row [column] space of A coincides with $\mathbb{R}^n$.
l. The rows [columns] of A form a linearly independent set in $\mathbb{R}^n$.

In each of these conditions, replacement of A by A^t yields an equivalent condition.

If A is an $m \times n$ matrix with $m < n$, the system $AX = B$ has either no solutions or many solutions, so it cannot have a unique solution. If $m > n$, the system $AX = B$ has a unique solution if and only if it is consistent and the rank of A equals n.

Proofs for the various conditions in Theorem 4.13 are scattered throughout Chapters 1–4. (See also Exercise 29.)

Consistency Conditions

If a system $AX = B$ does not have a unique solution, it may have many or none, depending on B. From here on, A will denote an $m \times n$ (i.e., not necessarily square) matrix.

In Chapter 1, a consistency test for linear systems was based on properties of Gaussian elimination:

> $AX = B$ is consistent if and only if no row echelon matrix equivalent to the augmented matrix of the system has a row with the first n entries 0 and the final entry 1.

This form is independent of the particular row echelon matrix, so

> $AX = B$ is consistent if and only if at least one row echelon matrix equivalent to the augmented matrix has no row with the first n entries 0 and the final entry 1.

This test can be reformulated in terms of rank:

> $AX = B$ is consistent if and only if the rank of the coefficient matrix equals the rank of the augmented matrix.

The column space interpretation in Section 4.5 provides another consistency condition. For if $A = [a_{ij}]_{n \times n}$ and the entries of X are $x_1, \ldots, x_n$, then the matrix product AX may be expressed as a linear combination of the columns of A:

$$AX = x_1 \begin{bmatrix} a_{11} \\ \vdots \\ a_{m1} \end{bmatrix} + \cdots + x_n \begin{bmatrix} a_{1n} \\ \vdots \\ a_{mn} \end{bmatrix}.$$

Hence

> $AX = B$ is consistent if and only if B lies in the column space of A.

Since the column space of A coincides with the row space of A^t,

> $AX = B$ is consistent if and only if B^t lies in the row space of A^t.

The following summary includes one characterization of consistency of a linear system $AX = B$ not noted above. Examples will be given to illustrate this condition.

THEOREM 4.14

The following statements are equivalent for a linear system $AX = B$ with an $m \times n$ coefficient matrix A.

a. The system is consistent.

b. For every row echelon matrix equivalent to the augmented matrix of the system, if the first n entries equal 0, then the $(n + 1)$st entry must also equal 0.

c. For at least one row echelon matrix equivalent to the augmented matrix, if the first n entries equal 0, then the $(n + 1)$st entry must also equal 0.

d. The rank of the augmented matrix equals the rank of the coefficient matrix.
e. B lies in the column space of A.
f. B^t lies in the row space of A^t.
g. B is orthogonal to every solution of $A^tY = 0$, that is, B is orthogonal to the nullspace of A^t.

The next examples illustrate condition (g) in Theorem 4.14. A proof that (g) is equivalent to (f) is given after Theorem 5.7 in Section 5.1.

Example 1

Verify that the system $AX = B$ given by

$$\begin{aligned} x_1 + 2x_2 + x_3 &= 1 \\ x_1 + 3x_2 - x_3 &= 3 \\ 3x_1 + 8x_2 - x_3 &= 7 \\ x_2 - 2x_3 &= 2 \end{aligned}$$

is consistent, and that B is orthogonal to every solution of the system $A^tY = 0$.

Solution

Apply Gaussian elimination to the augmented matrix:

$$\left[\begin{array}{ccc|c} 1 & 2 & 1 & 1 \\ 1 & 3 & -1 & 3 \\ 3 & 8 & -1 & 7 \\ 0 & 1 & -2 & 2 \end{array}\right] \to \left[\begin{array}{ccc|c} 1 & 2 & 1 & 1 \\ 0 & 1 & -2 & 2 \\ 0 & 2 & -4 & 4 \\ 0 & 1 & -2 & 2 \end{array}\right] \to \left[\begin{array}{ccc|c} 1 & 2 & 1 & 1 \\ 0 & 1 & -2 & 2 \\ 0 & 0 & 0 & 0 \\ 0 & 0 & 0 & 0 \end{array}\right].$$

The final matrix shows that the given system is consistent. For the second requirement, apply Gaussian elimination to A^t:

$$\begin{bmatrix} 1 & 1 & 3 & 0 \\ 2 & 3 & 8 & 1 \\ 1 & -1 & -1 & -2 \end{bmatrix} \to \begin{bmatrix} 1 & 1 & 3 & 0 \\ 0 & 1 & 2 & 1 \\ 0 & -2 & -4 & -2 \end{bmatrix} \to \begin{bmatrix} 1 & 1 & 3 & 0 \\ 0 & 1 & 2 & 1 \\ 0 & 0 & 0 & 0 \end{bmatrix}.$$

From this form, basis vectors for the nullspace of A^t can be found by back substitution, first with $(-1, 0)$, then with $(0, 1)$ assigned to (y_3, y_4):

$$Y_1 = \begin{bmatrix} 1 \\ 2 \\ -1 \\ 0 \end{bmatrix}, \qquad Y_2 = \begin{bmatrix} 1 \\ -1 \\ 0 \\ 1 \end{bmatrix}.$$

Then

$$B^tY_1 = 1(1) + 3(2) + 7(-1) + 2(0) = 0,$$

$$B^tY_2 = 1(1) + 3(-1) + 7(0) + 2(1) = 0.$$

Since $\{Y_1, Y_2\}$ is a basis for the solution space, any solution Y is expressible as a linear combination of Y_1, Y_2, i.e.,

$$Y = c_1Y_1 + c_2Y_2,$$

so by properties of matrix multiplication (from Chapter 2),

$$B^tY = B^t[c_1Y_1 + c_2Y_2] = c_1B^tY_1 + c_2B^tY_2 = c_1 \cdot 0 + c_2 \cdot 0 = 0,$$

hence B is orthogonal to Y, as required. ☐

Example 2

Let

$$A = \begin{bmatrix} 1 & 1 & 2 \\ 1 & 2 & 1 \\ 3 & 5 & 4 \end{bmatrix} \quad \text{and} \quad B = \begin{bmatrix} 1 \\ 0 \\ 1 \end{bmatrix}.$$

Use Theorem 4.14 to show that if the solution space of the system

$$\begin{aligned} x + y + 3z &= 0 \\ x + 2y + 5z &= 0 \\ 2x + y + 4z &= 0 \end{aligned}$$

coincides with the span of the vector $(1, 2, -1)$, then the linear system $AX = B$ is consistent.

Solution

Let $\mathbf{b} = (1, 0, 1)$ be the row vector representation of B, and let $\mathbf{u} = (1, 2, -1)$. Then for any scalar c,

$$\langle \mathbf{u}, c\mathbf{b} \rangle = (1)(c) + (2)(0) + (-1)(c) = 0.$$

Thus B is orthogonal to the solution space of the transposed homogeneous system $A^tX = 0$, so the system $AX = B$ is consistent by Theorem 4.14. ☐

We offer no separate summary for homogeneous systems since most of the information needed may be obtained from Theorem 4.13 and the "solution concepts" summaries in Chapter 1. There is one result based on the ideas in Chapter 1 that we formulate in terms of vector space concepts. This result is presented next because of its close relationship to Theorem 4.14. A proof will be given immediately following the proof of Theorem 5.6 in Section 5.1.

THEOREM 4.15

Let A be an $m \times n$ matrix of rank r, and let V be the nullspace of A. Then $\dim(V) = n - r$.

REVIEW CHECKLIST

1. Relate the uniqueness of the solution of a square linear system to properties of Gaussian elimination, matrix inversion, determinants, and vector spaces.
2. Apply the conditions from Theorem 4.13 to draw inferences concerning linear systems.
3. Relate the consistency of a linear system to properties of Gaussian elimination, matrix algebra, and vector space concepts.
4. Justify the various consistency conditions.
5. Apply the conditions from Theorem 4.14 to draw inferences concerning linear systems.
6. For a given matrix A, state the relationship between the dimension of the nullspace of A, the rank of A, and the number of columns in A.

EXERCISES

In Exercises 1–3, verify that the given matrix satisfies conditions (b), (e), (g), and (i) of Theorem 4.13.

1. $\begin{bmatrix} 1 & -3 \\ 2 & -2 \end{bmatrix}$ **2.** $\begin{bmatrix} 2 & 1 \\ 4 & 3 \end{bmatrix}$ **3.** $\begin{bmatrix} 1 & 4 \\ 2 & -3 \end{bmatrix}$

In Exercises 4–8, verify that the given matrix satisfies conditions (g), (i), and (j) of Theorem 4.13.

4. $\begin{bmatrix} 1 & -1 & 0 \\ 2 & 1 & 1 \\ 3 & 3 & 4 \end{bmatrix}$ **5.** $\begin{bmatrix} 1 & 2 & 1 \\ 4 & 7 & 1 \\ 3 & 2 & -8 \end{bmatrix}$ **6.** $\begin{bmatrix} 2 & 4 & 0 \\ 3 & 1 & -1 \\ 1 & 2 & 4 \end{bmatrix}$

7. $\begin{bmatrix} 1 & 1 & 0 & 2 \\ 2 & 3 & 1 & 4 \\ 3 & 3 & 0 & 7 \\ 2 & 2 & 1 & 1 \end{bmatrix}$ **8.** $\begin{bmatrix} 1 & 1 & 0 & 1 \\ 1 & 0 & 0 & 0 \\ 2 & 2 & 1 & 1 \\ 3 & 2 & -1 & -1 \end{bmatrix}$

In Exercises 9–11, let

$$A = \begin{bmatrix} 1 & 1 & -1 \\ 2 & 4 & -5 \\ 3 & -1 & 3 \end{bmatrix}, \quad \text{and} \quad B = \begin{bmatrix} 1 \\ -1 \\ 2 \end{bmatrix}.$$

9. Find the solution space of the system $AX = 0$.

10. Verify that B is orthogonal to every vector in the solution space of $AX = 0$.

11. What conclusion may we draw from the result of Exercise 10 concerning the system $A^tX = B$? Justify your answer.

In Exercises 12–14, let

$$A = \begin{bmatrix} 1 & 4 & 1 & 2 \\ 1 & 6 & 3 & 0 \\ 2 & 7 & 4 & -1 \\ 1 & 7 & 4 & -1 \end{bmatrix}, \quad \text{and} \quad B = \begin{bmatrix} 1 \\ 5 \\ 2 \\ 1 \end{bmatrix}.$$

12. Find the solution space of the system $AX = 0$.

13. Verify that B is orthogonal to every vector in this solution space.

14. What conclusion may we draw from the result of Exercise 13 concerning the system $A^tX = 0$? Justify your answer.

15. If A is an invertible $n \times n$ matrix, determine the relationship between $\mathbb{R}^n$ and the row space of A^t, and justify your answer.

16. If A is an $n \times n$ matrix for which $\det(A^t) \neq 0$, must the row vectors from A form a basis for $\mathbb{R}^n$? Justify your answer.

17. If A is an $n \times n$ matrix whose column space has dimension n, find the dimension of the row space of A.

18. If A is an $n \times n$ matrix and A^t is invertible, what may we conclude about $\det(A)$ and why?

In Exercises 19–22, $A = [a_{ij}]_{3\times 3}$ is a matrix for which

$$a_{i1} + a_{i2} + a_{i3} = 0 \qquad \text{for } i = 1, 2, 3$$

(i.e., the sum of the entries in each row equals 0).

19. Show that the column vector $X = [1 \quad 1 \quad 1]^t$ is in the nullspace of A.

20. What conclusion may be drawn from the result of Exercise 19 concerning the invertibility of A and why?

21. Generalize the results of Exercises 19 and 20 to any $n \times n$ matrix.

22. What conclusion, if any, may be drawn from the result of Exercise 21 concerning the determinant of a matrix for which the sum of the entries in each *column* is 0? Justify your answer.

In Exercises 23–26, V is the plane in $\mathbb{R}^3$ determined by

$$x - y + 2z = 0.$$

23. Show that if $\mathbf{x}_1 = (1, 1, 0)$ and $\mathbf{x}_2 = (2, 0, -1)$ then the set $\{\mathbf{x}_1, \mathbf{x}_2\}$ is a basis for V.

24. Show that every vector in the span of the set $\{(1, -1, 2)\}$ is orthogonal to every vector in V.

25. From the result of Exercise 23, $\mathbf{b} = (b_1, b_2, b_3)$ lies in V if and only if there exist scalars c_1, c_2 such that

$$c_1\mathbf{x}_1 + c_2\mathbf{x}_2 = \mathbf{b}.$$

Express this vector equation as a linear system in the matrix form

$$AC = B.$$

26. Use the form in Exercise 25 to relate this example to Theorem 4.14. Which conditions are directly involved?

27. If A and B are $n \times n$ matrices such that AB is a diagonal matrix with each main diagonal entry equal to 2, find the row space of A and the column space of B.

28. If A and B are $n \times n$ matrices for which $\det(AB) = 10$, what conclusion can be drawn concerning linear dependence or independence of the set of row vectors of A and why?

29. Prove that conditions (j), (k), and (l) are equivalent to conditions (a)–(j) in Theorem 4.13.

REVIEW EXERCISES

1. Give an example of a set S in $\mathbb{R}^3$ that is closed under multiplication by scalars but is not a vector space.

In Exercises 2–5, determine whether the set of all vectors $\mathbf{x} = (x_1, x_2)$ in $\mathbb{R}^2$ that satisfy the given condition is a vector space.

2. $2x_1 + 3x_2 = 0$

3. $x_1^2 + 3x_2 = 0$

4. $x_1 + x_2 = 1$

5. $x_1 = x_2$

In Exercises 6–8, determine whether the given set is linearly independent in $\mathbb{R}^3$.

6. $\{(1, 2, 2), (2, 1, 2), (1, 1, 0)\}$

7. $\{(1, 2, 2), (2, 1, 2), (3, -6, -2)\}$

8. $\{(4, 2, 1), (1, 1, 2), (5, 1, -4)\}$

In Exercises 9–11, if possible, express the given vector as a linear combination of the vectors $(1,1,0)$, $(1,2,1)$, and $(1,-1,-1)$ in $\mathbb{R}^3$.

9. $(2,4,3)$ **10.** $(2,7,5)$ **11.** $(1,-1,-1)$

In Exercises 12–18, let

$$A = \begin{bmatrix} 1 & -1 & 1 \\ 2 & 0 & 3 \\ 2 & 4 & 5 \end{bmatrix}.$$

12. Find a basis for the row space of A.

13. Find the dimension of the row space of A.

14. Find a basis for the column space of A.

15. Find the dimension of the column space of A.

16. Find a basis for the column space of A consisting of column vectors of A.

17. Find a basis for the nullspace of A.

18. Find the rank of A.

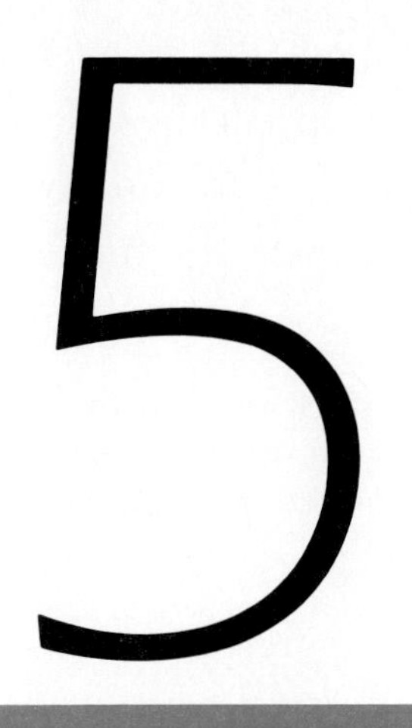

Inner Product Spaces

In Chapter 4, we studied the algebraic structure of vector spaces, emphasizing the spaces $\mathbb{R}^n$ and their relation to linear systems. In many vector spaces, including every $\mathbb{R}^n$, an inner product *enables us to achieve more than is possible or convenient using algebra alone, as condition (g) of Theorem 4.14 illustrates. In this chapter, we study vector spaces with inner products. One important application explored is the method of least squares for fitting curves to observed data.*

5.1 EUCLIDEAN SPACES

Properties of the spaces $\mathbb{R}^n$ relative to the inner product of Definition 2.3 are useful in the study of linear systems and more general inner product spaces. Most of the work in this section is related to orthogonality.

Orthonormal Bases

Each space $\mathbb{R}^n$ with the inner product of Definition 2.3 is called a **Euclidean space**. The inner product in a Euclidean space has the following characteristics.

Proposition

a. Positive definite For any $\mathbf{x}$ in $\mathbb{R}^n$, we have $\langle \mathbf{x}, \mathbf{x} \rangle \geq 0$, with equality if and only if $\mathbf{x} = \mathbf{0}$.

b. Symmetric For any $\mathbf{x}$ and $\mathbf{y}$ in $\mathbb{R}^n$,

$$\langle \mathbf{x}, \mathbf{y} \rangle = \langle \mathbf{y}, \mathbf{x} \rangle.$$

c. Linear For any vectors $\mathbf{x}_1$, $\mathbf{x}_2$, and $\mathbf{y}$ in $\mathbb{R}^n$ and any scalars c_1 and c_2,

$$\langle c_1\mathbf{x}_1 + c_2\mathbf{x}_2, \mathbf{y} \rangle = c_1\langle \mathbf{x}_1, \mathbf{y} \rangle + c_2\langle \mathbf{x}_2, \mathbf{y} \rangle.$$

For a proof of this proposition, see Exercise 38. In Exercise 39, a proof of (extended) linearity is assigned:

for any vectors $\mathbf{x}_1, \ldots, \mathbf{x}_k$, and $\mathbf{y}$ in $\mathbb{R}^n$ and any scalars $c_1, \ldots, c_k$, we have

$$\langle c_1\mathbf{x}_1 + \cdots + c_k\mathbf{x}_k, \mathbf{y}\rangle = c_1\langle \mathbf{x}_1, \mathbf{y}\rangle + \cdots + c_k\langle \mathbf{x}_k, \mathbf{y}\rangle.$$

In describing this relation, we often say that the inner product is *linear in the first variable*. By symmetry, the inner product is also linear in the second variable.

In $\mathbb{R}^n$, the use of a basis with certain special properties shortens the work of calculating coefficients.

DEFINITION 5.1

A set of vectors $\{\mathbf{x}_1, \ldots, \mathbf{x}_k\}$ in $\mathbb{R}^n$ is said to be **orthonormal** if $\|\mathbf{x}_i\| = 1$ for each i and $\langle \mathbf{x}_i, \mathbf{x}_j\rangle = 0$ whenever $i \neq j$.

(A unit vector is also called a **normalized** vector. Thus the vectors in an orthonormal set are pairwise *ortho*gonal and each of them is *normal*ized.)

Example 1

The set

$$\{(2/\sqrt{5}, 1/\sqrt{5}), (-1/\sqrt{5}, 2/\sqrt{5})\}$$

is orthonormal in $\mathbb{R}^2$ and the set

$$\{(1/3, 2/3, 2/3), (2/3, 1/3, -2/3), (2/3, -2/3, 1/3)\}$$

is orthonormal in $\mathbb{R}^3$. □

THEOREM 5.1

Any orthonormal set in $\mathbb{R}^n$ is linearly independent.

Proof Let $S = \{\mathbf{x}_1, \ldots, \mathbf{x}_k\}$ be an orthonormal set in $\mathbb{R}^n$ and suppose

$$c_1\mathbf{x}_1 + \cdots + c_k\mathbf{x}_k = \mathbf{0}.$$

Then for $i = 1, \ldots, k$,

$$\langle c_1\mathbf{x}_1 + \cdots + c_k\mathbf{x}_k, \mathbf{x}_i\rangle = \langle \mathbf{0}, \mathbf{x}_i\rangle = 0.$$

By linearity of the inner product in the first variable,

$$c_1\langle \mathbf{x}_1, \mathbf{x}_i \rangle + \cdots + c_i\langle \mathbf{x}_i, \mathbf{x}_i \rangle + \cdots + c_k\langle \mathbf{x}_k, \mathbf{x}_i \rangle = 0.$$

Since S is orthonormal, the left side equals c_i. Thus

$$c_i = 0$$

for $i = 1, \ldots, k$. By Definition 4.5, S is linearly independent. ☐

Since evaluating k inner products is generally easier than solving a $k \times k$ linear system, the next theorem shows how the use of an orthonormal basis saves effort.

THEOREM 5.2

Let $B = \{\mathbf{x}_1, \ldots, \mathbf{x}_k\}$ be an orthonormal set in $\mathbb{R}^n$. If $\mathbf{x}$ is in the span of B, then

$$\mathbf{x} = \langle \mathbf{x}, \mathbf{x}_1 \rangle \mathbf{x}_1 + \cdots + \langle \mathbf{x}, \mathbf{x}_k \rangle \mathbf{x}_k.$$

Proof Since $\mathbf{x}$ is in the span of B, $\mathbf{x}$ is expressible in the form

$$\mathbf{x} = c_1\mathbf{x}_1 + \cdots + c_k\mathbf{x}_k.$$

Thus for each $i = 1, \ldots, k$,

$$\begin{aligned}
\langle \mathbf{x}, \mathbf{x}_i \rangle &= \langle c_1\mathbf{x}_1 + \cdots + c_k\mathbf{x}_k, \mathbf{x}_i \rangle \\
&= c_1\langle \mathbf{x}_1, \mathbf{x}_i \rangle + \cdots + c_i\langle \mathbf{x}_i, \mathbf{x}_i \rangle + \cdots + c_k\langle \mathbf{x}_k, \mathbf{x}_i \rangle \\
&= c_i
\end{aligned}$$

as required. ☐

Example 2

Express the vector $\mathbf{x} = (3, 4)$ in $\mathbb{R}^2$ as a linear combination of $\mathbf{x}_1 = (2/\sqrt{5}, 1/\sqrt{5})$ and $\mathbf{x}_2 = (-1/\sqrt{5}, 2/\sqrt{5})$.

Solution

The vectors $\mathbf{x}_1$ and $\mathbf{x}_2$ form an orthonormal basis for $\mathbb{R}^2$. Since $\langle \mathbf{x}, \mathbf{x}_1 \rangle = 2\sqrt{5}$ and $\langle \mathbf{x}, \mathbf{x}_2 \rangle = \sqrt{5}$,

$$\begin{aligned}
\mathbf{x} &= \langle \mathbf{x}, \mathbf{x}_1 \rangle \mathbf{x}_1 + \langle \mathbf{x}, \mathbf{x}_2 \rangle \mathbf{x}_2 \\
&= 2\sqrt{5}\,\mathbf{x}_1 + \sqrt{5}\,\mathbf{x}_2
\end{aligned}$$

by Theorem 5.2. ☐

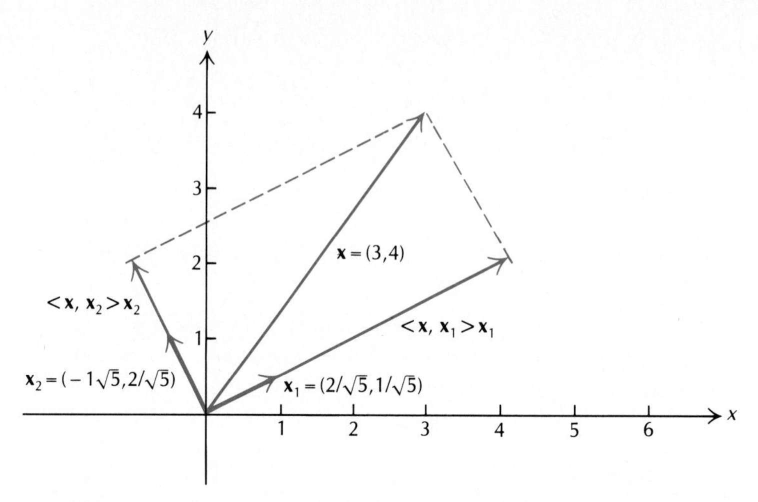

Figure 5.1 Resolution of the vector $(3, 4)$ into its components relative to the basis $\{(2/\sqrt{5}, 1/\sqrt{5}), (-1/\sqrt{5}, 2/\sqrt{5})\}$.

In Example 2 we *resolved* $\mathbf{x}$ *into its component vectors* relative to the basis $B = \{\mathbf{x}_1, \mathbf{x}_2\}$, i.e., we expressed $\mathbf{x}$ as the sum of its *orthogonal projections* onto the basis vectors (Definition 2.5). This resolution of $\mathbf{x}$ is shown in Fig. 5.1.

The use of an orthonormal basis also facilitates calculation of norms. A proof of the next result is assigned in Exercise 40.

THEOREM 5.3 (Generalized Pythagorean Theorem)

Let $B = \{\mathbf{x}_1, \ldots, \mathbf{x}_k\}$ be an orthonormal set in $\mathbb{R}^n$. If

$$\mathbf{x} = c_1\mathbf{x}_1 + \cdots + c_k\mathbf{x}_k,$$

then

$$\|\mathbf{x}\|^2 = c_1^2 + \cdots + c_k^2.$$

Example 3

For the vector $\mathbf{x} = (3, 4)$ of Example 2,

$$\|\mathbf{x}\| = \left((2\sqrt{5})^2 + (\sqrt{5})^2\right)^{1/2} = 5.$$

This value agrees with that obtained using the standard basis:

$$\|\mathbf{x}\| = \sqrt{3^2 + 4^2} = 5. \quad \square$$

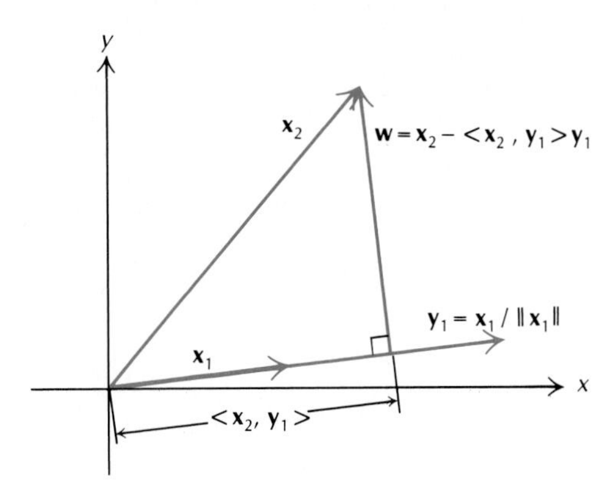

Figure 5.2 Gram-Schmidt procedure illustrated for two vectors in $\mathbb{R}^2$.

Gram-Schmidt* Procedure

The Gram-Schmidt procedure replaces any basis S of a subspace of $\mathbb{R}^n$ with an orthonormal basis T. To get an idea of how this procedure works, suppose S has two vectors: $S = \{\mathbf{x}_1, \mathbf{x}_2\}$. As illustrated in Fig. 5.2, let

$$\mathbf{y}_1 = \mathbf{x}_1/\|\mathbf{x}_1\|,$$

a unit vector in the direction in $\mathbf{x}_1$. Then the vector

$$\mathbf{w} = \mathbf{x}_2 - \langle \mathbf{x}_2, \mathbf{y}_1 \rangle \mathbf{y}_1$$

is orthogonal to $\mathbf{y}_1$ and is nonzero (Fig. 5.2). If

$$\mathbf{y}_2 = \mathbf{w}/\|\mathbf{w}\|,$$

then the set $T = \{\mathbf{y}_1, \mathbf{y}_2\}$ is an orthonormal basis for the span of S.

THEOREM 5.4 (Gram-Schmidt)

For any linearly independent set $S = \{\mathbf{x}_1, \ldots, \mathbf{x}_k\}$ in $\mathbb{R}^n$, define $T = \{\mathbf{y}_1, \ldots, \mathbf{y}_k\}$ as follows:

1. $\mathbf{y}_1 = \mathbf{x}_1/\|\mathbf{x}_1\|$;
2. with $\mathbf{y}_1, \ldots, \mathbf{y}_{i-1}$ determined, let

$$\mathbf{w}_i = \mathbf{x}_i - \langle \mathbf{x}_i, \mathbf{y}_1 \rangle \mathbf{y}_1 - \cdots - \langle \mathbf{x}_i, \mathbf{y}_{i-1} \rangle \mathbf{y}_{i-1},$$
$$\mathbf{y}_i = \mathbf{w}_i/\|\mathbf{w}_i\|.$$

Then T is an orthonormal basis for the span of S.

---

* **Jorgen Gram** (1850–1916) was a Danish actuary who was very much interested in mathematics. His contributions are widely used in studies related to linear algebra. **Erhard Schmidt** (1876–1905) was a German mathematician whose work on problems involving infinitely many variables resulted in the application of linear algebra to mathematical physics via the theory of integral equations.

Proof Since S is linearly independent, $\mathbf{x}_i \neq \mathbf{0}$ for $i = 1, \ldots, k$. In particular, the normalization of $\mathbf{x}_1$ is justified. The set $\{\mathbf{y}_1\}$ is orthonormal and its span coincides with the span of $\{\mathbf{x}_1\}$. Now suppose that for some i, where $1 \leq i \leq k$, the Gram-Schmidt procedure has produced an orthonormal basis $\{\mathbf{y}_1, \ldots, \mathbf{y}_{i-1}\}$ for the span of $\{\mathbf{x}_1, \ldots, \mathbf{x}_{i-1}\}$. Let $\mathbf{w}_i$ be defined as in (2), and let $1 \leq j \leq i-1$. Since $\{\mathbf{y}_1, \ldots, \mathbf{y}_{i-1}\}$ is orthonormal, and since the inner product is linear in the first variable,

$$\langle \mathbf{w}_i, \mathbf{y}_j \rangle = \langle \mathbf{x}_i, \mathbf{y}_j \rangle - \langle \mathbf{x}_i, \mathbf{y}_1 \rangle \langle \mathbf{y}_1, \mathbf{y}_j \rangle - \cdots - \langle \mathbf{x}_i, \mathbf{y}_{i-1} \rangle \langle \mathbf{y}_{i-1}, \mathbf{y}_j \rangle$$

$$= \langle \mathbf{x}_i, \mathbf{y}_j \rangle - \langle \mathbf{x}_i, \mathbf{y}_j \rangle = 0.$$

Thus $\mathbf{w}_i$ is orthogonal to each $\mathbf{y}_j$. Since each of the vectors $\mathbf{y}_1, \ldots, \mathbf{y}_{i-1}$ is a linear combination of $\mathbf{x}_1, \ldots, \mathbf{x}_{i-1}$ and the set $\{\mathbf{x}_1, \ldots, \mathbf{x}_i\}$ is linearly independent, $\mathbf{w}_i \neq \mathbf{0}$. Thus we may let $\mathbf{y}_i = \mathbf{w}_i / \|\mathbf{w}_i\|$, and the resulting set $B_i = \{\mathbf{y}_1, \ldots, \mathbf{y}_i\}$ is orthonormal. In addition, B_i is a basis for the span of $\{\mathbf{x}_1, \ldots, \mathbf{x}_i\}$ by Theorem 4.9. Since i was arbitrary between 1 and k, the process terminates after k steps with the required orthonormal basis $T = B_k$. ☐

Example 4

Find an orthonormal basis for the span of S in $\mathbb{R}^4$, if

$$S = \{(1, 0, -1, 0), (0, 1, 2, 1), (2, 1, -1, 0)\}.$$

Solution

That S is linearly independent can readily be verified. Denote the vectors from S in their given order by $\mathbf{x}_1$, $\mathbf{x}_2$, and $\mathbf{x}_3$. Let

$$\mathbf{y}_1 = (1/\sqrt{2})(1, 0, -1, 0),$$

and calculate $\mathbf{y}_2$ as in Theorem 5.1, step 2. This gives

$$\mathbf{w}_2 = \mathbf{x}_2 - \langle \mathbf{x}_2, \mathbf{y}_1 \rangle \mathbf{y}_1 = (0, 1, 2, 1) + (1, 0, -1, 0) = (1, 1, 1, 1),$$

from which

$$\mathbf{y}_2 = (1/2)(1, 1, 1, 1).$$

Similarly,

$$\begin{aligned} \mathbf{w}_3 &= \mathbf{x}_3 - \langle \mathbf{x}_3, \mathbf{y}_1 \rangle \mathbf{y}_1 - \langle \mathbf{x}_3, \mathbf{y}_2 \rangle \mathbf{y}_2 \\ &= (2,1,-1,0) - (3/2)(1,0,-1,0) - (1/2)(1,1,1,1) \\ &= (1/2)(0,1,0,-1), \end{aligned}$$

from which

$$\mathbf{y}_3 = (1/\sqrt{2})(0,1,0,-1).$$

The resulting orthonormal basis for the span of S is

$$T = \{(1/\sqrt{2})(1,0,-1,0), (1/2)(1,1,1,1), (1/\sqrt{2})(0,1,0,-1)\}.$$

A direct verification that the span of T coincides with the span of S is assigned in Exercise 41. □

Orthogonal Complement

The next examples illustrate orthogonality relations between sets in $\mathbb{R}^3$.

Example 5

Let V denote the subspace of $\mathbb{R}^3$ spanned by the vector $\mathbf{v} = (1, 2, -1)$. Describe geometrically the set W of all vectors in $\mathbb{R}^3$ that are orthogonal to every vector in V.

Solution

A vector $\mathbf{u}$ in $\mathbb{R}^3$ is orthogonal to every vector $c\mathbf{v}$ in V if and only if $\langle \mathbf{v}, \mathbf{u} \rangle = 0$. Thus W consists of all $\mathbf{u} = (x, y, z)$ such that

$$\langle \mathbf{v}, \mathbf{u} \rangle = x + 2y - z = 0.$$

Hence W coincides with the plane determined by this equation. □

Example 6

Let V denote the subspace of $\mathbb{R}^3$ spanned by the vectors $\mathbf{v}_1 = (1,3,0)$ and $\mathbf{v}_2 = (1,0,-3)$. Describe geometrically the set W of all vectors in $\mathbb{R}^3$ that are orthogonal to every vector in V.

Solution

A vector $\mathbf{u}$ in $\mathbb{R}^3$ is orthogonal to every vector $c_1\mathbf{v}_1 + c_2\mathbf{v}_2$ in V if and only if $\langle \mathbf{v}_1, \mathbf{u} \rangle = 0$ and $\langle \mathbf{v}_2, \mathbf{u} \rangle = 0$. Thus W consists of all solutions of the linear system

$$\begin{aligned} x + 3y \qquad &= 0 \\ x \qquad - 3z &= 0. \end{aligned}$$

Solving this linear system gives W as the span of the vector $(3, -1, 1)$. □

DEFINITION 5.2

The **orthogonal complement** of a set S in $\mathbb{R}^n$ is denoted by $S^\perp$ and is defined as the set of all vectors $\mathbf{y}$ in $\mathbb{R}^n$ that are orthogonal to *every* $\mathbf{x}$ in S.

THEOREM 5.5

The orthogonal complement of any set S in $\mathbb{R}^n$ is a subspace of $\mathbb{R}^n$.

A proof of Theorem 5.5 is assigned in Exercise 42.

THEOREM 5.6

If V is a *subspace* of $\mathbb{R}^n$, then every vector $\mathbf{x}$ in $\mathbb{R}^n$ is uniquely expressible in the form

$$\mathbf{x} = \mathbf{v} + \mathbf{u},$$

where $\mathbf{v}$ is in V and $\mathbf{u}$ in $V^\perp$. In addition, we have the Pythagorean relation

$$\|\mathbf{x}\|^2 = \|\mathbf{v}\|^2 + \|\mathbf{u}\|^2.$$

Proof If $V = \{\mathbf{0}\}$, then $V^\perp = \mathbb{R}^n$. If $V = \mathbb{R}^n$, then $V^\perp = \{\mathbf{0}\}$. In either of these cases, the conclusion follows from the defining condition for the additive identity,

$$\mathbf{x} = \mathbf{0} + \mathbf{x} = \mathbf{x} + \mathbf{0}.$$

If $V \neq \{\mathbf{0}\}$ and $V \neq \mathbb{R}^n$, then V is a **proper subspace** of $\mathbb{R}^n$ and $\dim(V) = r$, where $0 < r < n$. By Definition 4.7, V has a basis $S = \{\mathbf{v}_1, \ldots, \mathbf{v}_r\}$, which may be assumed orthonormal by Theorem 5.4. By Theorem 4.8b, there exists a basis B for

$\mathbb{R}^n$ containing S. By Theorem 5.4, B may be assumed orthonormal. Let

$$B = \{\mathbf{v}_1, \ldots, \mathbf{v}_r, \mathbf{u}_1, \ldots, \mathbf{u}_p\},$$

where $p = n - r$. If

$$\mathbf{v} = \langle \mathbf{x}, \mathbf{v}_1 \rangle \mathbf{v}_1 + \cdots + \langle \mathbf{x}, \mathbf{v}_r \rangle \mathbf{v}_r$$

and

$$\mathbf{u} = \langle \mathbf{x}, \mathbf{u}_1 \rangle \mathbf{u}_1 + \cdots + \langle \mathbf{x}, \mathbf{u}_p \rangle \mathbf{u}_p,$$

then by Theorem 5.2,

$$\mathbf{x} = \mathbf{v} + \mathbf{u}.$$

Since $\mathbf{v}$ is a linear combination of the vectors in S, $\mathbf{v}$ is in V. Since B is orthonormal, $\langle \mathbf{v}_i, \mathbf{u}_j \rangle = 0$ for each i and j, hence $\mathbf{u}$ is in $V^\perp$. Uniqueness of this expression follows from Theorem 4.6. The final relation,

$$\|\mathbf{x}\|^2 = \|\mathbf{v}\|^2 + \|\mathbf{u}\|^2,$$

follows from Theorem 5.3. $\square$

Proof of Theorem 4.15 For an $m \times n$ matrix A with row space V, if A has rank r, then $\dim(V) = r$. Let X be any $n \times 1$ (column) matrix. Since each entry in the product AX is the inner product of a row vector from A with X, the nullspace of A coincides with $V^\perp$. By the proof of Theorem 5.6 (which is independent of Theorem 4.15), $\dim(V^\perp) = n - r$. Hence the dimension of the nullspace of A is also $n - r$. $\square$

DEFINITION 5.3

Let V be a subspace of $\mathbb{R}^n$ and let $\mathbf{x}$ be a vector in $\mathbb{R}^n$. Then the vectors $\mathbf{v}$ in V and $\mathbf{u}$ in $V^\perp$ such that $\mathbf{x} = \mathbf{v} + \mathbf{u}$ are called the **orthogonal projections** of $\mathbf{x}$ onto V and $V^\perp$, respectively.

Note that Definition 5.3 is in harmony with Definition 2.5 when V (or $V^\perp$) is the span of a single nonzero vector.

The proof of Theorem 5.6 contains an explicit formula for the orthogonal projection of any vector $\mathbf{x}$ in $\mathbb{R}^n$ onto a subspace V in terms of an orthonormal basis for V. This proves the following result.

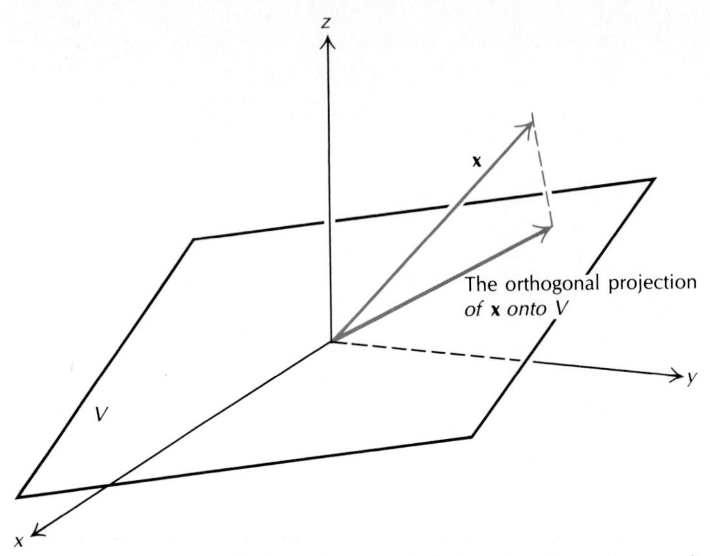

Figure 5.3 Illustration of the corollary to Theorem 5.6 showing the orthogonal projection of a vector $\mathbf{x}$ onto a subspace V of $\mathbb{R}^3$.

Corollary to Theorem 5.6 Let V be a subspace of $\mathbb{R}^n$ and let $\{\mathbf{v}_1, \ldots, \mathbf{v}_r\}$ be an orthonormal basis for V. If $\mathbf{x}$ is any vector in $\mathbb{R}^n$, then the orthogonal projection $\mathbf{v}$ of $\mathbf{x}$ on V is given by

$$\mathbf{v} = \langle \mathbf{x}, \mathbf{v}_1 \rangle \mathbf{v}_1 + \cdots + \langle \mathbf{x}, \mathbf{v}_r \rangle \mathbf{v}_r.$$

See Fig. 5.3 for an illustration of the corollary.

Example 7

Find the orthogonal projection of $\mathbf{x} = (4, 2, -1)$ on the subspace V of $\mathbb{R}^3$ spanned by $\{(1, 2, 2), (2, -2, 1)\}$.

Solution

Since the vectors in the given basis for V are orthogonal, an orthonormal basis may be obtained by normalization:

$$\mathbf{v}_1 = (1/3, 2/3, 2/3), \qquad \mathbf{v}_2 = (2/3, -2/3, 1/3).$$

Since $\langle \mathbf{x}, \mathbf{v}_1 \rangle = 2$ and $\langle \mathbf{x}, \mathbf{v}_2 \rangle = 1$, the orthogonal projection of $\mathbf{x}$ on V is given by

$$\mathbf{v} = 2(1/3, 2/3, 2/3) + 1(2/3, -2/3, 1/3) = (4/3, 2/3, 5/3). \qquad \square$$

For any set S in $\mathbb{R}^n$, denote $(S^\perp)^\perp$ by $S^{\perp\perp}$. The following result will be used to establish the equivalence of conditions (f) and (g) of Theorem 4.14.

THEOREM 5.7

For any set S in $\mathbb{R}^n$, the subspace $S^{\perp\perp}$ coincides with the span of S.

Proof Let V be the span of S. By the linearity properties of the inner product, a vector $\mathbf{u}$ in $\mathbb{R}^n$ is in $S^\perp$ if and only if it is in $V^\perp$. That is, $S^\perp = V^\perp$, and hence $S^{\perp\perp} = V^{\perp\perp}$. Since any $\mathbf{v}$ in V is orthogonal to $V^\perp$, V is contained in $V^{\perp\perp}$, hence also in $S^{\perp\perp}$.

Conversely, suppose $\mathbf{w}$ is in $S^{\perp\perp}$. By Theorem 5.6, $\mathbf{w}$ is uniquely expressible in the form

$$\mathbf{w} = \mathbf{v} + \mathbf{u},$$

where $\mathbf{v}$ is in V and $\mathbf{u}$ is in $V^\perp$. Since $\mathbf{w}$ is in $S^{\perp\perp}$ and $S^{\perp\perp} = V^{\perp\perp}$, $\mathbf{u} = \mathbf{0}$ by the corollary to Theorem 5.6. Hence $\mathbf{w} = \mathbf{v}$, and so $\mathbf{w}$ is in V, as required. □

Theorem 5.7 provides a convenient way to establish the final characterization of consistency of a linear system $AX = B$ in Theorem 4.14.

Proof of the equivalence of conditions (f) and (g) in Theorem 4.14 Let S be the set of row vectors of A^t. By condition (f), the system $AX = B$ is consistent if and only if B is in the column space of A, hence if and only if B is in the row space of A^t. That is, the system is consistent if and only if B lies in the span of S. By Theorem 5.7 (which is independent of Theorem 4.14), the span of S coincides with $S^{\perp\perp}$, which means that B is in the row space of A^t if and only if B is orthogonal to $S^\perp$, that is, if and only if B is orthogonal to the nullspace of A^t. □

For V a subspace of $\mathbb{R}^n$ and $\mathbf{x}$ a vector in $\mathbb{R}^n$, the orthogonal projection of $\mathbf{x}$ onto V is the "closest" vector to $\mathbf{x}$ in V, in the sense of the following theorem.

THEOREM 5.8

Let V be a subspace of $\mathbb{R}^n$, $\mathbf{x}$ be any vector in $\mathbb{R}^n$, and $\mathbf{v}$ be the orthogonal projection of $\mathbf{x}$ onto V. If $\mathbf{w}$ is any vector in V, then

$$\|\mathbf{x} - \mathbf{v}\| \leq \|\mathbf{x} - \mathbf{w}\|.$$

Proof By Theorem 5.6,

$$\mathbf{x} = \mathbf{v} + \mathbf{u},$$

where $\mathbf{u}$ is in $V^{\perp}$, hence

$$\|\mathbf{x} - \mathbf{v}\|^2 = \|\mathbf{u}\|^2.$$

Thus

$$\mathbf{x} - \mathbf{w} = \mathbf{v} + \mathbf{u} - \mathbf{w} = (\mathbf{v} - \mathbf{w}) + \mathbf{u},$$

and since $\mathbf{v} - \mathbf{w}$ is in V, Theorem 5.6 also implies that

$$\|\mathbf{x} - \mathbf{w}\|^2 = \|\mathbf{v} - \mathbf{w}\|^2 + \|\mathbf{u}\|^2 \geq \|\mathbf{u}\|^2 = \|\mathbf{x} - \mathbf{v}\|^2. \quad \square$$

In view of Theorem 5.8, the orthogonal projection of $\mathbf{x}$ onto V is also called the **least squares approximation** to $\mathbf{x}$ from V.

The ideas developed in this section have far-reaching implications for the theory and the application of mathematics. Some applications involving least squares approximation are given in Section 5.3. Many applications of these ideas can be found in studies of Fourier series and other orthogonal function expansions, mathematical physics, and engineering.

REVIEW CHECKLIST

1. State and prove the three basic properties of the inner product in $\mathbb{R}^n$.
2. Define an orthonormal set in $\mathbb{R}^n$.
3. Identify orthonormal sets as linearly independent.
4. Given a vector $\mathbf{x}$ in the span of an orthonormal set, find the coefficients in the expansion of $\mathbf{x}$ using the inner product.
5. Given a linearly independent set in $\mathbb{R}^n$, find an orthonormal set with the same span using the Gram-Schmidt procedure.
6. Define the orthogonal complement of a set in $\mathbb{R}^n$.
7. Define the orthogonal projection of a vector $\mathbf{x}$ onto a subspace V of $\mathbb{R}^n$.
8. Find the orthogonal projection of a given vector $\mathbf{x}$ in $\mathbb{R}^n$ onto a subspace V.
9. For any set S in $\mathbb{R}^n$, identify $S^{\perp\perp}$ as the span of S.
10. Identify the orthogonal projection of $\mathbf{x}$ on V as the least squares approximation to $\mathbf{x}$ from V.

EXERCISES

In Exercises 1–4, express the vector $\mathbf{x} = (3, 2)$ as a linear combination of the given vectors using Theorem 5.2, and then sketch.

1. $(1, 0), (0, 1)$
2. $(1/\sqrt{2}, 1/\sqrt{2}), (-1/\sqrt{2}, 1/\sqrt{2})$
3. $(3/5, 4/5), (-4/5, 3/5)$
4. $(\sqrt{3}/2, 1/2), (-1/2, \sqrt{3}/2)$
5. Using the results of Exercises 1–4, find the norm of the vector $(3, 2)$ in four distinct ways.

In Exercises 6–9, express the vector $\mathbf{x} = (1,1,-2)$ as a linear combination of the given vectors.

6. $(1,0,0), (0,1,0), (0,0,1)$
7. $(1/\sqrt{3}, 1/\sqrt{3}, 1/\sqrt{3}), (1/\sqrt{2}, 0, -1/\sqrt{2}), (1/\sqrt{6}, -2/\sqrt{6}, 1/\sqrt{6})$
8. $(2/7, 3/7, 6/7), (3/7, -6/7, 2/7), (-6/7, -2/7, 3/7)$
9. $(2/\sqrt{5}, 0, -1/\sqrt{5}), (1/\sqrt{6}, 1/\sqrt{6}, 2/\sqrt{6}), (1/\sqrt{30}, -5/\sqrt{30}, 2/\sqrt{30})$

In Exercises 10–12, apply the Gram-Schmidt procedure to the given vectors to obtain an orthonormal basis for $\mathbb{R}^2$.

10. $(1,2), (1,3)$ **11.** $(1,1), (1,0)$ **12.** $(1,0), (1,1)$

In Exercises 13–16, apply the Gram-Schmidt procedure to the given vectors to obtain an orthonormal basis for their span in $\mathbb{R}^3$.

13. $(3,0,4), (1,1,-1)$
14. $(1,1,-1), (3,0,4)$
15. $(2,2,1), (1,0,1), (1,1,5)$
16. $(1,0,1), (1,1,5), (2,2,1)$

In Exercises 17–20, apply the Gram-Schmidt procedure to the given vectors to obtain an orthonormal basis for their span in $\mathbb{R}^4$.

17. $(1,0,1,0), (1,1,1,1)$
18. $(1,1,1,1), (1,0,1,0)$
19. $(1,1,0,0), (0,1,1,0), (0,0,1,1)$
20. $(2,2,4,5), (-5,4,-2,2), (0,0,0,1)$

In Exercises 21–24, find the orthogonal projection of the vector $(1,-1,2)$ on the space spanned by the given vector(s) in $\mathbb{R}^3$.

21. $(2/\sqrt{6}, 1/\sqrt{6}, 1/\sqrt{6})$
22. $(1/\sqrt{2}, 0, 1/\sqrt{2}), (0,1,0)$
23. $(2,3,6), (6,2,-3)$
24. $(1,1,0), (0,1,1)$

In Exercises 25–27, find the orthogonal projection of the vector $(1,0,1,1)$ on the space spanned by the given vector(s) in $\mathbb{R}^4$.

25. $(2,-2,3,1)$
26. $(1/\sqrt{2}, 0, 1/\sqrt{2}, 0), (0, 1/\sqrt{2}, 0, 1/\sqrt{2})$
27. $(1,0,0,1), (1,1,1,0), (0,1,1,1)$
28. Find an orthonormal basis for the plane (subspace of $\mathbb{R}^3$) determined by

$$x - y + z = 0.$$

29. Find the projection of $(1,1,1)$ on the plane determined by

$$2x - z = 0.$$

30. Find an orthonormal basis for the nullspace of the matrix

$$A = \begin{bmatrix} 1 & -1 & 0 & 1 \\ 1 & 0 & 1 & 0 \\ -1 & 3 & 2 & -3 \\ 2 & -3 & -1 & 3 \end{bmatrix}.$$

31. Find the orthogonal projection of the vector $(1,1,1,1)$ onto the nullspace of the matrix A in Exercise 30.

32. Apply the Gram-Schmidt procedure to the set

$$S = \{(1,0,1),(0,1,1),(3,-2,1)\},$$

and explain the results.

33. Formulate a statement of the equivalence of conditions (a) and (g) in Theorem 4.14 in terms of orthogonal complements.

34. Find the distance from the point $(1,0,2)$ to the plane

$$2x + 3y + 6 = 0.$$

35. If $S = \{(1,-1,0)\}$, find an orthonormal basis for $S^{\perp}$ in $\mathbb{R}^3$.

36. Find an orthonormal basis for the orthogonal complement of the nullspace of the matrix A in Exercise 30.

37. Let $\mathbf{a}$ and $\mathbf{x}$ be vectors in $\mathbb{R}^n$. If V denotes the span of $\{\mathbf{a}\}$, use properties of the inner product to find an expression for k so that $\mathbf{x} - k\mathbf{a}$ will be in $V^{\perp}$.

38. Show that the inner product has properties (a)–(c) in the proposition of this section.

39. Establish the (extended) linearity of the inner product in the first variable in any Euclidean space for k vectors and k scalars.

40. Prove Theorem 5.3 by starting from the definition of $\|\mathbf{x}\|^2$, substituting the expression for $\mathbf{x}$, and applying properties of the inner product.

41. Verify directly that the span of the set T in Example 4, page 252, coincides with the span of S.

42. Prove Theorem 5.5.

5.2 GENERAL INNER PRODUCT SPACES

In Section 5.1, we derived a number of theorems for the Euclidean spaces using the symmetry, the linearity property, and the positive definiteness of the inner product of Definition 2.3. Since other functions have these same properties, we generalize the concept of inner product, *redefining* it in terms of these properties alone. The results of Section 5.1, as well as some results yet to come, will then apply to general *inner product spaces*.

General Inner Products

We formulate a general definition of an inner product.

DEFINITION 5.4

An **inner product** for a vector space is a function that assigns to each ordered pair of vectors $(\mathbf{x}, \mathbf{y})$ a scalar, denoted $\langle \mathbf{x}, \mathbf{y} \rangle$, with the following properties. For all vectors $\mathbf{x}$, $\mathbf{y}$, and $\mathbf{z}$, and all scalars c and k:

Positive definite $\langle \mathbf{x}, \mathbf{x} \rangle \geq 0$, with equality if and only if $\mathbf{x} = \mathbf{0}$;

Symmetric $\langle \mathbf{x}, \mathbf{y} \rangle = \langle \mathbf{y}, \mathbf{x} \rangle$;

Linear $\langle c\mathbf{x} + k\mathbf{y}, \mathbf{z} \rangle = \langle c\mathbf{x}, \mathbf{z} \rangle + \langle k\mathbf{y}, \mathbf{z} \rangle$.

An **inner product space** consists of a vector space V and an inner product for V.

In Exercise 24, we assign a proof of the general inner product space version of Theorem 5.1 to indicate how the results of Section 5.1 may be generalized.

Example 1

Every Euclidean space $\mathbb{R}^n$ is an inner product space relative to the **standard inner product** (Definition 2.3), as is any subspace of $\mathbb{R}^n$. □

Example 2

For any ordered pair of vectors $(\mathbf{x}, \mathbf{y})$ from $\mathbb{R}^2$, where $\mathbf{x} = (x_1, x_2)$ and $\mathbf{y} = (y_1, y_2)$, let

$$\langle \mathbf{x}, \mathbf{y} \rangle = 5x_1y_1 + 8x_2y_2.$$

This function satisfies the conditions in Definition 5.4. To see why, consider the linearity property, for instance. For scalars c and k and a vector $\mathbf{z} = (z_1, z_2)$,

$$c\mathbf{x} + k\mathbf{y} = (cx_1 + ky_1, cx_2 + ky_2).$$

Thus

$$\begin{aligned}\langle c\mathbf{x} + k\mathbf{y}, \mathbf{z} \rangle &= 5(cx_1 + ky_1)z_1 + 8(cx_2 + ky_2)z_2 \\ &= 5(cx_1)z_1 + 8(cx_2)z_2 + 5(ky_1)z_1 + 8(ky_2)z_2 \\ &= \langle c\mathbf{x}, \mathbf{z} \rangle + \langle k\mathbf{y}, \mathbf{z} \rangle,\end{aligned}$$

as required. □

Example 3

For any ordered pair of vectors $(\mathbf{x}, \mathbf{y})$ from $\mathbb{R}^2$, where $\mathbf{x} = (x_1, x_2)$ and $\mathbf{y} = (y_1, y_2)$, let

$$f(\mathbf{x}, \mathbf{y}) = x_1^2 y_1^2 + x_2 y_2.$$

Then f is positive definite and symmetric, but not linear in the first variable. For instance, let $\mathbf{x} = (2, 1)$, $\mathbf{y} = \mathbf{z} = (1, 1)$, and $c = k = 1$. Then

$$c\mathbf{x} + k\mathbf{y} = (3, 2),$$

from which

$$f(c\mathbf{x} + k\mathbf{y}, \mathbf{z}) = 3^2(1^2) + 2(1) = 11.$$

Since

$$f(c\mathbf{x}, \mathbf{z}) + f(k\mathbf{y}, \mathbf{z}) = [2^2(1^2) + 1(1)] + [1^2(1^2) + 1(1)] = 7,$$

f is not linear in the first variable, hence is not an inner product. □

Sigma notation for sums is especially useful for inner products. The upper case Greek sigma denotes a sum of terms a_i indexed by a variable i that assumes consecutive integer values:

$$\sum_{i=m}^{n} a_i = a_m + a_{m+1} + \cdots + a_n.$$

In this notation, the standard inner product may be written as

$$\langle x, y \rangle = \sum_{i=1}^{n} x_i y_i.$$

Example 4

Let $W = \{w_1, \ldots, w_n\}$ be any set of positive real numbers. For vectors $\mathbf{x}$ and $\mathbf{y}$ in $\mathbb{R}^n$, let

$$\langle \mathbf{x}, \mathbf{y} \rangle_W = \sum_{i=1}^{n} w_i x_i y_i.$$

This function satisfies the three conditions in Definition 5.4 (Exercise 25), hence is an inner product for $\mathbb{R}^n$ and for any subspace of $\mathbb{R}^n$. The numbers w_i are called **weights**, and the inner product is called a **weighted inner product** on $\mathbb{R}^n$. □

Example 5

(*Calculus based*) Let V be the subspace of all *continuous* real functions on the interval $[0, 1]$. For functions f and g from V, let

$$\langle f, g\rangle = \int_0^1 f(x)g(x)\,dx.$$

This form satisfies the three conditions in Definition 5.4 and hence is an inner product for V and for any subspace of V. Symmetry follows from the relation $f(x)g(x) = g(x)f(x)$ for $0 \le x \le 1$. Proofs for the other two conditions in Definition 5.4 involve only basic properties of the definite integral. (See Exercise 22.) □

The inner product of Example 5 is a continuous analogue of the standard (discrete) inner product for $\mathbb{R}^n$.

From the defining properties of an inner product follows an extremely important inequality that serves as a powerful problem solving tool and as a foundation for much mathematical theory.

THEOREM 5.9 (Cauchy-Schwarz*)

For any vectors $\mathbf{x}$ and $\mathbf{y}$ in an inner product space,

$$\langle \mathbf{x}, \mathbf{y}\rangle^2 \le \langle \mathbf{x}, \mathbf{x}\rangle\langle \mathbf{y}, \mathbf{y}\rangle,$$

with equality if and only if $\mathbf{x} = c\mathbf{y}$ or $\mathbf{y} = c\mathbf{x}$ for some scalar c.

Proof Since the inner product is positive definite,

$$\langle r\mathbf{x} + \mathbf{y}, r\mathbf{x} + \mathbf{y}\rangle \ge 0$$

for any scalar r. Expanding the expression on the left using symmetry and the linearity property of the inner product yields

$$0 \le \langle r\mathbf{x} + \mathbf{y}, r\mathbf{x} + \mathbf{y}\rangle = r^2\langle \mathbf{x}, \mathbf{x}\rangle + 2r\langle \mathbf{x}, \mathbf{y}\rangle + \langle \mathbf{y}, \mathbf{y}\rangle$$

* **Augustin-Louis Cauchy** (1789–1857) was a French mathematician who laid the foundations for a substantial part of real and complex analysis.
Hermann Amandus Schwarz (1843–1921) was a German mathematician whose work provides for much important interplay between linear algebra and mathematical analysis.

(Exercise 27). The graph of the function on the right is a parabola in the variable r. Since this graph does not cross the r-axis, the discriminant cannot be positive. That is,

$$(2\langle \mathbf{x}, \mathbf{y}\rangle)^2 - 4\langle \mathbf{x}, \mathbf{x}\rangle\langle \mathbf{y}, \mathbf{y}\rangle \leq 0,$$

which reduces to the desired inequality. A proof that equality is attained if and only if one of the vectors is a multiple of the other is assigned in Exercise 28. □

Example 6

With $\mathbf{x} = (x_1, \ldots, x_n)$ and $\mathbf{y} = (1, \ldots, 1)$ in $\mathbb{R}^n$, the general inequality

$$(x_1 + \cdots + x_n)^2 \leq n(x_1^2 + \cdots + x_n^2)$$

follows directly from the Cauchy-Schwarz inequality. □

The definition of *norm* in $\mathbb{R}^n$ may be extended to general inner product spaces.

DEFINITION 5.5

In any inner product space, the norm of a vector $\mathbf{x}$ is defined by

$$\|\mathbf{x}\| = \langle \mathbf{x}, \mathbf{x}\rangle^{1/2}.$$

The Cauchy-Schwarz inequality is often formulated in terms of the norm. For any x and y in an inner product space,

$$|\langle \mathbf{x}, \mathbf{y}\rangle| \leq \|\mathbf{x}\| \, \|\mathbf{y}\|.$$

THEOREM 5.10

In any inner product space V, the norm of Definition 5.5 satisfies the following conditions:

Positive definite for any $\mathbf{x}$ in V,

$$\|\mathbf{x}\| \geq 0,$$

with equality attained if and only if $\mathbf{x} = \mathbf{0}$;

Positive homogeneous for any $\mathbf{x}$ in V and any scalar c,

$$\|c\mathbf{x}\| = |c|\,\|x\|;$$

Triangle inequality for any $\mathbf{x}$ and $\mathbf{y}$ in V,

$$\|\mathbf{x} + \mathbf{y}\| \leq \|\mathbf{x}\| + \|\mathbf{y}\|.$$

Proof Positive definiteness of the norm follows from the corresponding property of the inner product. Positive homogeneity follows from the relation

$$\|c\mathbf{x}\|^2 = \langle c\mathbf{x}, c\mathbf{x}\rangle = c^2\|\mathbf{x}\|^2,$$

which, in turn, follows from linearity of the inner product in each variable. For the triangle inequality, we have

$$\|\mathbf{x} + \mathbf{y}\|^2 = \langle \mathbf{x} + \mathbf{y}, \mathbf{x} + \mathbf{y}\rangle = \langle \mathbf{x}, \mathbf{x}\rangle + 2\langle \mathbf{x}, \mathbf{y}\rangle + \langle \mathbf{y}, \mathbf{y}\rangle$$

by Definitions 5.5 and 5.4 (Exercise 29). By the Cauchy-Schwarz inequality,

$$\|\mathbf{x} + \mathbf{y}\|^2 \leq \|\mathbf{x}\|^2 + 2\|\mathbf{x}\|\,\|\mathbf{y}\| + \|\mathbf{y}\|^2 \leq (\|\mathbf{x}\| + \|\mathbf{y}\|)^2.$$

The desired result follows since the inequality is preserved by the square root function. □

Figure 5.4 shows the triangle inequality in $\mathbb{R}^2$: the shortest distance between two points in the plane is a straight line. The name "triangle inequality" is derived from this special case.

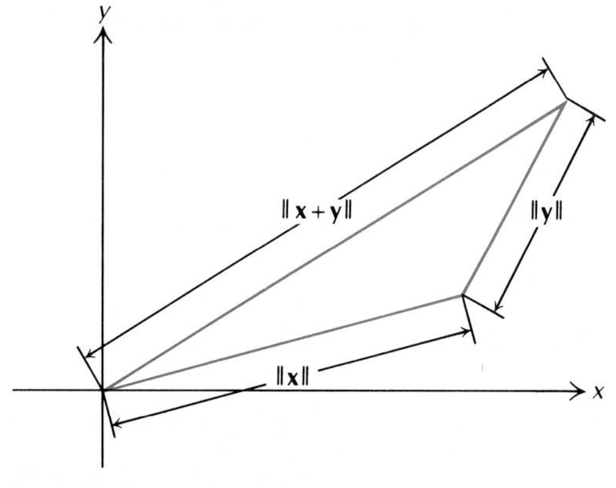

Figure 5.4 Illustration of the triangle inequality in $\mathbb{R}^2$.

Inner Products Determined by Matrices

Up to now we have primarily worked with row vectors, using an $n \times 1$ matrix, often denoted by X or B, when a column vector was needed. At this point, the relative frequency of use changes, and we shall primarily work with column vectors. The notation will shift accordingly: *from here on, we shall use* $\mathbf{x}$ *for a column vector and* $\mathbf{x}^t$ *for the corresponding row vector.*

If B is an invertible $n \times n$ matrix and

$$S = \{\mathbf{x}_1, \ldots, \mathbf{x}_n\}$$

a basis for $\mathbb{R}^n$, then

$$T = \{B\mathbf{x}_1, \ldots, B\mathbf{x}_n\}$$

is also a basis for $\mathbb{R}^n$ (Exercise 30). By properties of matrix multiplication, if

$$\mathbf{x} = \sum_{i=1}^{n} c_i \mathbf{x}_i,$$

then

$$B\mathbf{x} = \sum_{i=1}^{n} c_i B x_i.$$

Thus the replacement of $\mathbf{x}$ by $B\mathbf{x}$ represents a *change of basis* (which will be explored in Chapter 6).

For any $\mathbf{x}$ and $\mathbf{y}$ in $\mathbb{R}^n$, the *standard* inner product of $B\mathbf{x}$ and $B\mathbf{y}$ is given by

$$\langle B\mathbf{x}, B\mathbf{y}\rangle = (B\mathbf{x})^t(B\mathbf{y}).$$

The effect of this operation is the assignment of a scalar to any ordered pair $(\mathbf{x}, \mathbf{y})$ and we have

$$\langle B\mathbf{x}, B\mathbf{y}\rangle = (B\mathbf{x})^t(B\mathbf{y}) = (\mathbf{x}^t B^t)(B\mathbf{y}) = \mathbf{x}^t B^t B\mathbf{y} = \mathbf{x}^t A\mathbf{y},$$

where

$$A = B^t B.$$

Note that $A^t = A$ (why?) and that for every $\mathbf{x}$,

$$\mathbf{x}^t A\mathbf{x} = \langle B\mathbf{x}, B\mathbf{x}\rangle \geq 0.$$

These observations motivate following general concepts.

DEFINITION 5.6

An $n \times n$ matrix A is said to be **symmetric** if

$$A^t = A.$$

A symmetric matrix A is said to be **positive definite** if for every $\mathbf{x}$ in $\mathbb{R}^n$,

$$\mathbf{x}^t A\mathbf{x} \geq 0, \qquad \text{with equality if and only if} \qquad \mathbf{x} = \mathbf{0}.$$

If $A = [a_{ij}]_{n\times n}$, then A is symmetric if and only if $a_{ij} = a_{ji}$ for $i = 1, \ldots, n$ and $j = 1, \ldots, n$. The entries in a symmetric matrix are "reflected across the main diagonal."

Example 7

Show that the matrix

$$A = \begin{bmatrix} 2 & -2 \\ -2 & 5 \end{bmatrix}$$

is positive definite.

Solution

Since $a_{12} = a_{21}$, A is symmetric. In addition, for any $\mathbf{x} = [x_1 \quad x_2]^t$ in $\mathbb{R}^2$,

$$\mathbf{x}^t A\mathbf{x} = 2x_1^2 - 4x_1x_2 + 5x_2^2 = 2(x_1 - x_2)^2 + 3x_2^2,$$

which is nonnegative and zero if and only if $\mathbf{x} = \mathbf{0}$. Hence A is positive definite. □

Example 8

Show that the matrix

$$A = \begin{bmatrix} 2 & -1 & 1 \\ -1 & 4 & 1 \\ 1 & 1 & 4 \end{bmatrix}$$

is positive definite.

Solution

Since $a_{12} = a_{21}$, $a_{13} = a_{31}$, and $a_{23} = a_{32}$, A is symmetric. In addition, for any $\mathbf{x} = [x_1 \quad x_2 \quad x_3]^t$ in $\mathbb{R}^3$,

$$\begin{aligned} \mathbf{x}^t A\mathbf{x} &= 2x_1^2 - 2x_1x_2 + 4x_2^2 + 2x_1x_3 + 2x_2x_3 + 4x_3^2 \\ &= (x_1 - x_2)^2 + (x_1 + x_3)^2 + (x_2 + x_3)^2 + 2x_2^2 + 2x_3^2, \end{aligned}$$

which is nonnegative and zero if and only if $\mathbf{x} = \mathbf{0}$. Hence A is positive definite. ☐

Let A be an $n \times n$ matrix. For $\mathbf{x}$ and $\mathbf{y}$ in $\mathbb{R}^n$, we may assign a scalar $\langle \mathbf{x}, \mathbf{y} \rangle_A$ to the ordered pair $(\mathbf{x}, \mathbf{y})$ as follows:

$$\langle \mathbf{x}, \mathbf{y} \rangle_A = \mathbf{x}^t A \mathbf{y}.$$

The next examples illustrate that this assignment may determine an inner product for $\mathbb{R}^n$.

Example 9

For the matrix A of Example 7,

$$\langle \mathbf{x}, \mathbf{y} \rangle_A = \mathbf{x}^t A \mathbf{y} = \begin{bmatrix} x_1 & x_2 \end{bmatrix} \begin{bmatrix} 2 & -2 \\ -2 & 5 \end{bmatrix} \begin{bmatrix} y_1 \\ y_2 \end{bmatrix}$$

$$= 2x_1y_1 - 2x_1y_2 - 2x_2y_1 + 5x_2y_2.$$

Positive definiteness of this form was established in Example 7. Symmetry can be seen in the middle two terms of the expansion, since

$$-2x_1y_2 - 2x_2y_1 = -2y_1x_2 - 2y_2x_1.$$

This equality is related to the symmetry of A. Linearity in each variable follows from properties of matrix multiplication. Thus for this matrix A, $\langle \mathbf{x}, \mathbf{y} \rangle_A$ is an inner product for $\mathbb{R}^n$. ☐

Example 10

For the matrix A of Example 8,

$$\langle \mathbf{x}, \mathbf{y} \rangle_A = \mathbf{x}^t A \mathbf{y} = \begin{bmatrix} x_1 & x_2 & x_3 \end{bmatrix} \begin{bmatrix} 2 & -1 & 1 \\ -1 & 4 & 1 \\ 1 & 1 & 4 \end{bmatrix} \begin{bmatrix} y_1 \\ y_2 \\ y_3 \end{bmatrix}$$

$$= 2x_1y_1 - x_2y_1 - x_1y_2 + x_3y_1 + x_1y_3 + 4x_2y_2$$

$$+ x_3y_2 + x_2y_3 + 4x_3y_3.$$

An argument along the lines indicated in Example 9 shows that this form is an inner product for $\mathbb{R}^3$. ☐

THEOREM 5.11

If A is a positive definite $n \times n$ matrix, then the function defined by

$$\langle \mathbf{x}, \mathbf{y} \rangle_A = \mathbf{x}^t A \mathbf{y}$$

is an inner product for $\mathbb{R}^n$.

A proof is assigned in Exercise 31. The next theorem asserts that any inner product in $\mathbb{R}^n$ determined by a positive definite matrix may be calculated as a standard inner product relative to a suitable basis. This point will be clarified in the study of the change of basis problem in Chapter 6.

THEOREM 5.12

A square matrix A is positive definite if and only if

$$A = B^t B$$

for some invertible matrix B.

Example 11

We may verify by direct calculation that one factorization in the form $A = B^t B$ for the matrix A in Example 7 is given by

$$\begin{bmatrix} 2 & -2 \\ -2 & 5 \end{bmatrix} = \begin{bmatrix} 2/\sqrt{5} & -\sqrt{6}/\sqrt{5} \\ 1/\sqrt{5} & 2\sqrt{6}/\sqrt{5} \end{bmatrix} \begin{bmatrix} 2/\sqrt{5} & 1/\sqrt{5} \\ -\sqrt{6}/\sqrt{5} & 2\sqrt{6}/\sqrt{5} \end{bmatrix}.$$

(A procedure for finding such a matrix B will be given in Section 7.3.) The values of the inner product given by

$$\langle \mathbf{x}, \mathbf{y} \rangle_A = \mathbf{x}^t A \mathbf{y}$$

coincide with the values

$$\langle B\mathbf{x}, B\mathbf{y} \rangle = (B\mathbf{x})^t (B\mathbf{y}).$$

(See Exercise 32.) □

Example 12

Direct calculation also shows that one factorization in the form $A = B^tB$ for the matrix A in Example 8 is given by

$$\begin{bmatrix} 2 & -1 & 1 \\ -1 & 4 & 1 \\ 1 & 1 & 4 \end{bmatrix} = \begin{bmatrix} 2/\sqrt{6} & 2/\sqrt{3} & 0 \\ 1/\sqrt{6} & -2/\sqrt{3} & \sqrt{5}/\sqrt{2} \\ -1/\sqrt{6} & 2/\sqrt{3} & \sqrt{5}/\sqrt{2} \end{bmatrix}$$

$$\times \begin{bmatrix} 2/\sqrt{6} & 1/\sqrt{6} & -1/\sqrt{6} \\ 2/\sqrt{3} & -2/\sqrt{3} & 2/\sqrt{3} \\ 0 & \sqrt{5}/\sqrt{2} & \sqrt{5}/\sqrt{2} \end{bmatrix},$$

The values of the inner product given by

$$\langle \mathbf{x}, \mathbf{y} \rangle_A = \mathbf{x}^t A \mathbf{y}$$

coincide with the values

$$\langle B\mathbf{x}, B\mathbf{y} \rangle = (B\mathbf{x})^t (B\mathbf{y}). \quad \square$$

The "if" part of Theorem 5.12 follows from the discussion immediately preceding Definition 5.6. The converse will be proved in Section 7.3 in conjunction with the *eigenvalue problem* for positive definite matrices.

REVIEW CHECKLIST

1. Define inner product and inner product space.
2. Identify the standard inner product for $\mathbb{R}^n$.
3. Define weighted inner products and more general inner products defined by matrices.
4. State the Cauchy-Schwarz inequality and apply it to a given inner product.
5. Define the norm of a vector in an inner product space.
6. State the triangle inequality and interpret it in $\mathbb{R}^2$.
7. Define positive definite and symmetric matrices.
8. Test a given matrix for symmetry.
9. Test a given symmetric matrix for positive definiteness.
10. Identify the inner product determined by a positive definite matrix as the standard inner product relative to an invertible matrix (the matrix B of Theorem 5.12).

EXERCISES

In Exercises 1–4, decide whether the given form determines an inner product for $\mathbb{R}^3$.

1. $2x_1y_1 - x_2y_2 + 3x_3y_3$

2. $2x_1y_1 + x_2y_2 + 3x_3y_3$

3. $x_1y_2 + x_2y_1 + x_3y_3$

4. $x_1y_1 + x_2y_2$

5. State explicitly the Cauchy-Schwarz inequality for the $\mathbb{R}^3$ weighted inner product

$$\langle \mathbf{x}, \mathbf{y} \rangle_w = x_1y_1 + 2x_2y_2 + 3x_3y_3.$$

6. Sketch a graph to interpret the triangle inequality applied to $\mathbf{x} = (2, 1)$ and $\mathbf{y} = (-2, 2)$ in $\mathbb{R}^2$.

In Exercises 7–14, determine whether the given matrix is positive definite.

7. $\begin{bmatrix} 1 & 0 \\ 0 & 1 \end{bmatrix}$

8. $\begin{bmatrix} 1 & 1 \\ 1 & 1 \end{bmatrix}$

9. $\begin{bmatrix} 1 & -1 \\ 1 & 1 \end{bmatrix}$

10. $\begin{bmatrix} -1 & 2 & 1 \\ 2 & 1 & 4 \\ 1 & 4 & 2 \end{bmatrix}$

11. $\begin{bmatrix} 1 & 3 & 0 \\ 3 & 2 & 1 \\ 0 & 1 & 3 \end{bmatrix}$

12. $\begin{bmatrix} 1 & 1 & 2 \\ 2 & 3 & 0 \\ 1 & 1 & 2 \end{bmatrix}$

13. $\begin{bmatrix} 1 & 0 & 0 & 0 \\ 0 & 1 & 0 & 0 \\ 0 & 0 & 2 & 0 \\ 0 & 0 & 0 & 1 \end{bmatrix}$

14. $\begin{bmatrix} 1 & 1 & 2 & 2 \\ 1 & 2 & 1 & 0 \\ 2 & 1 & 3 & 2 \\ 2 & 0 & 2 & -1 \end{bmatrix}$

In Exercises 15–16, write out the expansion of $\mathbf{x}^t A \mathbf{y}$ for the given matrix and for general column vectors $\mathbf{x}$ and $\mathbf{y}$ of appropriate size, and show that A is positive definite.

15. $\begin{bmatrix} 2 & 1 \\ 1 & 4 \end{bmatrix}$

16. $\begin{bmatrix} 2 & 1 & 1 \\ 1 & 2 & 2 \\ 1 & 2 & 5 \end{bmatrix}$

17. Show that the standard inner product in $\mathbb{R}^n$ is determined by a matrix, i.e., find a matrix A for which

$$\langle \mathbf{x}, \mathbf{y} \rangle_A = \langle \mathbf{x}, \mathbf{y} \rangle.$$

Find the corresponding matrix for the general weighted inner product of Example 4, page 262.

In Exercises 18–19, let

$$A = \begin{bmatrix} 1 & -1 \\ -1 & 2 \end{bmatrix}, \quad \mathbf{x} = [1 \;\; 3]^t, \quad \text{and} \quad \mathbf{y} = [2 \;\; -1]^t.$$

18. Verify the Cauchy-Schwarz inequality for $\mathbf{x}, \mathbf{y}$, and the inner product determined by A.

19. Verify the triangle inequality for $\mathbf{x}, \mathbf{y}$, and the inner product determined by A.

In Exercises 20–21, let

$$A = \begin{bmatrix} 1 & 1 & 0 \\ 1 & 2 & -2 \\ 0 & -2 & 5 \end{bmatrix}, \qquad \mathbf{x} = [1 \quad 0 \quad 2]', \qquad \text{and} \qquad \mathbf{y} = [1 \quad 1 \quad 1]'.$$

20. Verify the Cauchy-Schwarz inequality for $\mathbf{x}, \mathbf{y}$, and the inner product determined by A.

21. Verify the triangle inequality for $\mathbf{x}, \mathbf{y}$, and the inner product determined by A.

22. (*Calculus based*) Show that the function

$$\langle f, g \rangle = \int_0^1 f(x)g(x)\,dx$$

satisfies the conditions for an inner product in Definition 5.4 for the vector space V of continuous real functions on the interval $[0, 1]$. (See Example 5, page 263.)

23. (*Calculus based*) Show that for $w(x) = x^2$, the function

$$\langle f, g \rangle_w = \int_0^1 x^2 f(x)g(x)\,dx$$

is also an inner product for the vector space V of continuous real functions on the interval $[0, 1]$.

24. Prove that an orthonormal set in an inner product space is linearly independent. (This generalizes Theorem 5.1.)

25. Show that the function defined in Example 4, page 262, satisfies all three conditions for an inner product.

26. For the inner product defined in Example 4, page 262, find a matrix A such that $\langle \mathbf{x}, \mathbf{y} \rangle_w = \mathbf{x}'A\mathbf{y}$ for all $\mathbf{x}$ and $\mathbf{y}$ in $\mathbb{R}^n$.

27. Apply the properties of the inner product in Definition 5.4 to show that for any $\mathbf{x}$ and $\mathbf{y}$ in $\mathbb{R}^n$ and any scalar r,

$$\langle r\mathbf{x} + \mathbf{y}, r\mathbf{x} + \mathbf{y} \rangle = r^2\langle \mathbf{x}, \mathbf{x} \rangle + 2r\langle \mathbf{x}, \mathbf{y} \rangle + \langle \mathbf{y}, \mathbf{y} \rangle.$$

28. Prove that equality is attained in the Cauchy-Schwarz inequality if and only if one of the vectors is a multiple of the other.

29. In the proof of the triangle inequality, show that

$$\langle \mathbf{x} + \mathbf{y}, \mathbf{x} + \mathbf{y} \rangle = \langle \mathbf{x}, \mathbf{x} \rangle + 2\langle \mathbf{x}, \mathbf{y} \rangle + \langle \mathbf{y}, \mathbf{y} \rangle.$$

30. Show that if B is an invertible $n \times n$ matrix and

$$S = \{\mathbf{x}_1, \ldots, \mathbf{x}_n\}$$

a basis for $\mathbb{R}^n$, then

$$T = \{B\mathbf{x}_1, \ldots, B\mathbf{x}_n\}$$

is also a basis for $\mathbb{R}^n$.

31. Prove Theorem 5.11.

32. Show that if $A = B^tB$, then for vectors $\mathbf{x}$ and $\mathbf{y}$,

$$\mathbf{x}^tA\mathbf{y} = (B\mathbf{x})^t(B\mathbf{y}).$$

33. Show that for any square matrix A, $A + A^t$ is symmetric.

34. Show that if A is positive definite, then A is invertible and A^{-1} is positive definite.

35. For a vector $\mathbf{x}$ in $\mathbb{R}^3$ and

$$A = \begin{bmatrix} 1 & 2 & -1 \\ 1 & 3 & 2 \\ 1 & -1 & 2 \end{bmatrix},$$

write out the expansions of $\mathbf{x}^tA\mathbf{x}$ and $\mathbf{x}^tA^t\mathbf{x}$. How are these expansions related? Generalize.

36. Identify the plane curve determined by the equation

$$\begin{bmatrix} x & y \end{bmatrix}\begin{bmatrix} 4 & 0 \\ 0 & 9 \end{bmatrix}\begin{bmatrix} x \\ y \end{bmatrix} = 36.$$

37. If $\mathbf{x} = [x_1 \quad x_2]^t$, find the symmetric matrix A such that

$$\mathbf{x}^tA\mathbf{x} = 2x_1^2 - 6x_1x_2 + 5x_2^2.$$

Is A positive definite?

38. Show that for x, y, and z lying between 0 and π and satisfying

$$\cos^2 x + \cos^2 y + \cos^2 z = 3/4,$$

the maximum value of

$$\cos x + \cos y + \cos z$$

is $3/2$, which is attained when $x = y = z = \pi/3$.

(o) 5.3 APPLICATIONS

Curve Fitting

Observation, recording, and analysis of data are frequent activities in application fields. Data may be obtained by devices, such as thermometers, strain gauges, and densitometers; by simple counting, as in studying occurrences of a specific event; or via governmental agencies, as in determining the gross national product.

Recorded data may be used to construct a model to explain some physical behavior. We also use data to validate a model derived by other means, such as application of a physical law. Still another use is model comparison—which of several valid models *best* describes a particular physical entity?

In many applications, one goal is to discover the *trend of the data*. For instance, do the points in a certain data set lie along a line, a parabola, or an exponential curve? Because of modeling and measurement errors, observed data points seldom lie exactly on a "nice" curve. Thus the problem is to *fit a curve to the data*, i.e., to find a curve of specified type that will be close enough to the data points to capture the trend. By specifying measures of closeness, we determine methods for finding curves to fit the data.

Least Squares Curve Fitting

The classical method for fitting a curve to a set of data points is called *least squares curve fitting*. Let

$$(x_i, y_i), \qquad i = 1, \ldots, m$$

be a given set of data points. We seek a curve $y = f(x)$ whose graph is "close" to the given points by some measure. The error in the approximation at the ith data point,

$$r_i = y_i - f(x_i),$$

is called the ***i*th residual** of the fit. (See Fig. 5.5.) Under the least squares method, the sum of the residuals squared describes closeness of the fit. Thus we seek

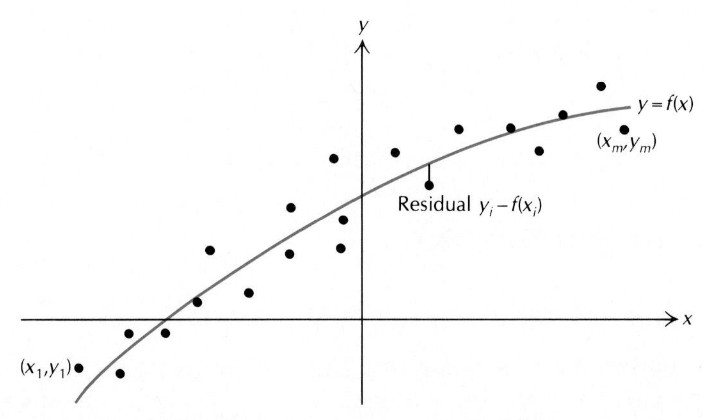

Figure 5.5 Fitting a curve to a set of data.

$f(x)$ for which

$$\sum_{i=1}^{m} (y_i - f(x_i))^2$$

is minimum.

For instance, suppose a graph indicates that the data points lie reasonably along a straight line. Unless the line is vertical, it is determined by a function of the form

$$f(x) = ax + b.$$

For a least squares fit, we seek values of a and b that minimize the sum

$$S = \sum_{i=1}^{m} (y_i - (ax_i + b))^2.$$

To solve this problem by linear algebra (methods based on calculus are also effective), let

$$B = \begin{bmatrix} x_1 & 1 \\ \vdots & 1 \\ x_m & 1 \end{bmatrix}, \qquad \mathbf{u} = \begin{bmatrix} a \\ b \end{bmatrix}, \qquad \text{and} \qquad \mathbf{y} = \begin{bmatrix} y_1 \\ \vdots \\ y_m \end{bmatrix}.$$

Let $\mathbf{r}$ be the **residual vector** corresponding to $\mathbf{u}$, i.e.,

$$\mathbf{r} = \mathbf{y} - B\mathbf{u},$$

with the residuals r_i as components. Then the linear least squares problem is to find the vector $\mathbf{u}$ for which $\|\mathbf{r}\|^2$ is minimum.

Thus we seek a vector $\mathbf{u}$ in $\mathbb{R}^2$ with residual

$$\mathbf{r} = \mathbf{y} - B\mathbf{u}$$

such that for all $\mathbf{v}$ in $\mathbb{R}^2$ with corresponding residual $\mathbf{s}$ given by

$$\mathbf{s} = \mathbf{y} - B\mathbf{v},$$

we have

$$\|\mathbf{r}\|^2 \leq \|\mathbf{s}\|^2.$$

Adding and subtracting $B\mathbf{u}$ in the expression for $\mathbf{s}$ gives

$$\mathbf{s} = \mathbf{y} - B\mathbf{u} + B\mathbf{u} - B\mathbf{v} = \mathbf{r} + B(\mathbf{u} - \mathbf{v}).$$

Hence

$$\begin{aligned}\|\mathbf{s}\|^2 &= [\mathbf{r} + B(\mathbf{u} - \mathbf{v})]^t[\mathbf{r} + B(\mathbf{u} - \mathbf{v})] \\ &= \|\mathbf{r}\|^2 + \mathbf{r}^tB(\mathbf{u} - \mathbf{v}) + [B(\mathbf{u} - \mathbf{v})]^t\mathbf{r} + \|B(\mathbf{u} - \mathbf{v})\|^2 \\ &= \|\mathbf{r}\|^2 + \|B(\mathbf{u} - \mathbf{v})\|^2 + (B^t\mathbf{r})^t(\mathbf{u} - \mathbf{v}) + (\mathbf{u} - \mathbf{v})^t(B^t\mathbf{r}).\end{aligned}$$

Suppose we can find $\mathbf{u}$ for which $B^t\mathbf{r} = \mathbf{0}$. Then for any $\mathbf{v}$,

$$\|\mathbf{s}\|^2 = \|\mathbf{r}\|^2 + \|B(\mathbf{u} - \mathbf{v})\|^2 \geq \|\mathbf{u}\|^2,$$

so this choice of $\mathbf{u}$ yields the required minimum. Since

$$B^t\mathbf{r} = B^t(\mathbf{y} - B\mathbf{u}) = B^t\mathbf{y} - B^tB\mathbf{u},$$

the minimizing vector $\mathbf{u}$ is the solution of the linear system

$$B^tB\mathbf{u} = B^t\mathbf{y}.$$

This system for $\mathbf{u}$ has a unique solution if the x_i are not all the same, i.e., the data do not all lie on one vertical line. We give a proof after formulating the procedure and illustrating it with an example. Our statement of the algorithm includes the term **regression**. This term is commonly used to describe least squares curve fitting.

Linear Least Squares Algorithm

To fit a straight line to points $(x_1, y_1), \ldots, (x_m, y_m)$, with the x_i not all equal, carry out the following steps.

1. Form the $m \times 2$ matrix B with the values x_i in column 1 and with each entry 1 in column 2, and form the column vector $\mathbf{y}$ with the y_i as components.
2. Calculate B^tB and $B^t\mathbf{y}$.
3. Solve the linear system

$$B^tB\mathbf{u} = B^t\mathbf{y}$$

for $\mathbf{u}$ with components a and b. The regression line is

$$y = ax + b.$$

Example 1

Find the linear least squares fit to the following data.

x	0.7	2.5	4.0	7.8	8.4
y	3.2	2.9	7.1	8.2	10.7

Solution

At step 1,

$$B = \begin{bmatrix} 0.7 & 1 \\ 2.5 & 1 \\ 4.0 & 1 \\ 7.8 & 1 \\ 8.4 & 1 \end{bmatrix} \quad \text{and} \quad \mathbf{y} = \begin{bmatrix} 3.2 \\ 2.9 \\ 7.1 \\ 8.2 \\ 10.7 \end{bmatrix}.$$

At step 2,

$$B^tB = \begin{bmatrix} 154.14 & 23.40 \\ 23.40 & 5 \end{bmatrix} \quad \text{and} \quad B^t\mathbf{y} = \begin{bmatrix} 191.73 \\ 32.10 \end{bmatrix}.$$

Solving the linear system

$$B^tB\mathbf{u} = B^t\mathbf{y}$$

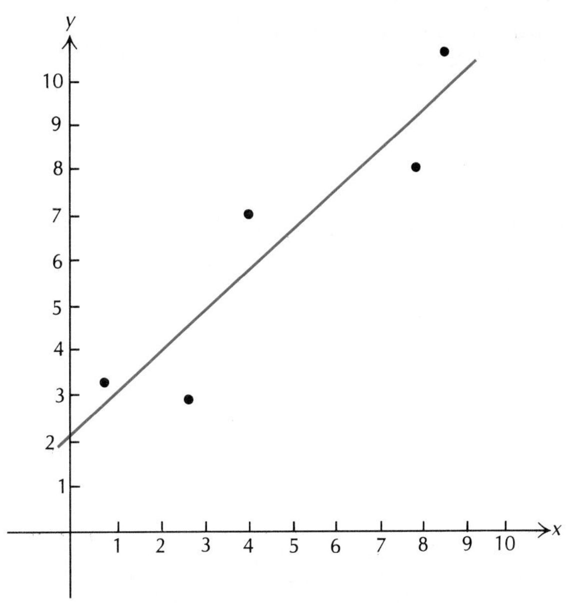

Figure 5.6 Linear least squares fit for the data in Example 1.

for **u** gives the equation of the regression line

$$y = 0.93x + 2.07.$$

(The actual values obtained may depend on the roundoff procedures used.) The residual values,

$$0.48, \ -1.49, 1.31, \ -1.12, 0.82,$$

may be calculated from the formula

$$r_i = y_i - (0.93x_i + 2.07)$$

for $i = 1, \ldots, 5$. The fit is illustrated in Fig. 5.6. □

To see why B^tB is invertible, let $\mathbf{x}$ denote the vector with entries x_i, and $\mathbf{1}$ the vector with each entry 1. Then

$$B^tB = \begin{bmatrix} \sum_{i=1}^{m} x_1^2 & \sum_{i=1}^{m} x_i \\ \sum_{i=1}^{m} x_i & m \end{bmatrix} = \begin{bmatrix} \langle \mathbf{x}, \mathbf{x} \rangle & \langle \mathbf{x}, \mathbf{1} \rangle \\ \langle \mathbf{x}, \mathbf{1} \rangle & \langle \mathbf{1}, \mathbf{1} \rangle \end{bmatrix},$$

from which

$$\det(B^tB) = \langle \mathbf{x}, \mathbf{x} \rangle \langle \mathbf{1}, \mathbf{1} \rangle - \langle \mathbf{x}, \mathbf{1} \rangle^2.$$

By the Cauchy-Schwarz inequality, this determinant is nonnegative and is nonzero since $\mathbf{x}$ is not a constant vector. (In fact, B^tB is positive definite.)

A similar procedure may be used for least squares curve fitting by polynomials of degree higher than one. For instance, suppose the problem is to fit a parabola of the form

$$f(x) = ax^2 + bx + c$$

to data points $(x_1, y_1), \ldots, (x_m, y_m)$. The residuals,

$$r_i = y_i - \left(ax_i^2 + bx_i + c\right)$$

for $i = 1, \ldots, m$ are the components of the vector $\mathbf{r}$ given by

$$\mathbf{r} = \mathbf{y} - B\mathbf{u},$$

where

$$B = \begin{bmatrix} x_1^2 & x_1 & 1 \\ \vdots & \vdots & \vdots \\ x_m^2 & x_m & 1 \end{bmatrix}, \qquad \mathbf{u} = \begin{bmatrix} a \\ b \\ c \end{bmatrix}, \qquad \text{and} \qquad \mathbf{y} = \begin{bmatrix} y_1 \\ \vdots \\ y_m \end{bmatrix}.$$

The argument given for linear least squares shows that the least squares fit for the data is obtained by using the solution of the linear system

$$B^t B\mathbf{u} = B^t\mathbf{y}.$$

The coefficient matrix of this linear system is

$$B^t B = \begin{bmatrix} \sum_{i=1}^m x_i^4 & \sum_{i=1}^m x_i^3 & \sum_{i=1}^m x_i^2 \\ \sum_{i=1}^m x_i^3 & \sum_{i=1}^m x_i^2 & \sum_{i=1}^m x_i \\ \sum_{i=1}^m x_i^2 & \sum_{i=1}^m x_i & m \end{bmatrix}.$$

An analog of the linear regression algorithm holds for *polynomial regression*. (See Exercise 3.) We illustrate the procedure by example.

Example 2

Find the quadratic least squares fit to the data in the following table.

x	0.5	2.2	3.7	5.3	6.5
y	3.1	1.9	1.3	2.7	5.5

Solution

First form B and $\mathbf{y}$ from the given data:

$$B = \begin{bmatrix} 0.25 & 0.5 & 1 \\ 4.84 & 2.2 & 1 \\ 13.69 & 3.7 & 1 \\ 28.09 & 5.3 & 1 \\ 42.25 & 6.5 & 1 \end{bmatrix} \qquad \text{and} \qquad \mathbf{y} = \begin{bmatrix} 3.1 \\ 1.9 \\ 1.3 \\ 2.7 \\ 5.5 \end{bmatrix}.$$

Next, with values rounded to 2 places,

$$B^tB = \begin{bmatrix} 2785.01 & 484.93 & 89.12 \\ 484.93 & 89.12 & 18.20 \\ 89.12 & 18.20 & 5.00 \end{bmatrix} \quad \text{and} \quad B^t\mathbf{y} = \begin{bmatrix} 335.99 \\ 60.60 \\ 14.50 \end{bmatrix}.$$

Solving the linear system

$$B^tB\mathbf{u} = B^t\mathbf{y}$$

for **u** produces the parabola

$$y = 0.32x^2 - 1.94x + 4.16.$$

The residuals,

$$-0.17, 0.46, -0.06, -0.17, 0.43,$$

may be calculated from the formula

$$r_i = y_i - \left(0.32x_i^2 - 1.94x_i + 4.16\right)$$

for $i = 1, \ldots, 5$. The fit is illustrated in Fig. 5.7. □

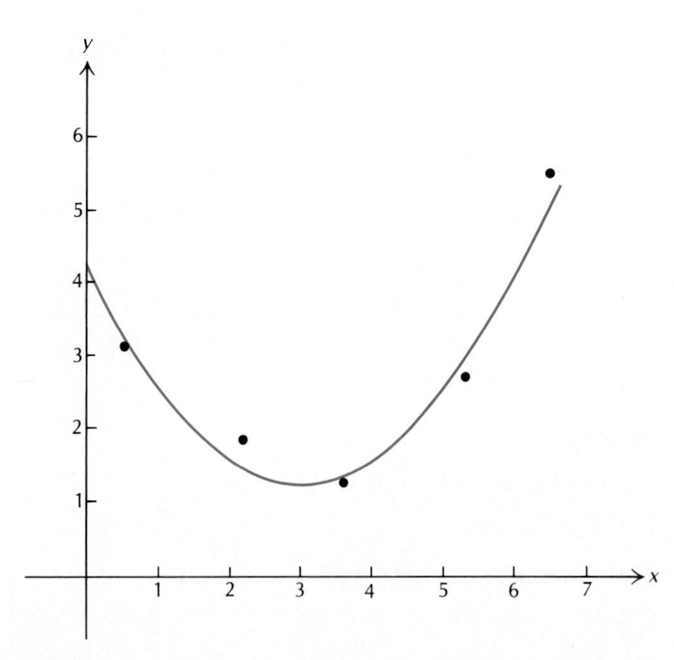

Figure 5.7 Quadratic least squares fit for the data in Example 2.

In fitting a parabola to a data set, we assume at least three distinct x-values. We show that if the data contain three distinct x-values, then B^tB is positive definite. Since

$$\mathbf{w}^t(B^tB)\mathbf{w} = (B\mathbf{w})^t(B\mathbf{w}) \geq 0$$

for any $\mathbf{w}$ in $\mathbb{R}^3$, we need only show that if $\mathbf{w} \neq 0$ then $B\mathbf{w} \neq \mathbf{0}$. Suppose (by changing subscripts, if necessary) that x_1, x_2, and x_3 are distinct. Then the corresponding components of $B\mathbf{w}$ are given by the product

$$\begin{bmatrix} x_1^2 & x_1 & 1 \\ x_2^2 & x_2 & 1 \\ x_3^2 & x_3 & 1 \end{bmatrix} \begin{bmatrix} w_1 \\ w_2 \\ w_3 \end{bmatrix}.$$

The determinant of this system is a *Vandermonde* determinant, and its value is given by

$$d = (x_2 - x_1)(x_3 - x_2)(x_1 - x_3).$$

(See Section 3.4, Exercise 42, page 171.) Since these x-values are distinct, $d \neq 0$, hence the system for these three components of $B\mathbf{w}$ has only the trivial solution. It follows that B^tB is invertible and, in fact, positive definite.

In many applications, the relation between the variables in question is transcendental, for instance, logarithmic or exponential. In such cases, it may be possible to generate a new data set by transforming the original data, and apply least squares to the transformed data.

Example 3

By a suitable transformation, fit a curve of the form

$$y = f(x) = ae^{bx}$$

to the data in the following table.

x	1	2	3	4	5	6
y	4	10	20	40	90	190

Solution

Take the natural logarithm of y and ae^{bx} to obtain

$$\ln y = bx + \ln a.$$

Calculate a table of values for $\ln y$, then fit a straight line to the points in the *transformed data* table.

x	1	2	3	4	5	6
$\ln y$	1.386	2.303	2.996	3.689	4.500	5.247

The linear least squares algorithm yields

$$B = \begin{bmatrix} 1 & 1 \\ 2 & 1 \\ 3 & 1 \\ 4 & 1 \\ 5 & 1 \\ 6 & 1 \end{bmatrix} \quad \text{and} \quad \ln \mathbf{y} = \begin{bmatrix} 1.386 \\ 2.303 \\ 2.996 \\ 3.689 \\ 4.500 \\ 5.247 \end{bmatrix},$$

from which

$$B^tB = \begin{bmatrix} 91 & 21 \\ 21 & 6 \end{bmatrix} \quad \text{and} \quad B^t\mathbf{y} = \begin{bmatrix} 83.718 \\ 20.121 \end{bmatrix}.$$

Solve the system

$$B^tB\mathbf{u} = B^t\mathbf{y}$$

for $\mathbf{u}$ to obtain the slope and the intercept of the regression line for the transformed data:

$$b = 0.760, \qquad \ln a = 0.695.$$

The required exponential curve is

$$y = 2.00e^{0.76}.$$

The residuals for this fit may be found from the relation

$$r_i = y_i - ae^x;$$

they are

$$-0.28, 0.86, 0.45, -1.81, 0.60, -1.17,$$

respectively. The fit is illustrated in Fig. 5.8. ▭

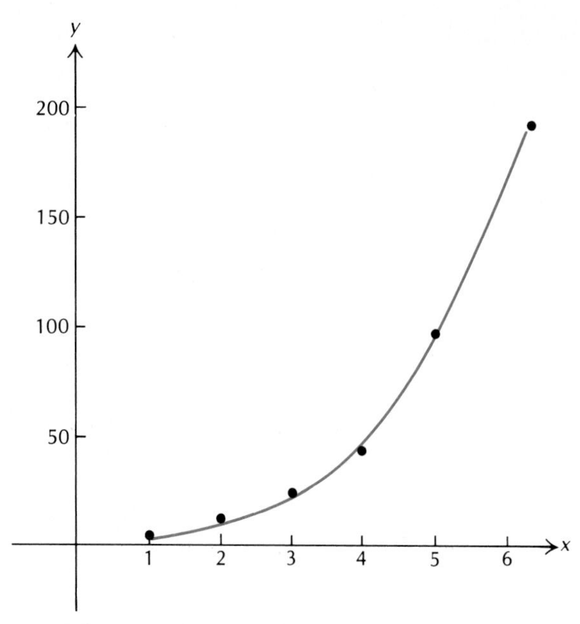

Figure 5.8 Exponential fit for the original data in Example 3 obtained from a linear least squares fit for a transformed data set.

We emphasize that the curve we found in Example 3 is *not the least squares fit* for the given data. The curve we found fits the given data, but not in the least squares sense. The least squares problem we solved was for the transformed data set.

Related Concepts and Computing

Many questions arise concerning inferences to be drawn from a particular least squares curve fitting and the level of confidence we may expect in predictions based on it. These are *statistical* questions, and this topic is often treated more thoroughly in probability and statistics.

In regression studies more generally, criteria for curve fitting are often described in terms of *measures of goodness of fit*. For least squares, the actual measure is called the **root mean square error** and is defined as

$$(\langle \mathbf{r}, \mathbf{r} \rangle / m)^{1/2},$$

where $\mathbf{r}$ is the residual vector and m the number of data points. This quantity represents one kind of "average error" and, although individual residuals may be large, the least squares procedure minimizes this error.

Another criterion used to fit a curve to a set of data points involves minimizing the maximum absolute residual. Under this method, which is often called Chebyshev curve fitting, we seek a curve of prescribed type that minimizes

$$\max\{|r_i| : i = 1, \ldots, m\}$$

A problem of this type may be regarded as a special *linear program*. Linear programming is studied in Chapter 8.

In practice, some observations may be "more important" than others in some sense, possibly because they are believed to be more reliable. For such problems, we may attach different weights to the residuals, thus minimizing the weighted least squares error. The theory of general inner products (Section 5.2) provides a conceptual framework for weighted least squares curve fitting.

In computing for least squares curve fitting, the implementation of algorithms based on the **normal equations**, i.e., the equations in the linear system

$$B^t B\mathbf{u} = B\mathbf{y},$$

is satisfactory for many problems, but certainly not all. For some problems, it may be better to base algorithms on the Gram-Schmidt procedure of Section 5.1. Still other algorithms use the *singular value decomposition*, which is usually treated in more advanced studies of applied linear algebra. Computing for least squares curve fitting is often included in courses on numerical analysis.

REVIEW CHECKLIST

1. Identify the criterion used for least squares curve fitting.
2. Formulate a statement of the polynomial least squares algorithm.
3. Fit polynomials of specified type to given data points.
4. Fit exponential and logarithmic curves to a given set of points by transforming the data.
5. Find the residual vector and the root mean square error of a polynomial fit.
6. Identify conditions under which the coefficient matrix of normal equations is positive definite.

EXERCISES

In Exercises 1–3, plot the points in the given data set, and find and plot the linear least squares fit.

1.

x	0	1	2	3
y	1.2	2.0	3.2	4.1

2.

x	1	3	4	9	10
y	2.5	2.4	6.1	11.3	11.0

3.

x	1.1	2.5	3.0	3.0	5.8	8.1	10.0
y	8.6	8.8	6.2	5.4	3.6	4.0	2.5

For Exercises 4–6, find the residual vector and the root mean square error for the data sets in Exercises 1–3.

7. Formulate a statement of a least squares algorithm for fitting polynomials of any degree to given data points.

In Exercises 8–10, plot the points in the given data set, and find and plot the quadratic least squares fit.

8.

x	1	2	3	4	5
y	4.3	5.0	4.0	3.4	1.0

9.

x	1	1	2	5	8	10
y	2	1	-1	-4	0	3

10.

x	0.5	1.5	2.5	3.5	4.5
y	5.0	2.2	0.1	-1.0	-1.5

For Exercises 11–13, find the residual vector and the root mean square error for the data sets in Exercises 8–10.

14. Write out a general expression for the normal equations for cubic least squares curve fitting.

In Exercises 15–16, plot the points in the given data set, and find and plot the cubic least squares fit.

15.

x	-2	-1	0	1	2
y	-20	-6	2	1	10

16.

x	0	1	2	3	4
y	-5.8	0.2	-0.1	0.1	5.5

For Exercises 17–18, find the residual vector and the root mean square error for the data sets in Exercises 15–16.

In Exercises 19–20, use a transformation of the data to fit a curve of the form

$$y = f(x) = ae^{bx}$$

to the data in the given table.

19.

x	0	2	4	6	8
y	1.2	1.1	1.5	2.5	3.0

20.

x	1.1	2.3	5.0	6.2	7.0
y	5.4	4.5	3.5	2.5	2.3

21. To test a belief that up to a point, sales volume is directly proportional to advertising expense, a retailer collects data for a full year. The values of x are the bimonthly advertising costs in thousands of dollars, and the corresponding values of y are the seasonally adjusted gross bimonthly sales figures in thousands. The figures, adjusted for inflation, are given in the following table.

x	1.0	1.2	1.4	1.6	1.8	2.0
y	50.1	55.6	52.3	64.0	58.6	68.0

Plot the data points on a graph, and find and plot the linear least squares fit.

22. A chemical used to control bacteria in a swimming pool is tested for effectiveness. In the following data table, x represents the number of gallons of the chemical mixture in the pool and y represents the corresponding bacteria count in thousands.

x	1	2	3	4	5	6
y	4.2	3.3	3.1	2.7	2.5	2.0

Plot the data points on a graph, and find and plot the linear least squares fit.

23. An electronic circuit component is tested for its response y to an input value x, with both parameters measured in suitable units. The data readings are recorded in the following table.

x	1.0	1.1	1.2	1.3	1.4	1.5
y	2.8	3.1	3.4	3.5	3.7	3.8

Plot the data points on a graph, and find and plot the linear least squares fit.

REVIEW EXERCISES

In Exercises 1–2, determine whether the given set is orthonormal in $\mathbb{R}^3$.

1. $\{(1/\sqrt{3}, 1/\sqrt{3}, 1/\sqrt{3}), (1/\sqrt{3}, -1/\sqrt{3}, -1/\sqrt{3}), (0, 1/\sqrt{2}, -1/\sqrt{2})\}$

2. $\{1/\sqrt{3}, 1/\sqrt{3}, 1/\sqrt{3}), (1/\sqrt{6}, -2/\sqrt{6}, 1/\sqrt{6}), (1/\sqrt{2}, 0, -1/\sqrt{2})\}$

In Exercises 3–4, express the vector $\mathbf{x} = (1, 2)$ in $\mathbb{R}^2$ as a linear combination of the vectors in the given set.

3. $\{(3/5, 4/5), (-4/5, 3/5)\}$ **4.** $\{(5/13, 12/13), (-12/13, 5/13)\}$

5. Calculate the norm of the vector $\mathbf{x} = (1, 2)$ in $\mathbb{R}^2$ in three distinct ways.

In Exercises 6–8, let $\mathbf{x} = (1, -1, 3)$, and let V be the span of the set $\{(1/\sqrt{3}, 1/\sqrt{3}, 1/\sqrt{3}), (1/\sqrt{2}, 0, -1/\sqrt{2})\}$ in $\mathbb{R}^3$.

6. Find the projection of $\mathbf{x}$ onto V.

7. Find the projection of $\mathbf{x}$ onto $V^\perp$.

8. Find the norm of $\mathbf{x}$ in two distinct ways.

In Exercises 9–10, apply the Gram-Schmidt procedure to the given set in $\mathbb{R}^3$.

9. $\{(1, 1, 0), (1, -2, 1), (1, 1, 1)\}$ **10.** $\{(1, -1, -1), (2, 1, 0), (1, 1, 2)\}$

11. Find an orthonormal basis for the nullspace of the matrix

$$A = \begin{bmatrix} 1 & -3 & 2 \\ 2 & -6 & 4 \\ -3 & 9 & -6 \end{bmatrix}.$$

Linear Transformations

Functions play a key role in the theory and the application of linear algebra. Special functions called linear transformations *are the main objects of study in a number of mathematical subjects. For us, they give additional insight into linear systems and complete the preparations for the eigenvalue problem. Linear transformations are also widely used in practice. In mechanics, they may describe the deformation of an elastic medium produced by external forces. In electronics, they may be used for signal selection and noise suppression in a communications system. In biology, they may describe changes in the distribution of genetic traits from one generation to the next. And in computer graphics, they may be used for the display of a surface on a video screen.*

6.1 LINEAR TRANSFORMATIONS, MATRICES, AND LINEAR SYSTEMS

In this section, we introduce the concept of a linear transformation, illustrate it by examples, and develop some of its basic properties. We also study the interplay between linear transformations and linear systems.

Concept and Basic Properties

We consider functions for which the domain and the codomain are vector spaces. The symbol T will often denote such a function.

DEFINITION 6.1

Let V and W be vector spaces. A function $T: V \to W$ is called a **linear transformation**, or a **linear mapping**, if for all vectors $\mathbf{u}$ and $\mathbf{v}$ in V and for every scalar c,

$$T(\mathbf{u} + \mathbf{v}) = T(\mathbf{u}) + T(\mathbf{v}) \qquad \text{and} \qquad T(c\mathbf{u}) = cT(\mathbf{u}).$$

If $W = V$, then T is called a **linear operator**. If $W = \mathbb{R}$, then the mapping is a **linear functional** and we use f instead of T.

Example 1

Show that the function defined for $\mathbf{u} = (u_1, u_2, u_3)$ by

$$T(\mathbf{u}) = (3u_2 - u_3, 2u_1 + 4u_2)$$

is a linear transformation from $\mathbb{R}^3$ into $\mathbb{R}^2$.

Solution

If $\mathbf{v} = (v_1, v_2, v_3)$ in $\mathbb{R}^3$, then

$$\mathbf{u} + \mathbf{v} = (u_1 + v_1, u_2 + v_2, u_3 + v_3).$$

Substituting into the formula for T gives

$$\begin{aligned} T(\mathbf{u} + \mathbf{v}) &= (3(u_2 + v_2) - (u_3 + v_3), 2(u_1 + v_1) + 4(u_2 + v_2)) \\ &= ((3u_2 - u_3) + (3v_2 - v_3), (2u_1 + 4u_2) + (2v_1 + 4v_2)) \\ &= (3u_2 - u_3, 2u_1 + 4u_2) + (3v_2 - v_3, 2v_1 + 4v_2) \\ &= T(\mathbf{u}) + T(\mathbf{v}). \end{aligned}$$

For any scalar c,

$$c\mathbf{u} = (cu_1, cu_2, cu_3).$$

Using the formula for T yields

$$\begin{aligned} T(c\mathbf{u}) &= (3(cu_2) - (cu_3), 2(cu_1) + 4(cu_2)) \\ &= c(3u_2 - u_3, 2u_1 + 4u_2) \\ &= cT(\mathbf{u}). \end{aligned}$$

Thus T is a linear transformation by Definition 6.1. □

Example 2

Find the range of the linear transformation T in Example 1.

Solution

The range of T consists of all vectors in $\mathbb{R}^2$ of the form

$$(3u_2 - u_3, 2u_1 + 4u_2),$$

where u_1, u_2, and u_3 are real numbers. For any (b_1, b_2) in $\mathbb{R}^2$, the equation

$$(3u_2 - u_3, 2u_1 + 4u_2) = (b_1, b_2)$$

is equivalent to the linear system

$$\begin{aligned} 3u_2 - u_3 &= b_1 \\ 2u_1 + 4u_2 \qquad &= b_2. \end{aligned}$$

This system is consistent (why?), so for any $\mathbf{b}$ in $\mathbb{R}^2$ there is a $\mathbf{u}$ in $\mathbb{R}^3$ for which $T(\mathbf{u}) = \mathbf{b}$. Thus the range of T coincides with $\mathbb{R}^2$. ☐

Example 3

By properties of the inner product, the function defined for any $\mathbf{u} = (u_1, u_2, u_3)$ in $\mathbb{R}^3$ by

$$f(\mathbf{u}) = 3u_1 - 2u_2 + u_3$$

is a linear functional on $\mathbb{R}^3$. This functional is also given by

$$f(\mathbf{u}) = \langle \mathbf{a}, \mathbf{u} \rangle,$$

where

$$\mathbf{a} = (3, -2, 1).$$

The range of f coincides with $\mathbb{R}$ because for any real number r, the equation

$$3u_1 - 2u_2 + u_3 = r$$

is consistent. ☐

Example 4

If T is linear from $\mathbb{R}^2$ to $\mathbb{R}^3$ and

$$T(1,0) = (4, -2, 3) \quad \text{and} \quad T(0,1) = (-2, -3, 1),$$

find a formula for $T(u_1, u_2)$, and describe the range of T.

Solution

By the standard decomposition of a vector in $\mathbb{R}^2$ and linearity of T,

$$\begin{aligned} T(u_1, u_2) &= T(u_1(1,0) + u_2(0,1)) \\ &= u_1T(1,0) + u_2T(0,1) \\ &= u_1(4,-2,3) + u_2(-2,-3,1) \\ &= (4u_1 - 2u_2, -2u_1 - 3u_2, 3u_1 + u_2). \end{aligned}$$

Since

$$(4u_1 - 2u_2, -2u_1 - 3u_2, 3u_1 + u_2) = u_1(4,-2,3) + u_2(-2,-3,1),$$

the range of T coincides with the span of the set $\{(4,-2,3),(-2,-3,1)\}$ in $\mathbb{R}^3$. This transformation is illustrated in Fig. 6.1. □

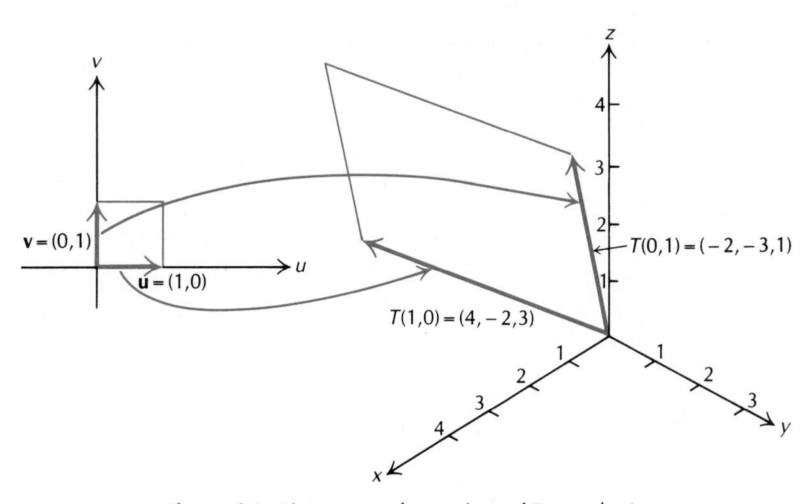

Figure 6.1 Linear transformation of Example 4.

Example 5

Let V be the vector space of all real polynomials, and define $T\colon V \to V$ as the operator corresponding to multiplication by x:

$$T(a_0 + a_1x + \cdots + a_n) = a_0x + a_1x^2 + \cdots + a_nx^{n+1}.$$

Then T is a linear operator on V (Exercise 40), and the range of T consists of all elements of V except the nonzero constants (Exercise 41). □

We make an important observation based on Definition 6.1 and Examples 1–4.

Proposition 1 The range R of any linear transformation $T: V \to W$ is a subspace of W.

Proof Let $\mathbf{u}$ and $\mathbf{v}$ be in V, so $T(\mathbf{u})$ and $T(\mathbf{v})$ are arbitrary vectors in R. Since T is a linear transformation,

$$T(\mathbf{u}) + T(\mathbf{v}) = T(\mathbf{u} + \mathbf{v}).$$

Since V is a vector space, $\mathbf{u} + \mathbf{v}$ is in V. Consequently, $T(\mathbf{u}) + T(\mathbf{v})$ is in R. For any scalar c,

$$cT(\mathbf{u}) = T(c\mathbf{u}).$$

Since V is a vector space, $c\mathbf{u}$ is in V and hence $cT(\mathbf{u})$ is in R. Therefore R is closed under the vector operations, so R is a subspace of W. □

The two conditions defining a linear transformation are equivalent to the single condition

$$T(c_1\mathbf{u}_1 + c_2\mathbf{u}_2) = c_1T(\mathbf{u}_1) + c_2T(\mathbf{u}_2)$$

for all vectors $\mathbf{u}_1$ and $\mathbf{u}_2$ in V and all scalars c_1 and c_2 (Exercise 32). This condition may be extended by induction to sums of any finite number of vectors (Exercise 33).

Proposition 2 A mapping $T: V \to W$ is linear if and only if

$$T(c_1\mathbf{u}_1 + \cdots + c_n\mathbf{u}_n) = c_1T(\mathbf{u}_1) + \cdots + c_nT(\mathbf{u}_n)$$

for all $\mathbf{u}_1, \ldots, \mathbf{u}_n$ in V and all scalars $c_1, \ldots, c_n$.

Proposition 2 has an important corollary that was illustrated in Example 4. Suppose the images $T(\mathbf{u}_1), \ldots, T(\mathbf{u}_n)$ of the vectors in a basis for V are known. Then the image of any

$$\mathbf{u} = c_1\mathbf{u}_1 + \cdots + c_n\mathbf{u}_n$$

in V is given by

$$T(\mathbf{u}) = c_1T(\mathbf{u}_1) + \cdots + c_nT(\mathbf{u}_n).$$

Corollary to Proposition 2 A linear transformation T is determined by the images of the vectors in a basis for the domain of T.

Linear Transformations and Matrices

If A is an $m \times n$ matrix, then for any $\mathbf{u}$ in $\mathbb{R}^n$ in column form, $A\mathbf{u}$ is in $\mathbb{R}^m$. This multiplication associates a unique vector in $\mathbb{R}^m$ with any given vector in $\mathbb{R}^n$.

Example 6

If

$$A = \begin{bmatrix} 3 & 1 \\ -1 & 2 \\ 0 & 1 \end{bmatrix} \quad \text{and} \quad \mathbf{u} = \begin{bmatrix} u_1 \\ u_2 \end{bmatrix},$$

then

$$A\mathbf{u} = \begin{bmatrix} 3 & 1 \\ -1 & 2 \\ 0 & 1 \end{bmatrix} \begin{bmatrix} u_1 \\ u_2 \end{bmatrix} = \begin{bmatrix} 3u_1 + u_2 \\ -u_1 + 2u_2 \\ u_2 \end{bmatrix}.$$

Defining $T: \mathbb{R}^2 \to \mathbb{R}^3$ by

$$T(\mathbf{u}) = A\mathbf{u},$$

we show that T is a linear transformation. For any $\mathbf{u}$ and $\mathbf{v}$ in $\mathbb{R}^2$,

$$T(\mathbf{u} + \mathbf{v}) = A(\mathbf{u} + \mathbf{v}) = A\mathbf{u} + A\mathbf{v} = T(\mathbf{u}) + T(\mathbf{v}).$$

The center equality follows from the left distributive law of matrix multiplication. In addition, for any $\mathbf{u}$ in $\mathbb{R}^2$ and any scalar c,

$$T(c\mathbf{u}) = A(c\mathbf{u}) = cA\mathbf{u} = cT(\mathbf{u}).$$

The center equality holds because the scalar c may be distributed to each component of $\mathbf{u}$ either before or after the multiplication by A, by the commutative and associative laws of multiplication in $\mathbb{R}$ and the distributive law in R. □

The converse of the principle embodied in Example 6 is also true.

Example 7

Find the matrix A of the linear mapping $T: \mathbb{R}^3 \to \mathbb{R}^2$ defined by

$$T(\mathbf{u}) = T(u_1, u_2, u_3) = (2u_1 - u_2 + 7u_3, 4u_2 + 2u_3).$$

Solution

The matrix of T can be obtained by inspection:

$$A = \begin{bmatrix} 2 & -1 & 7 \\ 0 & 4 & 2 \end{bmatrix}.$$

In column form,

$$T\begin{bmatrix} u_1 \\ u_2 \\ u_3 \end{bmatrix} = A\mathbf{u} = \begin{bmatrix} 2 & -1 & 7 \\ 0 & 4 & 2 \end{bmatrix}\begin{bmatrix} u_1 \\ u_2 \\ u_3 \end{bmatrix} = \begin{bmatrix} 2u_1 - u_2 + 7u_3 \\ 4u_2 + 2u_3 \end{bmatrix}. \quad \square$$

THEOREM 6.1

A mapping T from $\mathbb{R}^n$ to $\mathbb{R}^m$ is a linear transformation if and only if there is an $m \times n$ matrix A such that for every column vector $\mathbf{u}$ in $\mathbb{R}^n$,

$$T(\mathbf{u}) = A\mathbf{u}.$$

Proof For a given A, the linearity of T follows from the argument given in Example 6.

Let T be a linear transformation from $\mathbb{R}^n$ to $\mathbb{R}^m$. If $\mathbf{e}_1, \ldots, \mathbf{e}_n$ are the standard basis vectors for $\mathbb{R}^n$ in column form, then for any $\mathbf{u} = [u_1 \quad \cdots \quad u_n]^t$ in $\mathbb{R}^n$,

$$\mathbf{u} = u_1\mathbf{e}_1 + \cdots + u_n\mathbf{e}_n.$$

Hence if

$$\mathbf{v}_j = T(\mathbf{e}_j)$$

for $j = 1, \ldots, n$, then by Proposition 2,

$$T(\mathbf{u}) = T(u_1\mathbf{e}_1 + \cdots + u_n\mathbf{e}_n) = u_1\mathbf{v}_1 + \cdots + u_n\mathbf{v}_n.$$

Thus if A is the matrix with columns $\mathbf{v}_1, \ldots, \mathbf{v}_n$, then by the "column space interpretation" from Section 4.5,

$$T(\mathbf{u}) = A\mathbf{u}. \quad \square$$

Example 8 (Orthogonal Projection)

Find the matrix representation of the orthogonal projection of $\mathbb{R}^2$ onto the line $y = 2x$. (See Fig. 6.2.)

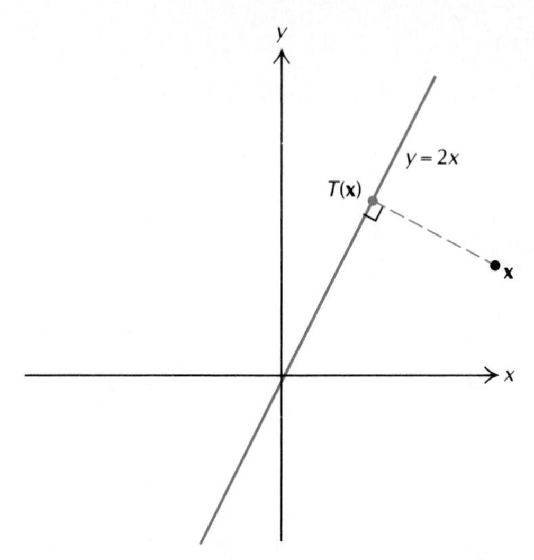

Figure 6.2 Linear transformation of Example 8.

Solution

The given line is spanned by the unit vector $\mathbf{v} = (1/\sqrt{5}, 2/\sqrt{5})$, hence by Definition 5.3, the image of $\mathbf{x} = (x_1, x_2)$ in $\mathbb{R}^2$ is given by

$$P(\mathbf{x}) = \langle \mathbf{x}, \mathbf{v} \rangle \mathbf{v} = \left(x_1/\sqrt{5} + 2x_2/\sqrt{5}\right)\left(1/\sqrt{5}, 2/\sqrt{5}\right)$$

$$= \left(x_1/5 + 2x_2/5, 2x_1/5 + 4x_2/5\right).$$

Thus the matrix of P is:

$$A = \begin{bmatrix} \frac{1}{5} & \frac{2}{5} \\ \frac{2}{5} & \frac{4}{5} \end{bmatrix}. \qquad \square$$

We note in Example 8 that

$$T\begin{bmatrix} 1 \\ 0 \end{bmatrix} = \begin{bmatrix} \frac{1}{5} \\ \frac{2}{5} \end{bmatrix} \quad \text{and} \quad T\begin{bmatrix} 0 \\ 1 \end{bmatrix} = \begin{bmatrix} \frac{2}{5} \\ \frac{4}{5} \end{bmatrix}.$$

That is, each column of A is the image of a standard basis vector. This idea provides a useful way to construct A in practical applications. This idea was used in the proof of Theorem 6.1, and it is used again in the next result to find another connection between linear transformations and linear systems.

THEOREM 6.2

The range of a linear transformation T from $\mathbb{R}^n$ into $\mathbb{R}^m$ coincides with the column space of the matrix of T.

Proof Let $\{\mathbf{e}_1, \ldots, \mathbf{e}_n\}$ be the standard basis for $\mathbb{R}^n$. Then for any vector $\mathbf{u} = (u_1, \ldots, u_n)$ in $\mathbb{R}^n$,

$$T(\mathbf{u}) = T(u_1\mathbf{e}_1 + \cdots + u_n\mathbf{e}_n) = u_1T(\mathbf{e}_1) + \cdots + u_nT(\mathbf{e}_n).$$

Let A be the matrix of T. Since each $T(\mathbf{e}_i)$ is a column of A, $T(\mathbf{u})$ is in the column space of A. Since $\mathbf{u}$ is arbitrary in $\mathbb{R}^n$, the range of T is contained in the column space of A.

Conversely, if $\mathbf{v}$ is in the column space of A, then

$$\mathbf{v} = c_1T(\mathbf{e}_1) + \cdots + c_nT(\mathbf{e}_n) = T(c_1\mathbf{e}_1 + \cdots + c_n\mathbf{e}_n),$$

and $\mathbf{v}$ is in the range of T. □

Linear Transformations and Linear Systems

Theorem 6.2 suggests a connection between linear transformations and linear systems, since the column space of a coefficient matrix coincides with the range of the mapping determined by A.

DEFINITION 6.2

The **kernel**, or **nullspace**, of a linear transformation T is the set of all $\mathbf{u}$ in the domain of T for which $T(\mathbf{u}) = \mathbf{0}$.

Example 9

Find the kernel of the linear transformation $T: \mathbb{R}^2 \to \mathbb{R}^2$ defined by

$$T(u_1, u_2) = (2u_1 - u_2, 4u_1 - 2u_2).$$

Solution

The vector equation $T(\mathbf{u}) = \mathbf{0}$ takes the form

$$(2u_1 - u_2, 4u_1 - 2u_2) = (0, 0),$$

which is equivalent to the homogeneous linear system

$$\begin{aligned} 2u_1 - u_2 &= 0 \\ 4u_1 - 2u_2 &= 0. \end{aligned}$$

The kernel of T, which is the solution space of this linear system, is the span of the vector $(1, 2)$. □

Example 9 illustrates the following general principle. A proof is assigned in Exercise 44.

Proposition 3 The kernel of a linear transformation $T: V \to W$ is a subspace of V.

We use the following result in proving the next theorem. A proof is assigned in Exercise 36.

Proposition 4 For a linear tranformation $T: V \to W$, the image of the additive identity in V is the additive identity in W:

$$T(\mathbf{0}) = \mathbf{0}.$$

THEOREM 6.3

The following are equivalent for a matrix A and the linear transformation T determined by A:

a. the system $A\mathbf{x} = \mathbf{0}$ has only the trivial solution;
b. T is one-to-one;
c. the kernel of T consists of $\mathbf{0}$ alone.

Proof

(a) implies (b). Suppose the system $A\mathbf{x} = \mathbf{0}$ has only the trivial solution. Then for any $\mathbf{u}$ and $\mathbf{v}$ in the domain of T,

$$A(\mathbf{u} - \mathbf{v}) = T(\mathbf{u} - \mathbf{v}) = T(\mathbf{u}) - T(\mathbf{v}).$$

Thus if $T(\mathbf{u}) = T(\mathbf{v})$, then

$$A(\mathbf{u} - \mathbf{v}) = \mathbf{0}.$$

Since only the trivial solution exists, $\mathbf{u} - \mathbf{v} = \mathbf{0}$, hence $\mathbf{u} = \mathbf{v}$.

(b) implies (c). Since $T(\mathbf{0}) = \mathbf{0}$ by Proposition 4, $\mathbf{0}$ is in the kernel of T. Since $T(\mathbf{0}) = \mathbf{0}$, if T is one-to-one and $\mathbf{u} \neq \mathbf{0}$, then $T(\mathbf{u}) \neq T(\mathbf{0})$, so the kernel of T consists of $\mathbf{0}$ alone.

(c) implies (a). Since $T(\mathbf{u}) = A\mathbf{u}$ for all $\mathbf{u}$ in the domain of T, if $\mathbf{0}$ is the only vector $\mathbf{u}$ for which $T(\mathbf{u}) = \mathbf{0}$, it is the only $\mathbf{u}$ for which $A\mathbf{u} = \mathbf{0}$. □

Other concepts from linear systems may be formulated in terms of linear transformations. For instance, suppose T is the linear operator determined by a square matrix A. By Theorems 6.3 and 4.13, A is invertible if and only if T is one-to-one. Thus the equivalence of conditions (a) and (c) in Theorem 4.14 may be reformulated in terms of linear transformations.

Proposition 5 Let A be the matrix of a linear operator T on $\mathbb{R}^n$. Then the linear system

$$A\mathbf{x} = \mathbf{b}$$

has a unique solution if and only if T is one-to-one.

As Theorem 6.2 suggests, the dimension of the range of a linear transformation is a useful quantity for linear systems.

DEFINITION 6.3

The **rank of a linear transformation** T is the dimension of the range of T.

Definition 6.3 allows a useful reformulation of Theorem 4.15.

Proposition 6 Let $T: \mathbb{R}^n \rightarrow \mathbb{R}^m$ be a linear transformation. If r is the rank of T and k is the dimension of the kernel of T, then $r + k = n$.

Proof Let A be the matrix of T. Since the range of T coincides with the column space of A by Theorem 6.2, r equals the column rank of A, hence also the rank by Theorem 4.12. Since $T(\mathbf{x}) = \mathbf{0}$ if and only if $A\mathbf{x} = \mathbf{0}$, the kernel of T coincides with the nullspace of A, so the dimension of the nullspace of A equals k. By Theorem 4.15, $r + k = n$. ☐

Example 10

The matrix of the linear transformation T defined by

$$T(x_1, x_2, x_3) = (x_1 + 2x_2 - x_3, 2x_1 + 3x_2 + x_3, 2x_1 + x_2 + 7x_3)$$

is

$$A = \begin{bmatrix} 1 & 2 & -1 \\ 2 & 3 & 1 \\ 2 & 1 & 7 \end{bmatrix}.$$

The kernel of T, which coincides with the nullspace of A, is the line in $\mathbb{R}^3$

spanned by the vector $(5, -3, -1)$. Hence the kernel of T has dimension 1. The range of T, which coincides with the column space of A, is spanned by the vectors

$$\mathbf{x} = \begin{bmatrix} 1 \\ 2 \\ 2 \end{bmatrix} \quad \text{and} \quad \mathbf{y} = \begin{bmatrix} 2 \\ 3 \\ 1 \end{bmatrix},$$

since the third column of A equals $5\mathbf{x} - 3\mathbf{y}$. The set $\{\mathbf{x}, \mathbf{y}\}$ is linearly independent, so T has rank 2. This result verifies the relation $r + k = n$ in Proposition 6. □

Additional relationships between linear transformations and linear systems are explored in the exercises.

REVIEW CHECKLIST

1. Formulate a definition of linear transformation.
2. Characterize a mapping as linear in terms of preserving linear combinations with finitely many summands.
3. Given the values of a linear transformation on a basis, find the image of any element in the domain.
4. Given a linear transformation T, find the matrix of T, and vice versa.
5. Given a geometric description of a linear transformation (such as an orthogonal projection), find its formula.
6. Identify the range of a linear transformation T with the column space of the matrix of T.
7. Define the rank and the kernel of a linear transformation.
8. Characterize a linear transformation T as one-to-one in terms of the kernel of T.
9. Formulate conditions relating to the solution of a linear system in terms of linear transformation concepts.
10. Relate the rank of a linear transformation to the dimensions of its domain and its range.

EXERCISES

In Exercises 1–5, use Definition 6.1 to determine whether or not the given function T is a linear transformation.

1. $T(u_1, u_2) = (u_1 - 4u_2, u_2)$
2. $T(u_1, u_2) = (u_1 + u_2, 0, u_1 - u_2)$
3. $T(u_1, u_2, u_3) = (1, u_1 + u_2 - u_3, 4u_2 + 2u_3)$
4. $T(u_1, u_2) = (u_1^2 + u_2, u_1)$
5. $T(u_1, u_2) = (u_1 + 2, u_1 - 2u_2, u_2)$

In Exercises 6–10, find the matrix representing the given linear transformation relative to the usual basis for the domain.

6. $T(x_1, x_2) = (2x_1 - x_2, 0)$

7. $T(x_1, x_2, x_3) = x_1 - x_2$

8. $T(x_1, x_2) = (4x_1 + 3x_2, 2x_2 - 3x_1, x_2)$

9. $T(x_1, x_2, x_3) = (2x_1 + 5x_2 - x_3, 3x_1 - x_2 + 2x_3, 6x_1 + 5x_2 + x_3)$

10. $T(x_1, x_2) = (x_2, x_1)$

In Exercises 11–15, find a row vector formula for the linear transformation defined by the given matrix.

11. $\begin{bmatrix} 2 \\ 1 \end{bmatrix}$

12. $\begin{bmatrix} 2 & -1 \\ 1 & 3 \end{bmatrix}$

13. $\begin{bmatrix} -1 & 4 & 3 \\ 2 & 1 & -3 \end{bmatrix}$

14. $\begin{bmatrix} 3 & 0 & 3 \\ 2 & 3 & -2 \\ 1 & -1 & 0 \end{bmatrix}$

15. $\begin{bmatrix} -4 & 1 & 0 & 3 \\ 0 & 2 & 1 & 5 \\ -2 & -3 & -1 & 0 \end{bmatrix}$

In Exercises 16–18, find $T(2, 1)$ if T is a linear transformation and has the given values.

16. $T(1, 0) = (2, 4)$, $T(0, 1) = (5, 1)$

17. $T(-8, -4) = (12, 4, -2)$

18. $T(1, 1) = (2, 4)$, $T(1, 2) = (2, 3)$

In Exercises 19–23, find the matrix representing the operator on $\mathbb{R}^2$ defined by the given geometric condition. Sketch.

19. Reflection in the x-axis $T(x_1, x_2) = (x_1, -x_2)$

20. Reflection in the y-axis $T(x_1, x_2) = (-x_1, x_2)$

21. Reflection in the line y = x $T(x_1, x_2) = (x_2, x_1)$.

22. Rotation through 45° ($\pi/4$ radians).

23. Stretching of both coordinates by a factor of 2.

In Exercises 24–26, find the rank of the given transformation.

24. $T(x_1, x_2) = (x_1 + x_2, 3x_1 + 3x_2)$

25. $T(x_1, x_2, x_3) = (4x_1 - 2x_2, -6x_1 + 3x_2, 2x_1 - x_2)$

26. $T(x_1, x_2, x_3) = (x_1 - x_2, x_3 - x_2, x_3 + 2x_1)$

In Exercises 27–30, find the nullspace of the given linear transformation.

27. $T(x_1, x_2) = (2x_1 - 5x_2, -4x_1 + 10x_2)$

28. $T(x_1, x_2) = (0, 0, 0)$

29. $T(x_1, x_2, x_3) = (x_1 - 2x_2 + x_3, 2x_1 - 3x_2 + x_3)$

30. $T(x_1, x_2, x_3) = (x_1 - x_3, x_1 + x_2 - 2x_3, 2x_1 + x_2 - x_3)$

31. Let T be the linear operator on $\mathbb{R}^2$ such that

$$T(1,1) = (-1,1), \qquad T(1,2) = (1,1).$$

Find the matrix of T.

32. Show that a function T from a vector space V to a vector space W is linear if and only if

$$T(c_1\mathbf{u}_1 + c_2\mathbf{u}_2) = c_1T(\mathbf{u}_1) + c_2T(\mathbf{u}_2)$$

for arbitrary $\mathbf{u}_1$ and $\mathbf{u}_2$ in V and scalars c_1 and c_2.

33. Prove Proposition 1, extending the result of Exercise 32 to sums of any finite number of vectors.

34. If $T(x_1, x_2, x_3) = (x_1 - x_2, x_3 - x_2)$ and $S(y_1, y_2) = (y_1 + y_2, 2y_1, y_2 - 3y_1)$, find the matrix of the composite transformation $S \circ T(\mathbf{x}) = S(T(\mathbf{x}))$.

35. Generalize the result of Exercise 34. That is, show that if T: $\mathbb{R}^n \to \mathbb{R}^m$ and S: $\mathbb{R}^m \to \mathbb{R}^p$ are linear transformations, and if B and A are the matrices of T and S respectively, then AB is the matrix of $S \circ T$.

36. By evaluating $T(\mathbf{0} + \mathbf{0})$ in two ways, show that for any linear transformation T, $T(\mathbf{0}) = \mathbf{0}$.

In Exercises 37 and 38,

$$A = \begin{bmatrix} 2/\sqrt{5} & 1/\sqrt{5} \\ -1/\sqrt{5} & 2/\sqrt{5} \end{bmatrix},$$

and T is the linear operator in $\mathbb{R}^2$ determined by A.

37. Show that if $\mathbf{x} \perp \mathbf{y}$, then $T(\mathbf{x}) \perp T(\mathbf{y})$.

38. Show that for every $\mathbf{x}$ and $\mathbf{y}$ in $\mathbb{R}^2$, $\|T(\mathbf{x}) - T(\mathbf{y})\| = \|\mathbf{x} - \mathbf{y}\|$.

39. Let V be the space of all polynomials of degree 3 or less and W those of degree 2 or less. Define T: $V \to W$ by

$$T\left(a_0 + a_1x + a_2x^2 + a_3x^3\right) = a_1 + 2a_2x + 3a_3x^2.$$

Show that T is a linear transformation and find its range.

40. Show that the function T in Example 5, page 292, is a linear operator on the space of all real polynomials.

41. Show that the range of the linear operator in Example 5 consists of all polynomials except the nonzero constants.

42. (*Calculus based*) Let V be the vector space of all continuous real functions on the interval $[0, 1]$. Show that the function defined by

$$f(x) = \int_0^1 x(t)\,dt$$

is a linear functional on V.

43. Find a basis for the range of the linear mapping defined by

$$T(x_1, x_2, x_3) = (x_1 - x_3, x_1 + x_2 + x_3, -x_1 + 2x_2 + 6x_3).$$

44. Show that the kernel of a linear transformation T is a subspace of the domain of T.

45. Let V be a subspace of $\mathbb{R}^n$, and define P on $\mathbb{R}^n$ by

$$P(\mathbf{x}) = \mathbf{v},$$

where $\mathbf{v}$ is the orthogonal projection of $\mathbf{x}$ onto V (Definition 5.3). Show that P is a linear operator on $\mathbb{R}^n$.

46. Find the matrix representation of the orthogonal projection of $\mathbb{R}^3$ onto the subspace spanned by $\{(1,1,0),(1,0,1)\}$.

47. Find the matrix representation of the orthogonal projection of $\mathbb{R}^4$ onto the solution space of the system

$$\begin{aligned} x_1 \qquad\qquad + x_3 - x_4 &= 0 \\ x_1 + x_2 + 3x_3 \qquad\quad &= 0 \\ -x_1 + 2x_2 + 3x_3 + 3x_4 &= 0 \\ 2x_1 - x_2 \qquad\qquad - 3x_4 &= 0. \end{aligned}$$

48. Show that if $m < n$, then a linear transformation from $\mathbb{R}^n$ to $\mathbb{R}^m$ cannot be one-to-one.

49. Let A be an $n \times n$ matrix. Show that $\det(A) = 0$ if and only if the range of the linear operator determined by A coincides with $\mathbb{R}^n$.

50. Let A be an $m \times n$ matrix and $\mathbf{b}$ a column vector in $\mathbb{R}^m$. Show that the linear system $A\mathbf{x} = \mathbf{b}$ is consistent if and only if $\mathbf{b}$ is in the orthogonal complement of the kernel of the linear transformation determined by A^t.

51. If T and S are linear transformations from $\mathbb{R}^n$ to $\mathbb{R}^m$ and c is a scalar, define $S + T$ and cT on $\mathbb{R}^n$ by

$$(S + T)(\mathbf{x}) = S(\mathbf{x}) + T(\mathbf{x}), \qquad (cT)(\mathbf{x}) = cT(\mathbf{x}).$$

Show that $S + T$ and cT are linear transformations from $\mathbb{R}^n$ to $\mathbb{R}^m$.

52. Show that the set of all linear transformations from $\mathbb{R}^n$ to $\mathbb{R}^m$ forms a vector space (Definition 4.1) under the operations in Exercise 51.

53. Prove that if A is an $m \times n$ matrix and T is defined by

$$T(\mathbf{u}) = A(\mathbf{u}),$$

then T is a linear transformation from $\mathbb{R}^n$ to $\mathbb{R}^m$.

6.2 CHANGE OF BASIS

Some bases may be more useful than others for a linear transformation of a vector space. For instance, the problem of simplifying the matrix representation of a linear operator on $\mathbb{R}^n$, which is studied in Chapter 7, involves finding a suitable basis for that space. In applications, certain bases may be associated with special physical properties. In this section we study problems related to a change of the basis in a vector space.

Changes Involving the Standard Basis in $\mathbb{R}^n$

If $B = \{\mathbf{u}_1, \ldots, \mathbf{u}_n\}$ is a basis for a vector space V, then by Theorem 4.6, any $\mathbf{x}$ in V is uniquely expressible in the form

$$\mathbf{x} = c_1\mathbf{u}_1 + \cdots + c_n\mathbf{u}_n. \tag{1}$$

The scalars $c_1, \ldots, c_n$ are called the **components of x relative to B**. If the vectors in B are kept in their given order, we may represent Eq. (1) by an n-tuple using the *position* to each c_i to indicate which vector it multiplies. We shall also use parentheses for a row vector and brackets for a column vector. Thus Eq. (1) is denoted by either

$$(\mathbf{x})_B = (c_1, \ldots, c_n) \qquad \text{or} \qquad [\mathbf{x}]_B = [c_1 \ \cdots \ c_n]^t,$$

as appropriate. Most of the work is with column vectors.

Example 1

For the basis

$$B = \left\{\begin{bmatrix}1\\1\end{bmatrix}, \begin{bmatrix}1\\3\end{bmatrix}\right\}$$

in $\mathbb{R}^2$ and the vector $\mathbf{x}$ given by

$$[\mathbf{x}]_B = \begin{bmatrix}3\\-2\end{bmatrix}$$

relative to B, find $\mathbf{x}$ relative to the standard basis.

Solution

In the notation of Eq. (1),

$$c_1 = 3, \qquad c_2 = -2, \qquad \mathbf{u}_1 = \begin{bmatrix}1\\1\end{bmatrix}, \qquad \text{and} \qquad \mathbf{u}_2 = \begin{bmatrix}1\\3\end{bmatrix}.$$

Substitute these values into Eq. (1) and simplify to find

$$\mathbf{x} = 3\begin{bmatrix}1\\1\end{bmatrix} - 2\begin{bmatrix}1\\3\end{bmatrix} = \begin{bmatrix}1\\-3\end{bmatrix}.$$

This expansion of $\mathbf{x}$ is illustrated in Fig. 6.3. □

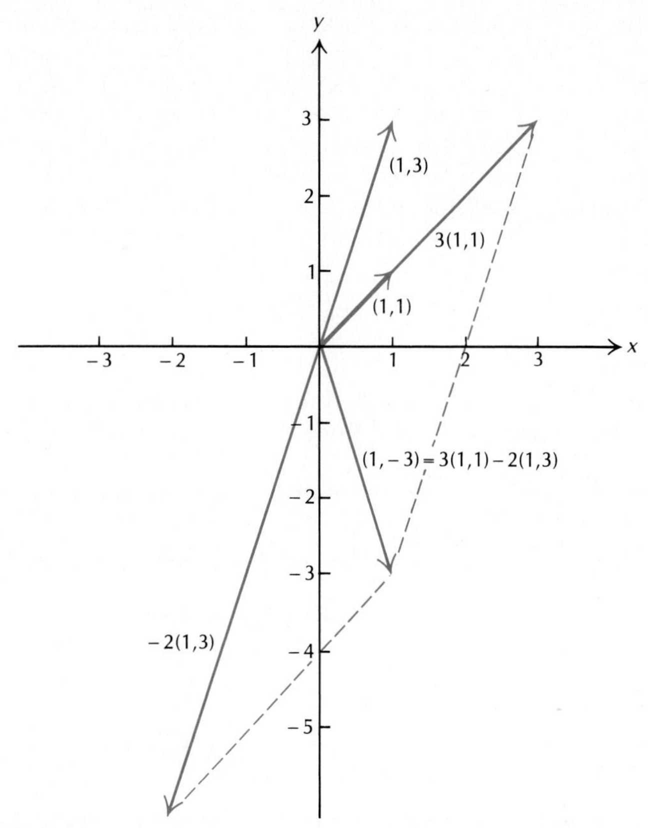

Figure 6.3 Expansion of the vector $\mathbf{x} = (1, -3)$ relative to the (ordered) basis $B = \{(1, 1), (1, 3)\}$ in $\mathbb{R}^2$.

For a given basis B of $\mathbb{R}^n$, we also need to be able to find $[\mathbf{x}]_B$ when $\mathbf{x}$ is known relative to the standard basis.

Example 2

Find $[\mathbf{x}]_B$ in $\mathbb{R}^2$ if

$$B = \left\{\begin{bmatrix} 2 \\ 1 \end{bmatrix}, \begin{bmatrix} 1 \\ 4 \end{bmatrix}\right\} \quad \text{and} \quad \mathbf{x} = \begin{bmatrix} 5 \\ -1 \end{bmatrix}.$$

Solution

Let c_1 and c_2 denote the components of $\mathbf{x}$ relative to B. Substitute $\mathbf{x}$, $\mathbf{u}_1$, and $\mathbf{u}_2$ in Eq. (1) to obtain

$$\begin{bmatrix} 5 \\ -1 \end{bmatrix} = c_1\begin{bmatrix} 2 \\ 1 \end{bmatrix} + c_2\begin{bmatrix} 1 \\ 4 \end{bmatrix} = \begin{bmatrix} 2c_1 + c_2 \\ c_1 + 4c_2 \end{bmatrix}.$$

This vector equation is equivalent to the linear system

$$\begin{aligned} 2c_1 + c_2 &= 5 \\ c_1 + 4c_2 &= -1. \end{aligned}$$

The solution of this system is $c_1 = 3$ and $c_2 = -1$, from which

$$[\mathbf{x}]_B = \begin{bmatrix} 3 \\ -1 \end{bmatrix}. \qquad \square$$

We may also use a matrix form of Eq. (1) to change bases. Let Q be the matrix with columns $\mathbf{u}_1, \ldots, \mathbf{u}_n$. Since the components of $[\mathbf{x}]_B$ are $c_1, \ldots, c_n$,

$$c_1\mathbf{u}_1 + \cdots + c_n\mathbf{u}_n = Q[\mathbf{x}]_B$$

by the *column space interpretation* of Section 4.5. Thus

$$\mathbf{x} = Q[\mathbf{x}]_B. \tag{2}$$

Example 3

For the basis B and the vector $[\mathbf{x}]_B$ in Example 1, use matrix algebra to find $\mathbf{x}$ relative to the standard basis.

Solution

Using the given basis

$$B = \left\{ \begin{bmatrix} 1 \\ 1 \end{bmatrix}, \begin{bmatrix} 1 \\ 3 \end{bmatrix} \right\},$$

form the matrix Q with the vectors from B as *columns*:

$$Q = \begin{bmatrix} 1 & 1 \\ 1 & 3 \end{bmatrix}.$$

By Eq. (2),

$$\mathbf{x} = Q[\mathbf{x}]_B = \begin{bmatrix} 1 & 1 \\ 1 & 3 \end{bmatrix} \begin{bmatrix} 3 \\ -2 \end{bmatrix} = \begin{bmatrix} 1 \\ -3 \end{bmatrix}. \qquad \square$$

Example 4

For the basis B and the vector $\mathbf{x}$ in Example 2, use matrix algebra to find $[\mathbf{x}]_B$.

Solution

Using the given basis

$$B = \left\{ \begin{bmatrix} 2 \\ 1 \end{bmatrix}, \begin{bmatrix} 1 \\ 4 \end{bmatrix} \right\},$$

form the matrix Q with the vectors from B as *columns*:

$$Q = \begin{bmatrix} 2 & 1 \\ 1 & 4 \end{bmatrix}.$$

Since B is linearly independent, Q is invertible. Find Q^{-1} and apply it to both sides of Eq. (2):

$$[\mathbf{x}]_B = Q^{-1}\mathbf{x} = \frac{1}{7}\begin{bmatrix} 4 & -1 \\ -1 & 2 \end{bmatrix}\begin{bmatrix} 5 \\ -1 \end{bmatrix} = \begin{bmatrix} 3 \\ -1 \end{bmatrix}. \quad \square$$

Figure 6.4 shows the change of basis in Example 4.

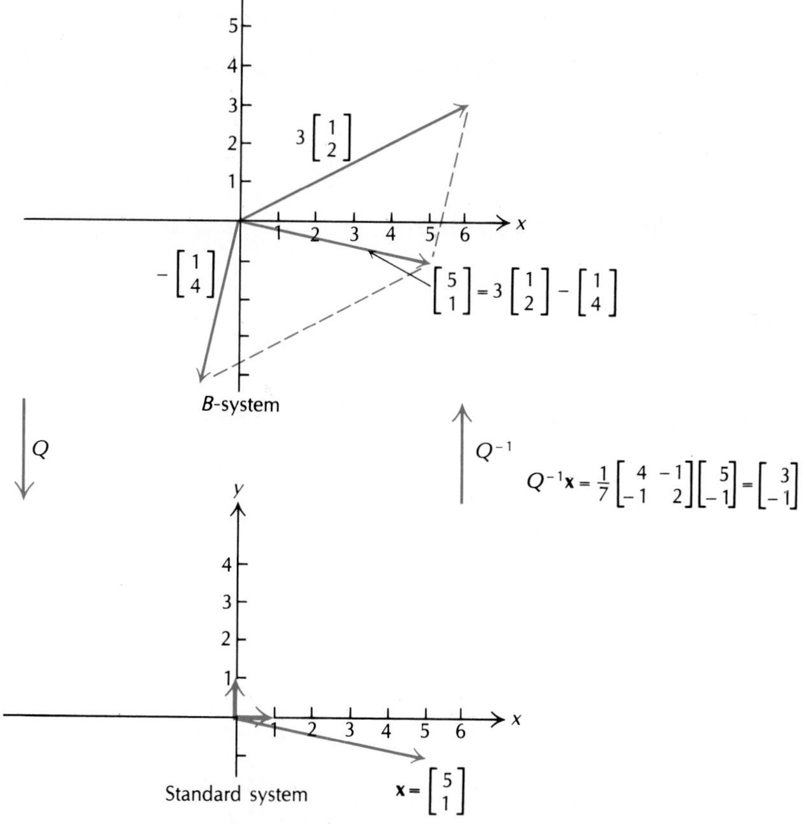

Figure 6.4 Change of basis in Example 4.

Changing Between Any Two Bases in $\mathbb{R}^n$

As above, let $B = \{\mathbf{u}_1, \ldots, \mathbf{u}_n\}$ be a basis for a vector space V. We fix the order of the vectors in B and represent any $\mathbf{x}$ in V as $[\mathbf{x}]_B$ in accordance with Eq. (1). If $B' = \{\mathbf{v}_1, \ldots, \mathbf{v}_n\}$ is also a basis for V, then for any $\mathbf{x}$ in V, we also have

$$\text{(3)} \qquad \mathbf{x} = k_1\mathbf{v}_1 + \cdots + k_n\mathbf{v}_n.$$

If the order of the vectors in B' is fixed, then we may use $[\mathbf{x}]_{B'} = [k_1 \;\cdots\; k_n]^t$ for Eq. (3). We also have

$$\text{(4)} \qquad \mathbf{x} = Q'[\mathbf{x}]_{B'},$$

where Q' is the matrix with columns $\mathbf{v}_1, \ldots, \mathbf{v}_n$. For a change of basis, the problem is to find $[\mathbf{x}]_{B'}$ for a given $[\mathbf{x}]_B$, or vice versa.

Example 5

Find $[\mathbf{x}]_{B'}$, if

$$B = \left\{\begin{bmatrix} 1 \\ -1 \end{bmatrix}, \begin{bmatrix} 2 \\ 1 \end{bmatrix}\right\}, \quad B' = \left\{\begin{bmatrix} 1 \\ 1 \end{bmatrix}, \begin{bmatrix} 1 \\ 2 \end{bmatrix}\right\}, \quad \text{and} \quad [\mathbf{x}]_B = \begin{bmatrix} 4 \\ 3 \end{bmatrix}.$$

Solution

Substitute the given vectors

$$\mathbf{u}_1 = \begin{bmatrix} 1 \\ -1 \end{bmatrix}, \quad \mathbf{u}_2 = \begin{bmatrix} 2 \\ 1 \end{bmatrix}, \quad \mathbf{v}_1 = \begin{bmatrix} 1 \\ 1 \end{bmatrix}, \quad \mathbf{v}_2 = \begin{bmatrix} 1 \\ 2 \end{bmatrix}, \quad \text{and} \quad \mathbf{x} = \begin{bmatrix} 4 \\ 3 \end{bmatrix}$$

into Eqs. (1) and (3) and equate the expressions for $\mathbf{x}$:

$$4\begin{bmatrix} 1 \\ -1 \end{bmatrix} + 3\begin{bmatrix} 2 \\ 1 \end{bmatrix} = k_1\begin{bmatrix} 1 \\ 1 \end{bmatrix} + k_2\begin{bmatrix} 1 \\ 2 \end{bmatrix}.$$

Simplify both sides to obtain the vector equation

$$\begin{bmatrix} k_1 + k_2 \\ k_1 + 2k_2 \end{bmatrix} = \begin{bmatrix} 10 \\ -1 \end{bmatrix}.$$

Solving the linear system corresponding to this equation gives

$$[\mathbf{x}]_{B'} = \begin{bmatrix} 21 \\ -11 \end{bmatrix}. \quad \square$$

Example 6

Carry out the change of basis in Example 5 by matrix algebra.

Solution

Equate the expressions for $\mathbf{x}$ in Eqs. (2) and (4) to obtain

$$Q[\mathbf{x}]_B = Q'[\mathbf{x}]_{B'}.$$

Since Q' is invertible,

$$[\mathbf{x}]_{B'} = Q'^{-1}Q[\mathbf{x}]_B.$$

Form Q and Q' from the given B and B':

$$Q = \begin{bmatrix} 1 & 2 \\ -1 & 1 \end{bmatrix}, \quad \text{and} \quad Q' = \begin{bmatrix} 1 & 1 \\ 1 & 2 \end{bmatrix}.$$

Then

$$Q'^{-1}Q = \begin{bmatrix} 2 & -1 \\ -1 & 1 \end{bmatrix}\begin{bmatrix} 1 & 2 \\ -1 & 1 \end{bmatrix} = \begin{bmatrix} 3 & 3 \\ -2 & -1 \end{bmatrix},$$

and

$$[\mathbf{x}]_{B'} = Q'^{-1}Q[\mathbf{x}]_B = \begin{bmatrix} 3 & 3 \\ -2 & -1 \end{bmatrix}\begin{bmatrix} 4 \\ 3 \end{bmatrix} = \begin{bmatrix} 21 \\ -11 \end{bmatrix},$$

in agreement with the result of Example 5. □

The next theorem summarizes the results above.

THEOREM 6.4

If $B = \{\mathbf{u}_1, \ldots, \mathbf{u}_n\}$ and $B' = \{\mathbf{v}_1, \ldots, \mathbf{v}_n\}$ are bases for $\mathbb{R}^n$, let Q and Q' be the matrices with column vectors from B and B', respectively. Then Q and Q' are invertible and if $P = Q^{-1}Q'$, then

$$[\mathbf{x}]_B = P[\mathbf{x}]_{B'} \quad \text{and} \quad [\mathbf{x}]_{B'} = P^{-1}[\mathbf{x}]_B.$$

Proof The matrices Q and Q' are invertible by Theorem 4.13. Equating the expressions for $\mathbf{x}$ in Eqs. (2) and (4) gives

$$Q[\mathbf{x}]_B = Q'[\mathbf{x}]_{B'}.$$

Multiplying by Q^{-1} or by Q'^{-1} yields the desired relations. □

DEFINITION 6.4

Let B, B', Q, Q', and P be as in Theorem 6.4. Then P is called the **transition matrix from B′ to B**, and P^{-1} is the **transition matrix from B to B′**.

We may also find P without a matrix inversion. Both methods will be justified after an illustrative example.

Two methods of finding a transition matrix If $B = \{\mathbf{u}_1, \ldots, \mathbf{u}_n\}$ and $B' = \{\mathbf{v}_1, \ldots, \mathbf{v}_n\}$ are bases for $\mathbb{R}^n$, let P be the transition matrix from B' to B.

Method 1 Form the matrices Q and Q' with column vectors from B and B', respectively, and find Q^{-1}. Then

$$P = Q^{-1}Q'.$$

Method 2 For $j = 1, \ldots, n$, form the vector equation

$$c_1\mathbf{u}_1 + \cdots + c_n\mathbf{u}_n = \mathbf{v}_j,$$

and solve the resulting linear system for a column vector $\mathbf{c}_j$. Then form P using $\mathbf{c}_j$ as the jth column.

Example 7

Find the transition matrix from B' to B in $\mathbb{R}^3$, if

$$B = \left\{ \begin{bmatrix} 1 \\ 1 \\ 1 \end{bmatrix}, \begin{bmatrix} 1 \\ 2 \\ 0 \end{bmatrix}, \begin{bmatrix} 0 \\ 2 \\ -1 \end{bmatrix} \right\} \quad \text{and} \quad B' = \left\{ \begin{bmatrix} 1 \\ 0 \\ 0 \end{bmatrix}, \begin{bmatrix} 1 \\ 1 \\ 0 \end{bmatrix}, \begin{bmatrix} 1 \\ 1 \\ 1 \end{bmatrix} \right\}.$$

Solution

Method 1. Since

$$Q = \begin{bmatrix} 1 & 1 & 0 \\ 1 & 2 & 2 \\ 1 & 0 & -1 \end{bmatrix} \quad \text{and} \quad Q' = \begin{bmatrix} 1 & 1 & 1 \\ 0 & 1 & 1 \\ 0 & 0 & 1 \end{bmatrix},$$

we have

$$P = Q^{-1}Q' = \begin{bmatrix} -2 & 1 & 2 \\ 3 & -1 & -2 \\ -2 & 1 & 1 \end{bmatrix} \begin{bmatrix} 1 & 1 & 1 \\ 0 & 1 & 1 \\ 0 & 0 & 1 \end{bmatrix} = \begin{bmatrix} -2 & -1 & 1 \\ 3 & 2 & 0 \\ -2 & -1 & 0 \end{bmatrix}.$$

Method 2. Form the vector equation

$$c_1\begin{bmatrix}1\\1\\1\end{bmatrix} + c_2\begin{bmatrix}1\\2\\0\end{bmatrix} + c_3\begin{bmatrix}0\\2\\-1\end{bmatrix} = \begin{bmatrix}1\\0\\0\end{bmatrix}$$

and solve the resulting linear system

$$\begin{aligned} c_1 + c_2 &= 1 \\ c_1 + 2c_2 + 2c_3 &= 0 \\ c_1 - c_3 &= 0 \end{aligned}$$

to obtain the first column of P:

$$\mathbf{c}_1 = \begin{bmatrix}-2\\3\\-2\end{bmatrix}.$$

In a similar way, solve the remaining two systems

$$\begin{aligned} c_1 + c_2 &= 1 \\ c_1 + 2c_2 + 2c_3 &= 1 \\ c_1 - c_3 &= 0 \end{aligned} \qquad \text{and} \qquad \begin{aligned} c_1 + c_2 &= 1 \\ c_1 + 2c_2 + 2c_3 &= 1 \\ c_1 - c_3 &= 1 \end{aligned}$$

to obtain the second and third columns

$$\mathbf{c}_2 = \begin{bmatrix}-1\\2\\-1\end{bmatrix} \qquad \text{and} \qquad \mathbf{c}_3 = \begin{bmatrix}1\\0\\0\end{bmatrix}.$$

Form P (given above) with $\mathbf{c}_1$, $\mathbf{c}_2$, and $\mathbf{c}_3$ as columns. ☐

Justification of the Methods for Finding a Transition Matrix Method 1 is based directly on Definition 6.4. To see why Method 2 is valid, suppose $\mathbf{c}_j$ is the solution of the jth vector equation formed under this method. Then

$$\mathbf{v}_j = c_1\mathbf{u}_1 + \cdots + c_n\mathbf{u}_n = Q\mathbf{c}_j,$$

from which

$$\mathbf{c}_j = Q^{-1}\mathbf{v}_j = Q^{-1}(Q'[\mathbf{v}_j]_{B'}) = (Q^{-1}Q')[\mathbf{v}_j]_{B'} = P[\mathbf{v}_j]_{B'}.$$

Since the jth component of $[\mathbf{v}_j]_{B'}$ equals 1 and the rest equal 0 (why?), the product on the right equals the jth column vector of P. ☐

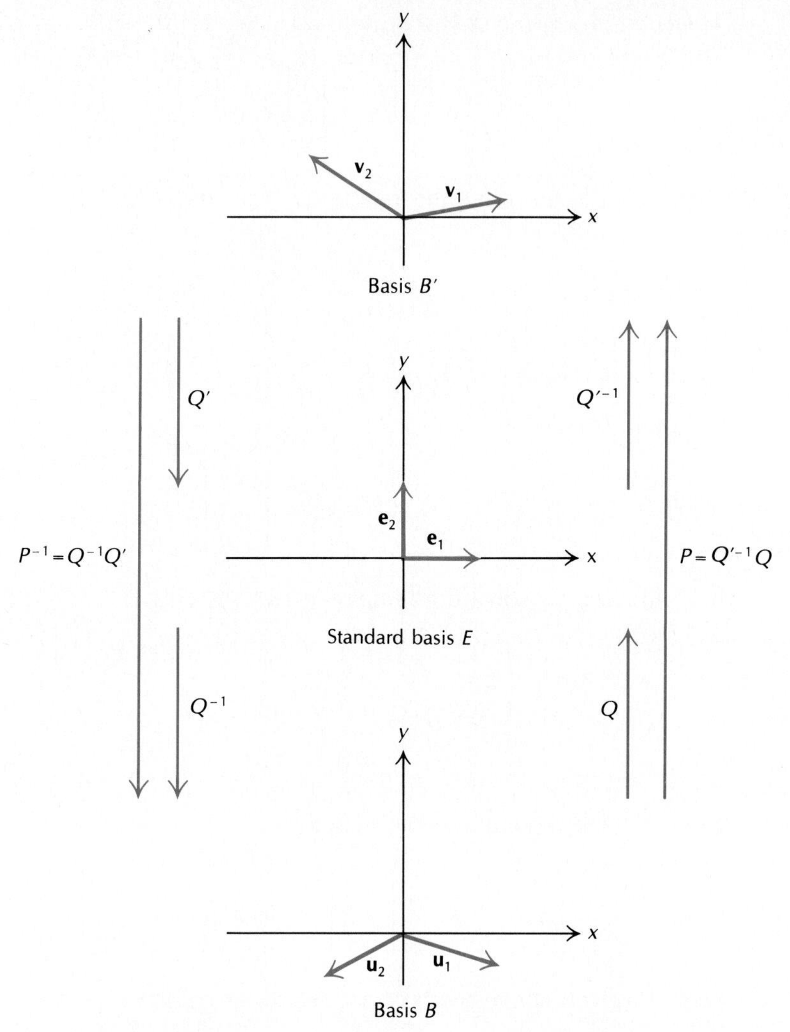

Figure 6.5 General change of basis problem.

We may regard the transition from B' to B as the composite of the transition from B' to the standard basis E with the transition from E to B, as shown in Fig. 6.5.

Change of Basis and Linear Transformations

In Section 6.1, we studied the linear transformation $T: \mathbb{R}^n \to \mathbb{R}^m$ determined by an $m \times n$ matrix A. This mapping is given by

$$T(\mathbf{x}) = A\mathbf{x},$$

where $\mathbf{x}$ and $A\mathbf{x}$ are expressed as column vectors relative to the standard bases for $\mathbb{R}^n$ and $\mathbb{R}^m$. This idea can be extended to other bases.

Example 8

Describe the linear transformation T determined by the matrix

$$A = \begin{bmatrix} 2 & -1 \\ 1 & 1 \\ 4 & 0 \end{bmatrix}$$

relative to the bases

$$B = \left\{ \begin{bmatrix} 2 \\ 1 \end{bmatrix}, \begin{bmatrix} -2 \\ 1 \end{bmatrix} \right\} \quad \text{and} \quad C = \left\{ \begin{bmatrix} 1 \\ 2 \\ 1 \end{bmatrix}, \begin{bmatrix} 2 \\ 1 \\ 0 \end{bmatrix}, \begin{bmatrix} 1 \\ 0 \\ 2 \end{bmatrix} \right\}$$

for $\mathbb{R}^2$ and $\mathbb{R}^3$, respectively.

Solution

Let $\mathbf{x}$ be any vector in $\mathbb{R}^2$. Since A is 3×2 and $[\mathbf{x}]_B$ is 2×1, the product

$$A[\mathbf{x}]_B$$

is a column vector with three components. This vector determines an element in $\mathbb{R}^3$ relative to the basis C. This procedure determines a linear transformation $T\colon \mathbb{R}^2 \to \mathbb{R}^3$, and the image $T(\mathbf{x})$ of $\mathbf{x}$ in $\mathbb{R}^2$ is determined by the relation

$$[T(\mathbf{x})]_C = A[\mathbf{x}]_B.$$

For instance, if

$$[\mathbf{x}]_B = \begin{bmatrix} 1 \\ 3 \end{bmatrix},$$

then

$$[T(\mathbf{x})]_C = \begin{bmatrix} 2 & -1 \\ 1 & 1 \\ 4 & 0 \end{bmatrix} \begin{bmatrix} 1 \\ 3 \end{bmatrix} = \begin{bmatrix} -1 \\ 4 \\ 4 \end{bmatrix}.$$

We may find $\mathbf{x}$ and $T(\mathbf{x})$ relative to the standard bases in $\mathbb{R}^2$ and $\mathbb{R}^3$ using Eq. (1) in both spaces:

$$\mathbf{x} = 1\begin{bmatrix} 2 \\ 1 \end{bmatrix} + 3\begin{bmatrix} -2 \\ 1 \end{bmatrix} = \begin{bmatrix} -4 \\ 4 \end{bmatrix} \quad \text{and} \quad T(\mathbf{x}) = -1\begin{bmatrix} 1 \\ 2 \\ 1 \end{bmatrix} + 4\begin{bmatrix} 2 \\ 1 \\ 0 \end{bmatrix} + 4\begin{bmatrix} 1 \\ 0 \\ 2 \end{bmatrix} = \begin{bmatrix} 11 \\ 2 \\ 7 \end{bmatrix}.$$

□

In general, let B be a basis for $\mathbb{R}^n$ and C a basis for $\mathbb{R}^m$. If A is an $m \times n$ matrix, then the relation

$$[T(\mathbf{x})]_C = A[\mathbf{x}]_B$$

determines a linear transformation T from $\mathbb{R}^n$ to $\mathbb{R}^m$. The converse is also true: for any linear transformation T: $\mathbb{R}^n \to \mathbb{R}^m$, there is a matrix A satisfying this relation. To see why, let A_1 be the matrix of T relative to the standard basis, i.e.,

$$T(\mathbf{x}) = A_1(\mathbf{x}).$$

Applying Eq. (4) in the codomain and Eq. (2) in the domain of T gives

$$Q'[T(\mathbf{x})]_C = A_1Q[\mathbf{x}]_B,$$

from which

$$[T(\mathbf{x})]_C = (Q'^{-1}A_1Q)[\mathbf{x}]_B = A[\mathbf{x}]_B.$$

The matrix A is called the matrix of T relative to B and C and is denoted by A_{BC}. In this notation, the relation that determines T becomes

$$[T(\mathbf{x})]_C = A_{BC}[\mathbf{x}]_B.$$

The remainder of this section is devoted to the relationship between the matrix of a linear transformation and the change of basis problem.

Example 9

Let B and B' be bases for $\mathbb{R}^2$ and let C and C' be bases for $\mathbb{R}^3$, where

$$B = \left\{\begin{bmatrix}3\\2\end{bmatrix}, \begin{bmatrix}4\\3\end{bmatrix}\right\}, \qquad C = \left\{\begin{bmatrix}1\\1\\0\end{bmatrix}, \begin{bmatrix}1\\0\\1\end{bmatrix}, \begin{bmatrix}0\\1\\1\end{bmatrix}\right\},$$

$$B' = \left\{\begin{bmatrix}1\\-1\end{bmatrix}, \begin{bmatrix}-2\\1\end{bmatrix}\right\}, \quad \text{and} \quad C' = \left\{\begin{bmatrix}1\\-1\\1\end{bmatrix}, \begin{bmatrix}0\\1\\-1\end{bmatrix}, \begin{bmatrix}1\\1\\0\end{bmatrix}\right\}.$$

Let

$$A = \begin{bmatrix}2 & 4\\-1 & 1\\2 & -2\end{bmatrix}.$$

If $A = A_{B'C'}$ find the matrix A' of T relative to B' and C'.

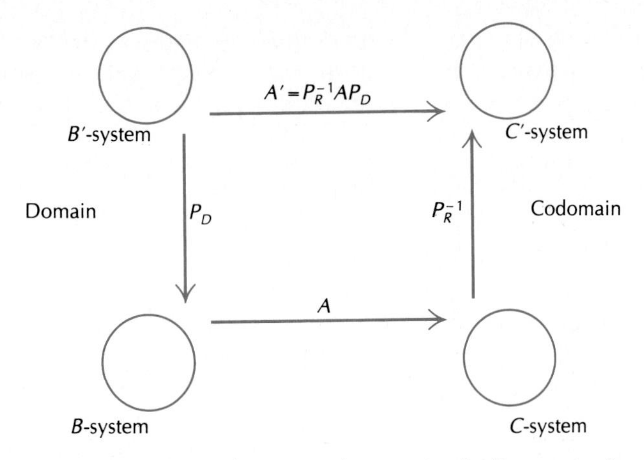

Figure 6.6 Effect of a basis change on the matrix of a linear transformation.

Solution

The solution is shown in Fig. 6.6. Let P_D and P_R^{-1} be the transition matrices from B' to B and C to C', respectively. For $[\mathbf{x}]_{B'}$ in $\mathbb{R}^2$, find $[\mathbf{x}]_B$ via P_D, apply A to find $[T(\mathbf{x})]_C$, then change to $[T(\mathbf{x})]_{C'}$ via P_R^{-1}. Thus

$$A' = P_R^{-1}AP_D.$$

Find P_R and P_D^{-1} as in Example 3. Since

$$\begin{bmatrix} 3 & 4 \\ 2 & 3 \end{bmatrix}^{-1} = \begin{bmatrix} 3 & -4 \\ -2 & 3 \end{bmatrix} \quad \text{and} \quad \begin{bmatrix} 1 & 0 & 1 \\ -1 & 1 & 1 \\ 1 & -1 & 0 \end{bmatrix}^{-1} = \begin{bmatrix} 1 & -1 & -1 \\ 1 & -1 & -2 \\ 0 & 1 & 1 \end{bmatrix},$$

it follows that

$$P_D = \begin{bmatrix} 3 & -4 \\ -2 & 3 \end{bmatrix}\begin{bmatrix} 1 & -2 \\ -1 & 1 \end{bmatrix} = \begin{bmatrix} 7 & -10 \\ -5 & 7 \end{bmatrix}$$

and

$$P_R^{-1} = \begin{bmatrix} 1 & -1 & -1 \\ 1 & -1 & -2 \\ 0 & 1 & 1 \end{bmatrix}\begin{bmatrix} 1 & 1 & 0 \\ 1 & 0 & 1 \\ 0 & 1 & 1 \end{bmatrix} = \begin{bmatrix} 0 & 0 & -2 \\ 0 & -1 & -3 \\ 1 & 1 & 2 \end{bmatrix}.$$

Thus

$$A' = P_R^{-1}AP_D = \begin{bmatrix} 0 & 0 & -2 \\ 0 & -1 & -3 \\ 1 & 1 & 2 \end{bmatrix}\begin{bmatrix} 2 & 4 \\ -1 & 1 \\ 2 & -2 \end{bmatrix}\begin{bmatrix} 7 & -10 \\ -5 & 7 \end{bmatrix} = \begin{bmatrix} -48 & 68 \\ -60 & 85 \\ 30 & -43 \end{bmatrix}. \quad \square$$

For a linear *operator* T, the domain coincides with the codomain and any matrix of T is square. If $B = C$ and $B' = C'$, then the problem in Example 9 is to find the matrix of T relative to B' given the matrix of T relative to B.

Example 10

Let

$$B = \left\{\begin{bmatrix} 2 \\ 1 \end{bmatrix}, \begin{bmatrix} 3 \\ 2 \end{bmatrix}\right\}, \quad B' = \left\{\begin{bmatrix} 2 \\ -1 \end{bmatrix}, \begin{bmatrix} -1 \\ -2 \end{bmatrix}\right\}, \quad \text{and} \quad A = \begin{bmatrix} 2 & -1 \\ 1 & 4 \end{bmatrix}.$$

If T is the linear operator on $\mathbb{R}^2$ whose matrix relative to B is A, find the matrix A' of T relative to B'.

Solution

The solution is shown in Fig. 6.7. Let P be the transition matrix from B' to B. For any given $[\mathbf{x}]_{B'}$ in $\mathbb{R}^2$, find $[\mathbf{x}]_B$ via P, apply A to find $[T(\mathbf{x})]_{B}$, then find $[T(\mathbf{x})]_{B'}$ via P^{-1}. That is,

$$A' = P^{-1}AP.$$

Find P from the vectors in B, and invert it. The results are

$$P = \begin{bmatrix} 2 & 3 \\ 1 & 2 \end{bmatrix} \quad \text{and} \quad P^{-1} = \begin{bmatrix} 2 & -3 \\ -1 & 2 \end{bmatrix},$$

from which

$$P^{-1}AP = \begin{bmatrix} 2 & -3 \\ -1 & 2 \end{bmatrix}\begin{bmatrix} 2 & -1 \\ 1 & 4 \end{bmatrix}\begin{bmatrix} 2 & 3 \\ 1 & 2 \end{bmatrix} = \begin{bmatrix} -12 & -25 \\ 9 & 18 \end{bmatrix}. \quad \square$$

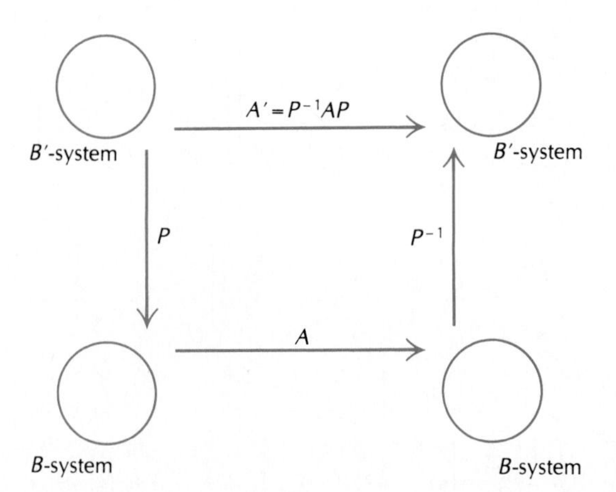

Figure 6.7 Effect of a basis change on the matrix of a linear operator.

Example 11

If T is defined on $\mathbb{R}^2$ by

$$T(x_1, x_2) = (x_1 + 3x_2, 4x_1 + 2x_2),$$

find the matrix A' of T relative to the basis

$$B' = \left\{ \begin{bmatrix} 2 \\ 3 \end{bmatrix}, \begin{bmatrix} 3 \\ 5 \end{bmatrix} \right\}.$$

Solution

This time use the images of the basis vectors to find A'. If the vectors in B' are denoted by $\mathbf{v}_1, \mathbf{v}_2$, then

$$T(\mathbf{v}_1) = (11, 14) \quad \text{and} \quad T(\mathbf{v}_2) = (18, 22).$$

Find $[T(\mathbf{v}_1)]_{B'}$ and $[T(\mathbf{v}_2)]_{B'}$ by solving the linear systems corresponding to the vector equations

$$c_1\begin{bmatrix} 2 \\ 3 \end{bmatrix} + c_2\begin{bmatrix} 3 \\ 5 \end{bmatrix} = \begin{bmatrix} 11 \\ 14 \end{bmatrix} \quad \text{and} \quad k_1\begin{bmatrix} 2 \\ 3 \end{bmatrix} + k_2\begin{bmatrix} 3 \\ 5 \end{bmatrix} = \begin{bmatrix} 18 \\ 22 \end{bmatrix}.$$

The solutions are given by

$$[T(\mathbf{v}_1)]_{B'} = \begin{bmatrix} 13 \\ -5 \end{bmatrix} \quad \text{and} \quad [T(\mathbf{v}_2)]_{B'} = \begin{bmatrix} 24 \\ -10 \end{bmatrix}.$$

Form A' using these vectors as columns:

$$A' = \begin{bmatrix} 13 & 24 \\ -5 & -10 \end{bmatrix}. \quad \square$$

We close this section by relating two important ideas. Since a change of basis corresponds to a square matrix which, in turn, corresponds to a linear operator, a *change of basis corresponds to a linear operator*. Intuitively, "holding a space fixed and moving a basis" (a change of basis) has the same effect as "holding a basis fixed and moving the space" (a linear operator). In view of this observation, we call special attention to the result assigned in Exercise 35 of Section 6.1:

> the matrix of the composite of two linear transformations is the product of the matrices of the transformations.

This result explains the product $P_R^{-1}AP_D$ in Example 9 as the matrix of the composite of three linear transformations. A similar remark applies to the product $P^{-1}AP$ in Example 10.

REVIEW CHECKLIST

1. Interpret a vector $[\mathbf{x}]_B$ as a linear combination of the basis vectors with the components as coefficients.
2. Interpret and use the notation for a vector relative to a basis in either row or column form.
3. For bases B and B' of $\mathbb{R}^n$, find the transition matrices from B to B' and from B' to B by two methods.
4. Draw symbolic diagrams to show the change of basis for any bases B and B' and the construction of the transition matrices by matrix products.
5. Formulate a relation between $\mathbf{x}$, $T(\mathbf{x})$, and A that defines the matrix representation of T by A relative to bases in the domain and the codomain of T.
6. Given a linear transformation T, two bases in the domain and in the codomain, and the matrix of T relative to either pair of bases, find the matrix of A relative to the other pair.
7. Specialize the general problem in (6) to linear operators, and interpret the meaning of $A' = P^{-1}AP$.
8. Draw symbolic diagrams to show the effect of a change of basis on the matrix representation of a linear transformation or a linear operator.

EXERCISES

In Exercises 1–3, given a basis B for $\mathbb{R}^2$ or $\mathbb{R}^3$ and a vector $[\mathbf{x}]_B$, find the standard components of $\mathbf{x}$.

1. $B = \left\{\begin{bmatrix} 5 \\ 1 \end{bmatrix}, \begin{bmatrix} 2 \\ 3 \end{bmatrix}\right\}, \quad [\mathbf{x}]_B = \begin{bmatrix} 1 \\ -1 \end{bmatrix}$

2. $B = \left\{\begin{bmatrix} 2 \\ 4 \end{bmatrix}, \begin{bmatrix} 1 \\ -1 \end{bmatrix}\right\}, \quad [\mathbf{x}]_B = \begin{bmatrix} 1 \\ 3 \end{bmatrix}$

3. $B = \left\{\begin{bmatrix} 1 \\ -1 \\ 2 \end{bmatrix}, \begin{bmatrix} 1 \\ 1 \\ 3 \end{bmatrix}, \begin{bmatrix} 0 \\ 1 \\ 4 \end{bmatrix}\right\}, \quad [\mathbf{x}]_B = \begin{bmatrix} 1 \\ -2 \\ 1 \end{bmatrix}$

In Exercises 4–6, given a basis B for $\mathbb{R}^2$ or $\mathbb{R}^3$ and a vector $\mathbf{x}$, find $[\mathbf{x}]_B$.

4. $B = \left\{\begin{bmatrix} 4 \\ 5 \end{bmatrix}, \begin{bmatrix} 3 \\ 4 \end{bmatrix}\right\}, \quad \mathbf{x} = \begin{bmatrix} -1 \\ 1 \end{bmatrix}$

5. $B = \left\{\begin{bmatrix} 2 \\ 2 \end{bmatrix}, \begin{bmatrix} 2 \\ 3 \end{bmatrix}\right\}, \quad \mathbf{x} = \begin{bmatrix} 2 \\ -1 \end{bmatrix}$

6. $B = \left\{\begin{bmatrix} 1 \\ -1 \\ 2 \end{bmatrix}, \begin{bmatrix} 1 \\ 0 \\ 4 \end{bmatrix}, \begin{bmatrix} 0 \\ 1 \\ 3 \end{bmatrix}\right\}, \quad \mathbf{x} = \begin{bmatrix} 1 \\ -2 \\ 1 \end{bmatrix}$

In Exercises 7–9, given B, B', and $[\mathbf{x}]_{B'}$, find $[\mathbf{x}]_B$, and check by converting both expressions for $\mathbf{x}$ to standard form.

7. $B = \left\{\begin{bmatrix}1\\3\end{bmatrix}, \begin{bmatrix}-1\\2\end{bmatrix}\right\}, \quad B' = \left\{\begin{bmatrix}1\\0\end{bmatrix}, \begin{bmatrix}1\\1\end{bmatrix}\right\}, \quad [\mathbf{x}]_{B'} = \begin{bmatrix}1\\-1\end{bmatrix}$

8. $B = \left\{\begin{bmatrix}1\\1\end{bmatrix}, \begin{bmatrix}1\\-1\end{bmatrix}\right\}, \quad B' = \left\{\begin{bmatrix}3\\5\end{bmatrix}, \begin{bmatrix}2\\4\end{bmatrix}\right\}, \quad [\mathbf{x}]_{B'} = \begin{bmatrix}2\\0\end{bmatrix}$

9. $B = \left\{\begin{bmatrix}1\\1\\-1\end{bmatrix}, \begin{bmatrix}1\\-1\\1\end{bmatrix}, \begin{bmatrix}-1\\1\\1\end{bmatrix}\right\}, \quad B' = \left\{\begin{bmatrix}1\\1\\1\end{bmatrix}, \begin{bmatrix}2\\3\\3\end{bmatrix}, \begin{bmatrix}3\\2\\3\end{bmatrix}\right\}, \quad [\mathbf{x}]_{B'} = \begin{bmatrix}1\\-1\\0\end{bmatrix}$

In Exercises 10–12, given B, B', and $[\mathbf{x}]_B$, find $[\mathbf{x}]_{B'}$, and check by converting both expressions for $\mathbf{x}$ to standard form.

10. $B = \left\{\begin{bmatrix}4\\1\end{bmatrix}, \begin{bmatrix}3\\1\end{bmatrix}\right\}, \quad B' = \left\{\begin{bmatrix}2\\5\end{bmatrix}, \begin{bmatrix}1\\3\end{bmatrix}\right\}, \quad [\mathbf{x}]_B = \begin{bmatrix}2\\-1\end{bmatrix}$

11. $B = \left\{\begin{bmatrix}2\\3\end{bmatrix}, \begin{bmatrix}3\\2\end{bmatrix}\right\}, \quad B' = \left\{\begin{bmatrix}2\\-3\end{bmatrix}, \begin{bmatrix}1\\3\end{bmatrix}\right\}, \quad [\mathbf{x}]_B = \begin{bmatrix}1\\-2\end{bmatrix}$

12. $B = \left\{\begin{bmatrix}1\\-1\\1\end{bmatrix}, \begin{bmatrix}2\\1\\-1\end{bmatrix}, \begin{bmatrix}0\\1\\1\end{bmatrix}\right\}, \quad B' = \left\{\begin{bmatrix}1\\1\\2\end{bmatrix}, \begin{bmatrix}3\\2\\8\end{bmatrix}, \begin{bmatrix}4\\2\\13\end{bmatrix}\right\}, \quad [\mathbf{x}]_B = \begin{bmatrix}1\\0\\0\end{bmatrix}$

13. Let B be a basis for $\mathbb{R}^3$ and $C = \{\mathbf{w}_1, \mathbf{w}_2\}$ a basis for $\mathbb{R}^2$. If $T: \mathbb{R}^3 \to \mathbb{R}^2$ is the linear transformation defined for $(\mathbf{x})_B = (c_1, c_2, c_3)$ by

$$T(\mathbf{x}) = (2c_1 - c_2 + c_3)\mathbf{w}_1 + (4c_1 + c_2 - c_3)\mathbf{w}_2,$$

find the matrix A_{BC} of T relative to B and C.

In Exercises 14–17, we are given bases B and B' for $\mathbb{R}^2$ or $\mathbb{R}^3$, and the matrix A of a linear operator T relative to B. Find the matrix A' of T relative to B'.

14. $B = \left\{\begin{bmatrix}1\\0\end{bmatrix}, \begin{bmatrix}0\\1\end{bmatrix}\right\}, \quad B' = \left\{\begin{bmatrix}2\\-3\end{bmatrix}, \begin{bmatrix}1\\-1\end{bmatrix}\right\}, \quad A = \begin{bmatrix}4 & 2\\-3 & -1\end{bmatrix}$

15. $B = \left\{\begin{bmatrix}1\\2\end{bmatrix}, \begin{bmatrix}2\\5\end{bmatrix}\right\}, \quad B' = \left\{\begin{bmatrix}2\\-3\end{bmatrix}, \begin{bmatrix}3\\-4\end{bmatrix}\right\}, \quad A = \begin{bmatrix}1 & 0\\-2 & 2\end{bmatrix}$

16. $B = \left\{\begin{bmatrix}1\\0\\0\end{bmatrix}, \begin{bmatrix}-3\\-2\\3\end{bmatrix}, \begin{bmatrix}0\\-1\\2\end{bmatrix}\right\}, \quad B' = \left\{\begin{bmatrix}1\\1\\0\end{bmatrix}, \begin{bmatrix}-3\\-2\\1\end{bmatrix}, \begin{bmatrix}-3\\0\\2\end{bmatrix}\right\},$

$A = \begin{bmatrix}1 & -2 & 0\\1 & 0 & -1\\2 & 1 & 1\end{bmatrix}$

17. $B = \left\{\begin{bmatrix}1\\0\\0\end{bmatrix}, \begin{bmatrix}0\\1\\0\end{bmatrix}, \begin{bmatrix}0\\0\\1\end{bmatrix}\right\}, \quad B' = \left\{\begin{bmatrix}1\\-1\\1\end{bmatrix}, \begin{bmatrix}2\\1\\-1\end{bmatrix}, \begin{bmatrix}0\\1\\1\end{bmatrix}\right\}, \quad A = \begin{bmatrix}2 & -1 & 1\\-1 & 4 & 1\\1 & 1 & 4\end{bmatrix}$

In Exercises 18–21, we are given bases B and B' for the domain, bases C and C' for the codomain, and the matrix A of a linear transformation T relative to B and C. Find the matrix A' of T relative to B' and C'. Use A' to find $[T(\mathbf{x})]_{C'}$ for the given $[\mathbf{x}]_{B'}$, and check by finding $[T(\mathbf{x})]_C$ in two ways.

18. $B = \left\{ \begin{bmatrix} 2 \\ 1 \\ 1 \end{bmatrix}, \begin{bmatrix} -1 \\ 0 \\ -1 \end{bmatrix}, \begin{bmatrix} -1 \\ -1 \\ 2 \end{bmatrix} \right\}$, $B' = \left\{ \begin{bmatrix} 1 \\ 0 \\ 0 \end{bmatrix}, \begin{bmatrix} -1 \\ -2 \\ -1 \end{bmatrix}, \begin{bmatrix} 2 \\ 3 \\ 2 \end{bmatrix} \right\}$, $A = \begin{bmatrix} 1 & 0 & -1 \\ 0 & -1 & 1 \end{bmatrix}$

$C = \begin{bmatrix} 1 & 3 \\ 2 & 5 \end{bmatrix}$, $C' = \begin{bmatrix} 1 & -2 \\ 2 & -3 \end{bmatrix}$, $[\mathbf{x}]_{B'} = \begin{bmatrix} 1 \\ 1 \\ 0 \end{bmatrix}$

19. $B = \left\{ \begin{bmatrix} 1 \\ 0 \\ 0 \end{bmatrix}, \begin{bmatrix} 0 \\ 1 \\ 0 \end{bmatrix}, \begin{bmatrix} 1 \\ 0 \\ 1 \end{bmatrix} \right\}$, $B' = \left\{ \begin{bmatrix} 1 \\ 1 \\ 0 \end{bmatrix}, \begin{bmatrix} 0 \\ -2 \\ 3 \end{bmatrix}, \begin{bmatrix} 0 \\ -1 \\ 2 \end{bmatrix} \right\}$, $A = \begin{bmatrix} 1 & -1 & 1 \\ 3 & 2 & 1 \end{bmatrix}$

$C = \left\{ \begin{bmatrix} 1 \\ 0 \end{bmatrix}, \begin{bmatrix} 0 \\ 1 \end{bmatrix} \right\}$, $C' = \left\{ \begin{bmatrix} 3 \\ -1 \end{bmatrix}, \begin{bmatrix} 4 \\ -2 \end{bmatrix} \right\}$, $[\mathbf{x}]_{B'} = \begin{bmatrix} 0 \\ 2 \\ 1 \end{bmatrix}$

20. $B = \left\{ \begin{bmatrix} 1 \\ 0 \end{bmatrix}, \begin{bmatrix} 0 \\ 1 \end{bmatrix} \right\}$, $B' = \begin{bmatrix} 2 \\ 1 \end{bmatrix}, \begin{bmatrix} 5 \\ 2 \end{bmatrix}$, $A = \begin{bmatrix} 1 & -1 \\ 2 & 1 \\ 1 & -1 \end{bmatrix}$

$C = \left\{ \begin{bmatrix} 1 \\ 0 \\ 0 \end{bmatrix}, \begin{bmatrix} 0 \\ 1 \\ 0 \end{bmatrix}, \begin{bmatrix} 1 \\ 0 \\ 1 \end{bmatrix} \right\}$, $C' = \left\{ \begin{bmatrix} 1 \\ 1 \\ 1 \end{bmatrix}, \begin{bmatrix} 1 \\ -2 \\ -3 \end{bmatrix}, \begin{bmatrix} -1 \\ 1 \\ 2 \end{bmatrix} \right\}$, $[\mathbf{x}]_{B'} = \begin{bmatrix} 1 \\ 3 \end{bmatrix}$

21. $B = \left\{ \begin{bmatrix} 3 \\ 2 \end{bmatrix}, \begin{bmatrix} 3 \\ 4 \end{bmatrix} \right\}$, $B' = \left\{ \begin{bmatrix} -1 \\ 0 \end{bmatrix}, \begin{bmatrix} 3 \\ 1 \end{bmatrix} \right\}$, $A = \begin{bmatrix} 2 & 1 \\ -1 & 0 \\ 1 & -2 \end{bmatrix}$

$C = \left\{ \begin{bmatrix} 1 \\ -1 \\ -1 \end{bmatrix}, \begin{bmatrix} 2 \\ -2 \\ -1 \end{bmatrix}, \begin{bmatrix} 1 \\ 0 \\ 2 \end{bmatrix} \right\}$, $C' = \left\{ \begin{bmatrix} 1 \\ -2 \\ 0 \end{bmatrix}, \begin{bmatrix} 2 \\ -2 \\ 1 \end{bmatrix}, \begin{bmatrix} 2 \\ 1 \\ 2 \end{bmatrix} \right\}$, $[\mathbf{x}]_{B'} = \begin{bmatrix} 1 \\ -1 \end{bmatrix}$

22. Let P be the orthogonal projection of $\mathbb{R}^2$ onto the line spanned by the vector $\mathbf{u}_1 = (2,1)$, and let $\mathbf{u}_2 = (-1,2)$ and $B = \{\mathbf{u}_1, \mathbf{u}_2\}$. Find the matrix of P relative to B.

23. Let P be the orthogonal projection of $\mathbb{R}^3$ onto the plane spanned by $\mathbf{u}_1 = (1,2,-2)$ and $\mathbf{u}_2 = (2,1,2)$, and let $\mathbf{u}_3 = (-2,2,1)$ and $B = \{\mathbf{u}_1, \mathbf{u}_2, \mathbf{u}_3\}$. Find the matrix of P relative to B.

24. Let V be the vector space of all real polynomials of at most degree 2. Let B be the basis $\{p_0, p_1, p_2\}$, where

$$p_0 = 1, \qquad p_1 = x, \qquad p_2 = x^2.$$

Let $B' = \{q_0, q_1, q_2\}$ the basis defined by

$$q_0 = 1, \qquad q_1 = x, \qquad q_2 = 2x^2 - 1.$$

Representing a polynomial as an ordered triple, i.e., making the identification

$$a_0 + a_1 x + a_2 x^2 = (a_0, a_1, a_2),$$

find the transition matrix from B' to B and the transition matrix from B to B'.

25. For the vector space V and the bases B and B' of Exercise 24, let D be the linear transformation determined by:

$$D(p_0) = 0, \qquad D(p_1) = p_0, \qquad D(p_2) = 2p_1.$$

Find the matrix of D relative to B and the matrix of D relative to B'.

26. (*Calculus based*) Let V be the vector space of all real functions on R with continuous first derivatives and let W be the subspace with basis $B = \{f_1, f_2\}$, where

$$f_1(x) = e^x \cos x, \qquad f_2(x) = e^x \sin x.$$

Find the matrix of the differentiation operator D on V, defined by $D(f) = f'$.

6.3 SIMILARITY OF MATRICES

As we saw in Section 6.2, the matrix of a linear operator on $\mathbb{R}^n$ depends on the basis for $\mathbb{R}^n$, so distinct matrices may determine the same operator. In this section, we study matrices that determine the same linear operator, in final preparation for the eigenvalue problem in Chapter 7.

The Concept of Similarity

The effect of a basis change on the matrix of a linear operator T was found in Section 6.2: if A represents T relative to B and A' represents the same T relative to B', then

$$A' = P^{-1}AP, \qquad \text{or} \qquad PA' = AP,$$

where P is the transition matrix from B' to B. This relation gives rise to an important concept.

DEFINITION 6.5

Let A and A' be $n \times n$ matrices. We say **A is similar to A'**, and write $A \sim A'$, if $A' = P^{-1}AP$ for some invertible matrix P.

THEOREM 6.5

Two $n \times n$ matrices A and A' are similar if and only if they represent the same linear operator T on $\mathbb{R}^n$ relative to some bases B and B'.

Proof Figure 6.7 may help in understanding the outline of this proof. If $A \sim A'$, then by Definition 6.5 there is an invertible matrix P such that $A' = P^{-1}AP$. Let B be any basis for $\mathbb{R}^n$, and define T on $\mathbb{R}^n$ by

$$[T(\mathbf{x})]_B = A[\mathbf{x}]_B.$$

Let Q be the matrix whose columns are the vectors in B, and put

$$Q' = QP.$$

Since Q and P are invertible, Q' is invertible, its columns form a basis B' for $\mathbb{R}^n$, and

$$P = Q^{-1}Q'.$$

By Definition 6.4, P is the transition matrix from B' to B, hence for any $\mathbf{x}$ in $\mathbb{R}^n$,

$$[\mathbf{x}]_B = P[\mathbf{x}]_{B'} \quad \text{and} \quad [T(\mathbf{x})]_B = P[T(\mathbf{x})]_{B'}.$$

Therefore

$$[T(\mathbf{x})]_{B'} = P^{-1}[T(\mathbf{x})]_B = P^{-1}A[\mathbf{x}]_B = P^{-1}AP[\mathbf{x}]_{B'} = A'[\mathbf{x}]_{B'},$$

so A' represents T relative to B'.

Conversely, if A and A' represent the same linear operator T relative to bases B and B', respectively, let P be the transition matrix from B' to B. By Definition 6.4,

$$[\mathbf{x}]_B = P[\mathbf{x}]_{B'} \quad \text{and} \quad [T(\mathbf{x})]_B = P[T(\mathbf{x})]_{B'}$$

for any $\mathbf{x}$ in $\mathbb{R}^n$. From these relations and the representation of T by A,

$$[T(\mathbf{x})]_{B'} = P^{-1}[T(\mathbf{x})]_B = P^{-1}A[\mathbf{x}]_B = P^{-1}AP[\mathbf{x}]_{B'}.$$

Thus $P^{-1}AP$ represents T relative to B'. Since A' also represents T relative to B', we have for all $\mathbf{x}$ in $\mathbb{R}^n$,

$$A'[\mathbf{x}]_{B'} = P^{-1}AP[\mathbf{x}]_{B'}.$$

In particular, this relation holds when $\mathbf{x}$ is any vector in B'. But for $\mathbf{x} = \mathbf{v}_j$, the jth vector in B', the vectors $A'[\mathbf{x}]_{B'}$ and $P^{-1}AP[\mathbf{x}]_{B'}$ equal the jth column vector of A' and $P^{-1}AP$, respectively. Thus $A' = P^{-1}AP$ since their columns agree. $\square$

The "only if" statement is illustrated in Theorem 6.5 in the next example. The "if" statement was illustrated in Section 6.2 by Example 10, page 316, and by Exercises 14–17.

Example 1

Let

$$A = \begin{bmatrix} 1 & 3 \\ 2 & -2 \end{bmatrix}, \quad P = \begin{bmatrix} 3 & 4 \\ 2 & 3 \end{bmatrix}, \quad \text{and} \quad B = \left\{ \begin{bmatrix} 1 \\ 1 \end{bmatrix}, \begin{bmatrix} 1 \\ -1 \end{bmatrix} \right\}$$

in $\mathbb{R}^2$, and let T be the linear operator determined by A relative to B. If $A' = P^{-1}AP$, find a basis B' for $\mathbb{R}^2$ such that T is the linear operator determined by A' relative to B', and verify that it is by direct calculation.

Solution

Let Q be the matrix formed with the vectors in B as columns, and put

$$Q' = QP = \begin{bmatrix} 1 & 1 \\ 1 & -1 \end{bmatrix}\begin{bmatrix} 3 & 4 \\ 2 & 3 \end{bmatrix} = \begin{bmatrix} 5 & 7 \\ 1 & 1 \end{bmatrix}.$$

Form the basis B' from the columns of Q':

$$B' = \left\{ \begin{bmatrix} 5 \\ 1 \end{bmatrix}, \begin{bmatrix} 7 \\ 1 \end{bmatrix} \right\}.$$

Then

$$A' = P^{-1}AP = \begin{bmatrix} 3 & -4 \\ -2 & 3 \end{bmatrix}\begin{bmatrix} 1 & 3 \\ 2 & -2 \end{bmatrix}\begin{bmatrix} 3 & 4 \\ 2 & 3 \end{bmatrix} = \begin{bmatrix} 19 & 31 \\ -12 & -20 \end{bmatrix}.$$

Use this matrix to verify that for all $\mathbf{x}$ in $\mathbb{R}^2$,

$$[T(\mathbf{x})]_{B'} = A'[\mathbf{x}]_{B'}.$$

For any $\mathbf{x}$ in $\mathbb{R}^2$, let

$$[\mathbf{x}]_B = \begin{bmatrix} x_1 \\ x_2 \end{bmatrix}.$$

Since A is the matrix of T relative to B,

$$[T(\mathbf{x})]_B = A[\mathbf{x}]_B = \begin{bmatrix} 1 & 3 \\ 2 & -2 \end{bmatrix}\begin{bmatrix} x_1 \\ x_2 \end{bmatrix} = \begin{bmatrix} x_1 + 3x_2 \\ 2x_1 - 2x_2 \end{bmatrix}.$$

Since P^{-1} is the transition matrix from B to B',

$$[T(\mathbf{x})]_{B'} = P^{-1}[T(\mathbf{x})]_B = \begin{bmatrix} 3 & -4 \\ -2 & 3 \end{bmatrix}\begin{bmatrix} x_1 + 3x_2 \\ 2x_1 - 2x_2 \end{bmatrix} = \begin{bmatrix} -5x_1 + 17x_2 \\ 4x_1 - 12x_2 \end{bmatrix}.$$

Applying P^{-1} also to $[\mathbf{x}]_B$ gives

$$[\mathbf{x}]_{B'} = P^{-1}[\mathbf{x}]_B = \begin{bmatrix} 3 & -4 \\ -2 & 3 \end{bmatrix}\begin{bmatrix} x_1 \\ x_2 \end{bmatrix} = \begin{bmatrix} 3x_1 - 4x_2 \\ -2x_1 + 3x_2 \end{bmatrix},$$

from which

$$A'[\mathbf{x}]_{B'} = \begin{bmatrix} 19 & 31 \\ -12 & -20 \end{bmatrix} \begin{bmatrix} 3x_1 - 4x_2 \\ -2x_1 + 3x_2 \end{bmatrix} = \begin{bmatrix} -5x_1 + 17x_2 \\ 4x_1 - 12x_2 \end{bmatrix}.$$

Since the expressions for $A'[\mathbf{x}]_{B'}$ and $[T(\mathbf{x})]_{B'}$ agree, A' represents T relative to B'. □

Similarity is one key to understanding the structure of a linear operator T. If A is the matrix of T relative to the standard basis, we can learn much about T by finding a *diagonal matrix* D such that $A \sim D$. The problem of finding D and the corresponding basis (Theorem 6.5) is studied in Chapter 7.

Example 2

Let

$$A = \begin{bmatrix} 1 & 3 \\ 2 & 2 \end{bmatrix}, \quad P = \begin{bmatrix} 3 & 4 \\ -2 & 4 \end{bmatrix}.$$

Show that if A represents a linear operator T relative to the standard basis for $\mathbb{R}^2$ and $D = P^{-1}AP$, then D is a diagonal matrix representing T relative to the basis determined by P.

Solution

In the notation of Theorem 6.5, B is the standard basis for $\mathbb{R}^2$ and Q is the 2×2 identity matrix since its columns are the vectors in B. Since

$$P^{-1}AP = \begin{bmatrix} 4/20 & -4/20 \\ 2/20 & 3/20 \end{bmatrix} \begin{bmatrix} 1 & 3 \\ 2 & 2 \end{bmatrix} \begin{bmatrix} 3 & 4 \\ -2 & 4 \end{bmatrix} = \begin{bmatrix} -1 & 0 \\ 0 & 4 \end{bmatrix},$$

D is a diagonal matrix. By Theorem 6.5, D represents T relative to the basis B' determined by $Q' = QP = P$. That is, D is the matrix of T relative to the basis

$$B' = \left\{ \begin{bmatrix} 3 \\ -2 \end{bmatrix}, \begin{bmatrix} 4 \\ 4 \end{bmatrix} \right\}.$$ □

Orthogonal Matrices

Certain matrices, including all symmetric matrices, have useful properties relative to similarity. As shown in Chapter 7, if A is symmetric, then there is a diagonal matrix D and a transition matrix P with an orthonormal set of column vectors such that $P^{-1}AP = D$. In this case, P^{-1} is easy to find since $P^{-1} = P^t$ (Exercise 17). We formalize these ideas.

DEFINITION 6.6

A square matrix P is said to be **orthogonal** if its column vectors form an orthonormal set. If $A' = P^{-1}AP$ and P is orthogonal, A' is said to be **orthogonally similar** to A'.

THEOREM 6.6

A square matrix P is orthogonal if and only if P^{-1} exists and satisfies $P^{-1} = P^t$.

Example 3

If

$$A = \begin{bmatrix} 1 & 2 \\ 2 & 1 \end{bmatrix} \quad \text{and} \quad P = \begin{bmatrix} 1/\sqrt{2} & 1/\sqrt{2} \\ -1/\sqrt{2} & 1/\sqrt{2} \end{bmatrix},$$

then A is symmetric, and P is orthogonal since its column vectors form an orthonormal set. Moreover,

$$P^{-1}AP = P^tAP = \begin{bmatrix} 1/\sqrt{2} & -1/\sqrt{2} \\ 1/\sqrt{2} & 1/\sqrt{2} \end{bmatrix} \begin{bmatrix} 1 & 2 \\ 2 & 1 \end{bmatrix} \begin{bmatrix} 1/\sqrt{2} & 1/\sqrt{2} \\ -1/\sqrt{2} & 1/\sqrt{2} \end{bmatrix} = \begin{bmatrix} -1 & 0 \\ 0 & 3 \end{bmatrix},$$

so A is orthogonally similar to a diagonal matrix. □

Properties of Similar Matrices

In more advanced studies, properties shared by A and $P^{-1}AP$, where P is any invertible matrix, play an important role. Although an exposition is beyond the scope of this book, we explore a few of these properties in the exercises. For example, let A be a square matrix. Then $\det(A) = \det(P^{-1}AP)$ (Exercise 23). Moreover $P^{-1}AP$ is invertible if and only if A is invertible (Exercise 24). In addition, $\text{rank}(A) = \text{rank}(P^{-1}AP)$ (Exercise 25), and this relation has implications for linear systems associated with A and $P^{-1}AP$ (Exercise 26).

The **trace of a matrix** is the sum of its diagonal entries. For $A = [a_{ij}]_{n\times n}$, the trace of A is denoted by $\text{tr}(A)$, and

$$\text{tr}(A) = \sum_{i=1}^{n} a_{ii}.$$

The value of the trace is also the same for A and $P^{-1}AP$. That is, if $A \sim A'$, then

$$\text{tr}(A) = \text{tr}(A')$$

(Exercise 29). The examples above illustrate this relation. In Example 1, $\text{tr}(A) = \text{tr}(A') = -1$; in Example 2, $\text{tr}(A) = \text{tr}(D) = 3$; and in Example 3, $\text{tr}(A) = \text{tr}(A') = 2$.

Finally, for any n, let M_n denote the vector space of all $n \times n$ matrices under ordinary matrix addition and multiplication by a scalar. If P is any invertible $n \times n$ matrix, then the function $\mathcal{T}$ defined by

$$\mathcal{T}(A) = P^{-1}AP$$

is a linear operator on M_n (Exercise 30). (Be careful to distinguish $\mathcal{T}$ from a linear operator T on $\mathbb{R}^n$.) Implications of this idea and related concepts are often studied in more advanced courses in linear algebra.

REVIEW CHECKLIST

1. Define similarity of matrices.
2. Characterize similarity of matrices in terms of representation of linear operators.
3. Verify the assertion of Theorem 6.5 for specific examples.
4. Formulate definitions of an orthogonal matrix and orthogonally similar matrices.
5. Identify an orthogonal matrix as one whose inverse equals its transpose.
6. Define the trace of a square matrix.
7. Identify the value of the trace, the determinant, and the rank of a matrix as equal for similar matrices.

EXERCISES

In Exercises 1–8, verify the assertion of Theorem 6.5. That is, find $A' = P^{-1}AP$ and the corresponding basis B', and show that A and A' represent the same linear operator T by showing that $A[\mathbf{x}]_B = A'[\mathbf{x}]_{B'}$ for each $\mathbf{x}$ in B.

1. $A = \begin{bmatrix} 1 & 1 \\ -1 & -1 \end{bmatrix}$, $P = \begin{bmatrix} 1 & 1 \\ 2 & 3 \end{bmatrix}$, $B = \left\{ \begin{bmatrix} 1 \\ 0 \end{bmatrix}, \begin{bmatrix} 0 \\ 1 \end{bmatrix} \right\}$

2. $A = \begin{bmatrix} 1 & 1 \\ -1 & -1 \end{bmatrix}$, $P = \begin{bmatrix} 1 & 1 \\ 2 & 3 \end{bmatrix}$, $B = \left\{ \begin{bmatrix} 1 \\ 2 \end{bmatrix}, \begin{bmatrix} 1 \\ 1 \end{bmatrix} \right\}$

3. $A = \begin{bmatrix} 1 & 0 \\ 1 & -2 \end{bmatrix}$, $P = \begin{bmatrix} 2 & 1 \\ 3 & 2 \end{bmatrix}$, $B = \left\{ \begin{bmatrix} 1 \\ 0 \end{bmatrix}, \begin{bmatrix} 0 \\ 1 \end{bmatrix} \right\}$

4. $A = \begin{bmatrix} 1 & 0 \\ 1 & -2 \end{bmatrix}$, $P = \begin{bmatrix} 2 & 1 \\ 3 & 2 \end{bmatrix}$, $B = \left\{ \begin{bmatrix} 2 \\ -3 \end{bmatrix}, \begin{bmatrix} -1 \\ 2 \end{bmatrix} \right\}$

5. $A = \begin{bmatrix} 2 & 1 \\ 1 & 2 \end{bmatrix}$, $P = \begin{bmatrix} 1 & 1 \\ 1 & -1 \end{bmatrix}$, $B = \left\{ \begin{bmatrix} 5 \\ 4 \end{bmatrix}, \begin{bmatrix} 4 \\ 3 \end{bmatrix} \right\}$

6. $A = \begin{bmatrix} 3 & 2 & -2 \\ 1 & 4 & -2 \\ 1 & 2 & 0 \end{bmatrix}$, $P = \begin{bmatrix} 2 & 2 & 1 \\ 0 & -1 & 1 \\ 1 & 0 & 1 \end{bmatrix}$, $B = \left\{ \begin{bmatrix} 1 \\ 0 \\ 0 \end{bmatrix}, \begin{bmatrix} 0 \\ 1 \\ 1 \end{bmatrix}, \begin{bmatrix} 0 \\ 0 \\ 1 \end{bmatrix} \right\}$

7. $A = \begin{bmatrix} 3 & 2 & -2 \\ 1 & 4 & -2 \\ 1 & 2 & 0 \end{bmatrix}$, $P = \begin{bmatrix} 2 & 2 & 1 \\ 0 & -1 & 1 \\ 1 & 0 & 1 \end{bmatrix}$, $B = \left\{ \begin{bmatrix} 1 \\ 1 \\ 0 \end{bmatrix}, \begin{bmatrix} 1 \\ 0 \\ 1 \end{bmatrix}, \begin{bmatrix} 0 \\ 1 \\ 1 \end{bmatrix} \right\}$

8. $A = \begin{bmatrix} 2 & 1 & -1 \\ 1 & 4 & 1 \\ -1 & 1 & 4 \end{bmatrix}$, $P = \begin{bmatrix} 2 & 1 & 0 \\ -1 & 1 & 1 \\ 1 & -1 & 1 \end{bmatrix}$, $B = \left\{ \begin{bmatrix} 1 \\ 0 \\ 0 \end{bmatrix}, \begin{bmatrix} 1 \\ 1 \\ 0 \end{bmatrix}, \begin{bmatrix} 1 \\ 1 \\ 1 \end{bmatrix} \right\}$

In Exercises 9–12, show that the given matrix is orthogonal and verify that its inverse equals its transpose.

9. $\begin{bmatrix} 3/5 & -4/5 \\ 4/5 & 3/5 \end{bmatrix}$

10. $\begin{bmatrix} 1/2 & -\sqrt{3}/2 \\ \sqrt{3}/2 & 1/2 \end{bmatrix}$

11. $\begin{bmatrix} 1/3 & -2/3 & 2/3 \\ 2/3 & 2/3 & 1/3 \\ -2/3 & 1/3 & 2/3 \end{bmatrix}$

12. $\begin{bmatrix} 1/\sqrt{3} & 1/\sqrt{6} & 1/\sqrt{2} \\ 1/\sqrt{3} & 1/\sqrt{6} & -1/\sqrt{2} \\ 1/\sqrt{3} & -2/\sqrt{6} & 0 \end{bmatrix}$

In Exercises 13–16, show that the matrices A and A' are orthogonally similar via the given P.

13. $A = \begin{bmatrix} 4 & 2 \\ 2 & 2 \end{bmatrix}$, $A' = \begin{bmatrix} 5 & -1 \\ -1 & 1 \end{bmatrix}$, $P = \begin{bmatrix} 1/\sqrt{2} & -1/\sqrt{2} \\ 1/\sqrt{2} & 1/\sqrt{2} \end{bmatrix}$

14. $A = \begin{bmatrix} 5 & 2 \\ 2 & 2 \end{bmatrix}$, $A' = \begin{bmatrix} 1 & 0 \\ 0 & 6 \end{bmatrix}$, $P = \begin{bmatrix} 1/\sqrt{5} & 2/\sqrt{5} \\ -2/\sqrt{5} & 1/\sqrt{5} \end{bmatrix}$

15. $A = \begin{bmatrix} 3 & 0 & 0 \\ 0 & 5 & 2 \\ 0 & 2 & 2 \end{bmatrix}$, $A' = \begin{bmatrix} 3 & 0 & 0 \\ 0 & 6 & 0 \\ 0 & 0 & 1 \end{bmatrix}$, $P = \begin{bmatrix} 1 & 0 & 0 \\ 0 & 2/\sqrt{5} & -1/\sqrt{5} \\ 0 & 1/\sqrt{5} & 2/\sqrt{5} \end{bmatrix}$

16. $A = \begin{bmatrix} 4 & 1 & -1 \\ 1 & 3 & -2 \\ -1 & -2 & 3 \end{bmatrix}$, $A' = \begin{bmatrix} 6 & 0 & 0 \\ 0 & 3 & 0 \\ 0 & 0 & 1 \end{bmatrix}$,

$P = \begin{bmatrix} 1/\sqrt{3} & 2/\sqrt{6} & 0 \\ 1/\sqrt{3} & -1/\sqrt{6} & 1/\sqrt{2} \\ -1/\sqrt{3} & 1/\sqrt{6} & 1/\sqrt{2} \end{bmatrix}$

In Exercises 17–21, P is a square matrix.

17. Show that P is orthogonal if and only if P^{-1} exists and satisfies $P^{-1} = P^t$.

18. If P is orthogonal, find the possible value(s) of $\det(P)$.

19. Show that P is orthogonal if and only if P^t is orthogonal.

20. Show that P is orthogonal if and only if P^{-1} exists and is orthogonal.

21. Let Q be a square matrix of the same dimension as is P. Show that if any two of P, Q, and PQ are orthogonal, then so is the third.

22. Show that if $A \sim A'$, then for every positive integer k, $A^k \sim A'^k$.

23. Show that if $A \sim A'$, then $\det(A) = \det(A')$.

24. Show that if $A \sim A'$, then A is invertible if and only if A' is invertible.

25. Show that if $A \sim A'$, then $\operatorname{rank}(A) = \operatorname{rank}(A')$.

26. If $A \sim A'$ how is the dimension of the solution space of the linear system $A\mathbf{x} = \mathbf{0}$ related to the dimension of the solution space of $A'\mathbf{x} = \mathbf{0}$?

27. If $A' = P^{-1}AP$, determine the relationship between the solution spaces of $A\mathbf{x} = \mathbf{0}$ and $A'\mathbf{x} = \mathbf{0}$.

28. Show that an $n \times n$ matrix A is orthogonal if for every $\mathbf{x}$ in $\mathbb{R}^n$, we have $\|A\mathbf{x}\| = \|\mathbf{x}\|$.

29. Show that if $A \sim A'$, then $\operatorname{tr}(A) = \operatorname{tr}(A')$.

30. For any n, let M_n denote the vector space of all $n \times n$ matrices under ordinary matrix addition and multiplication by a scalar. Show that if P is any invertible $n \times n$ matrix, then the function $\mathscr{T}$ defined by

$$\mathscr{T}(A) = P^{-1}AP$$

is a linear operator on M_n.

31. Show that $\sim$ is an *equivalence relation* on the set of all $n \times n$ matrices. That is, show that for $n \times n$ matrices A, B, and C, $\sim$ is:

a. *reflexive* $A \sim A$;

b. *symmetric* if $A \sim B$, then $B \sim A$; and

c. *transitive* if $A \sim B$ and $B \sim C$, then $A \sim C$.

(o) 6.4 APPLICATIONS AND COMPUTING

3-D Graphics

One application of computer graphics is the video screen display of a surface determined by an equation $z = f(x, y)$. Such a display involves representing a 3-dimensional object in a 2-dimensional medium. Figure 6.8a shows the projection of a spatial coordinate system onto an image plane by rays from a point source and Fig. 6.8b shows projection by parallel rays.

For rays from a source at a point P, the image plane is orthogonal to the line through P and the spatial origin O, as in Fig. 6.8a. The image plane may be behind O, as in Fig. 6.8a, or between O and P. For parallel rays, the image plane passes through O, as in Fig. 6.8b. For this method, the rays are orthogonal to the image plane.

The size of a point source image is controlled by the ratio d/ρ, where d is the distance between the image plane and P, and ρ is the distance between O and P. The value of ρ is often just large enough to prevent distortion, and d is

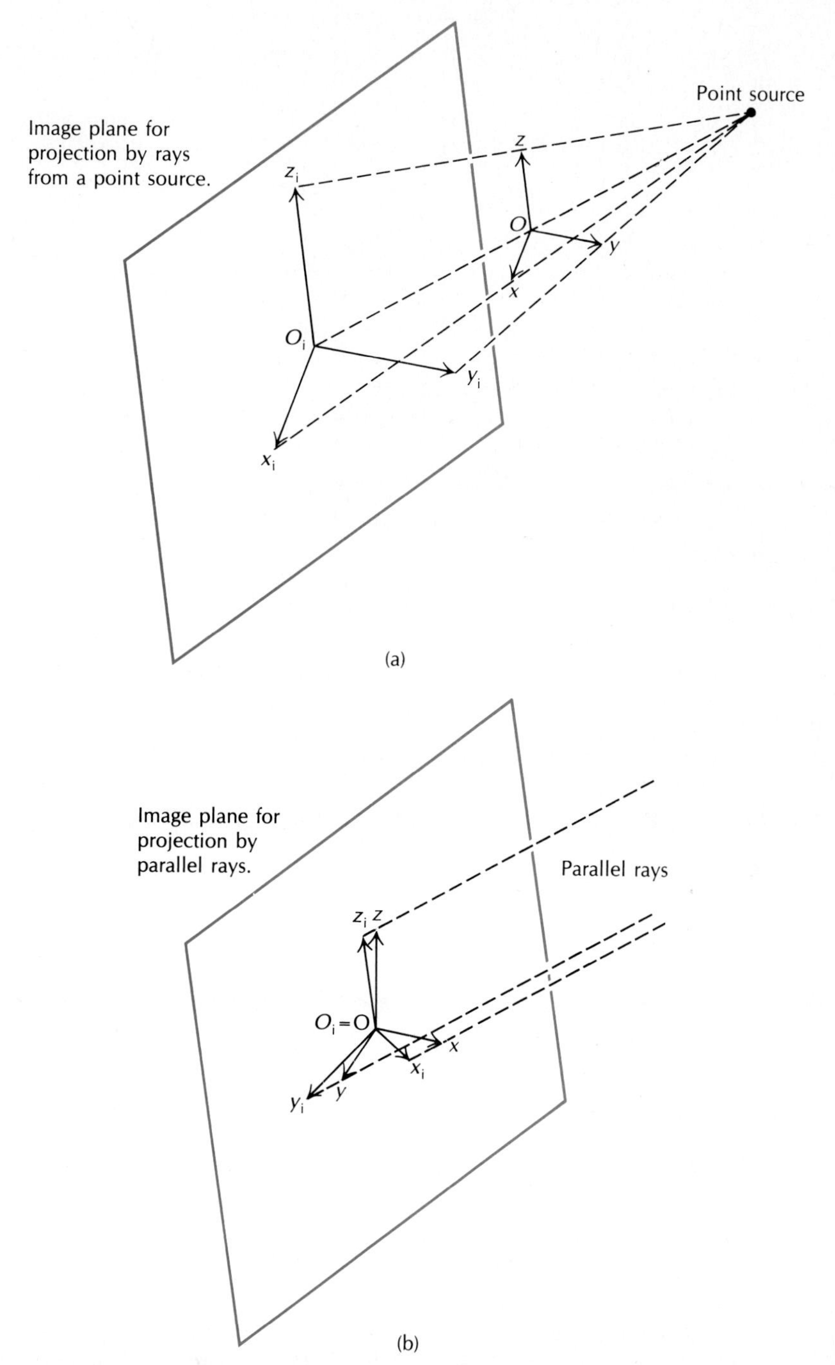

Figure 6.8 Projection onto an image plane by: (a) rays from a point source; (b) parallel rays.

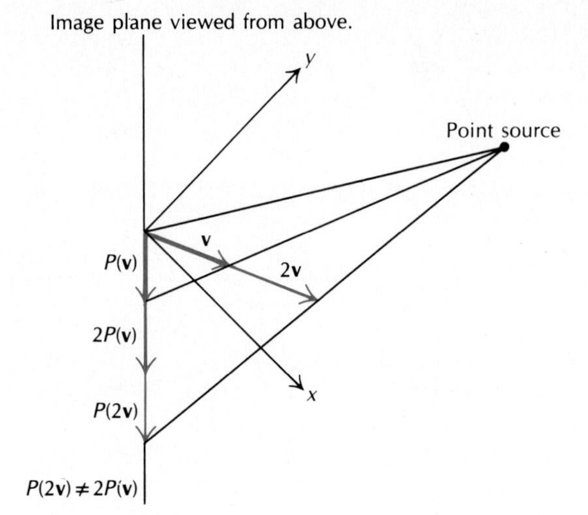

Figure 6.9 Point source projection is nonlinear. For the vector **v** shown here, we have $P(2\mathbf{v}) \neq 2P(\mathbf{v})$.

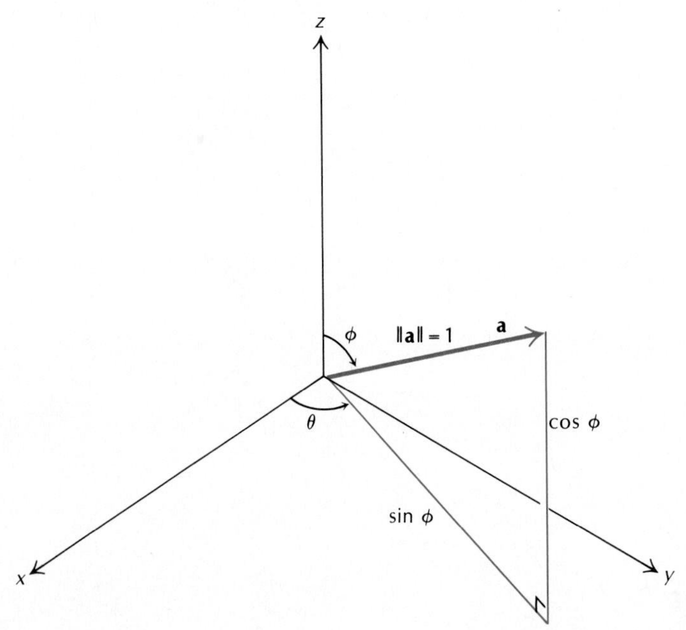

Figure 6.10 Interpretation of the variables θ and ϕ used to specify the direction of a parallel projection.

then adjusted to produce a visible image. With a little practice, a user can learn to set values of d and ρ for satisfactory displays. Under this method, the image resembles what is seen by a viewer with an eye at the ray source. However, this "projection" is *nonlinear*, as illustrated in Fig. 6.9, and we shall not study it here. A readable treatment of this method, complete with BASIC programs, can be found in *Microcomputer Graphics* (IBM or Apple version) by Roy E. Myers, Addison-Wesley Publishing Company.

Projection by parallel rays *is* a linear transformation, and it corresponds to the orthogonal projection of Definition 5.3. Under this method, the image plane passes through the origin O as in Fig. 6.8b, and P is located at a distance 1 from O. Then P corresponds to a unit vector $\mathbf{a}$, and either P or $\mathbf{a}$ determines the projection operator. One convenient way to specify $\mathbf{a}$ is by *spherical coordinates* (Fig. 6.10). If ϕ is the angle formed by $\mathbf{a}$ and the positive z-axis, and θ is the angle formed by the orthogonal projection of $\mathbf{a}$ on the xy-plane and the positive x-axis as shown in Fig. 6.10, then

$$\mathbf{a} = (\sin\phi\cos\theta, \sin\phi\sin\theta, \cos\phi).$$

With $\mathbf{a}$ specified, we find the orthogonal projection of $\mathbb{R}^3$ onto the plane through the origin perpendicular to $\mathbf{a}$. Let $E = \{\mathbf{e}_1, \mathbf{e}_2, \mathbf{e}_3\}$ be the standard basis for $\mathbb{R}^3$. We construct a basis $B = \{\mathbf{v}_1, \mathbf{v}_2\}$ in the image space. Let $\mathbf{v}_2$ be the unit

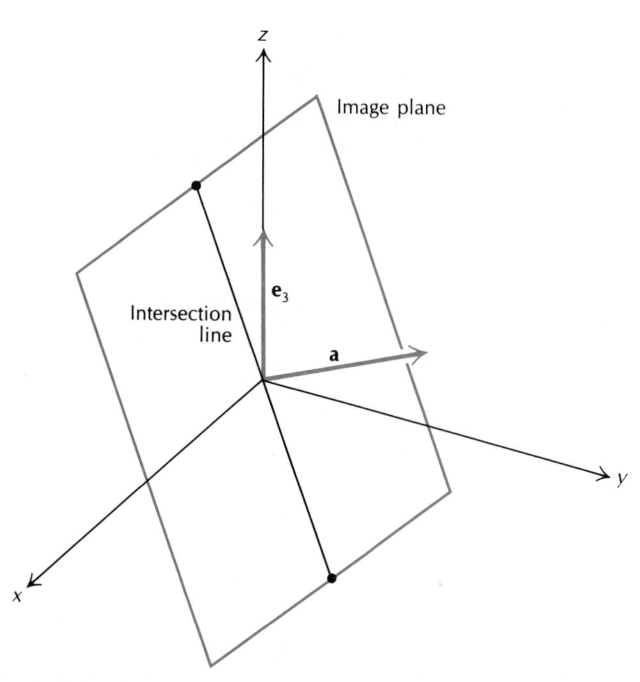

Figure 6.11 Intersection of the image plane with the span of $\mathbf{a}$ and $\mathbf{e}_3$.

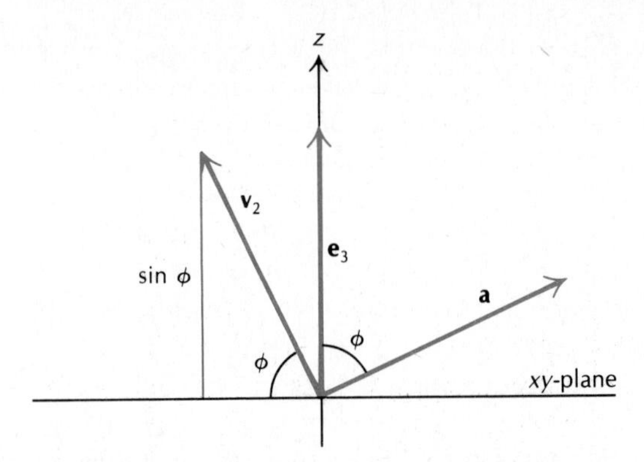

Figure 6.12 Side view of the span of **a** and $\mathbf{e}_3$.

vector with positive *z*-component in the intersection of the image plane with the plane spanned by **a** and $\mathbf{e}_3$ (see Fig. 6.11). A side view of the span of $\{\mathbf{a}, \mathbf{e}_3\}$ (Fig. 6.12) shows that $\mathbf{v}_2$ can be readily found. For $\mathbf{v}_1$, take a unit vector orthogonal to the plane of $\mathbf{v}_2$ and **a**. Then $\mathbf{v}_1$ is calculated either as the cross product of $\mathbf{v}_2$ and **a** (Section 3.4) or as a unit vector in the *xy*-plane with the first two components orthogonal to those of **a**. Either way,

$$\mathbf{v}_1 = (-\sin\theta, \cos\theta, 0), \qquad \text{and}$$

$$\mathbf{v}_2 = (-\cos\theta\cos\phi, -\sin\theta\sin\phi, \sin\phi).$$

The image $P(\mathbf{x})$ of any point $\mathbf{x} = (x_1, x_2, x_3)$ in $\mathbb{R}^3$ may be found by the corollary to Theorem 5.6, Section 5.1:

$$P(\mathbf{x}) = \langle \mathbf{x}, \mathbf{v}_1 \rangle \mathbf{v}_1 + \langle \mathbf{x}, \mathbf{v}_2 \rangle \mathbf{v}_2.$$

These inner products are given by

$$\langle \mathbf{x}, \mathbf{v}_1 \rangle = x_1(-\sin\theta) + x_2(\cos\theta)$$

and

$$\langle \mathbf{x}, \mathbf{v}_2 \rangle = x_1(-\cos\theta\cos\phi) + x_2(-\sin\theta\cos\phi) + x_3(\sin\phi).$$

Thus the matrix of P relative to the standard basis of $\mathbb{R}^3$ and the basis B in the image plane is:

$$A = \begin{bmatrix} -\sin\theta & \cos\theta & 0 \\ -\cos\theta\cos\phi & -\sin\theta\cos\phi & \sin\phi \end{bmatrix}.$$

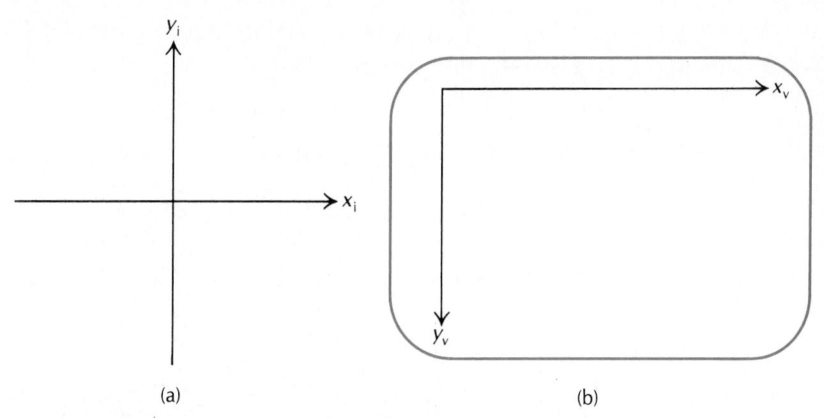

Figure 6.13 (a) Image plane coordinates; (b) viewing screen coordinates.

For BASIC implementation on a microcomputer, for instance, unless "window" commands are available the image plane coordinates (x_i, y_i) must be converted to viewing screen coordinates (x_v, y_v). The conversion is necessary if the origin of the latter system is often located at the upper left and y is directed positive downward. (See Fig. 6.13.) The conversion typically involves scaling, translation, and a reversal of vertical orientation. Denoting the scale factors by s_x and s_y, and the center of the screen by (c_x, c_y), we have

$$x_v = s_x x_i + c_x, \qquad y_v = -s_y y_i + c_y.$$

Substituting the expressions for x_i and y_i obtained above gives

$$x_v = s_x[x(-\sin\theta) + y(\cos\theta)] + c_x$$

and

$$y_v = s_y[x(-\cos\theta\cos\phi) + y(-\sin\theta\cos\phi) + z(\sin\phi)] + c_y$$

for a point (x, y, z) on a given surface.

A "plain vanilla" 3-dimensional plotter is included on the diskette that accompanies this book to illustrate an implementation of the formulas above. In the program, the perspective direction is set in variables THETA and PHI, the scale factors are set in SX, SY, and the center of the display screen is in CX, CY. The coefficient variables are set:

XC1 = −SX * SIN(THETA), XC2 = SX * COS(THETA),

YC1 = −SY * COS(THETA) * COS(PHI),

YC2 = −SY * SIN(THETA) * COS(PHI), and YC3 = SY * SIN(PHI).

For each point to be plotted, set the variables X and Y, and let Z = FNF(X, Y). The viewing screen coordinates are given by

$$XV = XC1 * X + XC2 * Y + CX$$
$$YV = -(YC1 * X + YC2 * Y + YC3 * Z) + CY.$$

These instructions implement the formulas derived above.

In general, good images of surfaces require hidden line removal. One method involves plotting front to back, updating the minimum and the maxi-

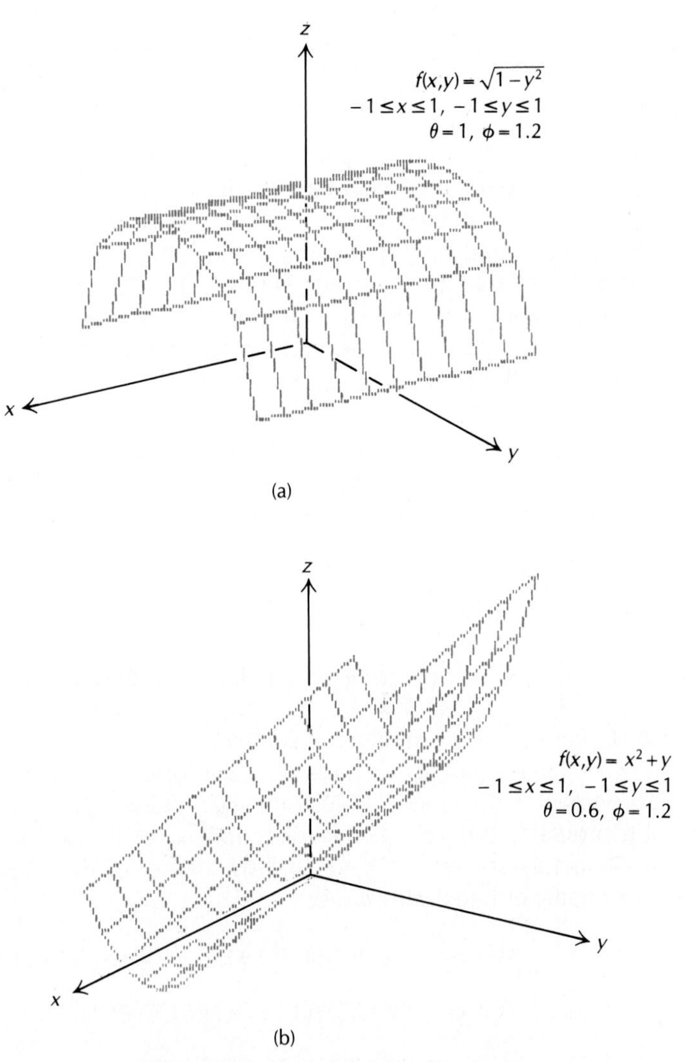

Figure 6.14 Sample plots of surfaces.

mum values of y_v plotted for x_v as we go. We ignore any (x_v, y_v) for which y_v lies between the extreme values already plotted. Graphers with hidden line removers are generally slow in interpreted BASIC, and the use of a compiler or a machine language subroutine is desirable to save time. A hidden lines option is included in our sample program, and the Myers book is recommended for additional details. Figure 6.14 shows sample plots from our IBM-PC.

Dynamic Computer Displays

Animation and computer-aided design involve dynamic relocation of objects on a viewing screen. Matrix multiplication is often used in computer programs to calculate new screen coordinates for the points to be moved.

One difficulty is that translation, which is defined by

$$T(x_1, x_2) = (x_1 + h, x_2 + k),$$

is nonlinear. For instance, since

$$T(2(x_1, x_2)) = T(2x_1, 2x_2) = (2x_1 + h, 2x_2 + k)$$

and

$$2T(x_1, x_2) = 2(x_1 + h, x_2 + k) = (2x_1 + 2h, 2x_2 + 2k),$$

$$T(2(x_1, x_2)) \neq 2T(x_1, x_2)$$

in general, so T is nonlinear.

For this problem, we use **homogeneous coordinates** $(x, y, 1)$ to represent a point (x, y) in the plane. Any linear transformation on $\mathbb{R}^2$ determined by the matrix

$$A = \begin{bmatrix} a_{11} & a_{12} \\ a_{21} & a_{22} \end{bmatrix}$$

relative to standard coordinates is determined by the matrix

$$B = \begin{bmatrix} a_{11} & a_{12} & 0 \\ a_{21} & a_{22} & 0 \\ 0 & 0 & 1 \end{bmatrix}$$

in homogeneous coordinates. But now translation may be represented by a matrix, since the product

$$\begin{bmatrix} 1 & 0 & h \\ 0 & 1 & k \\ 0 & 0 & 1 \end{bmatrix} \begin{bmatrix} x \\ y \\ 1 \end{bmatrix} = \begin{bmatrix} x + h \\ y + k \\ 1 \end{bmatrix}$$

produces the desired result.

The use of homogeneous coordinates and corresponding matrix representation is illustrated in the following examples.

Example 1

Find the matrix of a counterclockwise rotation through 30° followed by a horizontal translation 2 units to the right using homogeneous coordinates in the plane.

Solution

Rotating the plane through 30° with the standard basis fixed is equivalent to rotating the standard basis through −30° with the plane fixed. The rotated basis vectors,

$$\begin{bmatrix} \sqrt{3}/2 \\ -1/2 \end{bmatrix} \quad \text{and} \quad \begin{bmatrix} 1/2 \\ \sqrt{3}/2 \end{bmatrix},$$

are the column vectors of the rotation matrix

$$\begin{bmatrix} \sqrt{3}/2 & 1/2 \\ -1/2 & \sqrt{3}/2 \end{bmatrix}.$$

Form the rotation matrix in homogeneous coordinates as described above. The desired matrix is the product

$$\begin{bmatrix} 1 & 0 & 2 \\ 0 & 1 & 0 \\ 0 & 0 & 1 \end{bmatrix} \begin{bmatrix} \sqrt{3}/2 & -1/2 & 0 \\ 1/2 & \sqrt{3}/2 & 0 \\ 0 & 0 & 1 \end{bmatrix} = \begin{bmatrix} \sqrt{3}/2 & -1/2 & 2 \\ 1/2 & \sqrt{3}/2 & 0 \\ 0 & 0 & 1 \end{bmatrix}.$$

In this product, the second matrix (which is applied first) represents the rotation, the first the translation. □

Example 2

Find the matrix of a translation 3 units left and 1 unit up followed by a reflection in the x-axis for homogeneous coordinates in the plane.

Solution

Reflecting the plane in the x-axis with the standard basis fixed is equivalent to changing to the basis

$$B = \left\{ \begin{bmatrix} 1 \\ 0 \end{bmatrix}, \begin{bmatrix} 0 \\ -1 \end{bmatrix} \right\}$$

with the plane fixed. Thus the matrix of the reflection is

$$\begin{bmatrix} 1 & 0 \\ 0 & -1 \end{bmatrix}.$$

The required matrix is the product

$$\begin{bmatrix} 1 & 0 & 0 \\ 0 & -1 & 0 \\ 0 & 0 & 1 \end{bmatrix}\begin{bmatrix} 1 & 0 & h \\ 0 & 1 & k \\ 0 & 0 & 1 \end{bmatrix} = \begin{bmatrix} 1 & 0 & h \\ 0 & -1 & -k \\ 0 & 0 & 1 \end{bmatrix}. \quad \square$$

Other mappings used in computer graphics include *compressions*, *expansions*, and *shears*. These mappings are as follows:

x-compression $T(x, y) = (kx, y), \quad 0 < k < 1;$

y-compression $T(x, y) = (x, ky), \quad 0 < k < 1;$

x-expansion $T(x, y) = (kx, y), \quad k > 1;$

y-expansion $T(x, y) = (x, ky), \quad k > 1;$

x-shear $T(x, y) = (x + ky, y), \quad k \neq 0;$

y-shear $T(x, y) = (x, y + kx), \quad k \neq 0.$

Matrix representation and geometric interpretation of these transformations are explored in the exercises.

Example 3

The images of the standard basis vectors under the x-expansion with $k = 3$ are given by

$$T\left(\begin{bmatrix} 1 \\ 0 \end{bmatrix}\right) = \begin{bmatrix} 3 \\ 0 \end{bmatrix} \quad \text{and} \quad T\left(\begin{bmatrix} 0 \\ 1 \end{bmatrix}\right) = \begin{bmatrix} 0 \\ 1 \end{bmatrix}.$$

Thus the matrix of T relative to the standard basis is

$$\begin{bmatrix} 3 & 0 \\ 0 & 1 \end{bmatrix},$$

and the matrix of T relative to homogeneous coordinates is

$$\begin{bmatrix} 3 & 0 & 0 \\ 0 & 1 & 0 \\ 0 & 0 & 1 \end{bmatrix}. \quad \square$$

Example 4

The images of the standard basis vectors under the y-shear with $k = 4$ are given by

$$T\left(\begin{bmatrix}1\\0\end{bmatrix}\right) = \begin{bmatrix}1\\4\end{bmatrix} \quad \text{and} \quad T\left(\begin{bmatrix}0\\1\end{bmatrix}\right) = \begin{bmatrix}0\\1\end{bmatrix}.$$

Thus the matrix of T relative to the standard basis is

$$\begin{bmatrix}1 & 0\\4 & 1\end{bmatrix},$$

and the matrix of T relative to homogeneous coordinates is

$$\begin{bmatrix}1 & 0 & 0\\4 & 1 & 0\\0 & 0 & 1\end{bmatrix}. \quad \square$$

Integrating by Matrices (Calculus Based)

Linear operators provide a convenient way to integrate certain functions. Suppose f and g are defined on $\mathbb{R}$ by

$$f(x) = e^{ax}\cos bx \quad \text{and} \quad g(x) = e^{ax}\sin bx.$$

Then $B = \{f, g\}$ is linearly independent in the vector space V of all functions with continuous first derivatives (Exercise 22). Therefore B is the basis for a subspace W of V, and any h in W is expressible as a linear combination of f and g:

$$h = c_1 f + c_2 g.$$

We indicate this relation by writing h as a vector,

$$h = \begin{bmatrix}c_1\\c_2\end{bmatrix}.$$

By the product differentiation rule,

$$f'(x) = a(e^{ax}\cos bx) - b(e^{ax}\sin bx)$$

and

$$g'(x) = b(e^{ax}\cos bx) + a(e^{ax}\sin bx),$$

hence the derivative determines a linear operator D on W. The matrix of D relative to B may be obtained from the images of the basis vectors under D. Since

$$D(f) = \begin{bmatrix} a \\ -b \end{bmatrix} \quad \text{and} \quad D(g) = \begin{bmatrix} b \\ a \end{bmatrix},$$

the matrix of D relative to B is

$$A = \begin{bmatrix} a & b \\ -b & a \end{bmatrix}.$$

The inverse of A is the matrix of the integration operator:

$$A^{-1} = \frac{1}{a^2 + b^2}\begin{bmatrix} a & -b \\ b & a \end{bmatrix}.$$

The following example illustrates the use of this matrix.

Example 5

Find the indefinite integral

$$I = \int (2e^{3x}\cos 4x - e^{3x}\sin 4x)\, dx.$$

Solution

Let $B = \{f, g\}$, where

$$f(x) = e^{3x}\cos 4x \quad \text{and} \quad g(x) = e^{3x}\sin 4x.$$

Then B is a basis for a subspace W of the space of all real functions with a continuous first derivative and, as previously described, the matrix of the differentiation operator on W is

$$A = \begin{bmatrix} 3 & 4 \\ -4 & 3 \end{bmatrix}.$$

The desired integral may be obtained from the matrix product

$$A^{-1}\begin{bmatrix} 2 \\ -1 \end{bmatrix} = \frac{1}{3^2 + 4^2}\begin{bmatrix} 3 & -4 \\ 4 & 3 \end{bmatrix}\begin{bmatrix} 2 \\ -1 \end{bmatrix} = \begin{bmatrix} 10/25 \\ 5/25 \end{bmatrix}.$$

This gives

$$I = (2/5)(e^{3x}\cos 4x) + (1/5)(e^{3x}\sin 4x) + C,$$

where C is a constant of integration. This result may be checked by differentiation. □

We may also integrate the product of an exponential by a polynomial function using matrices. If

$$f(x) = x^k e^{ax},$$

then for any $k \neq 0$,

$$f'(x) = ax^k e^{ax} + kx^{k-1}e^{ax}.$$

Suppose $f_0, f_1, \ldots, f_n$ are defined on $\mathbb{R}$ by

$$f_0(x) = e^{ax}, f_1(x) = xe^{ax}, \ldots, f_n(x) = x^n e^{ax}.$$

Then $B = \{f_0, f_1, \ldots, f_n\}$ is linearly independent in the vector space V of real functions with continuous first derivatives (Exercise 23). Thus B is a basis for a subspace W of V. Differentiation determines a linear operator D on W, and the derivatives of these basis vectors can be used to form the matrix A of D relative to B. For instance, for $n = 3$,

$$A = \begin{bmatrix} a & 1 & 0 & 0 \\ 0 & a & 2 & 0 \\ 0 & 0 & a & 3 \\ 0 & 0 & 0 & a \end{bmatrix}.$$

The inverse of A can be found essentially by inspection. Start from the bottom row and work upward to obtain

$$A^{-1} = \begin{bmatrix} 1/a & -1/a^2 & 2/a^3 & -6/a^4 \\ 0 & 1/a & -2/a^2 & 6/a^3 \\ 0 & 0 & 1/a & -3/a^2 \\ 0 & 0 & 0 & 1/a \end{bmatrix}.$$

The following example illustrates the use of this technique.

Example 6

Find the indefinite integral

$$I = \int e^{2x}(1 - 2x + x^2 + 3x^3)\,dx.$$

Solution

Here $a = 2$, so the differentiation matrix is given by

$$A = \begin{bmatrix} 2 & 1 & 0 & 0 \\ 0 & 2 & 2 & 0 \\ 0 & 0 & 2 & 3 \\ 0 & 0 & 0 & 2 \end{bmatrix}.$$

The desired integral may be obtained from the matrix product

$$A^{-1}\begin{bmatrix}1\\-2\\1\\3\end{bmatrix}=\begin{bmatrix}1/2 & -1/4 & 1/4 & -3/8\\0 & 1/2 & -1/2 & 3/4\\0 & 0 & 1/2 & -3/4\\0 & 0 & 0 & 1/2\end{bmatrix}\begin{bmatrix}1\\-2\\1\\3\end{bmatrix}=\begin{bmatrix}1/8\\3/4\\-7/4\\3/2\end{bmatrix}.$$

That is,

$$I = e^{2x}(1/8 + 3x/4 - 7x^2/4 + 3x^3/2) + C. \quad \square$$

The method illustrated in Examples 5 and 6 may also be applied to the product of a polynomial function by a sine or a cosine. (Some specific illustrations are assigned in Exercises 24–27.)

REVIEW CHECKLIST

1. Describe geometrically the projection of a surface in space onto an image plane using point source rays and the projection by parallel rays.
2. Identify point source projection as nonlinear and parallel ray projection as linear.
3. Find image plane and viewing screen coordinates of a point projected by parallel rays when the direction is specified in spherical coordinates.
4. Describe the use of homogeneous coordinates to represent points in a plane.
5. Describe the use of matrices to represent transformations of the plane, including translations, when points are given in homogeneous coordinates.
6. Find matrices for compressions, expansions, shears, reflections, and composites of these transformations relative to homogeneous coordinates.
7. (*Calculus based*) Identify several types of functions whose first derivatives span finite-dimensional subspaces in the space of real functions with continuous first derivatives.
8. (*Calculus based*) Identify differentiation as a linear operator on the subspaces in item 7, represent this operator by a matrix, and obtain indefinite integrals by matrix inversion.

EXERCISES

1. For the parallel ray projection with $\theta = \pi/4$, $\phi = \pi/4$, $s_x = 30$, $s_y = 30$, $c_x = 159$, $c_y = 99$, find the image plane and viewing screen coordinates of the projection of the point $(1, 3, 2)$.

Exercises 2–4 require an IBM† PC or compatible microcomputer. If one is available, enter and run the program "3dgraph" to obtain images of the indicated surfaces.

† IBM is a registered trademark of International Business Machines Corporation.

2. $z = 1 - x - y, 0 \leq y \leq 1 - x, 0 \leq x \leq 1$. Use scale factors 150 for x, 90 for y, and no hidden line removal.
3. $z = 1 - x^2, -1 \leq x \leq 1, 0 \leq y \leq 2$. Use scale factors 125 for x, 75 for y, and remove hidden lines.
4. $z = y^2 - x^2, -1 \leq x \leq 1, -1 \leq y \leq 1$. Use scale factors 150 for x, 90 for y, and remove hidden lines.

In Exercises 5–8, find the matrices representing the indicated transformation of the plane in Cartesian coordinates and in homogeneous coordinates.

5. The x-expansion with $k = 2$.
6. The x-shear with $k = 3$.
7. The x-shear with $k = 2$ followed by a y-compression with $k = 1/2$.
8. The **scaling** with x-factor 1/2 and y-factor 3/2, which is composed of an x-compression with $k = 1/2$ and a y-expansion with $k = 3/2$.

In Exercises 9–15, sketch a graph showing the image of the unit square $0 \leq x \leq 1$, $0 \leq y \leq 1$ under the indicated transformation.

9. The x-compression with $k = 1/3$.
10. The y-expansion with $k = 4$.
11. The x-shear with $k = 2$.
12. The x-shear with $k = -2$.
13. The x-shear with $k = 3$ followed by reflection in the y-axis.
14. The x-shear with $k = 2$ followed by the y-shear with $k = 2$.
15. The scaling of Exercise 8.
16. If an IBM PC is available, run the program "movem" on the diskette for examples of moving objects on a video display.

In Exercises 17–21 (*calculus based*), find an indefinite integral by the matrix method illustrated in Examples 5 and 6.

17. $\int e^x(\cos x + \sin x)\,dx$
18. $\int e^{2x}(\cos 3x - 3\sin 3x)\,dx$
19. $\int e^x(1 + x)\,dx$
20. $\int e^{2x}(1 - x^2)\,dx$
21. $\int e^{3x}(5 + 3x + 3x^2 - 2x^3 + 3x^4)\,dx$
22. Show that the set of real functions $\{f, g\}$, where

$$f(x) = e^{ax}\cos bx,\ g(x) = e^{ax}\sin bx,$$

is linearly independent when the functions are defined on any real interval of positive length. That is, show that if

$$c_1(e^{ax}\cos bx) + c_2(e^{ax}\sin bx) = 0$$

for all x in an interval, then $c_1 = 0$ and $c_2 = 0$. (*Hint:* We may multiply and divide the expression

$$c_1 \cos bx + c_2 \sin bx$$

by $c_1^2 + c_2^2$ and convert it to the form $A\cos(bx - \theta)$ by applying a trigonometric identity.)

23. Show that for any n, the set of real functions $\{f_0, f_1, \ldots, f_n\}$, where

$$f_0(x) = e^{ax} \quad \text{and} \quad f_k(x) = x^k e^{ax} \quad \text{for } k = 1, \ldots, n,$$

is linearly independent on any interval of positive length. (*Hint:* A polynomial of degree n has at most n real zeros.)

In Exercises 24–27 (*calculus based*), assume that for any n, the set of real functions $\{f_0, f_1, \ldots, f_{2n}, f_{2n+1}\}$, where

$$f_0(x) = \cos bx, \qquad f_k(x) = x^k \cos bx$$

for $k = 1, \ldots, n$,

$$f_{n+1}(x) = \sin bx, \qquad f_k(x) = x^{k-n-1} \sin bx$$

for $k = n + 2, \ldots, 2n + 1$, is linearly independent. Obtain and use the inverse of the differentiation operator matrix to find the indefinite integral of the given function.

24. $\int (\cos x + x \sin x)\, dx$

25. $\int (2\cos x - x\cos x + \sin x - 2x \sin x)\, dx$

26. $\int (\cos 2x - x\cos 2x + \sin 2x)\, dx$

27. $\int (6x\cos 2x - \sin 2x - 4x\sin 2x + 8x^2 \sin 2x)\, dx$

REVIEW EXERCISES

In Exercises 1–10, let B and B' be the following (ordered) bases in $\mathbb{R}^2$:

$$B = \left\{ \begin{bmatrix} 3 \\ 2 \end{bmatrix}, \begin{bmatrix} 7 \\ 5 \end{bmatrix} \right\} \quad \text{and} \quad B' = \left\{ \begin{bmatrix} 3 \\ 4 \end{bmatrix}, \begin{bmatrix} 2 \\ 3 \end{bmatrix} \right\}.$$

1. Find the standard components of $\mathbf{x}$, if $(\mathbf{x})_B = (1, -1)$.

2. Find the standard components of $\mathbf{x}$, if $(\mathbf{x})_{B'} = (-2, 3)$.

3. If $\mathbf{x} = (1, 2)$, find $(\mathbf{x})_B$.

4. If $\mathbf{x} = (2, -1)$, find $(\mathbf{x})_{B'}$.

5. If $[\mathbf{x}]_B = [1 \quad -1]^t$, find $[\mathbf{x}]_{B'}$.

6. If $[\mathbf{x}]_{B'} = [2 \quad 2]^t$, find $[\mathbf{x}]_B$.

7. Find the transition matrix from B to the standard basis.

8. Find the transition matrix from the standard basis to B'.

9. Find the transition matrix from B to B'.

10. Find the transition matrix from B' to B.

In Exercises 11–13, find the matrix of the given linear transformation relative to the standard bases in $\mathbb{R}^2$ and $\mathbb{R}^3$.

11. $T(x_1, x_2) = (2x_1 + x_2, x_2, 3x_1 - 4x_2)$

12. $T(x_1, x_2, x_3) = (x_1 - x_2 + 2x_3, 3x_1 + x_3)$

13. $T(x_1, x_2) = (x_1 - x_2, x_1 + x_2)$

In Exercises 14–17, let B and B' be the bases for $\mathbb{R}^2$ from Exercises 1–10, let E be the standard basis for $\mathbb{R}^2$, and let

$$A = \begin{bmatrix} 1 & 2 \\ 1 & 3 \end{bmatrix}.$$

14. If A is the matrix of a linear operator T relative to E, find a formula for $T(x_1, x_2)$.

15. If A is the matrix of a linear operator T relative to E, find the matrix of T relative to B.

16. If A is the matrix of a linear operator T relative to B', find the matrix of T relative to B.

17. If A is the matrix of a linear operator T relative to B', find the matrix of T relative to E.

18. If A is an orthogonal 3×3 matrix whose first two row vectors are $(1/\sqrt{3}, 1/\sqrt{3}, 1/\sqrt{3})$ and $(1/\sqrt{2}, 0, -1/\sqrt{2})$, find all possibilities for the third row vector of A.

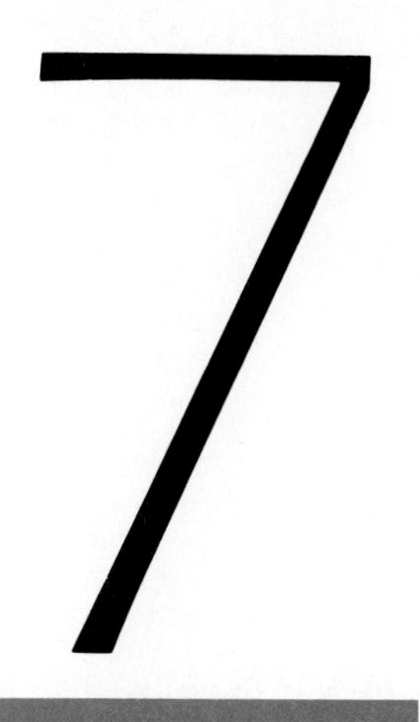

Eigenvalues and Eigenvectors

Eigenvalues *and* eigenvectors *are used widely in applications of linear algebra. For oscillatory motion, they describe the frequency and the mode of vibration. For a Leslie population model (Section 2.6), they describe the steady-state growth rate and age class distribution. In mechanics, they describe the principal stresses and the principal axes of stress in bodies acted upon by external forces.*

When the eigenvalue/eigenvector concept is extended from the finite dimensional setting of linear algebra to infinitely many dimensions, eigenvectors give way to eigenfunctions. *Eigenvalues and eigenfunctions are used to solve boundary value problems of mathematical physics. They also provide a major topic of study in an area of mathematics called functional analysis.*

7.1 EIGENVALUE PROBLEMS

For a square matrix A, any nonzero vector $\mathbf{v}$ such that $A\mathbf{v}$ lies on the line spanned by $\mathbf{v}$ is of special interest. In this section, we develop methods for determining when such vectors exist and finding them when they do.

The Characteristic Equation of a Matrix

For an $n \times n$ matrix A and a nonzero in $\mathbb{R}^n$, $A\mathbf{v}$ is in the span of $\mathbf{v}$ if and only if $A\mathbf{v}$ is a scalar multiple of $\mathbf{v}$. We denote the (unknown) scalar by a Greek lower case lambda (λ).

Example 1

For

$$A = \begin{bmatrix} 3 & 0 \\ 0 & 2 \end{bmatrix}, \qquad \lambda_1 = 3, \qquad \mathbf{v}_1 = \begin{bmatrix} 1 \\ 0 \end{bmatrix}, \qquad \lambda_2 = 2, \qquad \text{and} \qquad \mathbf{v}_2 = \begin{bmatrix} 0 \\ 1 \end{bmatrix},$$

we have $A\mathbf{v}_1 = \lambda_1\mathbf{v}_1$ and $A\mathbf{v}_2 = \lambda_2\mathbf{v}_2$, since

$$A\mathbf{v}_1 = \begin{bmatrix} 3 & 0 \\ 0 & 2 \end{bmatrix}\begin{bmatrix} 1 \\ 0 \end{bmatrix} = \begin{bmatrix} 3 \\ 0 \end{bmatrix} = 3\begin{bmatrix} 1 \\ 0 \end{bmatrix} = 3\mathbf{v}_1$$

and

$$A\mathbf{v}_2 = \begin{bmatrix} 3 & 0 \\ 0 & 2 \end{bmatrix}\begin{bmatrix} 0 \\ 1 \end{bmatrix} = \begin{bmatrix} 0 \\ 2 \end{bmatrix} = 2\begin{bmatrix} 0 \\ 1 \end{bmatrix} = 2\mathbf{v}_2. \quad \square$$

DEFINITION 7.1

Let A be an $n \times n$ matrix. Then λ is called an **eigenvalue of A** if there exists a *nonzero* vector $\mathbf{v}$ such that

$$A\mathbf{v} = \lambda\mathbf{v}.$$

Any nonzero $\mathbf{v}$ satisfying this equation is called an **eigenvector of A** corresponding to the eigenvalue λ.

Eigenvalues [eigenvectors] are also called **characteristic**, **proper**, or **latent** values [vectors]. Although the emphasis in this book is on real numbers and all given matrices have real entries, eigenvalues, eigenvectors, and matrices constructed from eigenvectors may be complex. A brief summary of complex number fundamentals is given in Appendix C.

Example 2

Show that for

$$A = \begin{bmatrix} 1 & -3 & 1 \\ -1 & 1 & 1 \\ 3 & -3 & -1 \end{bmatrix}, \qquad \lambda_1 = -2, \qquad \text{and} \qquad \mathbf{v}_1 = \begin{bmatrix} 3 \\ 2 \\ -3 \end{bmatrix},$$

λ_1 is an eigenvalue of A with corresponding eigenvector $\mathbf{v}_1$.

Solution

Verify the relation specified in Definition 7.1:

$$A\mathbf{v}_1 = \begin{bmatrix} 1 & -3 & 1 \\ -1 & 1 & 1 \\ 3 & -3 & -1 \end{bmatrix}\begin{bmatrix} 3 \\ 2 \\ -3 \end{bmatrix} = \begin{bmatrix} -6 \\ -4 \\ 6 \end{bmatrix} = -2\begin{bmatrix} 3 \\ 2 \\ -3 \end{bmatrix} = -2\mathbf{v}_1.$$

(Although not required in this illustration, note that for

$$\lambda_2 = 1, \qquad \mathbf{v}_2 = \begin{bmatrix} 3 \\ 1 \\ 3 \end{bmatrix}, \qquad \lambda_3 = 2, \qquad \text{and} \qquad \mathbf{v}_3 = \begin{bmatrix} 1 \\ 0 \\ 1 \end{bmatrix},$$

similar calculations show that $A\mathbf{v}_2 = \lambda_2\mathbf{v}_2$ and $A\mathbf{v}_3 = \lambda_3\mathbf{v}_3$.) □

The next example involves complex numbers.

Example 3

Let i denote the imaginary unit, $i = \sqrt{-1}$, and let

$$A = \begin{bmatrix} 1 & -1 \\ 1 & 1 \end{bmatrix}, \qquad \lambda_1 = 1 - i, \qquad \mathbf{v}_1 = \begin{bmatrix} 1 \\ i \end{bmatrix},$$

$$\lambda_2 = 1 + i, \qquad \text{and} \qquad \mathbf{v}_2 = \begin{bmatrix} 1 \\ -i \end{bmatrix}.$$

Show that $\mathbf{v}_1$ and $\mathbf{v}_2$ are eigenvectors of A for the eigenvalues λ_1 and λ_2, respectively.

Solution

Using $i^2 = -1$, verify the condition of Definition 7.1:

$$A\mathbf{v}_1 = \begin{bmatrix} 1 & -1 \\ 1 & 1 \end{bmatrix}\begin{bmatrix} 1 \\ i \end{bmatrix} = \begin{bmatrix} 1 - i \\ 1 + i \end{bmatrix} = (1 - i)\begin{bmatrix} 1 \\ i \end{bmatrix} = (1 - i)\mathbf{v}_1$$

and

$$A\mathbf{v}_2 = \begin{bmatrix} 1 & -1 \\ 1 & 1 \end{bmatrix}\begin{bmatrix} 1 \\ -i \end{bmatrix} = \begin{bmatrix} 1 + i \\ 1 - i \end{bmatrix} = (1 + i)\begin{bmatrix} 1 \\ -i \end{bmatrix} = (1 + i)\mathbf{v}_2. \quad \square$$

Since $\mathbf{v} = I\mathbf{v}$, where I is the identity matrix,

$$A\mathbf{v} = \lambda\mathbf{v} \qquad \text{if and only if} \qquad (\lambda I - A)\mathbf{v} = \mathbf{0}$$

for a given $n \times n$ matrix A. Thus by Definition 7.1, λ is an eigenvalue of A if and only if the linear system determined by the vector equation on the right has a nontrivial solution. By Theorem 4.13, this system has a nontrivial solution if and only if the determinant of the coefficient matrix equals zero. Thus λ is an eigenvalue of A if and only if

$$\det(\lambda I - A) = 0.$$

The expansion of the left side is a polynomial of degree n in the variable λ (Exercise 40).

DEFINITION 7.2

For a square matrix A, the equation

$$\det(\lambda I - A) = 0$$

is called the **characteristic equation** and the expansion of the left side is called the **characteristic polynomial**.

Example 4

Find the characteristic polynomial of the matrix

$$A = \begin{bmatrix} 2 & 4 \\ 1 & -1 \end{bmatrix}.$$

Solution

Form the matrix $\lambda I - A$:

$$\lambda I - A = \lambda \begin{bmatrix} 1 & 0 \\ 0 & 1 \end{bmatrix} - \begin{bmatrix} 2 & 4 \\ 1 & -1 \end{bmatrix} = \begin{bmatrix} \lambda - 2 & -4 \\ -1 & \lambda + 1 \end{bmatrix}.$$

The characteristic polynomial of A is given by

$$\begin{vmatrix} \lambda - 2 & -4 \\ -1 & \lambda + 1 \end{vmatrix} = (\lambda - 2)(\lambda + 1) - 4 = \lambda^2 - \lambda - 6. \quad \square$$

Example 5

Find the characteristic polynomial of the matrix

$$A = \begin{bmatrix} 1 & 0 & 3 \\ 3 & -2 & 0 \\ -1 & 1 & 2 \end{bmatrix}.$$

Solution

From the given matrix,

$$\lambda I - A = \lambda \begin{bmatrix} 1 & 0 & 0 \\ 0 & 1 & 0 \\ 0 & 0 & 1 \end{bmatrix} - \begin{bmatrix} 1 & 0 & 3 \\ 3 & -2 & 0 \\ -1 & 1 & 2 \end{bmatrix} = \begin{bmatrix} \lambda - 1 & 0 & -3 \\ -3 & \lambda + 2 & 0 \\ 1 & -1 & \lambda - 2 \end{bmatrix}.$$

Expanding the determinant along the first row gives

$$|\lambda I - A| = (\lambda - 1)\begin{vmatrix} \lambda + 2 & 0 \\ -1 & \lambda - 2 \end{vmatrix} - 0\begin{vmatrix} -3 & 0 \\ 1 & \lambda - 2 \end{vmatrix} + (-3)\begin{vmatrix} -3 & \lambda + 2 \\ 1 & -1 \end{vmatrix}$$

$$= (\lambda - 1)(\lambda + 2)(\lambda - 2) + (-3)[3 - (\lambda + 2)]$$

$$= \lambda^3 - \lambda^2 - \lambda + 1. \quad \square$$

Finding Eigenvalues and Eigenvectors

For small matrices, we often find eigenvalues and eigenvectors using the characteristic equation. Let λ be a root of this equation. Since $\det(\lambda I - A) = 0$, the matrix $\lambda I - A$ has rank $r < n$ by Theorem 4.13. Hence by Theorem 4.15, the dimension of the solution space of the system determined by the equation

$$(\lambda I - A)\mathbf{v} = \mathbf{0}$$

is $n - r$. This space is called the **eigenspace** corresponding to λ, and it consists of all eigenvectors corresponding to λ.

A Method for Finding Eigenvalues and Eigenvectors of a Matrix A

1. Form the characteristic equation

$$\det(\lambda I - A) = 0.$$

2. Find all roots of the characteristic equation.
3. For each λ found in step 2, solve the linear system determined by the equation

$$(\lambda I - A)\mathbf{v} = \mathbf{0}.$$

Example 6

Find the eigenvalues and corresponding eigenvectors of

$$A = \begin{bmatrix} 2 & 2 \\ 1 & 3 \end{bmatrix}.$$

Solution

Since

$$\det(\lambda I - A) = \begin{vmatrix} \lambda - 2 & -2 \\ -1 & \lambda - 3 \end{vmatrix} = \lambda^2 - 5\lambda + 4 = (\lambda - 1)(\lambda - 4),$$

the eigenvalues are 1 and 4. Let x_1 and x_2 denote the components of the unknown $\mathbf{v}$. For $\lambda_1 = 1$, solve

$$(\lambda I - A)\mathbf{v} = \mathbf{0},$$

or

$$\begin{bmatrix} 0 \\ 0 \end{bmatrix} = (\lambda I - A)\mathbf{v} = \begin{bmatrix} 1 - 2 & -2 \\ -1 & 1 - 3 \end{bmatrix} \begin{bmatrix} x_1 \\ x_2 \end{bmatrix} = \begin{bmatrix} -1 & -2 \\ -1 & -2 \end{bmatrix} \begin{bmatrix} x_1 \\ x_2 \end{bmatrix}.$$

The linear system determined by this vector equation reduces to the single equation

$$-x_1 - 2x_2 = 0.$$

Back substitution with $x_2 = t$ gives $x_1 = -2t$. Thus $\mathbf{v}_1$ is an eigenvector for $\lambda_1 = 1$ if and only if

$$\mathbf{v}_1 = \begin{bmatrix} -2t \\ t \end{bmatrix}$$

for some nonzero value of t. For $\lambda_2 = 4$, solve

$$\begin{bmatrix} 0 \\ 0 \end{bmatrix} = (\lambda I - A)\mathbf{v} = \begin{bmatrix} 4 - 2 & -2 \\ -1 & 4 - 3 \end{bmatrix} \begin{bmatrix} x_1 \\ x_2 \end{bmatrix} = \begin{bmatrix} 2 & -2 \\ -1 & 1 \end{bmatrix} \begin{bmatrix} x_1 \\ x_2 \end{bmatrix}.$$

The resulting linear system reduces to the equation

$$x_1 - x_2 = 0.$$

Back substitution with $x_2 = t$ yields $x_1 = t$. Thus $\mathbf{v}_2$ is an eigenvector for $\lambda_2 = 4$ if and only if

$$\mathbf{v}_2 = \begin{bmatrix} t \\ t \end{bmatrix}$$

for some nonzero value of t. ▭

Example 7

Find the eigenvalues and corresponding eigenvectors of

$$A = \begin{bmatrix} 5 & -8 \\ 2 & -3 \end{bmatrix}.$$

Solution

Since

$$\det(\lambda I - A) = \begin{vmatrix} \lambda - 5 & 8 \\ -2 & \lambda + 3 \end{vmatrix} = \lambda^2 - 2\lambda + 1 = (\lambda - 1)^2,$$

1 is the only eigenvalue of A, and its multiplicity is 2. For $\lambda = 1$, solve

$$\begin{bmatrix} 0 \\ 0 \end{bmatrix} = (\lambda I - A)\mathbf{v} = \begin{bmatrix} 1 - 5 & 8 \\ -2 & 1 + 3 \end{bmatrix}\begin{bmatrix} x_1 \\ x_2 \end{bmatrix} = \begin{bmatrix} -4 & 8 \\ -2 & 4 \end{bmatrix}\begin{bmatrix} x_1 \\ x_2 \end{bmatrix}.$$

The coefficient matrix has rank 1, so the dimension of the corresponding eigenspace is $2 - 1 = 1$. The linear system for x_1 and x_2 reduces to the single equation

$$-4x_1 + 8x_2 = 0.$$

Back substitution with $x_2 = t$ gives $x_1 = 2t$. Thus $\mathbf{v}$ is an eigenvector for $\lambda = 1$ if and only if

$$\mathbf{v} = \begin{bmatrix} 2t \\ t \end{bmatrix}$$

for some nonzero value of t. □

The next example involves complex conjugates (see Appendix C). *For a real matrix A,*

$$A\mathbf{x} = \lambda\mathbf{x} \qquad \text{if and only if} \qquad A\bar{\mathbf{x}} = \bar{\lambda}\bar{\mathbf{x}},$$

where the bar denotes the complex conjugate. Thus when λ is complex, we get the eigenvector $\bar{\mathbf{x}}$ for $\bar{\lambda}$ "for free."

Example 8

Find the eigenvalues and corresponding eigenvectors of

$$A = \begin{bmatrix} 1 & -2 \\ 1 & -1 \end{bmatrix}.$$

Solution

Since

$$\det(\lambda I - A) = \begin{vmatrix} \lambda - 1 & 2 \\ -1 & \lambda + 1 \end{vmatrix} = \lambda^2 + 1,$$

the eigenvalues of A are $\lambda_1 = -i$ and $\lambda_2 = i$. For $\lambda_1 = -i$, the linear system for the eigenspace reduces to

$$\begin{aligned} (-i - 1)x_1 + \qquad 2x_2 &= 0 \\ -x_1 + (-i + 1)x_2 &= 0. \end{aligned}$$

Since

$$-1(1 + i) = -i - 1 \qquad \text{and} \qquad (-i + i)(1 + i) = 2,$$

the first row of the coefficient matrix is a multiple of the second. Thus the system reduces to the single equation

$$(-i - 1)x_1 + 2x_2 = 0.$$

Back substitution with $x_2 = (1 + i)t$ yields $x_1 = 2t$. Thus $\mathbf{v}_1$ is an eigenvector for $\lambda_1 = -i$ if and only if

$$\mathbf{v}_1 = \begin{bmatrix} 2t \\ (1 + i)t \end{bmatrix},$$

where $t \neq 0$. For $\lambda = i$, $\mathbf{v}_2$ is the complex conjugate of $\mathbf{v}_1$:

$$\mathbf{v}_2 = \begin{bmatrix} 2t \\ (1 - i)t \end{bmatrix},$$

where $t \neq 0$. □

For a 3×3 matrix, the characteristic equation is cubic, and it may have three real and distinct roots, one real root of multiplicity one and one of multiplicity two, one real root of multiplicity three, or one real and two complex conjugate roots.

Example 9

Find all eigenvalues and eigenvectors of

$$A = \begin{bmatrix} 2 & 2 & 2 \\ -1 & 2 & 1 \\ 1 & -2 & -1 \end{bmatrix}.$$

Solution

Since

$$\begin{vmatrix} \lambda - 2 & -2 & -2 \\ 1 & \lambda - 2 & -1 \\ -1 & 2 & \lambda + 1 \end{vmatrix} = \lambda^3 - 3\lambda^2 + 2\lambda = \lambda(\lambda - 1)(\lambda - 2),$$

the eigenvalues of A are $0, 1, 2$. For $\lambda_1 = 0$, apply Gaussian elimination to the matrix $\lambda I - A$:

$$\begin{bmatrix} 0 - 2 & -2 & -2 \\ 1 & 0 - 2 & -1 \\ -1 & 2 & 0 + 1 \end{bmatrix} = \begin{bmatrix} -2 & -2 & -2 \\ 1 & -2 & -1 \\ -1 & 2 & 1 \end{bmatrix} \to \begin{bmatrix} 1 & 1 & 1 \\ 0 & -3 & -2 \\ 0 & 3 & 2 \end{bmatrix}.$$

Back substitution with $x_3 = 3t$ gives $x_2 = -2t$ and $x_1 = -t$. Thus $\mathbf{v}_1$ is an eigenvector for $\lambda = 0$ if and only if

$$\mathbf{v}_1 = \begin{bmatrix} -t \\ -2t \\ 3t \end{bmatrix}$$

for some nonzero value of t. Similarly, for $\lambda_2 = 1$

$$\begin{bmatrix} 1 - 2 & -2 & -2 \\ 1 & 1 - 2 & -1 \\ -1 & 2 & 1 + 1 \end{bmatrix} = \begin{bmatrix} -1 & -2 & -2 \\ 1 & -1 & -1 \\ -1 & 2 & 2 \end{bmatrix} \to \begin{bmatrix} 1 & 2 & 2 \\ 0 & -3 & -3 \\ 0 & 4 & 4 \end{bmatrix}.$$

Back substitution with $x_3 = t$ yields $x_2 = -t$ and $x_1 = 0$. Thus $\mathbf{v}_2$ is an eigenvector for $\lambda_2 = 1$ if and only if

$$\mathbf{v}_2 = \begin{bmatrix} 0 \\ t \\ -t \end{bmatrix}$$

for some nonzero value of t. For $\lambda_3 = 2$,

$$\begin{bmatrix} 2 - 2 & -2 & -2 \\ 1 & 2 - 2 & -1 \\ -1 & 2 & 2 + 1 \end{bmatrix} = \begin{bmatrix} 0 & -2 & -2 \\ 1 & 0 & -1 \\ -1 & 2 & 3 \end{bmatrix} \to \begin{bmatrix} 1 & 0 & -1 \\ 0 & 1 & 1 \\ 0 & 2 & 2 \end{bmatrix}.$$

Back substitution shows that $\mathbf{v}_3$ is an eigenvector for $\lambda_3 = 2$ if and only if

$$\mathbf{v}_3 = \begin{bmatrix} t \\ -t \\ t \end{bmatrix}$$

for some nonzero value of t. ☐

Example 10

Find the eigenvalues and corresponding eigenvectors of

$$A = \begin{bmatrix} 2 & 1 & -1 \\ 1 & 2 & -1 \\ 1 & 1 & 0 \end{bmatrix}.$$

Solution

Since

$$\begin{vmatrix} \lambda - 2 & -1 & 1 \\ -1 & \lambda - 2 & 1 \\ -1 & -1 & \lambda - 0 \end{vmatrix} = \lambda^3 - 4\lambda^2 + 5\lambda - 2 = (\lambda - 1)^2(\lambda - 2),$$

1 is an eigenvalue of multiplicity 2 and 2 is an eigenvalue of multiplicity 1. For $\lambda_1 = \lambda_2 = 1$,

$$\begin{bmatrix} 1 - 2 & -1 & 1 \\ -1 & 1 - 2 & 1 \\ -1 & -1 & 1 - 0 \end{bmatrix} = \begin{bmatrix} -1 & -1 & 1 \\ -1 & -1 & 1 \\ -1 & -1 & 1 \end{bmatrix} \to \begin{bmatrix} 1 & 1 & -1 \\ 0 & 0 & 0 \\ 0 & 0 & 0 \end{bmatrix}.$$

Back substituting with $x_3 = t$ and $x_2 = s$ gives $x_1 = t - s$. Thus $\mathbf{v}_1$ is an eigenvector for $\lambda_1 = 1$ if and only if

$$\mathbf{v}_1 = \begin{bmatrix} t - s \\ s \\ t \end{bmatrix},$$

where s and t are not both zero. For $\lambda_3 = 1$,

$$\begin{bmatrix} 2 - 2 & -1 & 1 \\ -1 & 2 - 2 & 1 \\ -1 & -1 & 2 - 0 \end{bmatrix} = \begin{bmatrix} 0 & -1 & 1 \\ -1 & 0 & 1 \\ -1 & -1 & 2 \end{bmatrix} \to \begin{bmatrix} 1 & 0 & -1 \\ 0 & -1 & 1 \\ 0 & -1 & 1 \end{bmatrix}.$$

Thus v_3 is an eigenvector for $\lambda_3 = 1$ if and only if

$$\mathbf{v}_3 = \begin{bmatrix} t \\ t \\ t \end{bmatrix}$$

for some nonzero value of t, again by back substitution. ☐

In Example 10, the rank of the coefficient matrix for $\lambda_1 = 1$ is 1, so the dimension of the eigenspace for λ_1 is 2. Thus we may also specify the eigenspace by finding any linearly independent set of two eigenvectors. For instance, setting $s = 0$ and $t = 1$, then $s = -1$ and $t = 0$ produces the basis

$$B = \left\{ \begin{bmatrix} 1 \\ 0 \\ 1 \end{bmatrix}, \begin{bmatrix} 1 \\ -1 \\ 0 \end{bmatrix} \right\}.$$

Example 11

Find the eigenvalues and corresponding eigenvectors of

$$A = \begin{bmatrix} 1 & -2 & 2 \\ 1 & -2 & 1 \\ -1 & 1 & 2 \end{bmatrix}.$$

Solution

The characteristic polynomial is

$$\lambda^3 - \lambda^2 - \lambda + 1 = (\lambda + 1)(\lambda - 1)^2.$$

For $\lambda_1 = -1$,

$$\begin{bmatrix} -2 & 2 & -2 \\ -1 & 1 & -1 \\ 1 & -1 & -3 \end{bmatrix} \to \begin{bmatrix} 1 & -1 & 1 \\ 1 & -1 & -3 \\ -1 & 1 & -1 \end{bmatrix} \to \begin{bmatrix} 1 & -1 & 1 \\ 0 & 0 & -4 \\ 0 & 0 & 0 \end{bmatrix}.$$

Then $x_3 = 0$, and back substitution with $x_2 = t$ gives $x_1 = t$. Thus $\mathbf{v}_1$ is an eigenvector for $\lambda_1 = -1$ if and only if

$$\mathbf{v}_1 = \begin{bmatrix} t \\ t \\ 0 \end{bmatrix}$$

for some nonzero value of t. Similarly, for $\lambda_2 = 1$,

$$\begin{bmatrix} 0 & 2 & -2 \\ -1 & 3 & -1 \\ 1 & -1 & -1 \end{bmatrix} \to \begin{bmatrix} 1 & -1 & -1 \\ -1 & 3 & -1 \\ 0 & 2 & -2 \end{bmatrix} \to \begin{bmatrix} 1 & -1 & -1 \\ 0 & 2 & -2 \\ 0 & 2 & -2 \end{bmatrix}.$$

This matrix has rank 1 (even though the multiplicity of λ_2 is 2). By back substitution, the general form of an eigenvector for $\lambda_2 = 1$ is

$$\mathbf{v}_2 = \begin{bmatrix} 2t \\ t \\ t \end{bmatrix}, \qquad t \neq 0. \qquad \square$$

Example 12

Find the eigenvalues and eigenvectors of

$$A = \begin{bmatrix} 1 & -2 & 0 \\ 1 & 1 & 1 \\ -2 & 0 & -1 \end{bmatrix}.$$

Solution

The characteristic polynomial is

$$\lambda^3 - \lambda^2 + \lambda - 1 = (\lambda - 1)(\lambda^2 + 1).$$

For $\lambda_1 = 1$, row reduce the coefficient matrix $I - A$:

$$\begin{bmatrix} 0 & 2 & 0 \\ -1 & 0 & -1 \\ 2 & 0 & 2 \end{bmatrix} \rightarrow \begin{bmatrix} 1 & 0 & 1 \\ 0 & 2 & 0 \\ 2 & 0 & 2 \end{bmatrix} \rightarrow \begin{bmatrix} 1 & 0 & 1 \\ 0 & 1 & 0 \\ 0 & 0 & 0 \end{bmatrix}.$$

Then $x_2 = 0$, and $x_1 = -x_3$. Thus $\mathbf{v}_1$ is an eigenvector for $\lambda_1 = 1$ if and only if

$$\mathbf{v}_1 = \begin{bmatrix} t \\ 0 \\ -t \end{bmatrix},$$

where $t \neq 0$. For $\lambda_2 = i$, row reduce the coefficient matrix $(i)I - A$, using the relation $i^2 = -1$:

$$\begin{bmatrix} i-1 & 2 & 0 \\ -1 & i-1 & -1 \\ 2 & 0 & i+1 \end{bmatrix} \rightarrow \begin{bmatrix} 1 & 1-i & 1 \\ -1+i & 2 & 0 \\ 2 & 0 & 1+i \end{bmatrix} \rightarrow$$

$$\begin{bmatrix} 1 & 1-i & 1 \\ 0 & 2(1-i) & 1-i \\ 0 & -2(1-i) & -(1-i) \end{bmatrix} \rightarrow \begin{bmatrix} 1 & 1-i & 1 \\ 0 & 2 & 1 \\ 0 & 0 & 0 \end{bmatrix}.$$

Back substitution with $x_3 = -2t$ yields $x_2 = t$ and

$$x_1 = -(1 - i)(t) - (-2t) = (1 + i)t.$$

Thus $\mathbf{v}_2$ is an eigenvector for $\lambda_2 = i$ if and only if

$$\mathbf{v}_2 = \begin{bmatrix} (1+i)t \\ t \\ -2t \end{bmatrix},$$

where $t \neq 0$. For $\lambda_3 = -i$, $\mathbf{v}_3$ is the complex conjugate of $\mathbf{v}_2$. That is, the general formula for an eigenvector for $\lambda_3 = -i$ is

$$\mathbf{v}_3 = \begin{bmatrix} (1-i)t \\ t \\ -2t \end{bmatrix},$$

where $t \neq 0$. ☐

For a 2×2 matrix, the characteristic equation may be solved by factoring or by the quadratic formula. For a 3×3 matrix, if the characteristic polynomial has integer coefficients and a root r is found among the divisors of the constant term, we may reduce the cubic to a quadratic by factoring $\lambda - r$. Newton's method (calculus based) can be used to approximate the first root. Alternatively, we may use known formulas found in standard mathematical reference books. Examples and exercises in this book involve relatively simple equations to keep the focus on eigenvalue concepts rather than on equation solving. Numerical methods for finding eigenvalues are discussed briefly in Section 7.4.

REVIEW CHECKLIST

1. Define an eigenvalue and an eigenvector of a matrix.
2. Identify the set of all eigenvectors for an eigenvalue as the nullspace of a certain matrix.
3. Define the characteristic polynomial and the characteristic equation of a matrix.
4. Describe the various possible types of eigenvalues possible for a 2×2 matrix and for a 3×3 matrix.
5. Solve the eigenvalue problem for a given 2×2 or 3×3 matrix. That is, find all eigenvalues of the matrix and for each, find the corresponding eigenspace.

EXERCISES

In Exercises 1–4, verify by direct calculation that $A\mathbf{v}_i = \lambda\mathbf{v}_i$.

1. $A = \begin{bmatrix} 2 & 1 \\ 2 & 3 \end{bmatrix}$, $\lambda_1 = 1$, $\mathbf{v}_1 = \begin{bmatrix} 1 \\ -1 \end{bmatrix}$, $\lambda_2 = 4$, $\mathbf{v}_2 = \begin{bmatrix} 1 \\ 2 \end{bmatrix}$

2. $A = \begin{bmatrix} 3 & 1 \\ -1 & 1 \end{bmatrix}$, $\lambda = 2$, $\mathbf{v} = \begin{bmatrix} 1 \\ -1 \end{bmatrix}$

3. $A = \begin{bmatrix} 1 & -2 & 0 \\ 0 & -1 & -2 \\ -1 & 1 & -1 \end{bmatrix}$, $\lambda = -1$, $\mathbf{v} = \begin{bmatrix} 1 \\ 1 \\ 0 \end{bmatrix}$

4. $A = \begin{bmatrix} 1 & 2 & 2 \\ -2 & 1 & 1 \\ 2 & -2 & -2 \end{bmatrix}$, $\lambda = -1$, $\mathbf{v} = \begin{bmatrix} 1 \\ 3 \\ -4 \end{bmatrix}$,

$\lambda_2 = 0$, $\mathbf{v}_2 = \begin{bmatrix} 0 \\ 1 \\ -1 \end{bmatrix}$, $\lambda_3 = 1$, $\mathbf{v}_3 = \begin{bmatrix} 1 \\ -2 \\ 2 \end{bmatrix}$

In Exercises 5–8, find the characteristic polynomial of the given matrix.

5. $\begin{bmatrix} 5 & -2 \\ 2 & 4 \end{bmatrix}$

6. $\begin{bmatrix} 3 & 6 \\ 4 & 5 \end{bmatrix}$

7. $\begin{bmatrix} 1 & 2 & 3 \\ 0 & 1 & 2 \\ 0 & -1 & -2 \end{bmatrix}$

8. $\begin{bmatrix} 1 & 3 & 2 \\ 2 & 0 & -1 \\ 1 & 2 & 2 \end{bmatrix}$

In Exercises 9–30, find all eigenvalues of the given matrix, and for each, find a basis for the corresponding eigenspace.

9. $\begin{bmatrix} 2 & 0 \\ 0 & 2 \end{bmatrix}$

10. $\begin{bmatrix} 2 & 0 \\ 0 & 5 \end{bmatrix}$

11. $\begin{bmatrix} 0 & 0 \\ 0 & 0 \end{bmatrix}$

12. $\begin{bmatrix} 1 & 0 \\ 0 & 0 \end{bmatrix}$

13. $\begin{bmatrix} 2 & -1 \\ 0 & 3 \end{bmatrix}$

14. $\begin{bmatrix} 3 & -1 \\ 2 & 0 \end{bmatrix}$

15. $\begin{bmatrix} 6 & 5 \\ -5 & -4 \end{bmatrix}$

16. $\begin{bmatrix} 3 & 1 \\ 1 & 3 \end{bmatrix}$

17. $\begin{bmatrix} 2 & -2 \\ 2 & 6 \end{bmatrix}$

18. $\begin{bmatrix} 3 & -2 \\ -1 & 2 \end{bmatrix}$

19. $\begin{bmatrix} 1 & 0 & 0 \\ 0 & 1 & 0 \\ 0 & 0 & 1 \end{bmatrix}$

20. $\begin{bmatrix} 2 & 0 & 0 \\ 0 & 1 & 0 \\ 0 & 0 & 3 \end{bmatrix}$

21. $\begin{bmatrix} 0 & 0 & 0 \\ 0 & 0 & 0 \\ 0 & 0 & 0 \end{bmatrix}$

22. $\begin{bmatrix} 1 & 2 & 3 \\ -1 & 1 & 0 \\ 0 & -2 & -2 \end{bmatrix}$

23. $\begin{bmatrix} 1 & -1 & -1 \\ 0 & -1 & -2 \\ -1 & 1 & 0 \end{bmatrix}$

24. $\begin{bmatrix} 2 & -1 & -1 \\ 1 & 0 & -1 \\ 1 & -1 & 2 \end{bmatrix}$

25. $\begin{bmatrix} 1 & -3 & 1 \\ 0 & 1 & 0 \\ 3 & 0 & -1 \end{bmatrix}$

26. $\begin{bmatrix} 1 & -3 & 1 \\ 0 & 1 & 0 \\ 3 & 1 & -1 \end{bmatrix}$

27. $\begin{bmatrix} 1 & -2 & 2 \\ -1 & -2 & 3 \\ -1 & -1 & 2 \end{bmatrix}$

28. $\begin{bmatrix} 0 & -2 & 2 \\ -3 & 1 & 3 \\ -1 & 1 & 3 \end{bmatrix}$

29. $\begin{bmatrix} 2 & 2 & 1 \\ 1 & 3 & 1 \\ 1 & 2 & 2 \end{bmatrix}$

30. $\begin{bmatrix} 1 & 0 & 0 \\ 0 & 1 & 1 \\ 0 & -1 & 1 \end{bmatrix}$

31. Show that if $A\mathbf{v} = \lambda\mathbf{v}$, then for every positive integer k, $A^k\mathbf{v} = \lambda^k\mathbf{v}$. (That is, if λ is an eigenvalue of A, then λ^k is an eigenvalue of A^k.)

32. Show that A is singular if and only if 0 is an eigenvalue of A.

33. Show that if A is invertible and λ is a nonzero eigenvalue of A, then $1/\lambda$ is an eigenvalue of A^{-1}.

34. Show that the characteristic polynomials of A and A^t are identical, so A and A^t have the same eigenvalues.

35. If $\mathbf{v}$ is an eigenvector of A corresponding to λ, is $\mathbf{v}$ necessarily also an eigenvector of A^t?

36. Show that the characteristic equation of the matrix

$$A = \begin{bmatrix} 1 & 2 \\ 0 & 1 \end{bmatrix}$$

is

$$\lambda^2 - 2\lambda + 1 = 0,$$

and that

$$A^2 - 2A + I = 0.$$

37. Show that the characteristic equation of the matrix

$$A = \begin{bmatrix} 1 & 2 & 2 \\ 0 & 1 & 0 \\ -1 & 0 & -2 \end{bmatrix}$$

is

$$\lambda^3 - \lambda = 0,$$

and that

$$A^3 - A = 0.$$

38. Formulate a conjecture based on the results of Exercises 36 and 37 and test it with some examples from this section.

39. Explain geometrically why the matrix of a rotation of the plane through an angle θ cannot have a real eigenvalue/eigenvector pair unless $\theta = n \cdot 180°$ for some integer n.

40. Show that for any $n \times n$ matrix A, the expansion of the determinant $\lambda I - A$ is a polynomial of degree n in λ.

7.2 DIAGONALIZATION

The eigenvectors of an $n \times n$ matrix A may or may not span $\mathbb{R}^n$. When a basis for $\mathbb{R}^n$ consists of eigenvectors of A, the basis vectors are "decoupled" for multiplication by A, and hence for finding the image of a vector under any linear operator determined by A. In this section, we learn how to determine whether such a basis exists for a given A and how to handle the change of basis problem when one does.

Linear Independence of Eigenvectors

We begin with an example to illustrate the central concept of this section.

Example 1

From Example 6 of Section 7.1, the eigenvalues of the matrix

$$A = \begin{bmatrix} 2 & 2 \\ 1 & 3 \end{bmatrix}$$

are $\lambda_1 = 1$ and $\lambda_2 = 4$ and the eigenvectors

$$\mathbf{v}_1 = \begin{bmatrix} 2 \\ -1 \end{bmatrix} \quad \text{and} \quad \mathbf{v}_2 = \begin{bmatrix} 1 \\ 1 \end{bmatrix}$$

span the corresponding eigenspaces. Since neither of these vectors is a scalar multiple of the other, the set $B = \{\mathbf{v}_1, \mathbf{v}_2\}$ is linearly independent and hence is a basis for $\mathbb{R}^2$. If

$$P = \begin{bmatrix} 2 & 1 \\ -1 & 1 \end{bmatrix},$$

the transition matrix from B to the standard basis, then

$$P^{-1}AP = \frac{1}{3}\begin{bmatrix} 1 & -1 \\ 1 & 2 \end{bmatrix}\begin{bmatrix} 2 & 2 \\ 1 & 3 \end{bmatrix}\begin{bmatrix} 2 & 1 \\ -1 & 1 \end{bmatrix} = \frac{1}{3}\begin{bmatrix} 1 & -1 \\ 1 & 2 \end{bmatrix}\begin{bmatrix} 2 & 4 \\ -1 & 4 \end{bmatrix} = \begin{bmatrix} 1 & 0 \\ 0 & 4 \end{bmatrix}.$$

Thus A is similar to the diagonal matrix $D = \text{diag}(\lambda_1, \lambda_2)$, and the transition matrix P has column vectors $\mathbf{v}_1$ and $\mathbf{v}_2$. □

Example 1 illustrates the following general result. If the eigenvectors of an $n \times n$ matrix A span $\mathbb{R}^n$, then A is similar to the diagonal matrix $D = \text{diag}(\lambda_1, \ldots, \lambda_n)$, and the columns of the transition matrix P are corresponding eigenvectors. We begin with an important result on linear independence of sets of eigenvectors.

THEOREM 7.1

If $\mathbf{v}_1, \ldots, \mathbf{v}_k$ are eigenvectors of a matrix A for *distinct* eigenvalues $\lambda_1, \ldots, \lambda_k$, respectively, then the set $\{\mathbf{v}_1, \ldots, \mathbf{v}_k\}$ is linearly independent.

Proof The singleton $\{\mathbf{v}_1\}$ is linearly independent, since $\mathbf{v}_1 \neq \mathbf{0}$. To see that $\{\mathbf{v}_1, \mathbf{v}_2\}$ is linearly independent, suppose

$$c_1\mathbf{v}_1 + c_2\mathbf{v}_2 = \mathbf{0}.$$

Multiply both sides of this equation first by λ_2 to obtain

$$c_1\lambda_2\mathbf{v}_1 + c_2\lambda_2\mathbf{v}_2 = \mathbf{0},$$

then by A to obtain

$$Ac_1\mathbf{v}_1 + Ac_2\mathbf{v}_2 = \mathbf{0}, \qquad \text{or} \qquad c_1\lambda_1\mathbf{v}_1 + c_2\lambda_2\mathbf{v}_2 = \mathbf{0}.$$

Subtracting these equations and factoring the result gives

$$c_1(\lambda_2 - \lambda_1)\mathbf{v}_1 = \mathbf{0}.$$

Since $\mathbf{v}_1 \neq \mathbf{0}$ and $\lambda_1 \neq \lambda_2$, $c_1 = 0$. Thus the first equation above reduces to

$$c_2\mathbf{v}_2 = \mathbf{0}.$$

Since $\mathbf{v}_2 \neq \mathbf{0}$, $c_2 = 0$. Hence $\{\mathbf{v}_1, \mathbf{v}_2\}$ is linearly independent. Similar arguments show that the set $\{\mathbf{v}_1, \mathbf{v}_2, \mathbf{v}_3\}$ is linearly independent (Exercise 21) and establish the inductive step for the extension to any k vectors. □

Corollary If an $n \times n$ matrix A has n distinct eigenvalues, then the eigenvectors of A form a basis of $\mathbb{R}^n$.

Example 2

If

$$A = \begin{bmatrix} 1 & -3 & 0 \\ -1 & 1 & -2 \\ -1 & 0 & -1 \end{bmatrix},$$

the eigenvalues and corresponding eigenvectors are:

$$\lambda_1 = -2, \quad \mathbf{v}_1 = \begin{bmatrix} 1 \\ 1 \\ 1 \end{bmatrix}, \quad \lambda_2 = 1, \quad \mathbf{v}_2 = \begin{bmatrix} 2 \\ 0 \\ -1 \end{bmatrix}, \quad \lambda_3 = 2, \quad \mathbf{v}_3 = \begin{bmatrix} 3 \\ -1 \\ -1 \end{bmatrix}.$$

By Theorem 4.13, these eigenvectors constitute a linearly independent set since they form a matrix of row rank 3:

$$\begin{bmatrix} 1 & 1 & 1 \\ 2 & 0 & -1 \\ 3 & -1 & -1 \end{bmatrix} \to \begin{bmatrix} 1 & 1 & 1 \\ 0 & -2 & -3 \\ 0 & -4 & -4 \end{bmatrix} \to \begin{bmatrix} 1 & 1 & 1 \\ 0 & 2 & 3 \\ 0 & 0 & 2 \end{bmatrix}.$$

Hence the eigenvectors of A constitute a basis for $\mathbb{R}^3$. □

A Diagonalization Procedure

The next theorem, which is also illustrated by Example 1, provides guidance for establishing a general method.

THEOREM 7.2

Let A be an $n \times n$ matrix. Then A is similar to a diagonal matrix D if and only if $\mathbb{R}^n$ has a basis consisting of eigenvectors of A.

Proof Suppose $\lambda_1, \ldots, \lambda_n$ are (not necessarily distinct) eigenvalues of A and the corresponding eigenvectors $\mathbf{v}_1, \ldots, \mathbf{v}_n$ form a linearly independent set. Let P be the matrix with ith column vector $[\mathbf{v}_i]$. By Theorem 4.13, P is invertible. We also have

$$AP = [[\lambda_1\mathbf{v}_1] \quad \cdots \quad [\lambda_n\mathbf{v}_n]],$$

by definition of eigenvalue and eigenvector. By the definition of the inverse of a matrix, $P^{-1}[\mathbf{v}_i]$ equals the ith column of the identity matrix, so $P^{-1}[\lambda_i\mathbf{v}_i]$ equals the ith column of the identity multiplied by λ_i. Therefore

$$P^{-1}AP = \operatorname{diag}(\lambda_1, \ldots, \lambda_n) = D.$$

Conversely, if there exists an invertible P for which

$$P^{-1}AP = D = \operatorname{diag}(d_1, \ldots, d_n),$$

then

$$AP = PD.$$

Hence if $[\mathbf{v}_1], \ldots, [\mathbf{v}_n]$ are the column vectors of P, then

$$[[A\mathbf{v}_1] \quad \cdots \quad [A\mathbf{v}_n]] = [[d_1\mathbf{v}_1] \quad \cdots \quad [d_n\mathbf{v}_n]].$$

By definition of matrix equality and vector equality,

$$A\mathbf{v}_i = d_1\mathbf{v}_i$$

for $i = 1, \ldots, n$. Since P is invertible, its column vectors are nonzero, hence d_i is an eigenvalue with corresponding eigenvector $\mathbf{v}_i$. Invertibility of P also implies linear independence of this set of eigenvectors by Theorem 4.13. ☐

The proof of Theorem 7.2 is especially important because it establishes a *diagonalization procedure*. We illustrate the procedure and details of the key calculations in the proof.

Example 3

For

$$A = \begin{bmatrix} 4 & -3 \\ 2 & -1 \end{bmatrix},$$

the characteristic equation method yields eigenvalues and corresponding eigenvectors

$$\lambda_1 = 1, \quad \mathbf{v}_1 = \begin{bmatrix} 1 \\ 1 \end{bmatrix}, \quad \lambda_2 = 2, \quad \mathbf{v}_2 = \begin{bmatrix} 3 \\ 2 \end{bmatrix}.$$

The matrix P in Theorem 7.2 is given by

$$P = \begin{bmatrix} 1 & 3 \\ 1 & 2 \end{bmatrix}.$$

The product AP can be written in a form that displays a key element in the proof of Theorem 7.2:

$$AP = \begin{bmatrix} 4 & -3 \\ 2 & -1 \end{bmatrix}\begin{bmatrix} 1 & 3 \\ 1 & 2 \end{bmatrix} = \begin{bmatrix} 1 & 6 \\ 1 & 4 \end{bmatrix} = \begin{bmatrix} 1\begin{bmatrix} 1 \\ 1 \end{bmatrix} & 2\begin{bmatrix} 3 \\ 2 \end{bmatrix} \end{bmatrix}.$$

Multiplying each column of AP by P^{-1} shows that the ith column of $P^{-1}AP$ equals the ith column of the identity multiplied by λ_i:

$$P^{-1}AP = \begin{bmatrix} -2 & 3 \\ 1 & -1 \end{bmatrix}\begin{bmatrix} 1\begin{bmatrix} 1 \\ 1 \end{bmatrix} & 2\begin{bmatrix} 3 \\ 2 \end{bmatrix} \end{bmatrix} = \begin{bmatrix} 1\begin{bmatrix} 1 \\ 0 \end{bmatrix} & 2\begin{bmatrix} 0 \\ 1 \end{bmatrix} \end{bmatrix} = \begin{bmatrix} 1 & 0 \\ 0 & 2 \end{bmatrix}. \quad \square$$

In the notation of Theorem 7.2, if the eigenvectors of A form a basis B for $\mathbb{R}^n$, then the matrix P is the transition matrix from B to the standard basis. The matrix A represents a linear operator T relative to the standard basis, and the diagonal matrix D is the matrix of T relative to B.

Ordinarily in a diagonalization problem, the calculations are not carried out in the detail of Example 3. Usually we first find the distinct eigenvalues of A and for each one, find a basis of the corresponding eigenspace. Then we determine whether a basis for $\mathbb{R}^n$ consisting of eigenvectors exists and if so, form the matrices P and D.

Example 4

For the matrix

$$A = \begin{bmatrix} 1 & -2 & -1 \\ -1 & 0 & -1 \\ 1 & 2 & 3 \end{bmatrix},$$

suppose the eigenvalues and eigenvectors are known:

$$\lambda_1 = 0, \quad \mathbf{v}_1 = \begin{bmatrix} 1 \\ 1 \\ -1 \end{bmatrix}, \quad \lambda_2 = \lambda_3 = 2, \quad \mathbf{v}_2 = \begin{bmatrix} 1 \\ 0 \\ -1 \end{bmatrix}, \quad \mathbf{v}_3 = \begin{bmatrix} 2 \\ -1 \\ 0 \end{bmatrix}.$$

Form the transition matrix P with the eigenvectors as columns:

$$P = \begin{bmatrix} 1 & 1 & 2 \\ 1 & 0 & -1 \\ -1 & -1 & 0 \end{bmatrix}.$$

The inverse of P may be obtained by the Gauss-Jordan procedure from Chapter 2 or by the adjoint method from Chapter 3. The diagonalization is given by

$$P^{-1}AP = \begin{bmatrix} 1/2 & 1 & 1/2 \\ -1/2 & -1 & -3/2 \\ 1/2 & 0 & 1/2 \end{bmatrix} \begin{bmatrix} 1 & -2 & -1 \\ -1 & 0 & -1 \\ 1 & 2 & 3 \end{bmatrix} \begin{bmatrix} 1 & 1 & 2 \\ 1 & 0 & -1 \\ -1 & -1 & 0 \end{bmatrix}$$

$$= \begin{bmatrix} 1/2 & 1 & 1/2 \\ -1/2 & -1 & -3/2 \\ 1/2 & 0 & 1/2 \end{bmatrix} \begin{bmatrix} 0 & 2 & 4 \\ 0 & 0 & -2 \\ 0 & -2 & 0 \end{bmatrix} = \begin{bmatrix} 0 & 0 & 0 \\ 0 & 2 & 0 \\ 0 & 0 & 2 \end{bmatrix}. \quad \square$$

Theorem 7.2 shows even more about diagonalization. Suppose A is diagonalizable and

$$P^{-1}AP = D,$$

where

$$D = \operatorname{diag}(\lambda_1, \ldots, \lambda_n)$$

and the corresponding eigenvectors form the columns of P. Since the eigenvalues of D are also $\lambda_1, \ldots, \lambda_n$, the characteristic polynomial of D in factored form is

$$(\lambda - \lambda_1)(\lambda - \lambda_2) \cdots (\lambda - \lambda_n).$$

Since similar matrices have the same characteristic polynomial (Exercise 22), this is also the characteristic polynomial of A. These observations establish an important theorem and lead to a general diagonalization procedure.

THEOREM 7.3

If an $n \times n$ matrix A is diagonalizable, then the multiplicity of any eigenvalue λ as a root of the characteristic equation, the **algebraic multiplicity of λ**, equals the dimension of the corresponding eigenspace, the **geometric multiplicity of λ**, and the sum of the geometric multiplicities of the eigenvalues of A equals n.

Corollary Let A be an $n \times n$ matrix. If the geometric multiplicity of any eigenvalue λ of A is less than the algebraic multiplicity of λ, then A is *not* diagonalizable.

A Method for Diagonalizing a Matrix

1. Solve the characteristic equation $\det(\lambda I - A) = 0$ for the eigenvalues of A.
2. For each distinct eigenvalue λ, find a basis for the eigenspace by solving $(\lambda I - A)\mathbf{v} = \mathbf{0}$. If for at least one λ, the geometric multiplicity is less than the algebraic multiplicity, then A is not diagonalizable.
3. If possible, form a basis B of $\mathbb{R}^n$ consisting of eigenvectors, and form the transition matrix P from the standard basis to B using these eigenvectors as columns.
4. The matrix $P^{-1}AP$ is a diagonal matrix with the eigenvalues on the main diagonal.

Example 5

Diagonalize the matrix

$$A = \begin{bmatrix} 1 & -1 & 1 \\ -1 & 1 & 1 \\ -1 & -1 & 3 \end{bmatrix}.$$

Solution

The characteristic polynomial and its factorization are given by

$$\det(\lambda I - A) = \lambda^3 - 5\lambda^2 + 8\lambda - 4 = (\lambda - 2)^2(\lambda - 1).$$

Reduce the coefficient matrix for the eigenspace of $\lambda_3 = 1$:

$$\begin{bmatrix} 0 & 1 & -1 \\ 1 & 0 & -1 \\ 1 & 1 & -2 \end{bmatrix} \rightarrow \begin{bmatrix} 1 & 0 & -1 \\ 0 & 1 & -1 \\ 0 & 1 & -1 \end{bmatrix}.$$

By back substitution, the eigenspace is spanned by

$$\mathbf{v}_3 = \begin{bmatrix} 1 \\ 1 \\ 1 \end{bmatrix}.$$

The coefficient matrix for the eigenspace of $\lambda_1 = \lambda_2 = 2$ is

$$\begin{bmatrix} 1 & 1 & -1 \\ 1 & 1 & -1 \\ 1 & 1 & -1 \end{bmatrix}.$$

This system has rank 1. Back substitution first with $x_2 = -1$ and $x_3 = 0$, and then with $x_2 = 0$ and $x_3 = 1$ yields

$$\mathbf{v}_1 = \begin{bmatrix} 1 \\ -1 \\ 0 \end{bmatrix} \quad \text{and} \quad \mathbf{v}_2 = \begin{bmatrix} 1 \\ 0 \\ 1 \end{bmatrix}.$$

For a basis of $\mathbb{R}^3$ consisting of eigenvectors, use

$$B = \left\{ \begin{bmatrix} 1 \\ -1 \\ 0 \end{bmatrix}, \begin{bmatrix} 1 \\ 0 \\ 1 \end{bmatrix}, \begin{bmatrix} 1 \\ 1 \\ 1 \end{bmatrix} \right\}.$$

For this choice of B, the transition matrix and its inverse are

$$P = \begin{bmatrix} 1 & 1 & 1 \\ -1 & 0 & 1 \\ 0 & 1 & 1 \end{bmatrix} \quad \text{and} \quad P^{-1} = \begin{bmatrix} 1 & 0 & -1 \\ -1 & -1 & 2 \\ 1 & 1 & -1 \end{bmatrix}.$$

It can then be verified that

$$P^{-1}AP = \operatorname{diag}(1, 2, 2). \quad \square$$

Example 6

Diagonalize the matrix

$$A = \begin{bmatrix} 1 & -2 & 1 \\ 1 & 1 & 2 \\ 1 & 0 & 3 \end{bmatrix}.$$

Solution

The characteristic polynomial is given by

$$\det(\lambda I - A) = \lambda^3 - 5\lambda^2 + 8\lambda - 4 = (\lambda - 1)(\lambda - 2)^2.$$

Reduce the coefficient matrix for the eigenspace of $\lambda_2 = 2$:

$$\begin{bmatrix} 1 & 2 & -1 \\ -1 & 1 & -2 \\ -1 & 0 & -1 \end{bmatrix} \rightarrow \begin{bmatrix} 1 & 2 & -1 \\ 0 & 3 & -3 \\ 0 & 2 & -2 \end{bmatrix}.$$

Since the rank of the system is 2, the eigenspace has dimension 1. Since the multiplicity of 2 as a root of the characteristic equation is 2, A is *not* diagonalizable. ☐

We conclude by illustrating a complex diagonalization.

Example 7

Example 3 of Section 7.1 gave

$$\lambda_1 = 1 - i, \quad \mathbf{v}_1 = \begin{bmatrix} 1 \\ i \end{bmatrix}, \quad \lambda_2 = 1 + i, \quad \text{and} \quad \mathbf{v}_2 = \begin{bmatrix} 1 \\ -i \end{bmatrix}$$

as eigenvalues and eigenvectors for the matrix

$$A = \begin{bmatrix} 1 & -1 \\ 1 & 1 \end{bmatrix}.$$

The transition matrix formed by the eigenvectors, i.e.,

$$P = \begin{bmatrix} 1 & 1 \\ i & -i \end{bmatrix},$$

gives

$$P^{-1} = \frac{1}{-2i}\begin{bmatrix} -i & -1 \\ -i & 1 \end{bmatrix} = \frac{i}{2}\begin{bmatrix} -i & -1 \\ -i & 1 \end{bmatrix} = \frac{1}{2}\begin{bmatrix} 1 & -i \\ 1 & i \end{bmatrix}.$$

Thus

$$P^{-1}AP = \frac{1}{2}\begin{bmatrix} 1 & -i \\ 1 & i \end{bmatrix}\begin{bmatrix} 1 & -1 \\ 1 & 1 \end{bmatrix}\begin{bmatrix} 1 & 1 \\ i & -i \end{bmatrix} = \frac{1}{2}\begin{bmatrix} 1 & -i \\ 1 & i \end{bmatrix}\begin{bmatrix} 1-i & 1+i \\ 1+i & 1-i \end{bmatrix}$$

$$= \frac{1}{2}\begin{bmatrix} 1-i-i+1 & 1+i-i-1 \\ 1-i+i-1 & 1+i+i+1 \end{bmatrix} = \begin{bmatrix} 1-i & 0 \\ 0 & 1+i \end{bmatrix}.$$

Therefore diagonalization of a matrix may be possible even when the eigenvalues and eigenvectors are complex. ☐

REVIEW CHECKLIST

1. Identify a set of eigenvectors corresponding to distinct eigenvalues as linearly independent.
2. Identify the existence of n distinct eigenvalues as a sufficient, but not necessary, condition for the diagonalizability of an $n \times n$ matrix.

3. Identify the existence of an eigenvalue with geometric multiplicity less than its algebraic multiplicity as sufficient to guarantee that A is not diagonalizable.
4. Characterize the diagonalizability of an $n \times n$ matrix A in terms of the existence of a linearly independent set of n eigenvectors.
5. Characterize the diagonalizability of A in terms of the sum of the dimensions of the eigenspaces of A.
6. For a diagonalizable $n \times n$ matrix A, find a basis B for $\mathbb{R}^n$ consisting of eigenvectors, form the transition matrix P from B to the standard basis, find P^{-1}, and describe in detail the results of the matrix multiplications AP and $P^{-1}AP$.

EXERCISES

In Exercises 1–20, determine whether the given matrix is diagonalizable, and diagonalize it when it is.

1. $\begin{bmatrix} 5 & 4 \\ -3 & -2 \end{bmatrix}$ **2.** $\begin{bmatrix} -1 & 0 \\ 3 & 1 \end{bmatrix}$ **3.** $\begin{bmatrix} 4 & 0 \\ 0 & 4 \end{bmatrix}$

4. $\begin{bmatrix} 1 & -2 \\ -1 & 0 \end{bmatrix}$ **5.** $\begin{bmatrix} 6 & 4 \\ -4 & -2 \end{bmatrix}$ **6.** $\begin{bmatrix} 2 & -2 \\ -1 & 3 \end{bmatrix}$

7. $\begin{bmatrix} 9 & -6 \\ 8 & -5 \end{bmatrix}$ **8.** $\begin{bmatrix} 1 & 5 \\ 1 & -3 \end{bmatrix}$ **9.** $\begin{bmatrix} 1 & -5 \\ 2 & -4 \end{bmatrix}$

10. $\begin{bmatrix} 1 & 2 & 2 \\ -1 & 1 & 0 \\ 0 & -3 & -2 \end{bmatrix}$ **11.** $\begin{bmatrix} 1 & -2 & 2 \\ -1 & 1 & -2 \\ 0 & 2 & -1 \end{bmatrix}$ **12.** $\begin{bmatrix} 1 & 1 & -1 \\ -2 & 4 & -2 \\ 1 & -1 & 3 \end{bmatrix}$

13. $\begin{bmatrix} 1 & 2 & 2 \\ 2 & 1 & 2 \\ 0 & -3 & -2 \end{bmatrix}$ **14.** $\begin{bmatrix} 1 & -2 & 3 \\ 1 & 2 & 0 \\ 0 & 2 & -1 \end{bmatrix}$ **15.** $\begin{bmatrix} 2 & 2 & -2 \\ 2 & 2 & 2 \\ -2 & 3 & 3 \end{bmatrix}$

16. $\begin{bmatrix} 1 & -2 & 2 \\ 0 & 1 & 2 \\ 3 & 1 & 2 \end{bmatrix}$ **17.** $\begin{bmatrix} 1 & 1 & 1 \\ -4 & -4 & -1 \\ 4 & 1 & -2 \end{bmatrix}$ **18.** $\begin{bmatrix} 1 & -2 & 0 \\ 1 & 1 & -2 \\ 1 & 0 & -1 \end{bmatrix}$

19. $\begin{bmatrix} 2 & 2 & -1 & 0 \\ 1 & 3 & -1 & 0 \\ 1 & 2 & 0 & 0 \\ 1 & 2 & -1 & 1 \end{bmatrix}$ **20.** $\begin{bmatrix} 1 & 1 & 2 & -1 \\ 0 & 3 & 4 & -2 \\ 1 & 0 & 2 & -1 \\ 2 & 0 & 2 & -1 \end{bmatrix}$

21. Show that if $\mathbf{v}_1$, $\mathbf{v}_2$, and $\mathbf{v}_3$ are eigenvectors of a matrix corresponding to the distinct eigenvalues λ_1, λ_2, and λ_3, and if $\{\mathbf{v}_1, \mathbf{v}_2\}$ is linearly independent, then $\{\mathbf{v}_1, \mathbf{v}_2, \mathbf{v}_3\}$ is also linearly independent. (This exercise illustrates the inductive step in the proof of Theorem 7.1.)

22. Show that if $A \sim A'$, then A and A' have the same characteristic polynomial.

23. If $A = \text{diag}(d_1, \ldots, d_n)$ and k is a positive integer, find A^k.

24. Show that if $P^{-1}AP = D$, then for any positive integer k, $P^{-1}A^kP = D^k$.

25. Find the matrix

$$\begin{bmatrix} -1 & 1 & 1 \\ 1 & -1 & 1 \\ 1 & 1 & -1 \end{bmatrix}^{50}.$$

7.3 SYMMETRIC MATRICES

Symmetric matrices often arise in mathematical and physical problems involving natural symmetries. In this section, we study some useful special properties of symmetric matrices relative to eigenvalues and eigenvectors.

Orthogonal Diagonalization

We illustrate two special properties of symmetric matrices.

Example 1

Verify that the eigenvalues of the symmetric matrix

$$A = \begin{bmatrix} 1 & 2 \\ 2 & 4 \end{bmatrix}$$

are real and that eigenvectors for distinct eigenvalues are orthogonal.

Solution

From the characteristic equation

$$\lambda^2 + 5\lambda = \lambda(\lambda + 5) = 0,$$

the distinct real eigenvalues are given by $\lambda_1 = 0$ and $\lambda_2 = -5$. The general forms of the corresponding eigenvectors are

$$\mathbf{v}_1 = \begin{bmatrix} 2s \\ -s \end{bmatrix}, \quad s \neq 0, \quad \text{and} \quad \mathbf{v}_2 = \begin{bmatrix} t \\ 2t \end{bmatrix}, \quad t \neq 0.$$

Since the inner product of any $\mathbf{v}_1$ and $\mathbf{v}_2$ is given by

$$\langle \mathbf{v}_1, \mathbf{v}_2 \rangle = (2s)(t) + (-s)(2t) = 0,$$

eigenvectors for the distinct eigenvalues are orthogonal. □

Lemma 1 If A is a symmetric $n \times n$ matrix, then for any column vectors $\mathbf{u}$ and $\mathbf{v}$ in $\mathbb{R}^n$,

$$\langle A\mathbf{u}, \mathbf{v} \rangle = \langle \mathbf{u}, A\mathbf{v} \rangle.$$

Proof The matrix form of the inner product $\langle \mathbf{u}, \mathbf{v} \rangle = \mathbf{u}^t\mathbf{v}$ gives

$$\langle A\mathbf{u}, \mathbf{v} \rangle = (A\mathbf{u})^t\mathbf{v} = (\mathbf{u}^tA^t)\mathbf{v} = (\mathbf{u}^tA)\mathbf{v} = \mathbf{u}^t(A\mathbf{v}) = \langle \mathbf{u}, A\mathbf{v} \rangle. \quad □$$

Lemma 2 Eigenvectors corresponding to distinct eigenvalues of a symmetric matrix A are orthogonal.

Proof Let $\mathbf{v}_1$ and $\mathbf{v}_2$ be eigenvectors for distinct eigenvalues λ_1 and λ_2, respectively. By properties of the inner product, the definition of eigenvalue and eigenvector, and Lemma 1,

$$\lambda_1\langle \mathbf{v}_1, \mathbf{v}_2\rangle = \langle \lambda_1\mathbf{v}_1, \mathbf{v}_2\rangle = \langle A\mathbf{v}_1, \mathbf{v}_2\rangle = \langle \mathbf{v}_1, A\mathbf{v}_2\rangle = \langle \mathbf{v}_1, \lambda_2\mathbf{v}_2\rangle = \lambda_2\langle \mathbf{v}_1, \mathbf{v}_2\rangle.$$

From the first and last of these expressions,

$$(\lambda_1 - \lambda_2)\langle \mathbf{v}_1, \mathbf{v}_2\rangle = 0.$$

Since $\lambda_1 \neq \lambda_2$,

$$\langle \mathbf{v}_1, \mathbf{v}_2\rangle = 0. \quad \square$$

Example 2

Since the characteristic polynomial of the symmetric matrix

$$A = \begin{bmatrix} 2 & 0 & -1 \\ 0 & 2 & -1 \\ -1 & -1 & 3 \end{bmatrix}$$

is

$$\lambda^3 - 7\lambda^2 + 14\lambda - 8 = (\lambda - 1)(\lambda - 2)(\lambda - 4),$$

A has the distinct real eigenvalues $\lambda_1 = 1$, $\lambda_2 = 2$, $\lambda_3 = 4$. The corresponding eigenvectors,

$$\mathbf{v}_1 = \begin{bmatrix} 1 \\ 1 \\ 1 \end{bmatrix}, \quad \mathbf{v}_2 = \begin{bmatrix} 1 \\ -1 \\ 0 \end{bmatrix}, \quad \text{and} \quad \mathbf{v}_3 = \begin{bmatrix} 1 \\ 1 \\ -2 \end{bmatrix},$$

are readily seen to be pairwise orthogonal. $\square$

If B is an *orthonormal* basis for $\mathbb{R}^n$, then the transition matrix P from B to the standard basis is *orthogonal*. Hence $P^{-1} = P^t$ by Theorem 6.6. In this case, A is orthogonally similar (Definition 6.6) to a diagonal matrix D, and A is said to be **orthogonally diagonalizable**.

THEOREM 7.4

The following are equivalent for an $n \times n$ matrix A with *real* entries:

a. A is symmetric;
b. there exists an orthonormal basis for $\mathbb{R}^n$ consisting of eigenvectors of A;
c. the (not necessarily distinct) eigenvalues $\lambda_1, \ldots, \lambda_n$ of A are real, and there exists an orthogonal matrix P with real entries such that $P^{-1}AP = \operatorname{diag}(\lambda_1, \ldots, \lambda_n)$.

Proof A proof that (a) $\Rightarrow$ (b) is given at the end of this section.
(b) $\Rightarrow$ (c). Let B be an orthonormal basis of $\mathbb{R}^n$ consisting of eigenvectors $\mathbf{v}_1, \ldots, \mathbf{v}_n$ of A for the eigenvalues $\lambda_1, \ldots, \lambda_n$, respectively. Let P be the matrix whose ith column is the vector $\mathbf{v}_i$. By the first part of the proof of Theorem 7.2,

$$P^{-1}AP = \operatorname{diag}(\lambda_1, \ldots, \lambda_n).$$

Since A is real and each $\mathbf{v}_i$ has real components, the relation $A\mathbf{v}_i = \lambda_i \mathbf{v}_i$ implies that λ_i is real. Since the columns of P form an orthonormal set, P is orthogonal by Definition 6.6.
(c) $\Rightarrow$ (a). If the diagonal matrix $\operatorname{diag}(\lambda_1, \ldots, \lambda_n)$ is denoted by D, then

$$P^{-1}AP = D,$$

from which

$$A = PDP^{-1}.$$

Since P is orthogonal, $P^{-1} = P^t$ by Theorem 6.6. Thus

$$A = PDP^t.$$

Since $D^t = D$,

$$A^t = (PDP^t)^t = PD^tP^t = PDP^t = A,$$

hence A is symmetric. $\square$

Example 3

The characteristic polynomial of the symmetric matrix

$$A = \begin{bmatrix} 2 & -1 & -1 \\ -1 & 2 & 1 \\ -1 & 1 & 2 \end{bmatrix}$$

is

$$\lambda^3 - 6\lambda^2 + 9\lambda - 4 = (\lambda - 1)^2(\lambda - 4).$$

The coefficient matrix for the eigenvectors for $\lambda_1 = \lambda_2 = 1$ is

$$\begin{bmatrix} -1 & 1 & 1 \\ 1 & -1 & -1 \\ 1 & -1 & -1 \end{bmatrix}.$$

Solving for two eigenvectors and using the Gram-Schmidt procedure yields an *orthonormal basis* for the eigenspace consisting of

$$\mathbf{v}_1 = \begin{bmatrix} 2/\sqrt{6} \\ 1/\sqrt{6} \\ 1/\sqrt{6} \end{bmatrix} \quad \text{and} \quad \mathbf{v}_2 = \begin{bmatrix} 0 \\ 1/\sqrt{2} \\ -1/\sqrt{2} \end{bmatrix}.$$

For $\lambda_3 = 4$ the coefficient matrix is

$$\begin{bmatrix} 2 & 1 & 1 \\ 1 & 2 & -1 \\ 1 & -1 & 2 \end{bmatrix} \rightarrow \begin{bmatrix} 1 & 2 & -1 \\ 0 & -3 & 3 \\ 0 & -3 & 3 \end{bmatrix}.$$

from which a normalized eigenvector is

$$\mathbf{v}_3 = \begin{bmatrix} 1/\sqrt{3} \\ -1/\sqrt{3} \\ -1/\sqrt{3} \end{bmatrix}.$$

We may readily verify that the matrix

$$P = \begin{bmatrix} 2/\sqrt{6} & 0 & 1/\sqrt{3} \\ 1/\sqrt{6} & 1/\sqrt{2} & -1/\sqrt{3} \\ 1/\sqrt{6} & -1/\sqrt{2} & -1/\sqrt{3} \end{bmatrix}$$

is orthogonal, and that

$$P^{-1}AP = P^tAP = \begin{bmatrix} 1 & 0 & 0 \\ 0 & 1 & 0 \\ 0 & 0 & 4 \end{bmatrix},$$

in illustration of Theorem 7.4. ▭

The wording of Theorem 7.4 was chosen with special care because studies involving complex numbers lead to *different* results. The standard inner product for vectors with n complex components is given by

$$\langle \mathbf{u}, \mathbf{v} \rangle = u_1\bar{v}_1 + \cdots + u_n\bar{v}_n.$$

Since $z\bar{z} = |z|^2$ for any complex number z, this definition preserves the norm relationship

$$\|\mathbf{u}\| = \sqrt{\langle \mathbf{u}, \mathbf{u} \rangle}.$$

The conclusion of Lemma 2 above holds for complex vectors, but *the equivalence asserted in Theorem 7.4 does not*, as the following example illustrates.

Example 4

Show that the nonsymmetric matrix

$$A = \begin{bmatrix} 0 & 1 \\ -1 & 0 \end{bmatrix}$$

can be diagonalized by an orthogonal complex matrix P.

Solution

The characteristic polynomial $\lambda^2 + 1$ yields the eigenvalues $\lambda_1 = -i$ and $\lambda_2 = i$. For corresponding eigenvectors, take

$$\mathbf{v}_1 = \begin{bmatrix} i \\ 1 \end{bmatrix} \quad \text{and} \quad \mathbf{v}_2 = \begin{bmatrix} 1 \\ i \end{bmatrix}.$$

Since $\bar{1} = 1$ and $\bar{i} = -i$,

$$\langle \mathbf{v}_1, \mathbf{v}_2 \rangle = (i)(1) + (1)(-i) = 0,$$

hence these eigenvectors are orthogonal. The transition matrix and its inverse are given by

$$P = \begin{bmatrix} i & 1 \\ 1 & i \end{bmatrix} \quad \text{and} \quad P^{-1} = \frac{1}{-2}\begin{bmatrix} i & -1 \\ -1 & i \end{bmatrix}.$$

The matrix P is orthogonal, and

$$P^{-1}AP = \begin{bmatrix} -i & 0 \\ 0 & i \end{bmatrix}$$

(Exercise 29). Thus A is orthogonally diagonalizable even though it is not symmetric. □

Further study of orthogonal diagonalization is beyond the scope of this book. We simply note without proof or clarification that although the matrix A in Example 4 is not symmetric, it does satisfy $AA^t = A^tA$. This condition characterizes *real* (but not complex in general) orthogonally diagonalizable matrices.

Positive Definite Matrices

In Definition 5.6, an $n \times n$ matrix A was defined to be *positive definite* if A is symmetric and for every nonzero column vector $\mathbf{v}$ in $\mathbb{R}^n$, $\mathbf{v}^t A\mathbf{v} > 0$. Theorem 5.12 asserted that A is positive definite if and only if there exists an invertible B such that

$$A = B^t B.$$

We are now ready to move toward a proof of Theorem 5.12.

Lemma 3 Every eigenvalue of a positive definite matrix is positive.

Proof Let A be positive definite, λ an eigenvalue of A, and $\mathbf{v}$ a corresponding eigenvector. Then

$$(\lambda I - A)\mathbf{v} = \mathbf{0},$$

hence

$$\mathbf{v}^t(\lambda I - A)\mathbf{v} = 0,$$

$$\lambda \mathbf{v}^t\mathbf{v} - \mathbf{v}^t A\mathbf{v} = 0.$$

Since $\mathbf{v}$ is an eigenvector of A, $\mathbf{v} \neq \mathbf{0}$, so $\mathbf{v}^t\mathbf{v} > 0$. We also have $\mathbf{v}^t A\mathbf{v} > 0$ because A is positive definite. It now follows from the last equation that $\lambda > 0$. □

Example 5

The symmetric matrix

$$A = \begin{bmatrix} 2 & 2 \\ 2 & 6 \end{bmatrix}$$

is positive definite, since for any $\mathbf{v}$ in $\mathbb{R}^2$,

$$\mathbf{v}^t A\mathbf{v} = \begin{bmatrix} x_1 & x_2 \end{bmatrix} \begin{bmatrix} 2 & 2 \\ 2 & 6 \end{bmatrix} \begin{bmatrix} x_1 \\ x_2 \end{bmatrix} = 2(x_1 + x_2)^2 + 4x_2^2.$$

The characteristic polynomial of A is

$$\lambda^2 - 7\lambda + 6 = (\lambda - 1)(\lambda - 6),$$

so the eigenvalues are 1 and 6, both positive as predicted by Lemma 3. □

Proof of Theorem 5.12 If A is expressible in the form

$$A = B^t B,$$

where B is invertible, note first that

$$A^t = (B^t B)^t = B^t (B^t)^t = B^t B = A,$$

hence A is symmetric. If $\mathbf{v}$ is a nonzero vector in $\mathbb{R}^n$, then $B\mathbf{v} \neq \mathbf{0}$ by Theorem 4.13 since B is invertible. Therefore

$$(B\mathbf{v})^t (B\mathbf{v}) > 0,$$

so A is positive definite.

For the converse, suppose A is positive definite. Then A is symmetric, so by Theorem 7.4, there is an orthogonal matrix P such that

$$P^{-1}AP = P^t AP = D,$$

where $D = \text{diag}(\lambda_1, \ldots, \lambda_n)$. Since A is positive definite, each $\lambda_i > 0$ by Lemma 3. Hence

$$D = D_1^2,$$

where

$$D_1 = \text{diag}\left(\sqrt{\lambda_1}, \ldots, \sqrt{\lambda_n}\right).$$

Thus since $D_1^t = D_1$,

$$A = PDP^t = (PD_1)(D_1^t P^t) = (PD_1)(PD_1)^t = B^t B,$$

where

$$B = (PD_1)^t.$$

Since P and D_1 are both invertible, so is B, hence this factorization of A has the required properties. □

We deferred the proof that (a) ⇒ (c) in Theorem 7.4 because it is rather long and involves complex numbers.

Proof that (a) ⇒ (c) in Theorem 7.4 Let A be a symmetric matrix. For the first part of the proof we show that if λ is a root of the characteristic equation and $\lambda = a + ib$, where $i = \sqrt{-1}$, then $b = 0$. Since

$$\det(\lambda I - A) = 0,$$

Gaussian elimination with complex arithmetic produces a (possibly complex) nonzero vector $\mathbf{v}$ such that

$$(\lambda I - A)\mathbf{v} = \mathbf{0}.$$

Then with $\bar{\lambda} = a - ib$, the complex conjugate of λ,

$$(\bar{\lambda} I - A)(\lambda I - A)\mathbf{v} = \mathbf{0},$$

$$[(aI - A) - ibI][(aI - A) + ibI]\mathbf{v} = \mathbf{0},$$

$$\left[(aI - A)^2 + b^2I\right]\mathbf{v} = \mathbf{0}.$$

If $\bar{\mathbf{v}}$ denotes the complex conjugate of $\mathbf{v}$, then

$$\bar{\mathbf{v}}^t\left[(aI - A)^2 + b^2I\right]\mathbf{v} = 0.$$

Since A is symmetric and real, $\overline{aI - A} = aI - A$, so

$$\left[\overline{(aI - A)\mathbf{v}}\right]^t[(aI - A)\mathbf{v}] + b^2\bar{\mathbf{v}}^t\mathbf{v} = 0.$$

Each inner product in the left side of this equality is the norm of a vector, hence is nonnegative. Therefore $b^2\bar{\mathbf{v}}^t\mathbf{v} = 0$. Since $\mathbf{v} \neq \mathbf{0}$, we have $\bar{\mathbf{v}}^t\mathbf{v} \neq 0$, hence $b = 0$.

For the second part of the proof, we show that if A is a symmetric $n \times n$ matrix, then $\mathbb{R}^n$ has an orthonormal basis consisting of eigenvectors of A. Let λ be an eigenvalue of A, as guaranteed by the first part of the proof, $\mathbf{v}$ a corresponding eigenvector of norm 1. Let M be the span of $\mathbf{v}$, N the orthogonal complement of M, that is, $N = M^\perp$. If $N = \{\mathbf{0}\}$, then the conclusion holds since $\{\mathbf{v}\}$ forms a basis for $\mathbb{R}^n = \mathbb{R}$, as required. Otherwise, if $\mathbf{y}$ is in N, then by Lemma 1,

$$\langle A\mathbf{y}, \mathbf{v}\rangle = \langle \mathbf{y}, A\mathbf{v}\rangle = \langle \mathbf{y}, \lambda\mathbf{v}\rangle = \lambda\langle \mathbf{v}, \mathbf{y}\rangle = 0,$$

so $A\mathbf{y}$ is in N. Thus if T is the linear operator on $\mathbb{R}^n$ represented by A, then the *restriction* of T to N is a linear operator on the subspace N. Denote this operator by T_1. Let $m = n - 1$, and let $B = \{\mathbf{y}_1, \ldots, \mathbf{y}_m\}$ be an orthonormal basis for N

(guaranteed by the Gram-Schmidt procedure). Let

$$\begin{aligned} T(\mathbf{y}_1) &= c_{11}\mathbf{y}_1 + \cdots + c_{m1}\mathbf{y}_m = A\mathbf{y}_1, \\ &\vdots \\ T(\mathbf{y}_m) &= c_{1m}\mathbf{y}_1 + \cdots + c_{mm}\mathbf{y}_m = A\mathbf{y}_m. \end{aligned}$$

If $A_1 = [c_{ij}]_{m\times m}$, then A_1 represents T_1 relative to B. In addition, A_1 is symmetric, since

$$c_{ij} = \langle A\mathbf{y}_i, \mathbf{y}_j\rangle = \langle \mathbf{y}_i, A\mathbf{y}_j\rangle = \langle A\mathbf{y}_j, \mathbf{y}_i\rangle = c_{ji}.$$

Thus A_1, hence also A, has an eigenvalue, and a normalized eigenvector $\mathbf{v}_2$ in N, and the entire argument may be repeated with the span of $\{\mathbf{v}_1, \mathbf{v}_2\}$ replacing N. The process terminates after a finite number of steps, yielding an orthonormal basis of $\mathbb{R}^n$ consisting of eigenvectors of A. ☐

REVIEW CHECKLIST

1. Formulate a definition of an orthogonally diagonalizable matrix.
2. Characterize symmetric matrices in terms of eigenvalue properties (Theorem 7.4).
3. State a necessary condition for the symmetry of A in terms of inner products with vectors **u** and **v** (Lemma 1).
4. Formulate the orthogonality relationship between eigenvectors corresponding to distinct eigenvalues of a symmetric matrix.
5. Identify the eigenvalues of a positive definite matrix as positive real numbers.
6. Factor a positive definite matrix A into as product of the form B^tB, where B is invertible.

EXERCISES

In Exercises 1–22, find the eigenvalues of the given matrix, find an orthonormal basis of $\mathbb{R}^n$ consisting of eigenvectors, form the transition matrix P, and verify that $P^tAP = D$, the corresponding diagonal matrix of eigenvalues.

1. $\begin{bmatrix} 1 & 1 \\ 1 & 1 \end{bmatrix}$ **2.** $\begin{bmatrix} 2 & 0 \\ 0 & 2 \end{bmatrix}$ **3.** $\begin{bmatrix} 2 & 1 \\ 1 & 2 \end{bmatrix}$

4. $\begin{bmatrix} 1 & 2 \\ 2 & 4 \end{bmatrix}$ **5.** $\begin{bmatrix} 1 & 2 \\ 2 & -2 \end{bmatrix}$ **6.** $\begin{bmatrix} -3 & 4 \\ 4 & 3 \end{bmatrix}$

7. $\begin{bmatrix} 1 & 2 \\ 2 & 1 \end{bmatrix}$ **8.** $\begin{bmatrix} 1 & 6 \\ 6 & 6 \end{bmatrix}$ **9.** $\begin{bmatrix} -4 & 6 \\ 6 & 5 \end{bmatrix}$

10. $\begin{bmatrix} 3 & 0 & 0 \\ 0 & 2 & -1 \\ 0 & -1 & 2 \end{bmatrix}$ **11.** $\begin{bmatrix} 2 & 0 & -1 \\ 0 & 2 & 1 \\ -1 & 1 & 3 \end{bmatrix}$ **12.** $\begin{bmatrix} 2 & 1 & 1 \\ 1 & 3 & 0 \\ 1 & 0 & 3 \end{bmatrix}$

13. $\begin{bmatrix} 2 & 1 & 1 \\ 1 & 2 & 1 \\ 1 & 1 & 4 \end{bmatrix}$ **14.** $\begin{bmatrix} 2 & -1 & 1 \\ -1 & 4 & 1 \\ 1 & 1 & 4 \end{bmatrix}$ **15.** $\begin{bmatrix} 1 & 2 & 1 \\ 2 & 1 & 1 \\ 1 & 1 & 2 \end{bmatrix}$

16. $\begin{bmatrix} 2 & 1 & -2 \\ 1 & 1 & 1 \\ -2 & 1 & 2 \end{bmatrix}$ **17.** $\begin{bmatrix} -1 & 1 & 2 \\ 1 & -1 & 2 \\ 2 & 2 & 2 \end{bmatrix}$ **18.** $\begin{bmatrix} 3 & -2 & -2 \\ -2 & 3 & 2 \\ -2 & 2 & 3 \end{bmatrix}$

19. $\begin{bmatrix} 1 & -1 & 1 & 0 \\ -1 & 1 & 1 & 0 \\ 1 & 1 & 1 & 0 \\ 0 & 0 & 0 & 2 \end{bmatrix}$ **20.** $\begin{bmatrix} 2 & -3 & 2 & -1 \\ -3 & 2 & 2 & 1 \\ 2 & 2 & 1 & 0 \\ -1 & 1 & 0 & 4 \end{bmatrix}$

21. $\begin{bmatrix} 2 & 0 & -1 & 0 \\ 0 & 2 & 0 & -1 \\ -1 & 0 & 2 & 0 \\ 0 & -1 & 0 & 2 \end{bmatrix}$ **22.** $\begin{bmatrix} 4 & -4 & 10 & -1 \\ -4 & 0 & -4 & 0 \\ 10 & -4 & 4 & 1 \\ -1 & 0 & 1 & -7 \end{bmatrix}$

In Exercises 23–24, let

$$A = \begin{bmatrix} 2 & 1 & 1 \\ 1 & 4 & 1 \\ 1 & 1 & 2 \end{bmatrix}.$$

23. Find a 3×3 matrix B such that $A = B^tB$.

24. Show that the matrix B found in Exercise 23 is invertible.

25. Show that if A is a symmetric $n \times n$ matrix, then there exists an orthonormal basis $\{\mathbf{v}_1, \ldots, \mathbf{v}_n\}$ for $\mathbb{R}^n$ such that for $i = 1, \ldots, n$,

$$\langle A\mathbf{v}_i, \mathbf{v}_i \rangle$$

is an eigenvalue of A.

26. Let A be a symmetric $n \times n$ matrix, $\lambda_1, \ldots, \lambda_n$ its eigenvalues, $\mathbf{v}_1, \ldots, \mathbf{v}_n$ corresponding eigenvectors forming an orthonormal basis of $\mathbb{R}^n$. Show that if λ is a scalar that is *not* an eigenvalue, then the solution of the linear system

$$\lambda \mathbf{v} - A\mathbf{v} = \mathbf{b}$$

is given by

$$\mathbf{v} = \sum_{i=1}^{n} \frac{\langle \mathbf{b}, \mathbf{v}_i \rangle}{\lambda - \lambda_i} \mathbf{v}_i.$$

27. Let A be a symmetric $n \times n$ matrix, T the linear operator on $\mathbb{R}^n$ represented by A relative to the standard basis, λ an eigenvalue of A, and P_λ the orthogonal projection operator projecting $\mathbb{R}^n$ onto the eigenspace corresponding to λ. Show that

$$TP_\lambda = P_\lambda T.$$

28. With A and T as in Exercise 27, let $\lambda_1, \ldots, \lambda_k$ be the distinct eigenvalues of A and P_i be the orthogonal projection operator for the eigenspace corresponding to λ_i. Show that

$$T = \sum_{i=1}^{k} \lambda_i P_i.$$

29. (*Complex number based*) For the matrix P and its inverse found in Example 4 (page 373), carry out the matrix multiplication to verify that $P^{-1}AP = \text{diag}(-i, i)$.

(o) 7.4 APPLICATIONS AND COMPUTING

Quadratic Forms

Eigenvalue theory can be used to simplify certain expressions arising in science and engineering studies. A **quadratic form** in n variables is an expression of the form

$$a_{11}x_1^2 + 2a_{12}x_1x_2 + a_{22}x_2^2 + 2a_{13}x_1x_3 + \cdots + a_{nn}x_n^2.$$

We often use $Q(\mathbf{x})$ for a quadratic form, where $\mathbf{x}$ is the vector with components $x_1, \ldots, x_n$. In an application, the path of a satellite or an electron may be described by an equation

$$Q(\mathbf{x}) = c,$$

where c is a constant and $\mathbf{x}$ is in $\mathbb{R}^2$. Similarly, an equipotential surface in a force field may be determined by an equation

$$Q(\mathbf{x}) = c,$$

where c is a constant and $\mathbf{x}$ is in $\mathbb{R}^3$.

Any quadratic form $Q(\mathbf{x})$ in n variables corresponds uniquely to a *symmetric* matrix $A = [a_{ij}]_{n \times n}$ via the relation

$$Q(\mathbf{x}) = \mathbf{x}^t A \mathbf{x}.$$

We illustrate the use of eigenvalues and eigenvectors in working with quadratic forms.

Example 1

(*Analytic geometry based*) Identify the plane curve determined by the equation

$$5x_1^2 - 4x_1x_2 + 8x_2^2 = 36.$$

Solution

The matrix of the given quadratic form may be obtained from the general relation $Q(\mathbf{x}) = \mathbf{x}^tA\mathbf{x}$, where $A = [a_{ij}]_{2\times 2}$. The product on the right is:

$$\mathbf{x}^tA\mathbf{x} = \begin{bmatrix} x_1 & x_2 \end{bmatrix}\begin{bmatrix} a_{11} & a_{12} \\ a_{12} & a_{22} \end{bmatrix}\begin{bmatrix} x_1 \\ x_2 \end{bmatrix} = a_{11}x_1^2 + 2a_{12}x_1x_2 + a_{22}x_2^2.$$

Comparing the righthand side with the given form shows that

$$A = \begin{bmatrix} 5 & -2 \\ -2 & 8 \end{bmatrix}.$$

The eigenvalues of A are 4 and 9, and a corresponding basis of normalized eigenvectors is given by

$$B = \{\mathbf{v}_1, \mathbf{v}_2\} = \left\{ \begin{bmatrix} 2/\sqrt{5} \\ 1/\sqrt{5} \end{bmatrix}, \begin{bmatrix} -1/\sqrt{5} \\ 2/\sqrt{5} \end{bmatrix} \right\}.$$

The transition matrix from B to the standard basis is

$$P = \begin{bmatrix} 2/\sqrt{5} & -1/\sqrt{5} \\ 1/\sqrt{5} & 2/\sqrt{5} \end{bmatrix}.$$

For any $\mathbf{x}$ in $\mathbb{R}^n$, let $\mathbf{x}' = [\mathbf{x}]_B$. Then

$$\mathbf{x} = P\mathbf{x}' \quad \text{and} \quad \mathbf{x}' = P^t\mathbf{x},$$

from which

$$\mathbf{x}^tA\mathbf{x} = (P\mathbf{x}')^tA(P\mathbf{x}') = \mathbf{x}'^t(P^tAP)\mathbf{x}' = \mathbf{x}'^tD\mathbf{x}',$$

where

$$D = \operatorname{diag}(4, 9).$$

Since

$$\mathbf{x}'^tD\mathbf{x}' = \begin{bmatrix} x_1' & x_2' \end{bmatrix}\begin{bmatrix} 4 & 0 \\ 0 & 9 \end{bmatrix}\begin{bmatrix} x_1' \\ x_2' \end{bmatrix},$$

the equation of the given curve relative to B is

$$4x_1'^2 + 9x_2'^2 = 36.$$

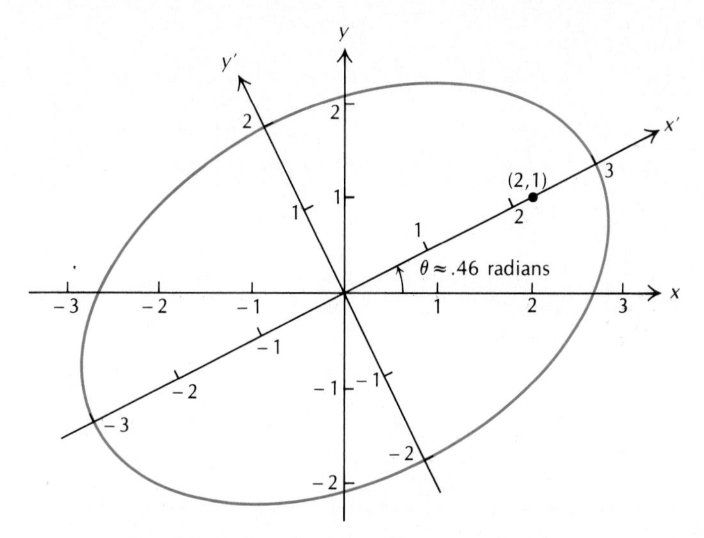

Figure 7.1 Graph of the ellipse in Example 1.

This is the equation of an ellipse in the coordinate system determined by the basis B, as shown in Fig. 7.1. ☐

Example 2

(*Analytic geometry based*) Identify the surface determined by the equation

$$2x_1^2 - 2x_1x_3 + 2x_2^2 - 2x_2x_3 + 3x_3^2 = 16.$$

Solution

Find the matrix A by comparing the given form with the general quadratic form in three variables

$$a_{11}x_1^2 + 2a_{12}x_1x_2 + a_{22}x_2^2 + 2a_{13}x_1x_3 + 2a_{23}x_2x_3 + a_{33}x_n^2:$$

$$A = \begin{bmatrix} 2 & 0 & -1 \\ 0 & 2 & -1 \\ -1 & -1 & 3 \end{bmatrix}.$$

From Example 2 of Section 7.3 (page 370), the eigenvalues of A are $\lambda_1 = 1$, $\lambda_2 = 2$, $\lambda_3 = 4$. The corresponding normalized eigenvectors from that example form the basis

$$B = \{\mathbf{v}_1, \mathbf{v}_2, \mathbf{v}_3\} = \left\{ \begin{bmatrix} 1/\sqrt{3} \\ 1/\sqrt{3} \\ 1/\sqrt{3} \end{bmatrix}, \begin{bmatrix} 1/\sqrt{2} \\ -1/\sqrt{2} \\ 0 \end{bmatrix}, \begin{bmatrix} 1/\sqrt{6} \\ 1/\sqrt{6} \\ -2/\sqrt{6} \end{bmatrix} \right\}.$$

Let P be the transition matrix from B to the standard basis. For any $\mathbf{x}$ in $\mathbb{R}^n$, let $\mathbf{x}' = [\mathbf{x}]_B$. Then

$$\mathbf{x} = P\mathbf{x}' \qquad \text{and} \qquad \mathbf{x}' = P^t\mathbf{x},$$

from which

$$\mathbf{x}^tA\mathbf{x} = (P\mathbf{x}')^tA(P\mathbf{x}') = \mathbf{x}'^t(P^tAP)\mathbf{x}' = \mathbf{x}'^tD\mathbf{x}',$$

where

$$D = \operatorname{diag}(1, 2, 4).$$

Since

$$\mathbf{x}'^tD\mathbf{x}' = [x_1' \quad x_2' \quad x_3'] \begin{bmatrix} 1 & 0 & 0 \\ 0 & 2 & 0 \\ 0 & 0 & 4 \end{bmatrix} \begin{bmatrix} x_1' \\ x_2' \\ x_3' \end{bmatrix},$$

the equation of the given curve relative to B is

$$x_1'^2 + 2x_2'^2 + 4x_3'^2 = 16.$$

This is the equation of an ellipsoid in the coordinate system determined by B. □

Diagonalization of quadratic forms in $\mathbb{R}^2$ and $\mathbb{R}^3$ is explored further in the exercises.

The Leslie Population Model Revisited

Section 2.6 contained a model in which a given female population is partitioned into n age brackets of equal time length t. The census was taken at intervals of length t, and constant birth and survival rates b_i and s_i were assumed for $i = 1, \ldots, n$. If $\mathbf{x}$ is the column vector whose components represent the class sizes at one census, the distribution at the next census was given by $L\mathbf{x}$, where L

is the corresponding Leslie matrix. One general problem is to predict **x** for future values of time.

We illustrate the solution of the steady-state distribution problem with $n = 3$ classes. The characteristic equation of

$$L = \begin{bmatrix} b_1 & b_2 & b_3 \\ s_1 & 0 & 0 \\ 0 & s_2 & 0 \end{bmatrix}$$

is

$$\lambda^3 = b_1\lambda^2 - b_2 s_1 \lambda - b_3 s_1 s_2 = 0.$$

For $\lambda \neq 0$, this equation may be written in the form

$$b_1/\lambda + b_2 s_1/\lambda^2 + b_3 s_1 s_2/\lambda^3 = 1.$$

When λ is near zero and positive, the value of the expression on the left is large and positive. As λ increases, the value steadily decreases toward zero. This function is continuous on the domain $(0, \infty)$, therefore it crosses the line $\lambda = 1$ at a unique point r. Thus r is the positive eigenvalue of A. To find a corresponding eigenvector, solve the vector equation

$$L\mathbf{v} = r\mathbf{v}.$$

This equation is equivalent to the system

$$\begin{aligned} b_1x_1 + b_2x_2 + b_3x_3 &= rx_1 \\ s_1x_1 \qquad\qquad &= rx_2 \\ s_2x_2 \qquad &= rx_3. \end{aligned}$$

We may assign any nonzero value to x_1 and solve the second and third equations for x_2 and x_3. With $x_1 = 1$,

$$x_2 = s_1/r \quad \text{and} \quad x_3 = s_2x_2/r = s_1s_2/r^2.$$

It can be shown that under reasonable conditions, such as the existence of an index $i < n$ for which

$$b_ib_{i+1} > 0,$$

the eigenvalue r is **dominant**, i.e., largest in magnitude. In this case, the limiting distribution is approximately given by the eigenvector **v**, and the limiting *growth factor* for consecutive censuses is approximately r. Thus r and **v** give the steady-state solution of the population growth problem.

Example 3

Suppose the females of a hypothetical animal population are divided into three age groups: 0–5, 5–10, and 10–15 years. If the Leslie matrix for this population is given by

$$L = \begin{bmatrix} 0.4 & 1.2 & 0.2 \\ 0.6 & 0 & 0 \\ 0 & 0.75 & 0 \end{bmatrix},$$

then the characteristic equation of L is

$$\lambda^3 - 0.4\lambda^2 - 0.72\lambda - 0.09 = 0.$$

The positive eigenvalue r can be found numerically. By Newton's method, for instance,

$$r = 1.1168.$$

The corresponding eigenvector with $x_1 = 1$ (obtained as described just prior to this example) is approximately

$$\mathbf{v} = \begin{bmatrix} 1 \\ 0.5372 \\ 0.3608 \end{bmatrix}.$$

The Leslie model predicts that the relative distribution of this female population will move toward the ratios determined by this eigenvector. These ratios may be found by dividing the individual components of **v** by the sum of the three. To four decimal places, these ratios are

$$.5269, \qquad .2830, \qquad \text{and} \qquad .1901.$$

In terms of percentages, the distribution is

53% in the 0–5 group, 28% in 5–10, and 19% in 10–15.

The steady-state growth rate for each class can be obtained from the eigenvalue r: each class will grow by approximately 11.68% during a census period.

Suppose that initially each class has 1000 females. Then the distributions in 50 and 55 years will be

4,198, 2,582, 1,513; and 4,692, 2,519, 1,694,

respectively. For 50 years, the *relative* distribution is

$$.5268, \qquad .2834, \qquad \text{and} \qquad .1898,$$

in close agreement with the values obtained from the eigenvector **v**. Moreover, the growth rate in each class from 50 to 55 years is approximately 11.68%. □

The Leslie model is explored further in the exercises.

The Power Method for Computing Eigenvalues

Let A be a diagonalizable $n \times n$ matrix with eigenvalues $\lambda_1, \ldots, \lambda_n$ and corresponding eigenvectors $\mathbf{v}_1, \ldots, \mathbf{v}_n$. Then any vector $\mathbf{x}$ in $\mathbb{R}^n$ may be expressed uniquely as

$$\mathbf{x} = c_1\mathbf{v}_1 + \cdots + c_n\mathbf{v}_n.$$

We may apply A to this expression iteratively to obtain

$$A^k\mathbf{x} = c_1\lambda_1^k\mathbf{v}_1 + \cdots + c_n\lambda_n^k\mathbf{v}_n$$

for any positive integer k. Suppose A has a dominant eigenvalue and we relabel it λ_1. That is, suppose

$$|\lambda_j| < |\lambda_1| \qquad \text{for } j = 2, \ldots, n.$$

Letting $b_k = |\lambda_1|^k$ gives

$$A^k\mathbf{x} = b_k\left[c_1\frac{\lambda_1^k}{b_1}\mathbf{v}_1 + \cdots + c_n\frac{\lambda_n^k}{b_n}\mathbf{v}_n\right].$$

As k increases, all terms except the first approach 0, so $A^k\mathbf{x}$ becomes approximately a scalar multiple of $\mathbf{v}_1$. Since

$$A(A^k\mathbf{x}) = A^{k+1}\mathbf{x},$$

$A(A^k\mathbf{x})$ and $A^k\mathbf{x}$ are both approximately scalar multiples of $\mathbf{v}_1$. Hence $A^k\mathbf{x}$ is approximately an eigenvector of A.

Under the power method, we choose an initial vector $\mathbf{x}$, often with all components 1, and compute $A\mathbf{x}$. Then write

$$A\mathbf{x} = c\mathbf{y},$$

where c is the component of $A\mathbf{x}$ that is largest in magnitude. At the next step we compute $A\mathbf{y}$ (rather than $A\mathbf{x}$), and repeat. If there exists a vector $\mathbf{v}$ such that the computed vectors approach $\mathbf{v}$ in each component, then $\mathbf{v}$ is an eigenvector of A. The corresponding eigenvalue is (approximately) c, the value factored at the last step. This eigenvalue is the dominant eigenvalue of A. We illustrate by an example.

Example 4

Let

$$L = \begin{bmatrix} 0.4 & 1.2 & 0.2 \\ 0.6 & 0 & 0 \\ 0 & 0.75 & 0 \end{bmatrix} \quad \text{and} \quad \mathbf{x} = \begin{bmatrix} 1 \\ 1 \\ 1 \end{bmatrix}.$$

Thus L is the Leslie matrix from Example 3 and $\mathbf{x}$ is the initial choice to start the iteration. Then

$$L\mathbf{x} = \begin{bmatrix} 0.4 & 1.2 & 0.2 \\ 0.6 & 0 & 0 \\ 0 & 0.75 & 0 \end{bmatrix} \begin{bmatrix} 1 \\ 1 \\ 1 \end{bmatrix} = \begin{bmatrix} 1.8 \\ 0.6 \\ 0.75 \end{bmatrix} = 1.800 \begin{bmatrix} 1 \\ 0.3333 \\ 0.4167 \end{bmatrix} = c\mathbf{y}.$$

At the next step,

$$L\mathbf{y} = \begin{bmatrix} 0.4 & 1.2 & 0.2 \\ 0.6 & 0 & 0 \\ 0 & 0.75 & 0 \end{bmatrix} \begin{bmatrix} 1 \\ 0.3333 \\ 0.4167 \end{bmatrix} = 0.8333 \begin{bmatrix} 1 \\ 0.6792 \\ 0.2830 \end{bmatrix}.$$

The results of the next three iterations are

$$1.2717 \begin{bmatrix} 1 \\ 0.4718 \\ 0.4006 \end{bmatrix}, \quad 1.0463 \begin{bmatrix} 1 \\ 0.5734 \\ 0.3382 \end{bmatrix}, \quad \text{and} \quad 1.156 \begin{bmatrix} 1 \\ 0.5191 \\ 0.3721 \end{bmatrix}.$$

For the 16th and subsequent iterations, the result is

$$1.1168 \begin{bmatrix} 1 \\ 0.5372 \\ 0.3608 \end{bmatrix}.$$

This expression displays the dominant eigenvalue of L and a corresponding eigenvector. □

Although useful for computing eigenvalues and eigenvectors, the power method does have limitations. Problems arise if the dominant eigenvalue has algebraic multiplicity greater than 1. Even if this multiplicity is 1, convergence may be very slow if the magnitudes of the two largest (in magnitude) eigenvalues are close.

In addition, the vector with each component equal to 1 may be a poor choice for starting the iteration.

In some applications, a variant of the power method may be used to find an eigenvalue other than the dominant one. For instance, the *inverse power method*, in which the power method is applied to A^{-1}, may be used to find the eigenvalue that is smallest in magnitude. Closely related is the *shifted power method*, in which a multiple of the identity is added to A before the power method is applied. If A is symmetric, we may use a technique called *deflation* to find eigenvalues and eigenvectors. These ideas are explored in the exercises.

Differential Equations (Calculus Based)

The general solution of a differential equation of the form

$$\ddot{y} = ay$$

(the dots indicate differentiation with respect to time) is:

$$y = \begin{cases} c_1 t + c_2 & \text{if} \quad a = 0 \\ c_1 e^{bt} + c_2 e^{-bt} & \text{if} \quad a = b^2 > 0 \\ c_1 \cos bt + c_2 \sin bt & \text{if} \quad a = -b^2 < 0. \end{cases}$$

The third solution is often converted to the form

$$y = A \sin(bt + c),$$

to display a *sinusoid of amplitude A*, *frequency b*, and *phase angle c*. Solutions of systems of the form

$$\begin{aligned} \ddot{y}_1 &= a_{11} y_1 + \cdots + a_{1n} y_n \\ &\vdots \\ \ddot{y}_n &= a_{n1} y_1 + \cdots + a_{nn} y_n \end{aligned}$$

are less apparent but can be obtained from the solutions above by application of eigenvalue theory.

Let

$$A = \begin{bmatrix} a_{11} & \cdots & a_{1n} \\ \vdots & & \vdots \\ a_{n1} & \cdots & a_{nn} \end{bmatrix}, \quad \mathbf{y} = \begin{bmatrix} y_1 \\ \vdots \\ y_3 \end{bmatrix}, \quad \ddot{\mathbf{y}} = \begin{bmatrix} \ddot{y}_1 \\ \vdots \\ \ddot{y}_3 \end{bmatrix}.$$

Then the system above may be written as

$$\ddot{\mathbf{y}} = A\mathbf{y}.$$

If A is diagonalizable, so that

$$P^{-1}AP = D = \operatorname{diag}(\lambda_1, \ldots, \lambda_n),$$

the system may be rewritten as

$$\ddot{\mathbf{y}} = PDP^{-1}\mathbf{y}, \qquad \text{or} \qquad P^{-1}\ddot{\mathbf{y}} = DP^{-1}\mathbf{y}.$$

Since differentiation is a linear operator, the substitution

$$\mathbf{x} = P^{-1}\mathbf{y}$$

reduces the system to the form

$$\ddot{\mathbf{x}} = D\mathbf{x}.$$

The n equations represented by this system may be solved for $x_1, \ldots, x_n$, and then $\mathbf{y}$ may be obtained from the relation

$$\mathbf{y} = P\mathbf{x}.$$

Example 5

We find the equations of motion for the spring-mass system of Fig. 7.2. By Hooke's law, the forces exerted on the mass m_1 by the first two springs are

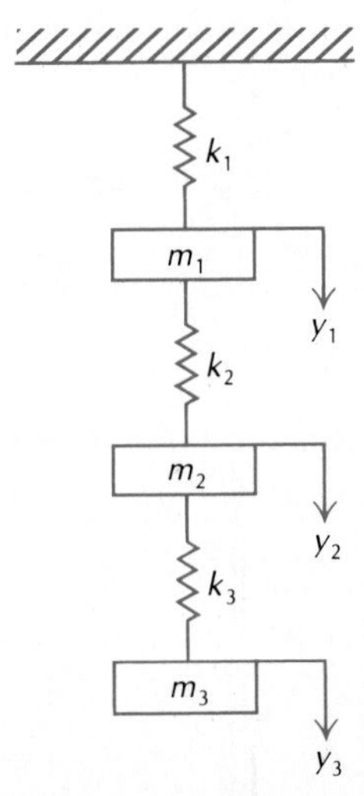

Figure 7.2 Spring-mass system of Example 5.

$-k_1y_1$ and $k_2(y_2 - y_1)$, where k_1 and k_2 are the spring constants. The terms in the remaining two equations are found in a similar way. By Newton's second law, each mass m_i times its acceleration $\ddot{y}_i$ equals the sum of the forces. If all other forces are neglected, the resulting system is

$$m_1\ddot{y}_1 = -k_1y_1 + k_2(y_2 - y_1)$$

$$m_2\ddot{y}_2 = -k_2(y_2 - y_1) + k_3(y_3 - y_2)$$

$$m_3\ddot{y}_3 = -k_3(y_3 - y_2).$$

With $m_1 = m_2 = m_3 = 2$ and $k_1 = k_3 = 2$, $k_2 = 4$, the system simplifies to

$$\ddot{\mathbf{y}} = A\mathbf{y},$$

where

$$A = \begin{bmatrix} -3 & 2 & 0 \\ 2 & -3 & 1 \\ 0 & 1 & -1 \end{bmatrix}.$$

The eigenvalues are -5.1249, -1.6367, -0.2384, and a corresponding matrix of eigenvectors

$$P = \begin{bmatrix} 0.6749 & 0.6189 & 0.4018 \\ -0.7171 & 0.4219 & 0.5548 \\ 0.1738 & -0.6626 & 0.7285 \end{bmatrix}.$$

As noted above, the substitution $\mathbf{x} = P^{-1}\mathbf{y}$ reduces the system to the form $\ddot{\mathbf{x}} = D\mathbf{x}$, or

$$\begin{bmatrix} \ddot{x}_1 \\ \ddot{x}_2 \\ \ddot{x}_3 \end{bmatrix} = \begin{bmatrix} -5.1249 & 0 & 0 \\ 0 & -1.6367 & 0 \\ 0 & 0 & -0.2384 \end{bmatrix} \begin{bmatrix} x_1 \\ x_2 \\ x_3 \end{bmatrix} = \begin{bmatrix} -5.1249x_1 \\ -1.6367x_2 \\ -0.2384x_3 \end{bmatrix}.$$

This vector equation is equivalent to the three equations

$$\ddot{x}_1 + 5.1249x_1 = 0, \qquad \ddot{x}_2 = 1.6367x_2 = 0, \qquad \ddot{x}_3 + 0.2384x_3 = 0.$$

We solve these equations to obtain

$$x_i(t) = A_i\sin(b_it + c_i) \qquad \text{for } i = 1, 2, 3,$$

where

$$b_1 = 2.26, \qquad b_2 = 1.28, \qquad b_3 = 0.49,$$

to two places. The constants A_i and c_i in $x_i(t)$ depend on the initial displacement and velocity of the ith mass. Then with

$$\mathbf{x} = \begin{bmatrix} x_1(t) \\ x_2(t) \\ x_3(t) \end{bmatrix},$$

the general solution for $\mathbf{y}$ is given by

$$\mathbf{y} = P\mathbf{x} = \begin{bmatrix} 0.67 & 0.62 & 0.40 \\ -0.72 & 0.42 & 0.55 \\ 0.17 & -0.66 & 0.73 \end{bmatrix} \begin{bmatrix} x_1(t) \\ x_2(t) \\ x_3(t) \end{bmatrix}.$$

By the *column interpretation* of this product,

$$\mathbf{y} = x_1(t)\begin{bmatrix} 0.67 \\ -0.72 \\ 0.17 \end{bmatrix} + x_2(t)\begin{bmatrix} 0.62 \\ 0.42 \\ -0.66 \end{bmatrix} + x_3(t)\begin{bmatrix} 0.40 \\ 0.55 \\ 0.73 \end{bmatrix}.$$

The vectors on the right determine the **normal modes** of **oscillation**. The mode for the lowest frequency is called the **fundamental mode**. As the expression for $\mathbf{y}$ indicates, the most general motion possible is a superposition (sum) of the three normal modes. □

The examples in this section convey only a hint of the utility of eigenvalues and eigenvectors. Many practical applications involve large matrices and require advanced numerical methods. One excellent method for finding eigenvalues of a general matrix is the "QR algorithm." The diskette accompanying this book contains a program for implementing this method.

REVIEW CHECKLIST

1. Find the matrix of a given quadratic form, and conversely, find the quadratic form corresponding to a given symmetric matrix.
2. Reduce a given quadratic form $Q(x)$ to one with square terms only, i.e., with cross terms removed, and identify the curve determined by $Q(\mathbf{x}) = c$, where c is a constant, in the plane or in space.
3. For a given Leslie matrix, find the eigenvalues and eigenvectors, and use them to describe steady-state population distribution and growth.
4. Describe and apply the power method and the inverse power method for finding the largest and the smallest eigenvalues of a matrix and their corresponding eigenvectors.
5. Find the normal modes of oscillation of a vibrating system, and describe the most general motion in terms of the normal modes.

EXERCISES

In Exercises 1–7, transform the given expression to a quadratic form with no "cross-terms" $x_i x_j$, $i \neq j$. Identify the curve or surface, and for Exercises 1–4, sketch.

1. $2x_1^2 + 4x_1x_2 + 5x_2^2 = 24$
2. $x_1^2 + 4x_1x_2 + 4x_2^2 = 25$
3. $x_1^2 + x_1x_2 + x_2^2 = 18$
4. $x_1^2 + 12x_1x_2 + 6x_2^2 = 15$
5. $x_1^2 - 8x_1x_2 + 7x_2^2 = 36$
6. $2x_1^2 + 2x_1x_2 + 2x_2^2 + 2x_1x_3 + 2x_2x_3 + 2x_3^2 = 16$
7. $x_1^2 - 2x_1x_2 + 3x_2^2 - 2x_1x_3 + 2x_2x_3 + x_3^2 = 36$
8. For a Leslie model with the female population partitioned into four age groups, use the method in the discussion preceding Example 3 to find a formula for the eigenvector once the dominant eigenvalue $\lambda = r$ is known.

In Exercises 9–11, find the steady-state age-class distribution and growth rate for the given Leslie matrix.

9. $\begin{bmatrix} 0.3 & 1.2 & 0.9 \\ 0.7 & 0 & 0 \\ 0 & 0.6 & 0 \end{bmatrix}$

10. $\begin{bmatrix} 0.2 & 0.8 & 0.9 & 0.4 \\ 0.6 & 0 & 0 & 0 \\ 0 & 0.6 & 0 & 0 \\ 0 & 0 & 0.5 & 0 \end{bmatrix}$

11. $\begin{bmatrix} 0.1 & 0.2 & 0.7 & 0.6 & 0.2 & 0.0 \\ 0.8 & 0 & 0 & 0 & 0 & 0 \\ 0 & 0.9 & 0 & 0 & 0 & 0 \\ 0 & 0 & 0.9 & 0 & 0 & 0 \\ 0 & 0 & 0 & 0.7 & 0 & 0 \\ 0 & 0 & 0 & 0 & 0.3 & 0 \end{bmatrix}$

In Exercises 12–19, apply the power method for the dominant eigenvalue and corresponding eigenvector, carrying out at least five iterations.

12. $\begin{bmatrix} 5 & 4 \\ 4 & -1 \end{bmatrix}$

13. $\begin{bmatrix} 7 & 3 \\ 3 & -1 \end{bmatrix}$

14. $\begin{bmatrix} 8 & -3 \\ 2 & 3 \end{bmatrix}$

15. $\begin{bmatrix} 2 & -1 & -1 \\ 1 & 0 & -1 \\ 1 & -1 & 2 \end{bmatrix}$

16. $\begin{bmatrix} 3 & 1 & 3 \\ -2 & 4 & 1 \\ 0 & -2 & -1 \end{bmatrix}$

17. $\begin{bmatrix} 1 & -3 & -3 \\ 0 & -2 & -3 \\ 1 & 0 & 2 \end{bmatrix}$

18. $\begin{bmatrix} 2 & 1 & -2 \\ 3 & 2 & 3 \\ -1 & 3 & 3 \end{bmatrix}$

19. $\begin{bmatrix} 2 & 2 & 1 \\ 1 & 2 & 1 \\ -2 & -2 & -1 \end{bmatrix}$

20. If I denotes the 3×3 identity, subtract $3I$ from the matrix of Exercise 18, apply the power method to the resulting matrix, and add 3 to the resulting eigenvalue to obtain an approximation to the smallest eigenvalue of the original matrix. (This technique is often called the *shifting method*.)
21. Apply the shifting method to the matrix in Exercise 19, shifting the matrix by $-2I$ and the resulting eigenvalue by 2 (see Exercise 20).

22. By considering the defining relation

$$(\lambda I - A)\mathbf{x} = \mathbf{0},$$

show that λ is an eigenvalue of A with corresponding eigenvector $\mathbf{x}$ if and only if $\lambda - c$ is an eigenvalue of $A - cI$ with corresponding eigenvector $\mathbf{x}$, thereby providing rationale for the shifting method.

23. Apply the power method to the inverse of the matrix in Exercise 14 to obtain an approximation to the smallest eigenvalue and corresponding eigenvector of the given matrix. (This technique is called the *inverse power method.*)

24. Use the defining relation

$$(\lambda I - A)\mathbf{x} = \mathbf{0}$$

to supply rationale for the inverse power method.

In Exercises 25–27, A is a real symmetric $n \times n$ matrix with eigenvalues $\lambda_1, \ldots, \lambda_n$ and corresponding eigenvectors $\mathbf{v}_1, \ldots, \mathbf{v}_n$ forming an orthonormal basis of $\mathbb{R}^n$, and

$$B = A - \lambda_1 \mathbf{v}_1 \mathbf{v}_1^t.$$

These exercises outline and justify the method of deflation, which is illustrated in Exercise 28.

25. Show that B is symmetric.

26. Show that the eigenvalues of B are $0, \lambda_2, \ldots, \lambda_n$, with corresponding eigenvectors $\mathbf{v}_1, \mathbf{v}_2, \ldots, \mathbf{v}_n$.

27. If A has n nonzero eigenvalues of distinct magnitudes, show that they and their corresponding eigenvectors may be found by repeated application of the power method.

28. Given the dominant eigenvalue $\lambda_1 = 6$ and corresponding eigenvector $[1 \quad -1 \quad -1]^t$ of

$$A = \begin{bmatrix} 3 & -1 & -2 \\ -1 & 4 & 1 \\ -2 & 1 & 3 \end{bmatrix},$$

apply the deflation method to find the second largest eigenvalue and corresponding eigenvector.

In Exercises 29–32 (calculus based), find the normal and fundamental modes of vibration of the given spring-mass system, and describe the most general motion possible. Use Fig. 7.2 (page 388) for Exercises 31 and 32.

29. $m_1 = m_2 = 2;\ k_1 = k_2 = 2$

30. $m_1 = 2,\ m_2 = 4;\ k_1 = k_2 = 2$

31. $m_1 = m_2 = m_3 = 1;\ k_1 = k_2 = k_3 = 2$

32. $m_1 = 1,\ m_2 = 2,\ m_3 = 2;\ k_1 = k_2 = 1,\ k_3 = 2$

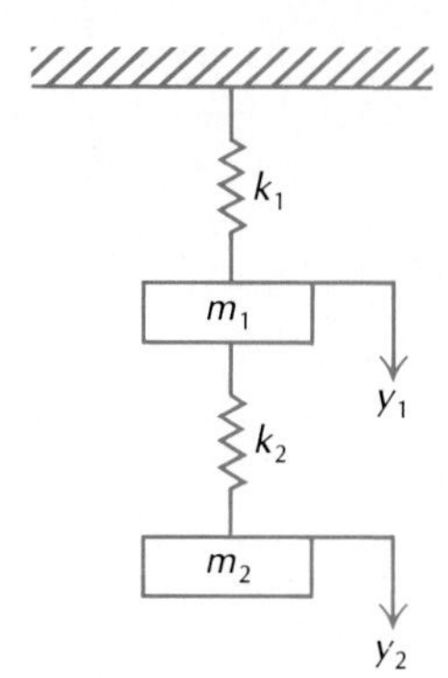

Spring-mass system for Exercises 29 and 30.

REVIEW EXERCISES

In Exercises 1–2, find the characteristic polynomial of the given matrix.

1. $\begin{bmatrix} 2 & 5 \\ -3 & -4 \end{bmatrix}$ **2.** $\begin{bmatrix} 1 & -2 & 0 \\ 1 & 1 & 2 \\ -1 & 0 & -1 \end{bmatrix}$

In Exercises 3–12, find all eigenvalues and eigenvectors of the given matrix, and determine whether the matrix is diagonalizable.

3. $\begin{bmatrix} 1 & -2 \\ 4 & -5 \end{bmatrix}$ **4.** $\begin{bmatrix} 1 & -1 \\ 1 & 3 \end{bmatrix}$ **5.** $\begin{bmatrix} 2 & -4 \\ 2 & -2 \end{bmatrix}$ **6.** $\begin{bmatrix} 2 & 0 \\ 0 & 2 \end{bmatrix}$

7. $\begin{bmatrix} 1 & -2 & 3 \\ -1 & 1 & 2 \\ -1 & 0 & 3 \end{bmatrix}$ **8.** $\begin{bmatrix} 1 & -3 & 0 \\ -1 & 1 & -2 \\ -1 & 0 & -1 \end{bmatrix}$ **9.** $\begin{bmatrix} 1 & -3 & 0 \\ 1 & -2 & 1 \\ 1 & 0 & 2 \end{bmatrix}$

10. $\begin{bmatrix} 3 & 0 & 0 \\ 0 & 3 & 0 \\ 0 & 0 & 3 \end{bmatrix}$ **11.** $\begin{bmatrix} 1 & 2 & 2 \\ -1 & 1 & 2 \\ 1 & -1 & -2 \end{bmatrix}$ **12.** $\begin{bmatrix} 2 & -2 & 1 \\ 2 & -3 & 2 \\ -1 & 2 & 0 \end{bmatrix}$

In Exercises 13–15, find the eigenvalues of the given matrix, and find an orthonormal basis for $\mathbb{R}^3$ consisting of eigenvectors of the matrix.

13. $\begin{bmatrix} 4 & 1 & 2 \\ 1 & 3 & -1 \\ 2 & -1 & 4 \end{bmatrix}$ **14.** $\begin{bmatrix} 2 & -1 & -2 \\ -1 & 2 & 2 \\ -2 & 2 & 5 \end{bmatrix}$ **15.** $\begin{bmatrix} 3 & -1 & -2 \\ -1 & 3 & 2 \\ -2 & 2 & 6 \end{bmatrix}$

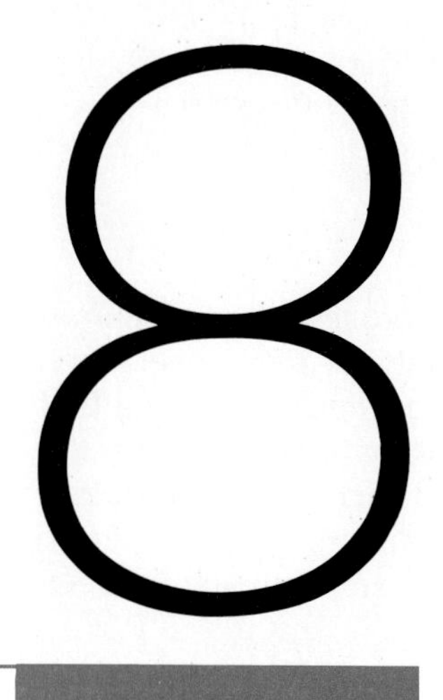

Linear Programming

We preface the presentation on linear programming by a note on our goals. This entire chapter is optional, and not an integral part of linear algebra. We describe the mathematical problem involved, indicate some application areas, provide some geometric insight, and describe an algorithm that shows how linear algebra is applied in solving the problem. The aim is a modest introduction to the subject, hopefully to inspire further study. References are provided at the end of Section 8.3 for readers who wish to learn more about linear programming or to begin with a more comprehensive treatment of the subject.

We use linear programming *in resource allocation problems to determine how to use limited assets to maximize gain or to minimize loss. Reasons for the spectacular growth of this field since World War II include an increased need to make optimal use of scarce resources, the availability of computers for solving large-scale problems, and the discovery of the* simplex algorithm *by George B. Dantzig in 1947. Applications abound in manufacturing, business, finance, agriculture, scheduling, medicine, transportation, communications, shipping, natural resource management, and engineering design. In addition to its practical utility, linear programming has been a fertile area for theoretical research, with new and exciting results on algorithms and computational complexity continuing to emerge.*

(o) 8.1 FORMULATION AND GEOMETRY OF A LINEAR PROGRAM

In this section we formulate the basic problem of linear programming, interpret it geometrically, and illustrate its application by examples.

Problem Formulation

Typically in linear programming, the mathematical model formulated directly from physical considerations is *not* a linear system, so the problems studied here are different from those we studied earlier. We begin with a simplified physical application to illustrate the type of mathematical problem to be solved.

Example 1

A firm manufactures products 1 and 2, each of which is processed on machines *A*, *B*, and *C*. For instance, the products may be distinctly shaped wheel axles, each of which is turned on a lathe and smoothed on a grinder, with lug bolt holes drilled and tapped on a drill press. The following table contains the machine processing times for each unit of product 1 and each unit of product 2, and the availability of each machine in an 8-hour work shift. All times are in hours. The table also gives the gross profit in dollars per unit for each product.

	machine A	**machine B**	**machine C**	**gross**
product 1	1/10	1/12	3/20	$55.00
product 2	1/10	1/6	1/20	$50.00
availability	5	7	6	—

Formulate a problem for the number of units of each product the firm should produce in each shift in order to maximize the gross.

Solution

Let x_1 and x_2 denote the number of units of products 1 and 2, respectively, to be produced in each 8-hour shift. The total gross profit in dollars per work shift, P, can be determined from the table. Since the gross profit for each unit of product 1 is $55 and x_1 units are produced, the total gross from this product is $55x_1$ dollars. Similarly, the gross profit from product 2 is $50x_2$ dollars. The grand total is the sum of these two, i.e.,

$$P = 55x_1 + 50x_2.$$

The machine availability *constraints* can be determined from the table in a similar way. For instance, the two products require a total of $x_1/10 + x_2/10$ hours on machine A, and this sum cannot exceed 5 hours. The respective inequality constraints for machines A, B, and C are

$$x_1/10 + x_2/10 \leq 5, \qquad x_1/12 + x_2/6 \leq 7, \qquad \text{and} \qquad 3x_1/20 + x_2/20 \leq 6,$$

each in hours. Taking into account that x_1 and x_2 must be nonnegative gives the following mathematical problem.

Maximize the function

$$P = 55x_1 + 50x_2$$

subject to the constraints

$$x_1 + 2x_2 \leq 84$$

$$x_1 + x_2 \leq 50$$

$$3x_1 + x_2 \leq 120$$

and the nonnegativity conditions

$$x_1 \geq 0 \qquad x_2 \geq 0. \quad \square$$

The problem formulated in Example 1 is one form of a *linear programming problem*, which is abbreviated to *linear program*, or simply LP. Linear programs may appear in other forms, for instance with the entire orientation reversed, as is illustrated next.

Example 2

A diet-conscious person wishes to derive at least 16 grams of protein (1/3 the Recommended Daily Allowance, or RDA) and .5 g. of calcium (1/2 the RDA) daily from milk and yogurt, but also wishes to minimize the number of calories consumed in the process. The protein, calcium, and calorie content of each of these products is given in the following table.

	protein	**calcium**	**calories**
1 oz. yogurt	1.2 g.	0.032 g.	14
1 oz. milk	1.0 g.	0.036 g.	15

Formulate a linear program to determine the daily amounts of milk and yogurt this person should consume.

Solution

Let x_1 and x_2 be the number of ounces of milk and yogurt, respectively, to be consumed each day. The number of calories consumed each day can be determined from the table. Since each ounce of yogurt has 14 calories and x_1 ounces are consumed, the individual takes in $14x_1$ calories with the yogurt each day. In a similar way, $15x_2$ calories are consumed with the milk each day, so the number of calories from these foods is given by

$$C = 14x_1 + 15x_2.$$

The nutrition requirement *constraints* can be determined from the table in a similar way. For instance, the two foods provide a total of $1.2x_1 + x_2$ grams of protein, and this sum must be at least 16 grams. The inequality constraints corresponding to the requirements for protein and calcium are

$$1.2x_1 + x_2 \geq 16 \quad \text{and} \quad .032x_1 + .036x_2 \geq .5,$$

respectively. Since x_1 and x_2 must both be nonnegative, the linear program may be formulated as follows.

Minimize

$$C = 14x_1 + 15x_2$$

subject to

$$6x_1 + 5x_2 \geq 80$$

$$32x_1 + 36x_2 \geq 500$$

and

$$x_1 \geq 0 \quad x_2 \geq 0. \quad \square$$

The function P in Example 1 is called a *payoff*, and C in Example 2 is called a *penalty*. The term *objective function* refers to either a payoff or a penalty. To *optimize* the objective function means to maximize or to minimize it, whichever is appropriate. We define these terms formally and describe a general linear program in Section 8.2. Meanwhile, we assume an LP involves optimizing a linear objective function subject to a set of *linear inequality constraints* and nonnegativity conditions, as illustrated by Examples 1 and 2.

Geometric Interpretation

Linear programs with two variables x_1 and x_2 may be pictured in the Cartesian plane. As indicated in Fig. 8.1, any linear equation of the form

$$ax_1 + bx_2 = c$$

determines two **closed half-planes**, one consisting of all points (x_1, x_2) such that

$$ax_1 + bx_2 \le c,$$

the other all points (x_1, x_2) for which

$$ax_1 + bx_2 \ge c.$$

Each value P_0 assumed by an objective function of the form

$$f(x_1, x_2) = px_1 + qx_2 + r$$

determines a line called a **level line**. The level line consists of all points (x_1, x_2) for which $f(x_1, x_2) = P_0$. Some level lines of the payoff function in Example 1 are shown in Fig. 8.2.

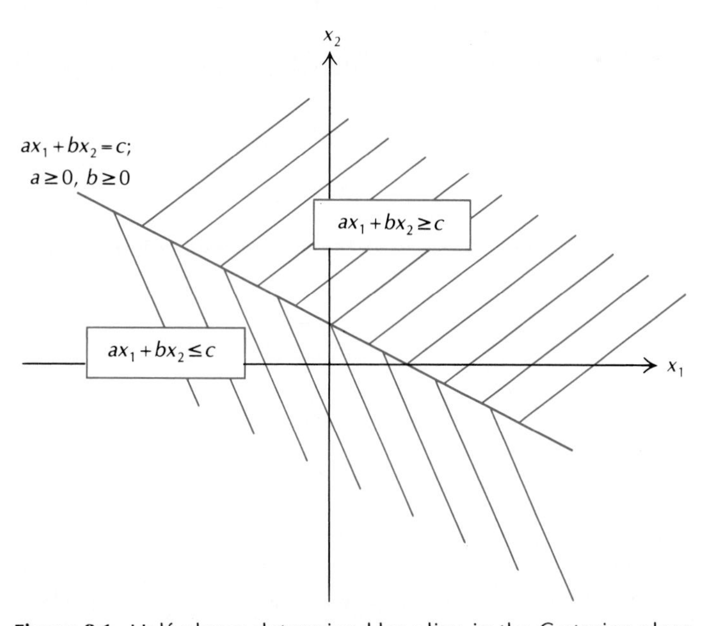

Figure 8.1 Half-planes determined by a line in the Cartesian plane.

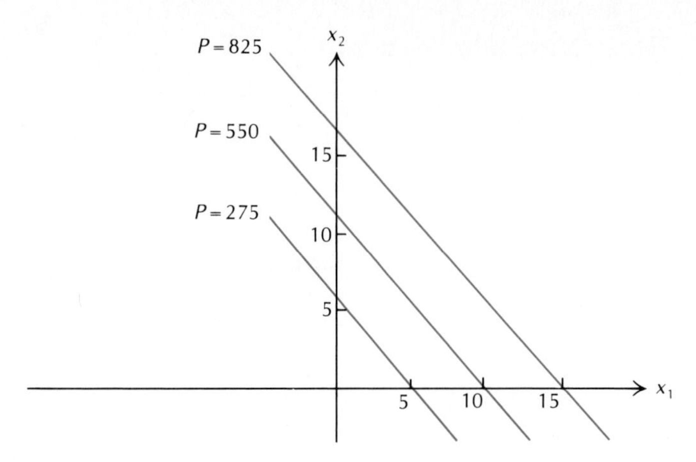

Figure 8.2 Some level lines of the payoff function $P = 55x_1 + 50x_2$ in Example 1.

Example 3

Solve the LP from Example 1 by using the geometric interpretation of linear inequality constraints and the level lines of the linear objective function.

Solution

Figure 8.3a shows the set of all points (x_1, x_2) that satisfy the three inequality constraints and the conditions $x_1 \geq 0$ and $x_2 \geq 0$ of Example 1. Each constraint is satisfied by the points in a *half-plane*, and the set of points satisfying all three constraints is the intersection of these half-planes.

Figure 8.3b shows the solution of the LP. The graph shows several level lines of the objective function. The level lines are mutually parallel. Each level line corresponds to a value of P and consists of all points (x_1, x_2) such that

$$55x_1 + 50x_2 = P.$$

The value of $P/50$ is the x_2-intercept of the level line.

The point (35, 15) satisfies the inequality constraints and produces a gross profit $P = 2675$. This point is the intersection of the lines determined by the *equations*

$$x_1 + x_2 = 50 \quad \text{and} \quad 3x_1 + x_2 = 120.$$

Any other point satisfying the inequality constraints lies on a level line corre-

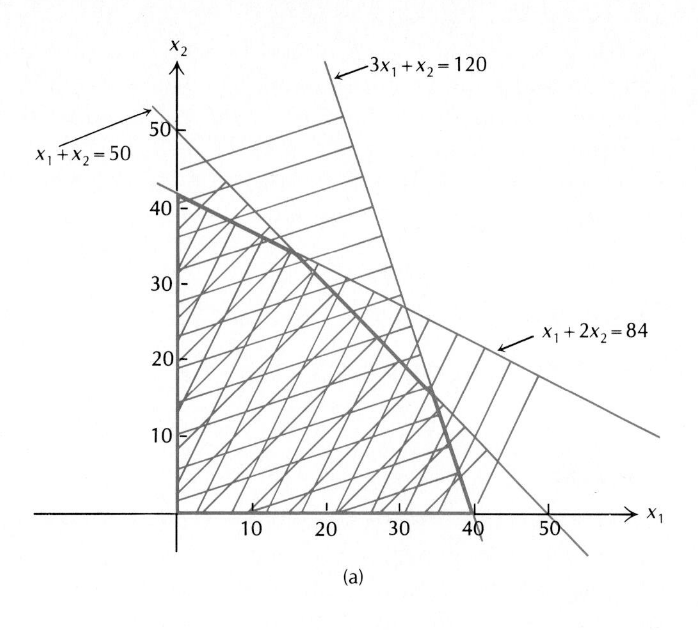

(a)

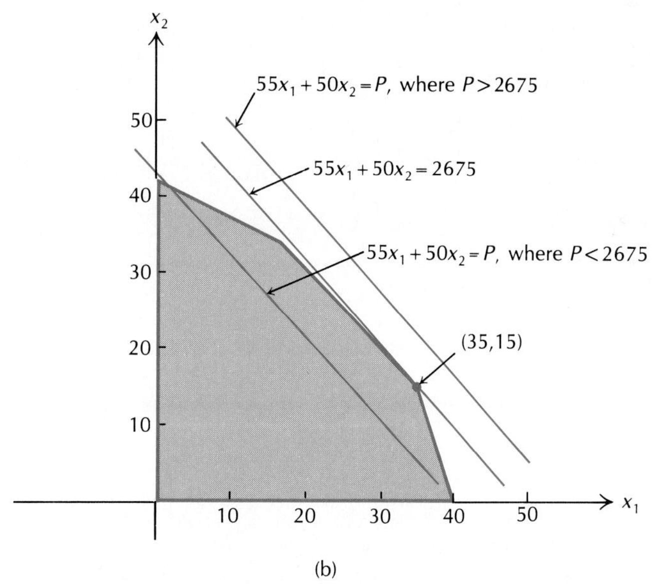

(b)

Figure 8.3 (a) The set of points satisfying the inequality constraints and the nonnegative conditions in Example 1. (b) The maximum of the objective function for these points is attained at the point (35, 15).

sponding to a value $P_1 < 2675$ and hence yields a smaller gross. Moreover, no point on a level line corresponding to a value $P > 2675$ satisfies the inequality constraints. Thus the maximum value of the objective function for points satisfying the inequality constraints is 2675, and this value is attained when $x_1 = 35$ and $x_2 = 15$. That is, the maximum gross profit, $2675, is attained for 35 units of product 1 and 15 of product 2 per shift. □

Example 4

Solve the LP from Example 2 geometrically.

Solution

Figure 8.4a shows the set of all points (x_1, x_2) satisfying the inequality constraints and the inequality conditions $x_1 \geq 0$ and $x_2 \geq 0$ of Example 2. Each constraint corresponds to a half-plane, and the points satisfying both inequality constraints constitute the intersection of these half-planes.

Figure 8.4b shows the solution of the LP. The graph shows several level lines of the objective function. Each level line corresponds to a value of C and consists of all points (x_1, y_1) for which

$$14x_1 + 15x_2 = C.$$

The lines bounding the inequality constraint regions,

$$6x_1 + 5x_2 = 80 \quad \text{and} \quad 32x_1 + 36x_2 = 500,$$

intersect at the point $(95/14, 55/7)$. This point satisfies both inequality constraints and corresponds to $C = 1490/7$. Any other point satisfying both inequality constraints lies on a level line corresponding to a value $C_2 > 1490/7$. Moreover, any point on a level line corresponding to a value $C_1 < 1490/7$ cannot satisfy both inequality constraints. Thus the minimum value of the objective function for points satisfying both inequality constraints is $1490/7$, and it is attained when $x_1 = 95/14$ and $x_2 = 55/7$. The calorie counter from Example 2 will likely use integer approximations, minimizing the daily intake of calories at 218 while meeting the protein and calcium goals by drinking 7 ounces of milk and eating 8 ounces of yogurt each day. □

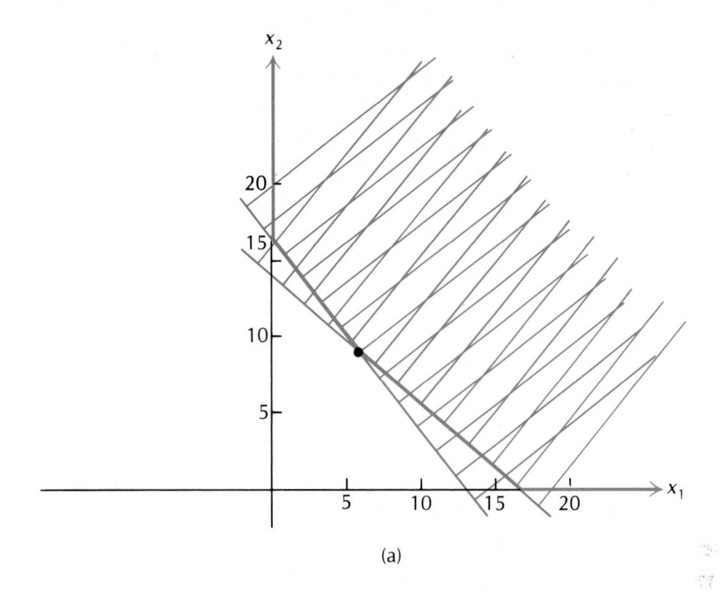

(a)

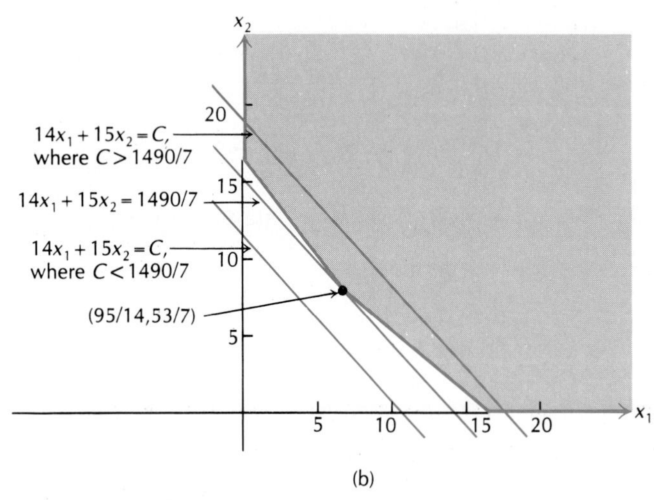

(b)

Figure 8.4 (a) The set of points satisfying the inequality constraints and the nonnegative conditions in Example 2. (b) The minimum of the objective function for these points is attained at the point $(95/14, 55/7)$.

As Examples 3 and 4 indicate, the nature of the set of points satisfying the inequality constraints is important in finding optimizing points in a linear program. Such sets are called *feasible sets* and their points *feasible points*. Before formulating the general LP, we develop additional insight by exploring feasible sets and their properties in the plane.

Two concepts related to sets in $\mathbb{R}^n$ are particularly important in linear programming. Rather than giving a general definition, we concentrate on $\mathbb{R}^2$ and

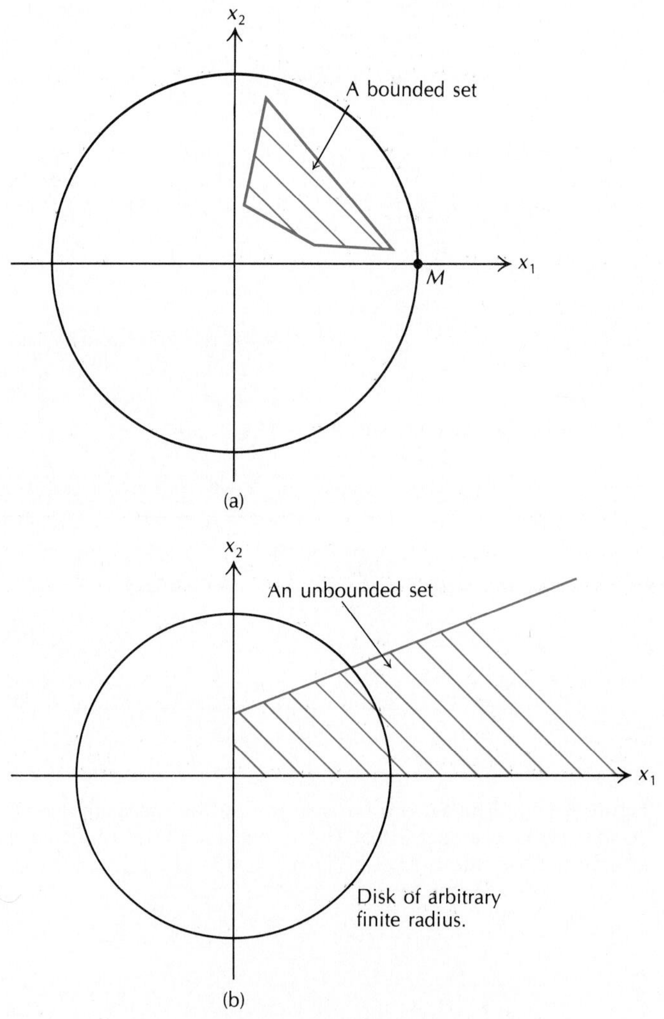

Figure 8.5 (a) A bounded set: it is contained in a disk of finite radius M. (b) An unbounded set: it is not contained in any disk of finite radius.

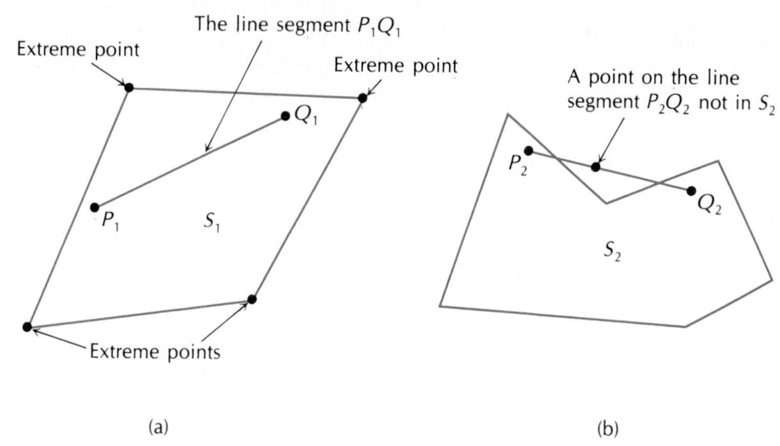

Figure 8.6 (a) The set S_1 is convex: the line segment between arbitrary points P_1 and Q_1 in S_1 is contained in S_1. This convex set has four extreme points. (b) The set S_2 is not convex: the points P_2 and Q_2 are in S_2 but the line segment between them is not contained in S_2.

formulate the concepts in terms of points in the Cartesian plane. A set S in the plane is said to be **bounded** if there exists a real number M such that S is contained in a disk of radius M centered at the origin. The feasible set in Example 1 is bounded, while the one in Example 2 is not. The concept is illustrated in Fig. 8.5.

Another concept used in linear programming is *convexity*. A plane set S is said to be **convex** if for every two points P and Q in S, the entire line segment PQ is contained in S. A point P is an **extreme point** of a convex set S in the plane if it does not lie strictly between two other points of S. These definitions are illustrated in Fig. 8.6.

Convexity is relevant because feasible sets are convex. A half-plane is a convex set and the intersection of a finite number of convex sets is convex, so any feasible set that coincides with the intersection of a finite number of half-planes is convex. It is also known that any real linear function that assumes a maximum or a minimum on a convex set does so at an extreme point. Thus in solving an LP, we look for solutions at the extreme points, or the "corners" of the feasible region.

Example 5

Find the maximum and the minimum of

$$f(x_1, x_2) = 2x_1 + x_2 + 1$$

subject to

$$x_1 + 3y_1 \geq 11$$

$$4x_1 + x_2 \geq 11$$

$$5x_1 + 4x_2 \leq 33$$

and

$$x_1 \geq 0, \qquad x_2 \geq 0.$$

Solution

The solution of this LP is shown in Fig. 8.7. The feasible points constitute the bounded and convex set with extreme points (1, 7), (2, 3), (5, 2). The desired minimum value of 8 is attained at (2, 3), the maximum 13 at (5, 2). ☐

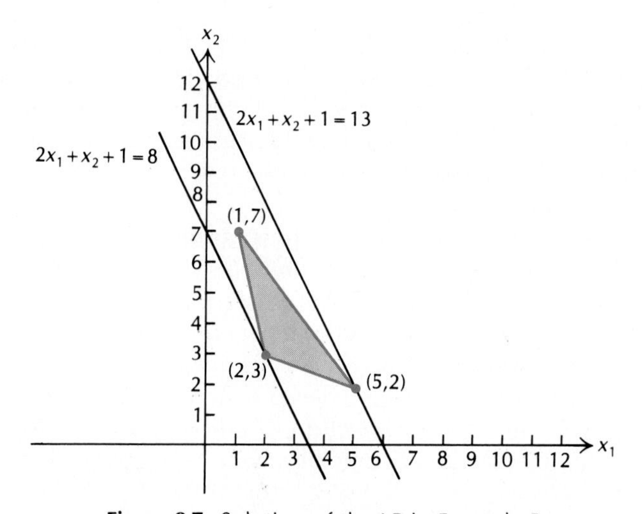

Figure 8.7 Solution of the LP in Example 5.

Example 6

Find all maxima and minima of

$$f(x_1, x_2) = 2x_1 + x_2$$

subject to

$$x_1 + x_2 \geq 20$$

and

$$x_1 \geq 0, \qquad x_2 \geq 0.$$

Solution

The solution is shown in Fig. 8.8. The feasible set is unbounded, and its extreme points are (20, 0), (0, 20). The minimum value of f on the feasible set, 20, is attained at (0, 20). The function f has no maximum on the feasible set. ☐

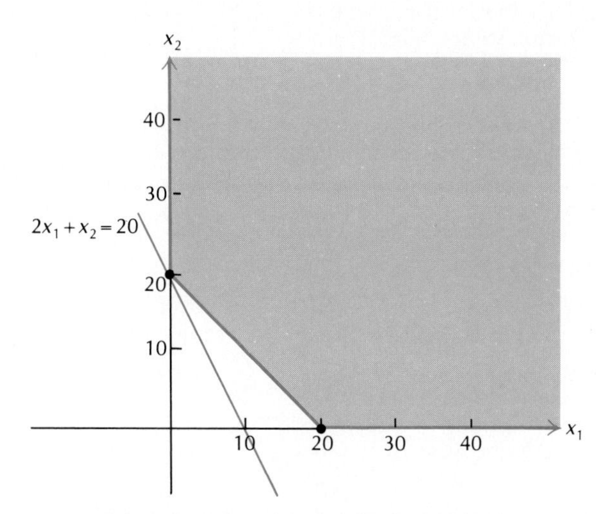

Figure 8.8 Solution of the LP in Example 6.

REVIEW CHECKLIST

1. Formulate linear programs from physical specifications.
2. Find feasible sets from inequality constraints.
3. Solve linear programs in the plane geometrically.
4. Define bounded set, unbounded set, and convex set.
5. Determine whether a given set is bounded and/or convex.
6. Identify feasible sets of linear programs as convex.

EXERCISES

In Exercises 1–4, sketch the region in the first quadrant determined by the given inequalities.

1. $x_1 + 2x_2 \geq 6$

2. $3x_1 + x_2 \geq 3$
$2x_1 + 3x_2 \leq 12$

3. $x_1 + x_2 \geq 4$
$x_1 - x_2 \leq 1$
$-x_1 + 4x_2 \leq 20$

4. $-3x_1 + 2x_2 \leq 6$
$-3x_1 + 2x_2 \geq -6$
$3x_1 + 2x_2 \leq 6$
$3x_1 + 2x_2 \geq 18$

Exercises 5–8. Find the extreme points of the convex sets in Exercises 1–4.

9. Show that the set determined in Exercise 2 is bounded by finding a disk of radius M centered at the origin and containing the set.

10. Sketch the level lines of the function $f(x_1, x_2) = 3x_1 + 2x_2$ corresponding to the function values 1, 2, 3, 4, 5.

In Exercises 11–18, find the maximum and the minimum of the function subject to the inequality constraints, assuming also the conditions $x_1 \geq 0$, $x_2 \geq 0$.

11. $f(x_1, x_2) = x_1 + x_2$;
$x_1 + 3x_2 \leq 15$
$4x_1 + 3x_2 \leq 24$

12. $f(x_1, x_2) = x_1 + 5x_2$; same constraints as Exercise 11

13. $f(x_1, x_2) = 3x_1 + x_2$; same constraints as Exercise 11

14. $f(x_1, x_2) = -x_1 + x_2$;
$-x_1 + 5x_2 \leq 10$

15. $f(x_1, x_2) = 4x_1 + 3x_2 - 1$
$7x_1 + x_2 \geq 14$
$2x_1 - x_2 \leq 10$
$x_1 + 2x_2 \leq 15$

16. $f(x_1, x_2) = 2x_1 + 6x_2$; same constraints as Exercise 15

17. $f(x_1, x_2) = 4x_1 - 4x_2$; same constraints as Exercise 15

18. $f(x_1, x_2) = 8x_1 + x_2$; same constraints as Exercise 15

In Exercises 19–22, formulate a linear program modeling the given physical problem, and solve geometrically.

19. A winery sells two blended wines to restaurants for "house" wines. Both blends are made from varietal wines A and B, with blending percentages, availability, and gross profits given in the following table.

	varietal A	**varietal B**	**gross**
blend 1	30%	70%	\$1.20/gal.
blend 2	60%	40%	\$1.00/gal.
availability	2000 gal.	3000 gal.	

Find the number of gallons of each blend the winery should produce to maximize the gross profit from them.

20. An investor has selected two securities for dividends and capital growth. This individual uses "risk factors" to account for relative risk—each investment is multiplied by the corresponding factor, and the total risk R is the sum of these two products. The dividend percentages, capital growth percentages, and risk factors are given in the following table.

	dividend	growth	risk factor
security 1	4%	12%	3
security 2	8%	7%	2

If the total amount to be invested is not to exceed $25,000, find the number of dollars to invest in these securities to minimize the total risk R while producing a dividend of at least $1200 and a capital growth of at least $2000.

21. One department in a furniture manufacturing firm handles assembly, sanding, staining, and finishing two types of wooden chairs. The number of hours each chair ties up these facilities, the gross profit per chair, and the number of hours per week the facilities are available are given in the following table.

	assembly	sanding	staining	finishing	profit
chair 1	1.00	2.00	1.50	2.50	$60
chair 2	1.25	2.25	1.75	3.00	$70
availability	15.00	24.00	30.00	40.00	—

Find the number of chairs of each type to be made each week to maximize gross profit while meeting the availability constraints.

22. A diet-conscious person wishes to prepare a dish of crabmeat and shad. Find the relative amounts of each to be used in the dish to maximize the total protein while keeping the calorie count under 300 and the total cholesterol under 0.25 grams, if the protein and cholesterol amounts and the calorie counts are given in the following table.

	protein	calories	cholesterol
1 oz. crabmeat	5 g.	25	0.04 g.
1 oz. shad	6 g.	62	0.02 g.

(o) 8.2 THE SIMPLEX METHOD

The linear programs in Section 8.1 involved only two variables, x_1 and x_2, the *decision variables*. In general, linear programs arising in practice may involve many decision variables. For instance, a feed producer may use twenty ingredients in a mix, an investor may maintain fifty positions in a portfolio, and an electronic circuit designer may work with a hundred component parameters in a sensitivity analysis. Large linear programs are usually solved by the *simplex method.* In this section, we describe the simplex method and illustrate its application to one particular form of a linear program.

The First Primal Problem and Its Standard Form

The type of LP defined next includes the problem of Example 1 in Section 8.1.

The First Primal Problem

Maximize an objective function

$$c_1x_1 + \cdots + c_nx_n$$

subject to the inequality constraints

$$\begin{array}{ccccc} a_{11}x_1 & + \cdots + & a_{1n}x_n & \leq & b_1 \\ \vdots & & \vdots & & \vdots \\ a_{m1}x_1 & + \cdots + & a_{mn}x_n & \leq & b_m \end{array}$$

and nonnegativity conditions

$$x_j \geq 0 \qquad \text{for } j = 1, \ldots, n.$$

The first primal form is adequate for many applications, but by no means all. We defer questions about other forms to Section 8.3 in order to reach our main objective, the simplex algorithm, as quickly as possible.

The first step in solving a linear program in first primal form is to convert each *inequality constraint*

$$a_{i1}x_1 + \cdots + a_{in}x_n \leq b_i$$

to an *equality constraint* by introducing a new variable:

$$x_{n+i} = b_i - a_{i1}x_1 - \cdots - a_{in}x_n.$$

These new variables are *nonnegative* and are called **slack variables**, because they "take up the slack" in the inequality constraints. The objective function may be regarded as a linear form in $m + n$ variables, with the coefficients of the m slack variables zero. We identify the form of an LP obtained this way.

Standard Form of the First Primal Problem

Maximize the objective function

$$c_1x_1 + \cdots + c_nx_n + 0x_{n+1} + \cdots + 0x_{n+m}$$

subject to the *equality* constraints

$$\begin{matrix} a_{11}x_1 + \cdots + a_{1n}x_n + x_{n+1} + 0x_{n+2} + \cdots + 0x_{n+m} = b_1 \\ a_{21}x_1 + \cdots + a_{2n}x_n + 0x_{n+1} + x_{n+2} + \cdots + 0x_{n+m} = b_2 \\ \vdots \\ a_{m1}x_1 + \cdots + a_{mn}x_n + 0x_{n+1} + 0x_{n+2} + \cdots + x_{n+m} = b_m \end{matrix}$$

and nonnegativity conditions

$$x_j \geq 0 \qquad \text{for } j = 1, \ldots, m + n.$$

Example 1

Write the following primal problem in standard form.

Maximize

$$2x_1 + 3x_2 - x_3$$

subject to

$$x_1 + 4x_2 - 3x_3 \leq 2$$

$$2x_1 - 5x_2 + 4x_3 \leq 6$$

and

$$x_i \geq 0 \qquad \text{for} \qquad i = 1, 2, 3.$$

Solution

If slack variables x_4 and x_5 are introduced in the inequality constraints and included in the objective function, the following problem in standard form is obtained.

Maximize

$$2x_1 + 3x_2 - x_3 + 0x_4 + 0x_5$$

subject to

$$x_1 + 4x_2 - 3x_3 + x_4 = 2$$

$$2x_1 - 5x_2 + 4x_3 + x_5 = 6$$

and

$$x_i \geq 0 \quad \text{for} \quad i = 1, \ldots, 5. \quad \square$$

The change to standard form produces an equivalent LP, i.e., one with the same solution as the original. The objective function remains the same since only zeros are added, and a point $(x_1, \ldots, x_n)$ satisfies the original constraints if and only if there exist $x_{n+1}, \ldots, x_{n+m}$ such that $(x_1, \ldots, x_n, x_{n+1}, \ldots, x_{n+m})$ satisfies the standard form constraints.

Why the change? The reason may be understood in terms of geometric ideas from Section 8.1. A **feasible solution** for any LP is any point satisfying all the constraints and nonnegativity conditions; the **feasible set** consists of all feasible solutions. The reason for the change to standard form is related to three observations: i) the objective function is maximized on the feasible set at an extreme point; ii) finding extreme points of the feasible set is easier for the standard problem than for the original; and iii) the extreme points of the original feasible set can be determined easily from those for the standard problem.

The simplex method is applied to problems in standard form with one additional simplification: *each constant on the righthand side of a constraint is nonnegative.* This convention will be in effect throughout the remainder of Chapter 8.

The Simplex Algorithm

A little more terminology will help to describe the simplex procedure. In solving a linear system with many solutions, we may assign values to some of the variables and then solve for the remaining ones. If the system is nonhomogeneous and all the assigned values equal 0, the assigned variables are called the **nonbasic variables**. In this case, the variables for which we solve are the **basic variables** and the resulting solution is called a **basic solution**. The following result is cited without proof.

THEOREM 8.1

The extreme points of the feasible region for the primal problem in standard form are precisely the basic solutions of the system of constraint equations.

If

$$z = c_1x_1 + \cdots + c_nx_n,$$

then the objective function is given by the equation

$$z - c_1x_1 - \cdots - c_nx_n = 0.$$

Then we use the linear system (of $m + 1$ equations in $m + n + 1$ unknowns)

$$\begin{array}{lcr} 0z + a_{11}x_1 + \cdots + x_{1n}x_n + x_{n+1} & & = b_1 \\ \vdots \quad\;\; \vdots \qquad\qquad\quad \vdots & & \vdots \\ 0z + a_{m1}x_1 + \cdots + a_{mn}x_n + & + x_{n+m} & = b_m \\ z - \;\; c_1x_1 - \cdots - \;\; c_nx_n & & = 0 \end{array}$$

to find the extreme points of the feasible set and the corresponding values of z.

We write the augmented matrix of this system, dropping the zeros in column 1, using z instead of the usual 1, and adding variable names to the columns. Such an array is determined at each step of the simplex procedure. Each of these arrays is called a **tableau**, and the first is called the **initial tableau**. We describe the method with an example.

Example 2

Solve the following linear program by the simplex method.

Maximize the objective function

$$55x_1 + 50x_2$$

subject to the inequality constraints

$$\begin{aligned} x_1 + 2x_2 &\le 84 \\ x_1 + x_2 &\le 50 \\ 3x_1 + x_2 &\le 120 \end{aligned}$$

and the nonnegativity conditions

$$x_1 \ge 0, \qquad x_2 \ge 0.$$

Solution

This problem, from Example 1 of Section 8.1, is in first primal form. The following is an equivalent LP in standard form.

Maximize

$$z = 55x_1 + 50x_2 + 0x_3 + 0x_4 + 0x_5, \text{ i.e.,}$$

$$z - 55x_1 - 50x_2 - 0x_3 - 0x_4 - 0x_5 = 0,$$

subject to

$$\begin{aligned} x_1 + 2x_2 + x_3 \qquad\qquad &= 84 \\ x_1 + x_2 \qquad + x_4 \qquad &= 50 \\ 3x_1 + x_2 \qquad\qquad + x_5 &= 120 \end{aligned}$$

and

$$x_1 \geq 0, \ldots, x_5 \geq 0.$$

The initial tableau is

	x_1	x_2	x_3	x_4	x_5	
	1	2	1	0	0	84
	1	1	0	1	0	50
	(3)	1	0	0	1	120
z	(−55)	−50	0	0	0	0

Step 1–exit test If all numeric entries in the bottom row, excluding the last, are nonnegative, then exit and return the last entry in the bottom row as the required maximum value. The 0 located at the lower right of the initial tableau is the value of the objective function at the primal extreme point $(0, 0)$. The point $(0, 0)$ is obtained from the first two coordinates of the standard extreme point $(0, 0, 84, 50, 120)$. This extreme point is found, in turn, with x_1 and x_2 as the nonbasic variables, and x_3, x_4, and x_5 as the basic variables (shown in the tableau by the shaded columns). In this example, the result of the exit test is negative for the initial tableau.

Step 2–pivot column selection Locate the negative entries that are in the bottom row but not in the rightmost column. From these entries, choose one that has the largest magnitude. (If there are two or more such entries, any choice may be made among them. This choice is made in the hope of maximizing the gain in the objective function among *adjacent* extreme points, i.e., those corresponding to a change of one variable in the basic set.) The column of this entry is the pivot column. In our example, the circled entry -55 in the initial tableau indicates the choice of the x_1 column.

Step 3–pivot row selection Choose the row of the positive entry in the pivot column, column j, for which b_i/a_{ij} is minimal. (If there are two or more such entries, any choice may be made among them. In this case, the procedure may fail in general, but not in practical problems. If the pivot column has no positive entries above the bottom row, the feasible set is unbounded.) Let row i be the pivot row. In this example, the least of 84/1, 50/1, and 120/3 is the 120/3, hence $i = 3$. The corresponding entry is circled in the initial tableau.

Step 4–new tableau calculation Multiply the pivot row by $1/a_{ij}$ and eliminate the nonzero entries in the pivot column. The result corresponds to an equivalent linear system with the pivot column variable as a new basic variable. The columns corresponding to the new basic variables are shaded in the resulting tableau as a reminder of the search for extreme points of the feasible set.

	x_1	x_2	x_3	x_4	x_5	
	0	5/3	1	0	−1/3	44
	0	(2/3)	0	1	−1/3	10
	1	1/3	0	0	1/3	40
z	0	(−95/3)	0	0	55/3	2200

At this step, the basic variables are x_1, x_3, and x_4. Assign 0 to the variables x_2 and x_5 and solve for x_1, x_3, and x_4. From the tableau, $x_1 = 40$, $x_3 = 44$, and $x_4 = 10$. This solution yields the extreme point $(40, 0, 44, 10, 0)$ of the standard feasible set. The first two coordinates of this point form the extreme point $(40, 0)$ of the primal feasible set, and the corresponding objective function value, $55(40) + 50(0) = 2200$, is the bottom right entry in the tableau.

Now repeat the steps. Since the bottom row has a negative entry, proceed to Step 2. The only negative entry in the bottom row is $-95/3$, so the x_2 column is the pivot column. In Step 3, find the pivot row by comparing $44/(5/3)$, $10/(2/3)$, and $40/(1/3)$. The least of these is $10/(2/3)$, so the pivot row is row 2 and the pivot is 2/3. The following tableau results from Step 4.

	x_1	x_2	x_3	x_4	x_5	
	0	0	1	−5/2	1/2	19
	0	1	0	3/2	−1/2	15
	1	0	0	−1/2	1/2	35
z	0	0	0	95/2	5/2	2675

Since all bottom row entries are nonnegative, the procedure terminates. With the basic variables x_1, x_2, and x_3, the standard extreme point (35, 15, 19, 0, 0) can be found from the tableau. The first two coordinates yield the extreme point (35, 15) of the original feasible set. The desired maximum of the objective function, 2675, is attained at this extreme point and is the bottom right entry in the final tableau. □

Example 3

Solve the following LP by the simplex method.

Maximize

$$4x_1 + 5x_2$$

subject to

$$\begin{aligned} -5x_1 + x_2 &\le 0 \\ -x_1 + 5x_2 &\le 24 \\ 3x_1 + 2x_2 &\le 30 \\ 3x_1 + x_2 &\le 27 \end{aligned}$$

and

$$x_1 \ge 0, \qquad x_2 \ge 0.$$

Solution

First write the given LP in standard form.

Maximize

$$z = 4x_1 + 5x_2, \qquad \text{or}$$

$$z - 4x_1 - 5x_2 = 0,$$

subject to

$$\begin{aligned} -5x_1 + x_2 + x_3 &= 0 \\ -x_1 + 5x_2 + x_4 &= 24 \\ 3x_1 + 2x_2 + x_5 &= 30 \\ 3x_1 + x_2 + x_6 &= 27 \end{aligned}$$

and

$$x_1 \geq 0, \qquad x_2 \geq 0.$$

The initial tableau is

	x_1	x_2	x_3	x_4	x_5	x_6	
	−5	(1)	1	0	0	0	0
	−1	5	0	1	0	0	24
	3	2	0	0	1	0	30
	3	1	0	0	0	1	27
z	−4	(−5)	0	0	0	0	0

We proceed through the simplex algorithm, marking the tableaus as above. At each step, try to calculate the entries in the new tableau before reading the results.

	x_1	x_2	x_3	x_4	x_5	x_6	
	−5	1	1	0	0	0	0
	(24)	0	−5	1	0	0	24
	13	0	−2	0	1	0	30
	8	0	−1	0	0	1	27
z	(−29)	0	−5	0	0	0	0

	x_1	x_2	x_3	x_4	x_5	x_6	
	0	1	−1/24	5/24	0	0	5
	1	0	−5/24	1/24	0	0	1
	0	0	(17/24)	−13/24	1	0	17
	0	0	2/3	−1/3	0	1	19
z	0	0	(−25/24)	29/24	0	0	29

	x_1	x_2	x_3	x_4	x_5	x_6	
	0	1	0	3/17	1/17	0	6
	1	0	0	−2/17	5/17	0	6
	0	0	1	−13/17	24/17	0	24
	0	0	0	3/17	−16/17	1	3
z	0	0	0	7/17	25/17	0	54

The maximum value $z = 54$ is attained at $(6, 6)$. □

Other forms of linear programs will be considered in Section 8.3. But first, we recap the algorithm for convenient reference and offer some exercises for familiarization with the procedure.

The Simplex Algorithm

For a first primal linear program in standard form, start from the initial tableau and perform the following steps.

Step 1–exit test End the procedure if in the current tableau, all of the bottom row entries are nonnegative except possibly the last one.

Step 2–pivot column selection Choose any column whose entry in the bottom row of the current tableau is minimal among the bottom row entries, excluding the last one.

Step 3–pivot row selection Denoting the pivot column as column j, choose any row i for which $a_{ij} > 0$ and the value of b_i/a_{ij} is minimal in column j. If no such a_{ij} exists, the LP has no solution and the procedure ends.

Step 4–next tableau calculation Make x_j a new basic variable by dividing row i by a_{ij} and eliminating the nonzero entries in column j. Go to Step 1.

REVIEW CHECKLIST

1. Describe the first primal form of a linear program.
2. Convert a given first primal LP to standard form.
3. Formulate a definition of a feasible solution and a feasible set.
4. Explain why a primal problem is converted to standard form.
5. Formulate a definition of basic and nonbasic variables in solutions of linear systems.
6. Relate the basic and nonbasic variables to the extreme points of feasible sets for linear programs.
7. Formulate the initial tableau for a first primal LP.
8. Apply the simplex method to find the maximum of the objective function of an LP and a point at which it is attained.
9. In each tableau generated in applying the simplex method, identify the basic and nonbasic variables, and find the extreme points of the standard and original feasible sets.

EXERCISES

In Exercises 1–13, write out the standard form of the given primal problem and solve the LP by the simplex method. In each exercise, all decision variables are required to be nonnegative.

1. Maximize
$x_1 + x_2$
subject to
$x_1 + 3x_2 \le 15$
$4x_1 + 3x_3 \le 24$

2. Maximize
$x_1 + 3x_2$
subject to the constraints in Exercise 1.

3. Maximize
$3x_1 + x_2$
subject to the constraints in Exercise 1.

4. Maximize
$x_1 + 2x_2$
subject to
$x_1 + 4x_2 \le 28$
$3x_1 + 4x_2 \le 36$
$3x_1 + x_2 \le 27$

5. Maximize
$2x_1 + x_2$
subject to the constraints in Exercise 4.

6. Maximize
$x_1 - x_2$
subject to the constraints in Exercise 4.

7. Maximize
$x_1 + 5x_2$
subject to the constraints in Exercise 4.

8. Maximize
$x_1 + 2x_2 + 3x_3$
subject to
$x_1 + x_2 + x_3 \le 5$
$x_2 + 2x_3 \le 6$
$4x_1 + x_2 + 6x_3 \le 18$
$5x_1 + 3x_2 + 4x_3 \le 19$

9. Maximize
$3x_1 + 4x_2 + 6x_3$
subject to the constraints in Exercise 8.

10. Maximize
$5x_1 + 5x_2 + 4x_3$
subject to the constraints in Exercise 8.

11. Maximize
$3x_1 + 4x_2 + 5x_3$
subject to
$8x_1 + 9x_2 + 14x_3 \le 156$
$32x_1 + 29x_2 + 14x_3 \le 372$
$x_1 + 2x_2 \le 16$

12. Maximize
$5x_1 - x_2 + 4x_3$
subject to the constraints in Exercise 11.

13. Maximize
$4x_1 + 4x_2 + 3x_3$
subject to the constraints in Exercise 11.

14. For the LP of Exercise 1, sketch the feasible region for the primal problem, identifying the extreme points. For each iteration in the simplex method, find the extreme point determined by the tableau, and identify the value of the objective function as a tableau entry.

(o) 8.3 MORE GENERAL LINEAR PROGRAMS

In Section 8.2, we used the simplex method to solve first primal linear programs. However, many practical linear programs appear in other forms. In this section, we adapt the simplex method to some other types of linear programs that arise often in practice.

The Second Primal Problem and Some Variations

The type of LP identified next includes the problem of Example 2 in Section 8.1.

The Second Primal Problem

Minimize the objective function

$$c_1x_1 + \cdots + c_nx_n$$

subject to the inequality constraints

$$\begin{array}{ccc} a_{11}x_1 + \cdots + a_{1n}x_n & \geq & b_1 \\ \vdots & & \vdots \\ a_{m1}x_1 + \cdots + a_{mn}x_n & \geq & b_m \end{array}$$

and nonnegativity conditions

$$x_j \geq 0 \quad \text{for} \quad j = 1, \ldots, n.$$

For each inequality constraint in a second primal problem, we *subtract* a **surplus variable** to "use up the surplus" and add an **artificial variable** related to the reversal of the sense of the inequality. These new variables are all nonnegative. In the resulting standard form, $x_{n+1}, \ldots, x_{n+m}$ are the surplus variables and $x_{n+m+1}, \ldots, x_{n+2m}$ are the artificial variables.

Standard Form of the Second Primal Problem

Minimize

$$c_1x_1 + \cdots + c_nx_n + 0x_{n+1} + \cdots + 0x_{n+2m}$$

subject to

$$\begin{array}{cccccc} a_{11}x_1 + \cdots + a_{1n}x_n & - x_{n+1} & & + x_{n+m+1} & & = b_1 \\ \vdots & & & & & \vdots \\ a_{m1}x_1 + \cdots + a_{mn}x_n & & - x_{n+m} & & + x_{n+2m} & = b_m \end{array}$$

and

$$x_i \geq 0, \qquad i = 1, \ldots, n + 2m.$$

The simplex algorithm must be modified for second primal problems. The modifications are described in the next example.

Example 2

Minimize

$$15x_1 + 7x_2$$

subject to

$$x_1 + x_2 \geq 2$$

$$3x_1 + x_2 \geq 3$$

and

$$x_1 \geq 0, \qquad x_2 \geq 0$$

using the simplex method.

Solution

Write the problem in standard form, subtracting the surplus variables x_3 and x_4, and adding in the artificial variables x_5 and x_6.

Minimize

$$15x_1 + 7x_2 + 0x_3 + 0x_4 + 0x_5 + 0x_6$$

subject to

$$\begin{aligned} x_1 + x_2 - x_3 \qquad + x_5 \qquad &= 2 \\ 3x_1 + x_2 \qquad - x_4 \qquad + x_6 &= 3 \end{aligned}$$

and

$$x_1 \geq 0, \ldots, x_6 \geq 0.$$

To implement the simplex method for a second primal problem, first use the objective function to form the equation

$$-z + 15x_1 + 7x_2 + 0x_3 + 0x_4 + Mx_5 + Mx_6 = 0,$$

where M denotes a large positive number. (Here the surplus variables are multiplied by 0 and the artificial variables are multiplied by M.) Add the equality constraints to obtain the equation

$$4x_1 + 2x_2 - x_3 - x_4 + x_5 + x_6 = 5.$$

Multiplying through this equation by M gives

$$4Mx_1 + 2Mx_2 - Mx_3 - Mx_4 + Mx_5 + Mx_6 = 5M.$$

The *objective function constraint* is formed by subtracting this equation from the equation formed above. The result, recorded on two lines, is

$$-z + 15x_1 + 7x_2 + 0x_3 + 0x_4 + 0x_5 + 0x_6$$

$$-4Mx_1 - 2Mx_2 + Mx_3 + Mx_4 + 0x_5 + 0x_6 = -5M.$$

Form the initial tableau with these two lines at the bottom. The M does not appear explicitly in each term, but the bottom row is labeled with an M as a reminder of its presence. The row above the bottom row is labeled with a $-z$ as a reminder that the *negative* of the objective function in the last entry of that row is being calculated. The initial tableau is

	x_1	x_2	x_3	x_4	x_5	x_6	
	1	1	−1	0	1	0	2
	(3)	1	0	−1	0	1	3
$-z$	15	7	0	0	0	0	0
M	(−4)	−2	1	1	0	0	−5

Carry out the steps of the simplex with one modification.

Modified Step 1–exit test If at least one of the M row entries, excluding the last one, in the current tableau is negative, go to Step 2 (pivot column selection). If these M row entries are all nonnegative, end the procedure if all of the z row entries above the zero entries in the M row are nonnegative.

The M row of the initial tableau has two negative entries other than the -5, so proceed to Step 2. Since the minimum of these two is -4, the pivot column corresponds to x_1. In Step 3, the lesser of 2/1 and 3/3 is 3/3, so pivot on the (2, 1)-entry in this tableau to obtain

	x_1	x_2	x_3	x_4	x_5	x_6	
	0	(2/3)	−1	1/3	1	−1/3	1
	1	1/3	0	−1/3	0	1/3	1
$-z$	0	2	0	5	0	−5	−15
M	0	(−2/3)	1	−1/3	0	4/3	−1

The basic variables are now x_1 and x_5. This tableau has two negative entries, excluding the -1, in the M row. Since the lesser of these two entries is $-2/3$, choose the x_2 column for the pivot in Step 2. Since the lesser of $1/(2/3)$ and $1/(1/3)$ is $1/(2/3)$, pivot on the $(1, 2)$-entry of this tableau to obtain

	x_1	x_2	x_3	x_4	x_5	x_6	
	0	1	$-3/2$	$1/2$	$3/2$	$-1/2$	$3/2$
	1	0	$1/2$	$-1/2$	$-1/2$	$1/2$	$1/2$
$-z$	0	0	3	4	-3	-4	-18
M	0	0	0	0	1	1	0

This tableau passes the exit test, since all M row entries are nonnegative, as are all z row entries above the zeros in the M row. The value of $-z$ is the final entry: $-z = -18$, from which the required minimum is 18. This value is achieved at $(1/2, 3/2)$, the feasible set extreme point obtained from the basic solution $(1/2, 3/2, 0, 0, 0, 0)$ reflected in rows 1 and 2. □

How do we handle mixed inequality constraints and objective functions containing a linear form and a constant term?

Example 3

Solve the following LP by the simplex method.

Maximize

$$2x_1 + x_2 + 1$$

subject to

$$5x_1 + 4x_2 \leq 33$$

$$x_1 + 3x_2 \geq 11$$

$$4x_1 + x_2 \geq 11$$

and

$$x_1 \geq 0, \qquad x_2 \geq 0.$$

Solution

This problem was solved in Example 5 of Section 8.1. (See Fig. 8.7.) A slack variable x_5 is added in the first constraint, and subtract surplus variables x_3, x_4, and artificial variables x_6, x_7 are added in the second and third. Since this is a maximization problem, reverse the sign of every entry in the z row, and mark this row with a z. Enter the constant 1 from the objective function in the last position of the z row in the initial tableau,

	x_1	x_2	x_3	x_4	x_5	x_6	x_7	
	5	4	0	0	1	0	0	33
	1	3	−1	0	0	1	0	11
	(4)	1	0	−1	0	0	1	11
z	−2	−1	0	0	0	0	0	1
M	(−5)	−4	1	1	0	0	0	−22

Since −5 is the most negative entry, excluding the −22, in the M row, the x_1 column is the pivot column. Since 11/4 is less than 11/1 and 33/5, pivot on the (3, 1)-entry of this tableau to obtain

	x_1	x_2	x_3	x_4	x_5	x_6	x_7	
	0	11/4	0	5/4	1	0	−5/4	77/4
	0	(11/4)	−1	1/4	0	1	−1/4	33/4
	1	1/4	0	−1/4	0	0	1/4	11/4
z	0	−1/2	0	−1/2	0	0	1/2	13/2
M	0	(−11/4)	1	−1/4	0	0	5/4	−33/4

Now −11/4 is the most negative entry, excluding the −33/4, in the M row, so the x_2 column is the pivot column. Since (33/4)/(11/4) is less than (77/4)/(11/4) and (11/4)/(1/4), pivot on the (2, 2)-entry of this tableau to find

	x_1	x_2	x_3	x_4	x_5	x_6	x_7	
	0	0	(1)	(1)	1	−1	−1	11
	0	1	−4/11	1/11	0	4/11	−1/11	3
	1	0	1/11	−3/11	0	−1/11	3/11	2
z	0	0	−2/11	(−5/11)	0	2/11	5/11	8
M	0	0	0	0	0	1	1	0

In this tableau, all entries in the M row are nonnegative. The x_4 column is the pivot column because of the −5/11 in the z row above a zero in the M row. Since 11/1 < 3/(1/11), use row 2 for the pivot to obtain

	x_1	x_2	x_3	x_4	x_5	x_6	x_7	
	0	0	1	1	1	−1	−1	11
	0	1	−5/11	0	−1/11	5/11	0	2
	1	0	4/11	0	3/11	−4/11	0	5
z	0	0	3/11	0	5/11	−3/11	0	13
M	0	0	1	0	0	1	1	0

This tableau passes the exit test, since all entries in the M row are nonnegative, as are all entries in the z row above the zeros in the M row. The maximum value, 13, is attained at (5, 2), which corresponds to the basic solution (5, 2, 0, 0, 0, 0) from rows 2 and 3. □

Example 4

Solve the following problem by the simplex method.

Minimize

$$2x_1 + x_2 + 1$$

subject to

$$5x_1 + 4x_2 \leq 33$$

$$x_1 + 3x_2 \geq 11$$

$$4x_1 + x_2 \geq 11$$

and

$$x_1 \geq 0, \qquad x_2 \geq 0.$$

Solution

This problem was also solved in Example 5 of Section 8.1. The initial tableau differs from that of Example 3 of this section only in the z row, where the sign of each entry is reversed:

	x_1	x_2	x_3	x_4	x_5	x_6	x_7	
	5	4	0	0	1	0	0	33
	1	3	−1	0	0	1	0	11
	(4)	1	0	−1	0	0	1	11
$-z$	2	1	0	0	0	0	0	−1
M	(−5)	−4	1	1	0	0	0	−22

The next two are similar to the corresponding tableau in Example 3, differing only in the z row. Thus

	x_1	x_2	x_3	x_4	x_5	x_6	x_7	
	0	11/4	0	5/4	1	0	−5/4	77/4
	0	(11/4)	−1	1/4	0	1	−1/4	33/4
	1	1/4	0	−1/4	0	0	1/4	11/4
$-z$	0	−1/2	0	1/2	0	0	−1/2	−13/2
M	0	(−11/4)	1	−1/4	0	0	5/4	−33/4

and

	x_1	x_2	x_3	x_4	x_5	x_6	x_7	
	0	0	1	1	1	−1	−1	11
	0	1	−4/11	1/11	0	4/11	−1/11	3
	1	0	1/11	−3/11	0	−1/11	3/11	2
$-z$	0	0	2/11	5/11	0	−2/11	−5/11	−8
M	0	0	0	0	0	1	1	0

This tableau passes the exit test, since all entries in the M row and all entries in the z row above M row zeros are nonnegative. The required minimum, 8, is determined from the last z row entry. The minimum is attained at $(3, 2)$, which is found in the basic solution $(3, 2, 0, 0, 11, 0, 0)$ from rows 1, 2, and 3. □

How do we allow for possibly negative decision variables?

Example 5

Solve the following problem by the simplex method.
Maximize

$$3x_1 + 4x_2$$

subject to

$$x_1 + 2x_2 \le 2$$

$$5x_1 + 4x_2 \le 16$$

and

$$x_1 \ge 0.$$

Solution

Introducing two *nonnegative* variables x_3, x_4, and replacing x_2 by $x_3 - x_4$ gives the following equivalent first primal problem.
Maximize

$$3x_1 + 4x_3 - 4x_4$$

subject to

$$x_1 + 2x_3 - 2x_4 \le 2$$

$$5x_1 + 4x_3 - 4x_4 \le 16$$

and

$$x_1 \ge 0, \qquad x_3 \ge 0, \qquad x_4 \ge 0.$$

This problem is in first primal form. The simplex method of Section 8.2 may be applied directly. The final tableau (shown) is reached after only three iterations.

	x_1	x_3	x_4	x_5	x_6	
	1	0	0	$-2/3$	$1/3$	4
	0	-1	1	$-5/6$	$1/6$	1
z	0	0	0	$4/3$	$1/3$	8

The required maximum value is 8, and it corresponds to the basic solution $(4, 0, 1, 0, 0)$. Thus $x_3 = 0$ and $x_4 = 1$ for the maximum, hence $x_2 = -1$. Since also $x_1 = 4$, the maximum 8 is attained at $(4, -1)$ in the original problem. □

One other modification of the simplex method is needed for second primal problems. Namely, if the final tableau indicates a nonzero artificial variable in the basic set, the LP has no solution. We suggest the references given at the end of this section for more information on this point.

Further Study in Linear Programming

One direction for further study is a *justification* of the principles and techniques used in the examples of this chapter. For instance, how can we prove Theorem 8.1, which relates extreme points of feasible regions to basic solutions of the constraint equations? How can it be proved that the objective function in a first primal problem with a bounded feasible set must have a maximum on this set? What does the inclusion of the "M" in the objective function constraints for second primal problems really mean? Although we tried to explain the simplex method in a reasonably meaningful way, we gave little justification, leaving much to do for anyone who wants a fundamental understanding of the method.

Applications represent a second direction for further study. Serious use of linear programming over the past forty years has saved untold millions of dollars for large institutions, such as the military services, governmental agencies, and manufacturing firms. Application areas include production planning and scheduling, shipping of goods and transportation of people, communications, inventory control, natural resource management, food production, and the management of investments. Many interesting and valuable uses of linear programming can be found in application fields, such as operations research.

The mathematical context of linear programming is another direction worth exploring. Linear programming is closely related to *game theory*, a relatively new branch of mathematics concerned with the behavior of people in competitive situations. This area is also relevant to the field of economics. Linear programming can be used for curve fitting when the criterion for the fit is to minimize the maximum absolute error at the data points. (Recall the related least squares curve fitting in Section 5.3 where we minimized the "average" error.) Finally, linear programming is only one of several "programming" studies. A real challenge and useful applications can be found in related areas such as integer, nonlinear, and dynamic programming.

For further study, we recommend M. S. Bazaraa and J. J. Jarvis, *Linear Programming and Network Flows*, John Wiley and Sons, New York (1977); G. B. Dantzig, *Linear Programming and Extensions*, Princeton University Press, Princeton, N.J. (1963); and S. I. Gass, *Linear Programming*, McGraw-Hill, New York (1969).

REVIEW CHECKLIST

1. Describe the second primal form of a linear program.
2. Formulate the standard form of the second primal problem.
3. Apply the simplex method to the standard form of the second primal problem and problems with mixed constraints.
4. Solve linear programs with inequality conditions absent by introducing variables to account for possible negative values.

EXERCISES

In Exercises 1–15, every decision variable x_i is assumed nonnegative.

In Exercises 1–3, minimize the given objective function subject to

$$x_1 + 3x_2 \geq 15$$

$$2x_1 + x_2 \geq 10.$$

1. $4x_1 + 3x_2$ **2.** $x_1 + 5x_2$ **3.** $6x_1 + x_2$

In Exercises 4–7, minimize the given objective function subject to

$$2x_1 + x_2 \geq 7$$

$$3x_1 + 4x_2 \leq 23$$

$$x_1 + 3x_2 \geq 11.$$

4. $2x_1 + 3x_2$ **5.** $4x_1 + 3x_2$ **6.** $6x_1 + x_2$ **7.** $x_1 + 6x_2$

In Exercises 8–11, minimize the given objective function subject to

$$-x_1 + x_2 \leq 2$$
$$x_1 + x_2 \geq 12$$
$$3x_1 - x_2 \leq 12.$$

8. $x_1 + x_2$ **9.** $x_1 + 2x_2$ **10.** $x_1 - 2x_2$ **11.** $-x_1 - x_2$

In Exercises 12–15, maximize the given objective function subject to

$$-x_1 + x_2 \leq 2$$
$$x_1 + 2x_2 \leq 10$$
$$-x_1 + 2x_2 \geq 2.$$

12. $-x_1 + 2x_2$ **13.** $x_1 + 2x_2$ **14.** $x_1 + x_2$ **15.** $-x_1 - x_2$

16. Maximize
$x_1 + x_2$
subject to
$5x_1 + 2x_2 \leq 2$
$x_1 + 2x_2 \leq 10$
and
$x_2 \geq 0.$

17. Minimize
$2x_1 + x_2$
subject to
$-x_1 - x_2 \leq 5$
$-3x_1 - x_2 \geq 9.$

REVIEW EXERCISES

In Exercises 1–3, let R be the diamond-shaped region of the plane bounded by the lines $y = 2x + 2$, $y = -2x + 2$, $y = 2x - 2$, and $y = -2x - 2$, including the bounding lines.

1. Is R a convex set?

2. If $z = 3x + 2y$, find the maximum value attained by z on R. Where is this maximum attained on R?

3. If $z = 2x - 3y$, find the minimum value attained by z on R. Where is this minimum attained on R?

In Exercises 4–8, solve the given LP by the simplex method. All decision variables are required to be nonnegative.

4. Maximize
$2x_1 + x_2$
subject to
$x_1 + 3x_2 \leq 11$
$2x_1 + x_2 \leq 12$

5. Maximize
$5x_1 + x_2$
subject to
$x_1 + 3x_2 \leq 11$
$2x_1 + x_2 \leq 12$

6. Maximize

$$x_1 + 2x_2$$

subject to

$$\begin{aligned} x_1 + \quad y_1 &\le 8 \\ -x_1 + 3x_2 &\ge 4 \\ 3x_1 - \quad x_2 &\ge 4 \end{aligned}$$

7. Minimize

$$x_1 + x_2$$

subject to

$$\begin{aligned} x_1 + 2x_2 &\ge 11 \\ 3x_1 + \quad x_2 &\ge 13 \end{aligned}$$

8. Minimize

$$3x_1 + 4x_2$$

subject to

$$\begin{aligned} x_1 + 2x_2 &\le 11 \\ 3x_1 + \quad x_2 &\le 13 \\ 2x_1 + 3x_2 &\ge \ 6 \end{aligned}$$

APPENDIX A

Mathematical Induction

Many formulas, like

$$1 + 2 + \cdots + n = \frac{n(n+1)}{2},$$

that hold for all positive integers n can be proved by applying an axiom called *the mathematical induction principle*. A proof that uses this axiom is called a proof by mathematical induction, or a proof by induction.

The steps in proving a formula by induction are

1. check that it holds for $n = 1$, and
2. prove that if it holds for any positive integer $n = k$, then it also holds for $n = k + 1$.

Once these steps are completed (the axiom says), we know the formula holds for all positive integers n. By step 1 it holds for $n = 1$. By step 2 it holds for $n = 2$, and therefore by step 2 also for $n = 3$, and by step 2 again for $n = 4$, and so on. If the first domino falls, and the kth domino always knocks over the $(k + 1)$-st when it falls, all the dominoes fall.

From another point of view, suppose we have a sequence of statements

$$S_1, S_2, \ldots, S_n, \ldots,$$

one for each positive integer. Suppose we can show that assuming any one of the statements to be true implies that the next statement in line is true. Suppose that we can also show that S_1 is true. Then we may conclude that all the statements are true from S_1 on.

Example 1

Show that

$$1 + 2 + \cdots + n = \frac{n(n+1)}{2}$$

for all positive integers n.

Solution

We carry out the two steps of mathematical induction.

1. The formula holds for $n = 1$ because

$$1 = \frac{1(1+1)}{2}.$$

2. If

$$1 + 2 + \cdots + k = \frac{k(k+1)}{2},$$

then

$$\begin{aligned} 1 + 2 + \cdots + k + (k+1) &= \frac{k(k+1)}{2} + (k+1) \\ &= \frac{k^2 + k + 2k + 2}{2} \\ &= \frac{(k+1)(k+2)}{2} \\ &= \frac{(k+1)((k+1)+1)}{2}. \end{aligned}$$

The last expression in this string of equalities is the expression $n(n+1)/2$ for $n = (k+1)$.

The mathematical induction principle now guarantees the formula for all positive integers n.

Note that all we have to do here is carry out steps 1 and 2. The mathematical induction principle does the rest. □

Example 2

Show that

$$\frac{1}{2^1} + \frac{1}{2^2} + \cdots + \frac{1}{2^n} = 1 - \frac{1}{2^n}$$

for all positive integers n.

Solution

We carry out the two steps of mathematical induction.

1. The formula holds for $n = 1$ because

$$\frac{1}{2^1} = 1 - \frac{1}{2^1}.$$

2. If

$$\frac{1}{2^1} + \frac{1}{2^2} + \cdots + \frac{1}{2^k} = 1 - \frac{1}{2^k},$$

then

$$\begin{aligned}\frac{1}{2^1} + \frac{1}{2^2} + \cdots + \frac{1}{2^k} + \frac{1}{2^{k+1}} &= 1 - \frac{1}{2^k} + \frac{1}{2^{k+1}} \\ &= 1 - \frac{1 \cdot 2}{2^k \cdot 2} + \frac{1}{2^{k+1}} \\ &= 1 - \frac{2}{2^{k+1}} + \frac{1}{2^{k+1}} \\ &= 1 - \frac{1}{2^{k+1}}.\end{aligned}$$

The mathematical induction principle now guarantees the formula for all positive integers n. □

Instead of starting at $n = 1$, some induction arguments start at another integer. The steps for such an argument are

1. check that the formula holds for $n = n_1$ (whatever the appropriate first integer is), and
2. prove that if it holds for any integer $n = k \geq n_1$, then it also holds for $n = k + 1$.

Once these steps are completed, the mathematical induction principle will guarantee the formula for all $n \geq n_1$.

Example 3

Show that $n! > 3^n$ for n sufficiently large.

Solution

How large? We experiment:

n	1	2	3	4	5	6	7
$n!$	1	2	6	24	120	720	5040
3^n	3	9	27	81	243	729	2187

It looks as if $n! > 3^n$ for $n \geq 7$. To be sure, we apply mathematical induction. We take $n_1 = 7$ in step 1, and try for step 2.

Suppose $k! > 3^k$ for some $k \geq 7$. Then

$$(k+1)! = (k+1)(k!) > (k+1)3^k > 8 \cdot 3^k > 3^{k+1}.$$

Thus, for $k \geq 7$,

$$k! > 3^k \Rightarrow (k+1)! > 3^{k+1}.$$

The mathematical induction principle now guarantees $n! \geq 3^n$ for all $n \geq 7$. ☐

EXERCISES

1. Assuming the triangle inequality $|a + b| \leq |a| + |b|$, show that

$$|x_1 + x_2 + \cdots + x_n| \leq |x_1| + |x_2| + \cdots + |x_n|$$

for any n numbers.

2. Show that if $r \neq 1$, then

$$1 + r + r^2 + \cdots + r^n = \frac{1 - r^{n+1}}{1 - r}$$

for all positive integers n.

3. (*Calculus based*) Use the product rule

$$\frac{d}{dx}(uv) = u\frac{dv}{dx} + v\frac{du}{dx}$$

and the fact that

$$\frac{d}{dx}(x) = 1$$

to show that

$$\frac{d}{dx}(x^n) = nx^{n-1}$$

for all positive integers n.

4. Suppose that a function $f(x)$ has the property that $f(x_1x_2) = f(x_1) + f(x_2)$ for any two positive numbers x_1 and x_2. Show that

$$f(x_1x_2 \cdots x_n) = f(x_1) + f(x_2) + \cdots + f(x_n)$$

for the product of any n positive numbers $x_1, x_2, \ldots, x_n$.

5. Show that

$$\frac{2}{3^1} + \frac{2}{3^2} + \cdots + \frac{2}{3^n} = 1 - \frac{1}{3^n}$$

for all positive integers n.

6. Show that $n! > n^3$ for n sufficiently large.

7. Show that $2^n > n^2$ for n sufficiently large.

8. Show that $2^n \geq \frac{1}{8}$ for $n \geq -3$.

APPENDIX B

Common Geometric Figures

A **line** in the plane is a set of points (x, y) such that (a) x is fixed and y is arbitrary, (b) y is fixed and x is arbitrary, or (c) x and y may both vary in such a way that a change in y is proportional to a change in x. An **equation of a line** L is an algebraic relation in x and y that is satisfied by a point $P(x, y)$ if and only if P lies on L.

Any vertical line L (Fig. A.1) has an equation of the form

$$x = x_1.$$

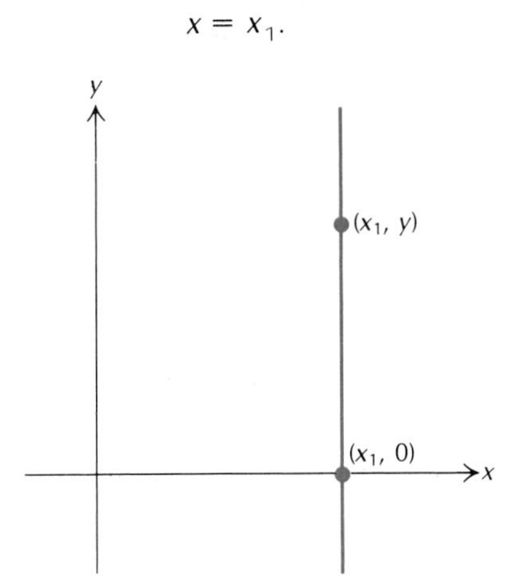

Figure A.1 The vertical line in the plane determined by the equation $x = x_1$.

The variable x is fixed at x_1, and y is arbitrary. A point (x, y) lies on L if and only if its coordinate representation is of the form (x_1, y).

Any horizontal line L (Fig. A.2) has an equation of the form

$$y = y_1.$$

The variable y is fixed at y_1, and x is arbitrary. A point (x, y) lies on L if and only if its coordinate representation is of the form (x, y_1).

If L is a line for which both variables may change, we may find an equation for L once we have a point (x_1, y_1) on L and the proportionality constant m for changes in x and y. A point $P(x, y)$ lies on L if and only if

$$y - y_1 = m(x - x_1),$$

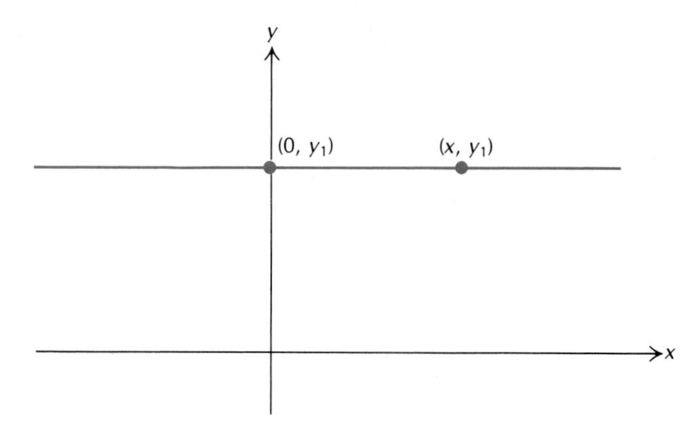

Figure A.2 The horizontal line in the plane determined by the equation $y = y_1$.

since $y - y_1$ is the change in y corresponding to the change $x - x_1$ in x. See Fig. A.3.

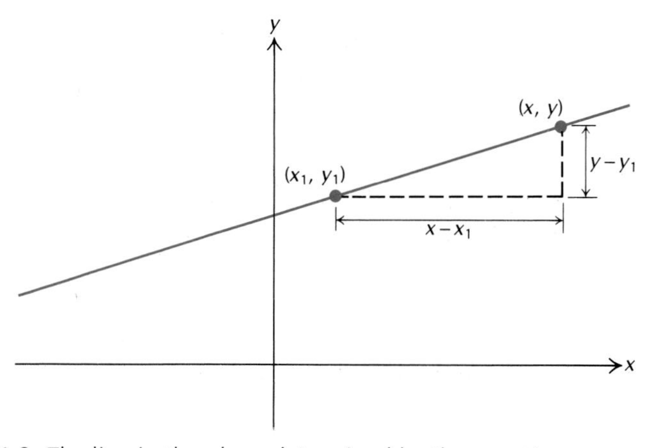

Figure A.3 The line in the plane determined by the equation $y - y_1 = m(x - x_1)$.

A single form that includes the three cases above is

$$Ax + By = C,$$

where A and B are not both zero. With $B = 0$, we obtain a vertical line. With $A = 0$, we obtain a horizontal line. With $A = m$, $B = -1$, and $C = mx_1 - y_1$, we obtain any other line, since the equation obtained above may be rearranged to

$$mx + (-1)y = (mx_1 - y_1).$$

Example 1

The equation of the vertical line through the point (1, 4) is

$$x = 1.$$

The equation of the line through the points (2, 1) and (5, 3) is

$$y - 1 = \frac{2}{3}(x - 2).$$

The constant $m = 2/3$ may be obtained from the proportionality condition

$$(3 - 1) = m(5 - 2). \quad \square$$

A **circle** in the plane is the set of all points located at a given distance r from a fixed point (h, k). The **center** of the circle is the point (h, k), and the **radius** is r. The equation of a circle C in the plane with center (h, k) and radius r may be obtained from the definition of a circle and the Pythagorean theorem (see Fig. A.4). A point $P(x, y)$ lies on C if and only if

$$(x - h)^2 + (y - k)^2 = r^2.$$

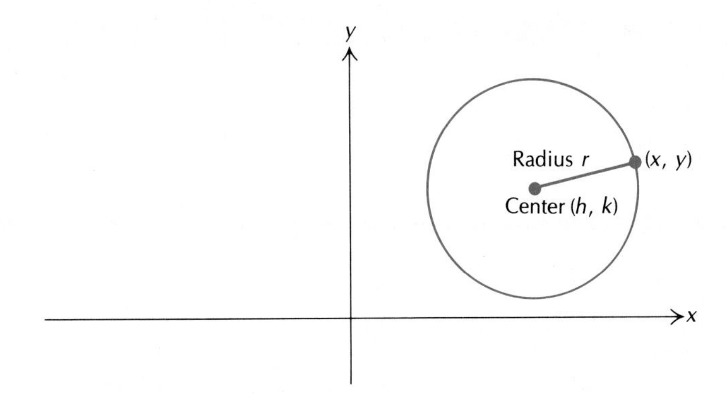

Figure A.4 The definition of a circle in the plane.

Squaring both sides and rearranging the terms, we may write this equation in the general form

$$x^2 + y^2 + Ax + By = C,$$

where $A = -2h$, $B = -2k$, and $C = r^2 - h^2 - k^2$. Given this form of a circle, we may find the center and the radius by completing the square on $x^2 + Ax$ and $y^2 + By$.

Example 2

The equation of the circle with center $(2, -1)$ and radius 4 is

$$(x-2)^2 + (y+1)^2 = 16.$$

This equation may also be expressed in the form

$$x^2 + y^2 - 4x + 2y = 11. \quad \square$$

A **parabola** with vertical axis of symmetry in standard position is the set of all points equidistant from the point $(0, p)$ and the line $y = -p$, where $p > 0$ (see Fig. A.5). The point $(0, p)$ is the **focus** of the parabola and the line $y = -p$ is the **directrix**. By definition, a point (x, y) lies on the parabola if and only if

$$x^2 + (y-p)^2 = y + p.$$

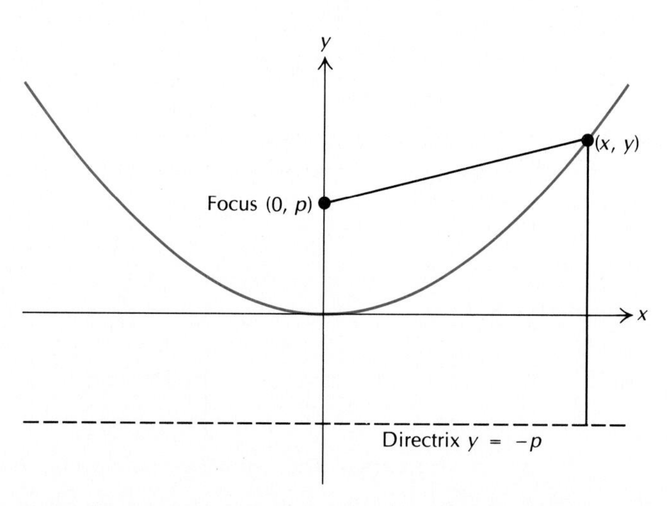

Figure A.5 The definition of a parabola in the plane with vertical axis of symmetry.

Squaring both sides and simplifying the resulting expression, we obtain the equation

$$x^2 = 4py.$$

If the parabola is centered at (h, k), with focus at the point $(h, k + p)$ and directrix with equation $y = k - p$, we may obtain the equation

$$(x - h)^2 = 4p(y - k)$$

by a similar analysis or by using horizontal and vertical translations. Expanding the left side of this equation and simplifying the result, we obtain the general form

$$y = Ax^2 + Bx + C,$$

where $A = 1/4p$, $B = -h/2p$, and $C = (h^2 + 4pk)/4p$.

The equation

$$y = x^2 - 2x + 4$$

determines a parabola with vertical axis of symmetry. Writing this equation in the form

$$y - 3 = (x - 1)^2,$$

we see that the *vertex* (lowest point) is at $(1, 3)$, the axis of symmetry is the line $x = 1$, and the *focal length* p is $1/4$. □

For a line L in xyz-space, any one or two of the variables may be fixed, but any variables that change must have the same proportionality property as a line in the plane, and the constant of proportionality between any two changing variables must be the same. Suppose (x_1, y_1, z_1) and (x_2, y_2, z_2) are distinct points on L. If we let

$$x_2 - x_1 = a, \qquad y_2 - y_1 = b, \qquad \text{and} \qquad z_2 - z_1 = c,$$

then a point (x, y, z) lies on L if and only if

$$x - x_1 = ta, \qquad y - y_1 = tb, \qquad \text{and} \qquad z - z_1 = tc,$$

since the proportionality constant t must be the same (see Fig. A.6). Thus, a line L through (x_1, y_1, z_1) in space is determined by equations

$$x = x_1 + at$$

$$y = y_1 + bt$$

$$z = z_1 + ct,$$

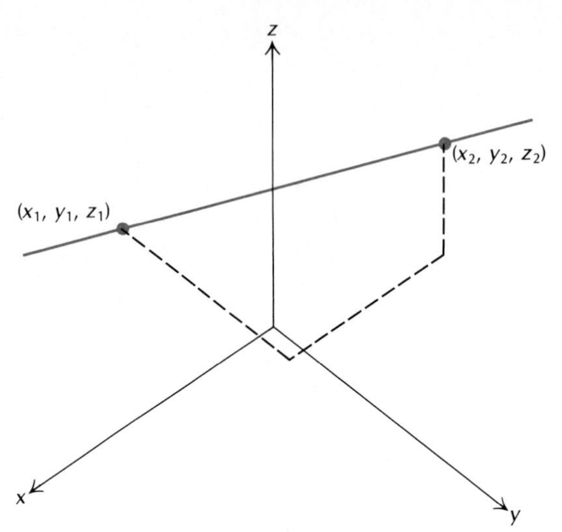

Figure A.6 The line through the points (x_1, y_1, z_1) and (x_2, y_2, z_2) in space.

where t is any real number, and a, b, and c may be found once a second point on L is known.

Example 4

The line through $(1, 2, 1)$ and $(-1, 3, 2)$ is given by

$$x = 1 - 2t$$

$$y = 2 + t$$

$$z = 1 + t.$$

Here we have taken $x_0 = 1$, $y_0 = 2$, and $z_0 = 1$. The values $-2, 1, 1$ may be obtained by subtracting respective coordinates:

$$-1 - 1 = -2, \quad 3 - 2 = 1, \quad \text{and} \quad 2 - 1 = 1. \quad \square$$

The ordered triple (a, b, c) may be regarded as a *vector* (Section 2.1). For any fixed point $P_1(x_1, y_1, z_1)$, any line through P_1 determines a unique vector as above and conversely, any vector determines a unique line through P_1. A **plane** in space through P_1 is defined as the set of all points P such that the vector determined by the directed line segment from P_1 to P is perpendicular to the vector (a, b, c). Thus, a point $P(x, y, z)$ lies on the plane if and only if the dot product (Section 2.1) of the two vectors is zero, i.e.,

$$a(x - x_1) + b(y - y_1) + c(z - z_1) = 0$$

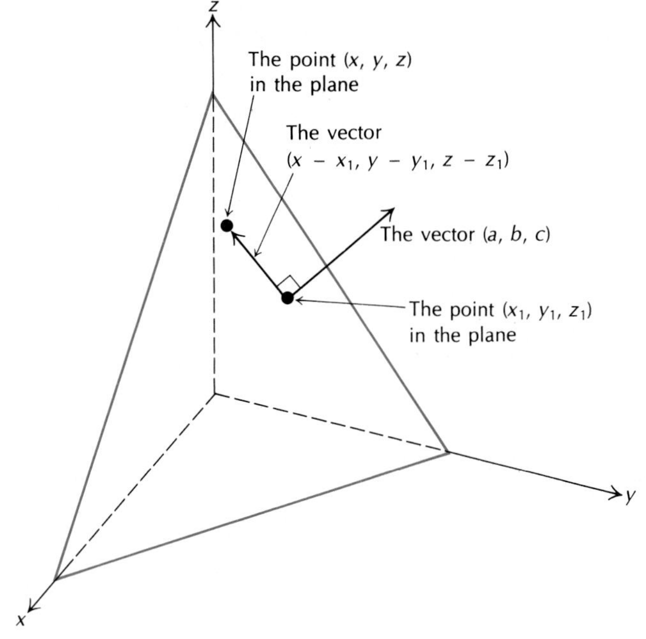

Figure A.7 The plane in space determined by the equation $a(x - x_1) + b(y - y_1) + c(z - z_1) = 0$.

(see Fig. A.7). Expanding and rearranging the terms, we may write this equation in the general form

$$Ax + By + Cz = D,$$

where $A = a$, $B = b$, $C = c$, and $D = ax_1 + by_1 + cz_1$.

Example 5

The equation of the plane through the point $(1, -2, 4)$ and perpendicular to the line through the origin and $(3, 1, -5)$ is

$$3(x - 1) + 1(y + 2) + (-5)(z - 4) = 0,$$

or

$$3x + y - 5z = -19. \quad \square$$

An **ellipse** is determined by two points, each called a **focus**, and a positive constant denoted by a. If we locate the foci in the plane at the points $(-c, 0)$

and $(c, 0)$, then the ellipse consists of all points (x, y) for which the sum of the distances to the foci equals $2a$ (see Fig. A.8). That is, (x, y) lies on the ellipse if and only if

$$\sqrt{(x+c)^2+y^2}+\sqrt{(x-c)^2+y^2}=2a.$$

We may simplify this expression by transposing either term on the left, squaring both sides, simplifying, isolating the resulting radical, and squaring again. The result is an equation of the form

$$\frac{x^2}{a^2}+\frac{y^2}{b^2}=1,$$

where $b^2 = a^2 - c^2$. As shown in Fig. A.8, a and b determine the intercepts of the ellipse on the coordinate axes. By relocating the foci in the plane, we obtain different forms for the equation of the resulting ellipse. More information on the ellipse can be found in books covering analytic geometry, for instance, *Calculus and Analytic Geometry, 7th ed.*, by G. B. Thomas, Jr. and Ross L. Finney.

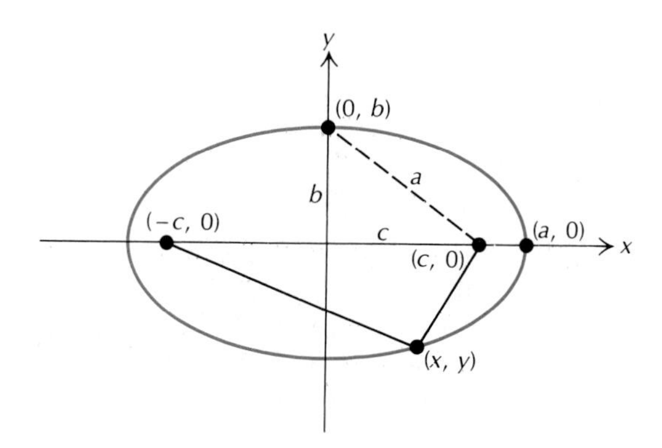

Figure A.8 The definition of an ellipse in the plane.

Example 6

The equation

$$4x^2+9y^2=36$$

has intercepts at $(\pm 3, 0)$ and $(0, \pm 2)$ (see Fig. A.9). The foci of this ellipse are the points $(\pm\sqrt{5}, 0)$, since $a^2 - b^2 = 9 - 4 = 5$. ▭

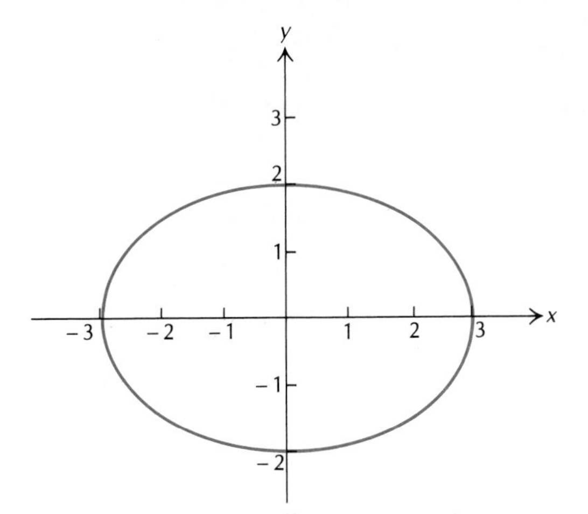

Figure A.9 The ellipse in Example 6.

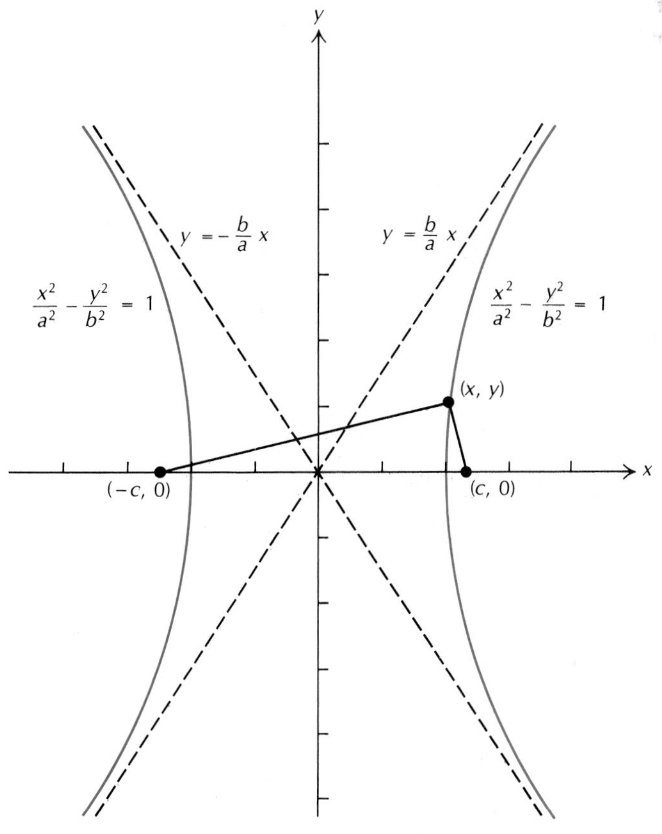

Figure A.10 The definition of a hyperbola in the plane.

A **hyperbola** is defined in much the same way as an ellipse, except that the *difference* between the distances to the foci is constant (see Fig. A.10). Calculations similar to those for an ellipse yield an equation of the form

$$\frac{x^2}{a^2} - \frac{y^2}{b^2} = 1$$

if the foci are located on the x-axis and are symmetric with respect to the origin. Again, information on other forms of a hyperbola may be obtained from references on analytic geometry.

Example 7

The graph of the equation

$$x^2 - 4y^2 = 4$$

intercepts the x-axis at ± 2. The foci are located at $(\pm\sqrt{5}, 0)$. These points may be obtained from the relation $c^2 = a^2 + b^2$. The graph of this hyperbola is shown in Fig. A.11. □

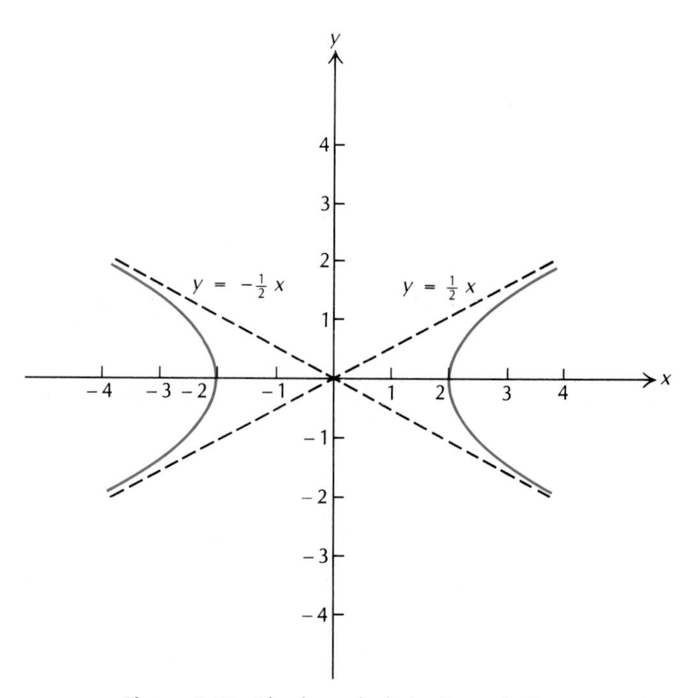

Figure A.11 The hyperbola in Example 7.

EXERCISES

1. Sketch the graph of the line determined by the equation $2x + 3y = 6$.
2. Find an equation for the horizontal line through the point $(3, -2)$ in the plane.
3. Sketch a graph of the parabola in the plane determined by the equation $y = x^2 - 2x + 4$.
4. Sketch a graph of the circle in the plane determined by the equation $(x - 3)^2 + (y + 2)^2 = 4$.
5. Sketch a graph of the circle in the plane determined by the equation $x^2 + y^2 + 2x - 4y = 0$.
6. Find a set of equations that determine the line through the points $(2, -1, 3)$ and $(1, 4, 1)$ in space.
7. Sketch a graph of the ellipse in the plane determined by the equation $x^2 + 9y^2 = 9$.
8. Sketch a graph of the hyperbola in the plane determined by the equation $4x^2 - 9y^2 = 36$.

APPENDIX C

Invented Number Systems. Complex Numbers

In this appendix we shall discuss complex numbers. These are expressions of the form $a + ib$, where a and b are "real" numbers and i is a symbol for $\sqrt{-1}$. Unfortunately, the words "real" and "imaginary" have connotations that somehow place $\sqrt{-1}$ in a less favorable position than $\sqrt{2}$ in our minds. As a matter of fact, a good deal of imagination, in the sense of *inventiveness*, has been required to construct the *real* number system, which forms the basis of the calculus. In this appendix we shall review the various stages of this invention. The further invention of a complex number system will then not seem so strange. It is fitting for us to study such a system, since modern engineering has found therein a convenient language for describing vibratory motion, harmonic oscillation, damped vibrations, alternating currents, and other wave phenomena.

The earliest stage of number development was the recognition of the *counting numbers* $1, 2, 3, \ldots$, which we now call the *natural numbers*, or the *positive integers*. Certain simple arithmetical operations can be performed with these numbers without getting outside the system. That is, the system of positive integers is *closed* with respect to the operations of *addition* and *multiplication*. By this we mean that if m and n are any positive integers, then

$$m + n = p \qquad \text{and} \qquad mn = q \tag{1}$$

are also positive integers. Given the two positive integers on the *left-hand side* of either equation in (1), we can find the corresponding positive integer on the right. More than this, we may sometimes specify the positive integers m and p and find a positive integer n such that $m + n = p$. For instance, $3 + n = 7$ can be *solved* when the only numbers we know are the positive integers. But the equation $7 + n = 3$ cannot be solved unless the number system is enlarged. The number concepts that we denote by zero and the *negative* integers were invented to solve equations like that. In a civilization that recognizes all the integers

$$\ldots, -3, -2, -1, 0, 1, 2, 3, \ldots, \tag{2}$$

an educated person may always find the missing integer that solves the equation $m + n = p$ when given the other two integers in the equation.

Suppose our educated people also know how to multiply any two integers of the set in (2). If, in Eqs. (1), they are given m and q, they discover that sometimes they can find n and sometimes they can't. If their *imagination* is still in good working order, they may be inspired to invent still more numbers and introduce fractions, which are just ordered pairs m/n of integers m and n. The number zero has special properties that may bother them for a while, but they ultimately discover that it is handy to have all ratios of integers m/n, excluding only those having zero in the denominator. This system, called the set of *rational numbers*, is now rich enough for them to perform the so-called *rational operations* of arithmetic:

1. a) addition
   b) subtraction

2. a) multiplication
   b) division

on any two numbers in the system, *except that they cannot divide by zero.*

The geometry of the unit square (Fig. A.12) and the Pythagorean Theorem showed that they could construct a geometric line segment which, in terms of some basic unit of length, has length equal to $\sqrt{2}$. Thus they could solve the equation

$$x^2 = 2$$

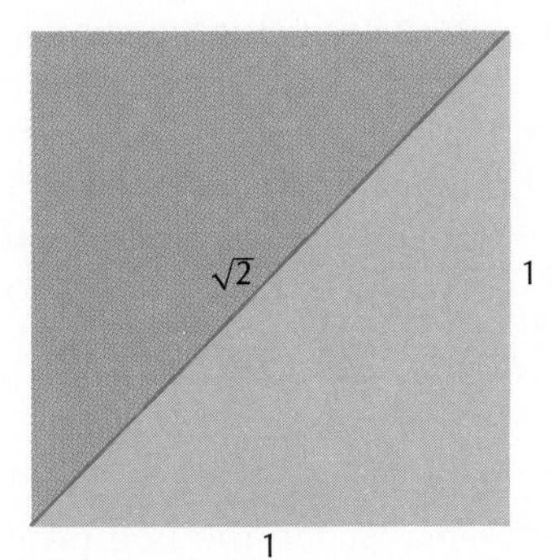

Figure A.12 Segments of irrational length can be constructed with straightedge and compass.

by a geometric construction. But then they discoverd that the line segment representing $\sqrt{2}$ and the line segment representing the unit of length 1 were incommensurable quantities. This means that the ratio $\sqrt{2}/1$ cannot be expressed as the ratio of two *integral* multiples of some other, presumably more fundamental, unit of length. That is, our educated people could not find a rational number solution of the equation $x^2 = 2$.

There is a nice algebraic argument that there is no rational number whose square is 2. Suppose that there were such a rational number. Then we could find integers p and q with no common factor other than 1, and such that

$$p^2 = 2q^2. \tag{3}$$

Since p and q are integers, p must then be even, say $p = 2p_1$, where p_1 is an integer. This leads to $2p_1^2 = q^2$, which says that q must also be even, say $q = 2q_1$, where q_1 is also an integer. But this is contrary to our choice of p and q as integers having no common factor other than 1. Hence there is no rational number whose square is 2.

Our educated people *could*, however, get a *sequence* of rational numbers

$$\frac{1}{1}, \frac{7}{5}, \frac{41}{29}, \frac{239}{169}, \ldots, \tag{4}$$

whose squares form a sequence

$$\frac{1}{1}, \frac{49}{25}, \frac{1681}{841}, \frac{57{,}121}{28{,}561}, \ldots, \tag{5}$$

that converges to 2 as its *limit*. This time their *imagination* suggested that they needed the concept of a *limit of a sequence* of rational numbers. If we accept the fact that an increasing sequence that is bounded from above always approaches a limit, and observe that the sequence in (4) has these properties, then we want it to have a limit L. This would also mean, from (5), that $L^2 = 2$, and hence L is *not* one of our rational numbers. If to the *rational* numbers we further add the *limits* of all bounded increasing sequences of rational numbers, we arrive at the system of all "real" numbers. The word *real* is placed in quotes because there is nothing that is either "more real" or "less real" about this system than there is about any other well-defined mathematical system.

Imagination was called upon at many stages during the development of the real number system from the system of positive integers. In fact, the art of invention was needed at least three times in constructing the systems we have discussed so far:

1. The *first invented* system; the set of all *integers* as constructed from the counting numbers.
2. The *second invented* system; the set of *rational* numbers m/n as constructed from the integers.
3. The *third invented* system; the set of all *"real"* numbers x as constructed from the rational numbers.

These invented systems form a hierarchy in which each system contains the previous system. Each system is also richer than its predecessor in that it permits additional operations to be performed without going outside the system. Expressed in algebraic terms, we may say that

1. In the system of all integers, we can solve all equations of the form

$$x + a = 0. \tag{6}$$

where a may be any integer.

2. In the system of all rational numbers, we can solve all equations of the form

$$ax + b = 0 \tag{7}$$

provided a and b are rational numbers and $a \neq 0$.

3. In the system of all real numbers, we can solve all of the Eqs. (6) and (7) and, in addition, all quadratic equations

$$ax^2 + bx + c = 0 \quad \text{having} \quad a \neq 0 \quad \text{and} \quad b^2 - 4ac \geq 0. \tag{8}$$

Every student of algebra is familiar with the formula that gives the solutions of (8), namely,

$$x = \frac{-b \pm \sqrt{b^2 - 4ac}}{2a}, \tag{9}$$

and familiar with the further fact that when the discriminant, $d = b^2 - 4ac$, is *negative*, the solutions in (9) do *not* belong to any of the systems discussed above. In fact, the very simple quadratic equation

$$x^2 + 1 = 0 \tag{10}$$

is impossible to solve if the only number systems that can be used are the three invented systems mentioned so far.

Thus we come to the *fourth invented* system, the set of all complex numbers $a + ib$. We could, in fact, dispense entirely with the symbol i and use a notation like (a, b). We would then speak simply of a pair of real numbers a and b. Since, under algebraic operations, the numbers a and b are treated somewhat differently, it is essential to keep the *order* straight. We therefore might say that the *complex number system consists of the set of all ordered pairs of real numbers* (a, b), together with the rules by which they are to be equated, added, multiplied, and so on, listed below. We shall use both the (a, b) notation and the notation $a + ib$. We call a the "real part" and b the "imaginary part" of (a, b). We make the following definitions.

Equality

$a + ib = c + id$
if and only if
$a = c$ and $b = d$.

Two complex numbers (a, b) and (c, d) are *equal* if and only if $a = c$ and $b = d$.

Addition

$(a + ib) + (c + id)$
$= (a + c) + i(b + d)$

The sum of the two complex numbers (a, b) and (c, d) is the complex number $(a + c, b + d)$.

Multiplication

$(a + ib)(c + id)$ $= (ac - bd) + i(ad + bc)$	The product of two complex numbers (a, b) and (c, d) is the complex number $(ac - bd, ad + bc)$.
$c(a + ib) = ac + i(bc)$	The product of a real number c and the complex number (a, b) is the complex number (ac, bc).

The set of all complex numbers (a, b) in which the second number is zero has all the properties of the set of ordinary "real" numbers a. For example, addition and multiplication of $(a, 0)$ and $(c, 0)$ give

$$(a, 0) + (c, 0) = (a + c, 0),$$

$$(a, 0) \cdot (c, 0) = (ac, 0),$$

which are numbers of the same type with "imaginary part" equal to zero. Also, if we multiply a "real number" $(a, 0)$ and the "complex number" (c, d), we get

$$(a, 0) \cdot (c, d) = (ac, cd) = a(c, d).$$

In particular, the complex number $(0, 0)$ plays the role of zero in the complex number system and the complex number $(1, 0)$ plays the role of unity.

The number pair $(0, 1)$, which has "real part" equal to zero and "imaginary part" equal to one, has the property that its square,

$$(0, 1)(0, 1) = (-1, 0),$$

has "real part" equal to minus one and "imaginary part" equal to zero. Therefore, in the system of complex numbers (a, b), there is a number $x = (0, 1)$ whose square can be added to unity $= (1, 0)$ to produce zero $= (0, 0)$; that is

$$(0, 1)^2 + (1, 0) = (0, 0).$$

The equation

$$x^2 + 1 = 0$$

therefore has a solution $x = (0, 1)$ in this new number system.

You are probably more familiar with the $a + ib$ notation than you are with the notation (a, b). And since the laws of algebra for the ordered pairs enable us to write

$$(a, b) = (a, 0) + (0, b) = a(1, 0) + b(0, 1),$$

while (1, 0) behaves like unity and (0, 1) behaves like a square root of minus one, we need not hesitate to write $a + ib$ in place of (a, b). The i associated with b is like a tracer element that tags the "imaginary part" of $a + ib$. We can pass at will from the realm of ordered pairs (a, b) to the realm of expressions $a + ib$, and conversely. But there is nothing less "real" about the symbol $(0, 1) = i$ than there is about the symbol $(1, 0) = 1$, once we have learned the laws of algebra in the complex number system (a, b).

To reduce any rational combination of complex numbers to a single complex number, we need only apply the laws of elementary algebra, replacing i^2 wherever it appears by -1. Of course, we cannot divide by the complex number $(0, 0) = 0 + i0$. But if $a + ib \neq 0$, then we may carry out a division as follows:

$$\frac{c + id}{a + ib} = \frac{(c + id)(a - ib)}{(a + ib)(a - ib)} = \frac{(ac + bd) + i(ad - bc)}{a^2 + b^2}.$$

The result is a complex number $x + iy$ with

$$x = \frac{ac + bd}{a^2 + b^2}, \qquad y = \frac{ad - bc}{a^2 + b^2},$$

and $a^2 + b^2 \neq 0$, since $a + ib = (a, b) \neq (0, 0)$.

The number $a - ib$ that is used as a multiplier to clear the i out of the denominator is called the *complex conjugate* of $a + ib$. It is customary to use $\bar{z}$ (read "z bar") to denote the complex conjugate of z; thus

$$z = a + ib, \qquad \bar{z} = a - ib.$$

Thus, we multiplied the numerator and denominator of the complex fraction $(c + id)/(a + ib)$ by the complex conjugate of the denominator. This will always replace the denominator by a real number.

Argand Diagrams. There are two geometric representations of the complex number $z = x + iy$:

a) as the point $P(x, y)$ in the xy-plane, and

b) as the vector **OP** from the origin to P.

In each representation, the x-axis is called the "axis of reals" and the y-axis is the "imaginary axis." Both representations are called *Argand diagrams* (Fig. A13).

In terms of the polar coordinates of x and y, we have

$$x = r\cos\theta, \qquad y = r\sin\theta,$$

and

$$z = x + iy = r(\cos\theta + i\sin\theta). \tag{11}$$

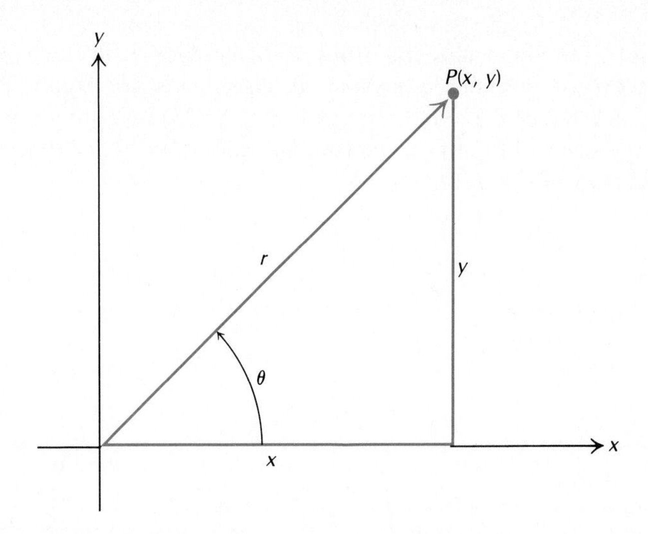

Figure A.13 An Argand diagram representing $z = x + iy$ both as a point $P(x, y)$ and as a vector **OP**.

We define the absolute value of a complex number $x + iy$ to be the length r of a vector **OP** from the origin to $P(x, y)$. We denote the absolute value by vertical bars, thus:

$$|x + iy| = \sqrt{x^2 + y^2}. \tag{12a}$$

If we always choose the polar coordinates r and θ so that r is nonnegative, then we have

$$r = |x + iy|. \tag{12b}$$

The polar angle θ is called the *argument* of z and is written $\theta = \arg z$. Of course, any integral multiple of 2π may be added to θ to produce another appropriate angle. The *principal* value of the argument will, in this book, be taken to be that value of θ for which $-\pi < \theta \leq +\pi$.

The following equation gives a useful formula connecting a complex number z, its conjugate $\bar{z}$, and its absolute value $|z|$, namely,

$$z \cdot \bar{z} = |z|^2. \tag{13}$$

The identity

$$e^{i\theta} = \cos\theta + i\sin\theta \tag{14}$$

leads to the following rules for calculating products, quotients, powers, and roots of complex numbers. It also leads to Argand diagrams for $e^{i\theta}$. Since $\cos\theta + i\sin\theta$ is what we get from Eq. (11) by taking $r = 1$, we can say that $e^{i\theta}$ is represented by a unit vector that makes an angle θ with the positive x-axis, as shown in Fig. A.14.

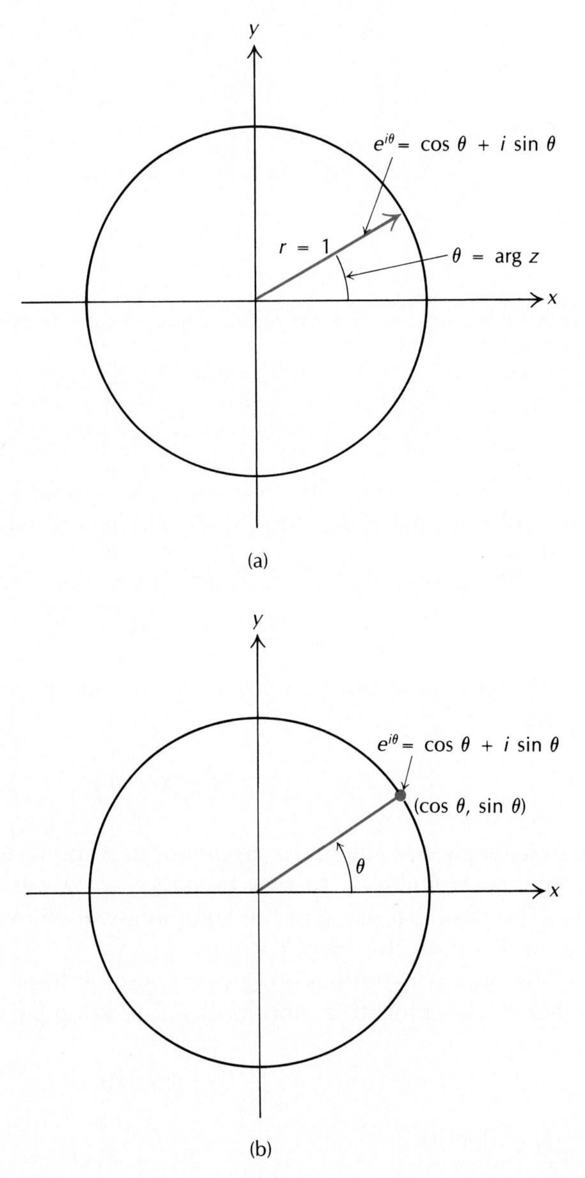

Figure A.14 Argand diagrams for $e^{i\theta} = \cos\theta + i\sin\theta$ (a) as a vector, (b) as a point.

Products. Let

$$z_1 = r_1 e^{i\theta_1}, \qquad z_2 = r_2 e^{i\theta_2} \tag{15}$$

so that

$$|z_1| = r_1, \quad \arg z_1 = \theta_1; \qquad |z_2| = r_2, \quad \arg z_2 = \theta_2. \tag{16}$$

Then

$$z_1 z_2 = r_1 e^{i\theta_1} \cdot r_2 e^{i\theta_2} = r_1 r_2 e^{i(\theta_1 + \theta_2)}$$

and hence

$$|z_1 z_2| = r_1 r_2 = |z_1| \cdot |z_2|,$$

$$\arg(z_1 z_2) = \theta_1 + \theta_2 = \arg z_1 + \arg z_2. \tag{17}$$

Thus the product of two complex numbers is represented by a vector whose length is the product of the lengths of the two factors and whose argument is the sum of their arguments (Fig. A.15). In particular, a vector may be rotated in the counterclockwise direction through an angle θ by simply multiplying it by $e^{i\theta}$. Multiplication by i rotates 90°, by -1 rotates 180°, by $-i$ rotates 270°, etc.

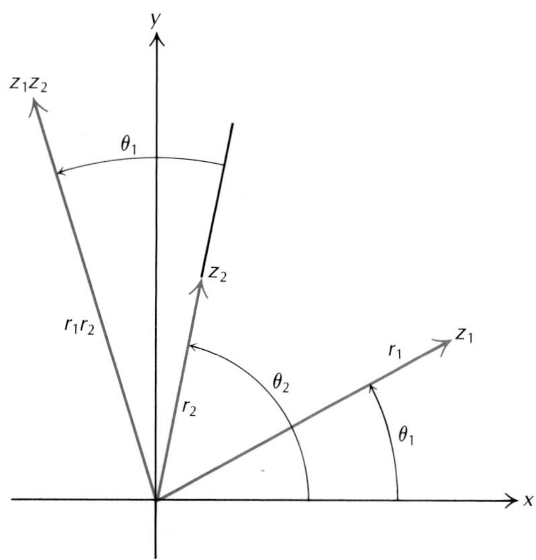

Figure A.15 When z_1 and z_2 are multiplied, $|z_1 z_2| = r_1 \cdot r_2$, and $\arg(z_1 z_2) = \theta_1 + \theta_2$.

Example 1

Let

$$z_1 = 1 + i, \qquad z_2 = \sqrt{3} - i.$$

We plot these complex numbers in an Argand diagram (Fig. A.16) from which we read off the polar representations

$$z_1 = \sqrt{2}\, e^{i\pi/4}, \qquad z_2 = 2e^{-i\pi/6}.$$

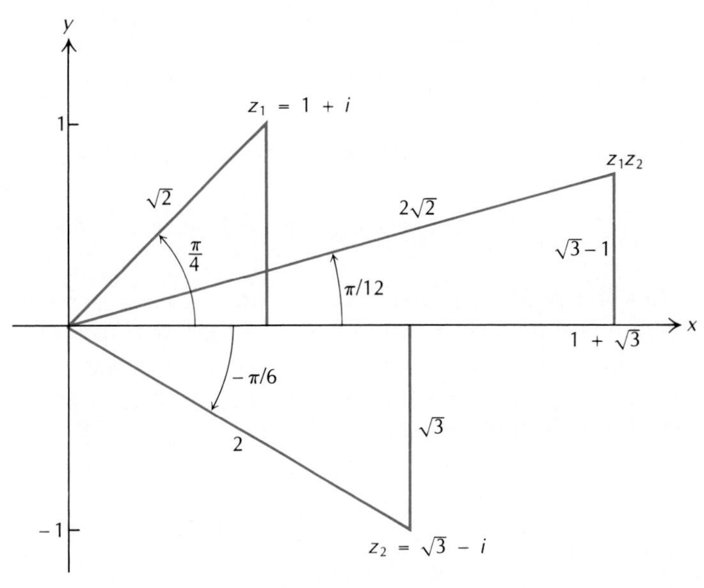

Figure A.16 To multiply two complex numbers, one multiplies their absolute values, and adds their arguments.

Then

$$z_1z_2 = 2\sqrt{2}\exp\left(\frac{i\pi}{4} - \frac{i\pi}{6}\right) = 2\sqrt{2}\exp\left(\frac{i\pi}{12}\right)$$

$$= 2\sqrt{2}\left(\cos\frac{\pi}{12} + i\sin\frac{\pi}{12}\right) \approx 2.73 + 0.73i. \qquad \square$$

Quotients. Suppose $r_2 \neq 0$ in Eq. (15). Then

$$\frac{z_1}{z_2} = \frac{r_1e^{i\theta_1}}{r_2e^{i\theta_2}} = \frac{r_1}{r_2}e^{i(\theta_1-\theta_2)}.$$

Hence

$$\left|\frac{z_1}{z_2}\right| = \frac{r_1}{r_2} = \frac{|z_1|}{|z_2|},$$

$$\arg\left(\frac{z_1}{z_2}\right) = \theta_1 - \theta_2 = \arg z_1 - \arg z_2.$$

That is, we divide lengths and subtract angles.

Example 2

Let $z_1 = 1 + i$ and $z_2 = \sqrt{3} - i$, as in Example 1. Then,

$$\begin{aligned}\frac{1+i}{\sqrt{3}-i} &= \frac{\sqrt{2}\,e^{i\pi/4}}{2e^{-i\pi/6}} = \frac{\sqrt{2}}{2}e^{5\pi i/12}\\ &\approx 0.707\left(\cos\frac{5\pi}{12} + i\sin\frac{5\pi}{12}\right)\\ &\approx 0.183 + 0.683i. \quad \square\end{aligned}$$

Powers. If n is a positive integer, we may apply the product formulas in (17) to find

$$z^n = z \cdot z \cdot \ldots \cdot z \quad (n \text{ factors}).$$

With $z = re^{i\theta}$, we obtain

$$\begin{aligned}z^n = \left(re^{i\theta}\right)^n &= r^n e^{i(\theta+\theta+\cdots+\theta)} \quad (n \text{ summands})\\ &= r^n e^{in\theta}.\end{aligned} \tag{18}$$

The length $r = |z|$ is raised to the nth power and the angle $\theta = \arg z$ is multiplied by n.

In particular, if we place $r = 1$ in Eq. (18), we obtain De Moivre's theorem.

DE MOIVRE'S THEOREM

$$(\cos\theta + i\sin\theta)^n = \cos n\theta + i\sin n\theta. \tag{19}$$

If we expand the left-hand side of this equation by the binomial theorem and reduce it to the form $a + ib$, we obtain formulas for $\cos n\theta$ and $\sin n\theta$ as polynomials of degree n in $\cos\theta$ and $\sin\theta$.

Example 4

If $n = 3$ in Eq. (19), we have

$$(\cos\theta + i\sin\theta)^3 = \cos 3\theta + i\sin 3\theta.$$

The left-hand side of this equation is

$$\cos^3\theta + 3i\cos^2\theta\sin\theta - 3\cos\theta\sin^2\theta - i\sin^3\theta.$$

The real part of this must equal $\cos 3\theta$ and the imaginary part must equal $\sin 3\theta$. Therefore,

$$\cos 3\theta = \cos^3\theta - 3\cos\theta\sin^2\theta.$$

$$\sin 3\theta = 3\cos^2\theta\sin\theta - \sin^3\theta. \quad \square$$

Roots. If $z = re^{i\theta}$ is a complex number different from zero and n is a positive integer, then there are precisely n different complex numbers $w_0, w_1, \ldots, w_{n-1}$, that are nth roots of z. To see why, let $w = \rho e^{i\alpha}$ be an nth root of $z = re^{i\theta}$, so that

$$w^n = z$$

or

$$\rho^n e^{in\alpha} = re^{i\theta}. \tag{20}$$

Then

$$\rho = \sqrt[n]{r} \tag{21}$$

is the real, positive, nth root of r. As regards the angle, although we cannot say that $n\alpha$ and θ must be equal, we can say that they may differ only by an integral multiple of 2π. That is,

$$n\alpha = \theta + 2k\pi, \qquad k = 0, \pm 1, \pm 2, \ldots \tag{22}$$

Therefore

$$\alpha = \frac{\theta}{n} + k\frac{2\pi}{n}.$$

Hence all the nth roots of $z = re^{i\theta}$ are given by

$$\sqrt[n]{re^{i\theta}} = \sqrt[n]{r}\exp\left(\frac{\theta}{n} + k\frac{2\pi}{n}\right), \qquad k = 0, \pm 1, \pm 2, \ldots. \tag{23}$$

There might appear to be infinitely many different answers corresponding to the infinitely many possible values of k. But one readily sees that $k = n + m$ gives the same answer as $k = m$ in Eq. (23). Thus we need only take n consecutive values for k to obtain all the different nth roots of z. For convenience, we may take

$$k = 0, 1, 2, \ldots, n - 1.$$

All the nth roots of $re^{i\theta}$ lie on a circle centered at the origin O and having radius equal to the real, positive nth root of r. One of them has argument $\alpha = \theta/n$. The others are uniformly spaced around the circle, each being separated from its neighbors by an angle equal to $2\pi/n$. Figure A.17 illustrates the placement of the three cube roots, w_0, w_1, w_2, of the complex number $z = re^{i\theta}$.

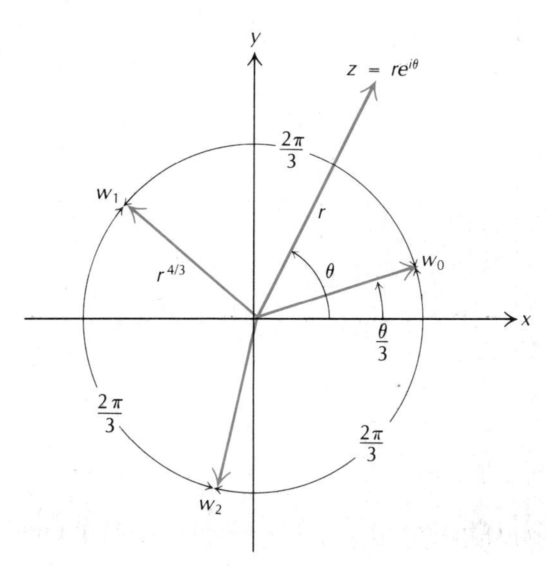

Figure A.17 The three cube roots of $z = re^{i\theta}$.

Example 5

Find the four fourth roots of -16.

Solution

As our first step, we plot the given number in an Argand diagram (Fig. A.18) and determine its polar representation $re^{i\theta}$. Here,

$$z = -16, \qquad r = +16, \qquad \theta = \pi.$$

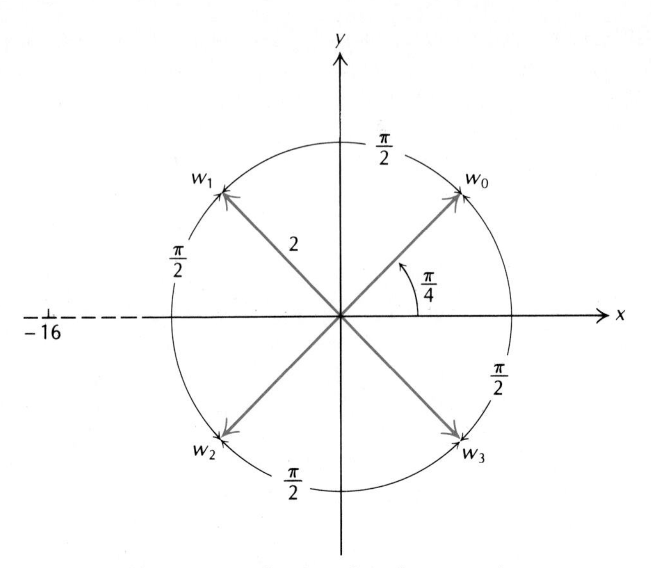

Figure A.18 The four fourth roots of -16.

One of the fourth roots of $16e^{i\pi}$ is $2e^{i\pi/4}$. We obtain others by successive additions of $2\pi/4 = \pi/2$ to the argument of this first one. Hence

$$\sqrt[4]{16 \exp i\pi} = 2 \exp i\left(\frac{\pi}{4}, \frac{3\pi}{4}, \frac{5\pi}{4}, \frac{7\pi}{4}\right),$$

and the four roots are

$$w_0 = 2\left[\cos\frac{\pi}{4} + i\sin\frac{\pi}{4}\right] = \sqrt{2}(1 + i),$$

$$w_1 = 2\left[\cos\frac{3\pi}{4} + i\sin\frac{3\pi}{4}\right] = \sqrt{2}(-1 + i),$$

$$w_2 = 2\left[\cos\frac{5\pi}{4} + i\sin\frac{5\pi}{4}\right] = \sqrt{2}(-1 - i),$$

$$w_3 = 2\left[\cos\frac{7\pi}{4} + i\sin\frac{7\pi}{4}\right] = \sqrt{2}(1 - i). \quad \square$$

The Fundamental Theorem of Algebra. One may well say that the invention of $\sqrt{-1}$ is all well and good and leads to a number system that is richer than the real number system alone; but where will this process end? Are we also going to invent still more systems so as to obtain $\sqrt[4]{-1}$, $\sqrt[6]{-1}$, and so on? By now it should be clear that this is not necessary. These numbers are already expressible in terms of the complex number system $a + ib$. In fact, the Fundamental Theorem of Algebra says that with the introduction of the complex numbers we now have enough numbers to factor every polynomial into a product of linear factors and hence enough numbers to solve every possible polynomial equation.

THE FUNDAMENTAL THEOREM OF ALGEBRA

Every polynomial equation of the form

$$a_0 z^n + a_1 z^{n-1} + a_2 z^{n-2} + \cdots + a_{n-1} z + a_n = 0,$$

in which the coefficients $a_0, a_1, \ldots, a_n$ are any complex numbers, whose degree n is greater than or equal to one, and whose leading coefficient a_0 is not zero, has exactly n roots in the complex number system, provided each multiple root of multiplicity m is counted as m roots.

This theorem is rather difficult to prove, and we can only state it here.

EXERCISES

1. Find $(a, b) \cdot (c, d) = (ac - bd, ad + bc)$.
   a) $(2, 3) \cdot (4, -2)$
   b) $(2, -1) \cdot (-2, 3)$
   c) $(-1, -2) \cdot (2, 1)$
   (*Note*: This is how complex numbers are multiplied by computers.)
2. Solve the following equations for the real numbers x and y.
   a) $(3 + 4i)^2 - 2(x - iy) = x + iy$
   b) $\left(\dfrac{1+i}{1-i}\right)^2 + \dfrac{1}{x + iy} = 1 + i$
   c) $(3 - 2i)(x + iy) = 2(x - 2iy) + 2i - 1$
3. Show with an Argand diagram that the law for adding complex numbers is the same as the parallelogram law for adding vectors.
4. How may the following complex numbers be obtained from $z = x + iy$ geometrically? Sketch.
   a) $\bar{z}$
   b) $\overline{(-z)}$
   c) $-z$
   d) $1/z$
5. Show that the conjugate of the sum (product, or quotient) of two complex numbers z_1 and z_2 is the same as the sum (product, or quotient) of their conjugates.
6. a) Extend the results of Problem 5 to show that

$$f(\bar{z}) = \overline{f(z)} \quad \text{if}$$

$$f(z) = a_0 z^n + a_1 z^{n-1} + \cdots + a_{n-1} z + a_n$$

   is a polynomial with real coefficients $a_0, \ldots, a_n$.

b) If z is a root of the equation $f(z) = 0$, where $f(z)$ is a polynomial with real coefficients as in part (a) above, show that the conjugate $\bar{z}$ is also a root of the equation. (*Hint*: Let $f(z) = u + iv = 0$; then both u and v are zero. Now use the fact that $f(\bar{z}) = \overline{f(z)} = u - iv$.)

7. Show that $|\bar{z}| = |z|$.

8. If z and $\bar{z}$ are equal, what can you say about the location of the point z in the complex plane?

9. Let $R(z)$, $I(z)$ denote respectively the real and imaginary parts of z. Show that

a) $z + \bar{z} = 2R(z)$, b) $z - \bar{z} = 2iI(z)$,

c) $|R(z)| \leq |z|$,

d) $|z_1 + z_2|^2 = |z_1|^2 + |z_2|^2 + 2R(z_1\bar{z}_2)$,

e) $|z_1 + z_2| \leq |z_1| + |z_2|$.

10. Show that the distance between the two points z_1 and z_2 in an Argand diagram is equal to $|z_1 - z_2|$.

In Problems 11–15, graph the points $z = x + iy$ that satisfy the given conditions.

11. a) $|z| = 2$ b) $|z| < 2$ c) $|z| > 2$

12. $|z - 1| = 2$

13. $|z + 1| = 1$

14. $|z + 1| = |z - 1|$

15. $|z + i| = |z - 1|$

Express the answer to each of the Problems 16–19 in the form $r^{i\theta}$, with $r \geq 0$ and $-\pi < \theta \leq \pi$. Sketch.

16. $(1 + \sqrt{-3})^2$

17. $\dfrac{1 + i}{1 - i}$

18. $\dfrac{1 + i\sqrt{3}}{1 - i\sqrt{3}}$

19. $(2 + 3i)(1 - 2i)$

20. Use De Moivre's theorem (Eq. 19) to express $\cos 4\theta$ and $\sin 4\theta$ as polynomials in $\cos\theta$ and $\sin\theta$.

21. Find the three cube roots of unity.

22. Find the two square roots of i.

23. Find the three cube roots of $-8i$.

24. Find the six sixth roots of 64.

25. Find the four roots of the equation $z^4 - 2z^2 + 4 = 0$.

26. Find the six roots of the equation $z^6 + 2z^3 + 2 = 0$.

27. Find all roots of the equation $x^4 + 4x^2 + 16 = 0$.

28. Solve: $x^4 + 1 = 0$.

APPENDIX D

Sample Programs

In this appendix, we give listings of BASIC programs for some vector and matrix operations. These programs may be keyed into and run on a microcomputer loaded with essentially any version of BASIC. The listings given here constitute a sample to indicate the possibilities of using the computer for matrix algebra. Since keying in programs can be tedious and time consuming, only very short programs are listed here. The longer programs in support of this book are contained on the enclosed diskette.

```
10 'Addition of two vectors
20 '
50 'Enter the number of components in line 900
60 'Enter the conponents of one vector in line 910,
70 'the other in line 920
80 '
100 READ N
120 DIM U(N), V(N), W(N)
140 FOR I=1 TO N
150 READ U(I)
160 NEXT I
170 FOR I=1 TO N
180 READ V(I)
190 NEXT I
200 FOR I=1 TO  N
210 W(I) = U(I)+V(I)
220 NEXT
300 PRINT
310 PRINT "(";
320 FOR I=1 TO N - 1
330 PRINT U(I) ",";
340 NEXT
350 PRINT U(N) ")" +" * ";
410 PRINT "(";
420 FOR I=1 TO N - 1
430 PRINT V(I) ",";
440 NEXT
450 PRINT V(N) ")" " = ";
510 PRINT "(";
520 FOR I=1 TO N - 1
530 PRINT W(I) ",";
540 NEXT
550 PRINT W(N) ")"
800 END
900 DATA 3
910 DATA 2,-1,5
920 DATA 4,3,-2
```

```
10 'Multiplication of a vector by a scalar
20 '
50 'Enter the number of components and the scalar in line 900
60 'Enter the components in line 910
70 '
100 READ N
110 READ C
120 DIM U(N), V(N)
150 FOR I=1 TO N
160 READ V(I)
170 NEXT I
200 FOR I=1 TO  N
210 U(I) = C*V(I)
220 NEXT
300 PRINT
310 PRINT C "* " "(";
320 FOR I=1 TO N - 1
330 PRINT V(I) ",";
340 NEXT
350 PRINT V(N) ")" +" = ";
410 PRINT "(";
420 FOR I=1 TO N - 1
430 PRINT U(I) ",";
440 NEXT
450 PRINT U(N) ")"
800 END
900 DATA 3, 4
910 DATA 2,-1,5
```

```
10 'Inner product of two vectors  u  and  v
20 '
50 'Enter the number of components in line 900
60 'Enter the conponents of  u  starting at line 910
70 'Enter the components of  v  after the components of  u
80 '
100 READ N
120 DIM U(N), V(N)
140 FOR I=1 TO N
150 READ U(I)
160 NEXT I
170 FOR I=1 TO N
180 READ V(I)
190 NEXT I
200 SUM = 0
210 FOR I=1 TO  N
220 SUM = SUM + U(I) * V(I)
230 NEXT
300 PRINT
310 PRINT "(";
320 FOR I=1 TO N - 1
330 PRINT U(I) ",";
340 NEXT
350 PRINT U(N) ")" +" . ";
410 PRINT "(";
420 FOR I=1 TO N - 1
430 PRINT V(I) ",";
```

```
440 NEXT
450 PRINT V(N) ")" " = ";
500 PRINT SUM
800 END
900 DATA 3
910 DATA 2,-1,5
920 DATA 4,3,-2
```

```
10 'Addition of matrices,  A + B
20 '
50 'Enter the number of rows and the number of columns in line 900
60 'Enter  A  by rows starting at line 910
70 'Enter  B  by rows after the entries in  A
80 '
100 READ M,N
110 DIM A(M,N),B(M,N),C(M,N)
120 FOR I=1 TO M
125 FOR J=1 TO N
130 READ A(I,J)
140 NEXT J
145 NEXT I
150 FOR I=1 TO M
155 FOR J=1 TO N
160 READ B(I,J)
170 NEXT J
175 NEXT I
200 FOR I=1 TO M
205 FOR J=1 TO N
210 C(I,J)=A(I,J)+B(I,J)
220 NEXT J
225 NEXT I
300 PRINT
310 PRINT "A + B ="
320 FOR I = 1 TO M
325 FOR J = 1 TO N
330 PRINT TAB(1 + 8 * (J - 1)) C(I,J);
340 NEXT J
345 PRINT
350 NEXT I
800 END
900 DATA 3,5
910 DATA 1,0,2,-4,3
920 DATA 5,1,-2,-3,1
930 DATA 4,4,1,-1,3
950 DATA 3,-1,2,2,4
960 DATA 4,-1,0,1,2
970 DATA -2,1,3,4,5
```

```
10 'Multiplication of a matrix  A  by a scalar  c
20 '
50 'Enter the number of rows of  A,  the number of columns of  A,
60 'and  c  in line 900
70 'Enter  A  by rows starting at line 910
90 '
100 READ M,N
```

```
110 READ C
120 DIM A(M,N),B(M,N)
150 FOR I=1 TO M
160 FOR J=1 TO N
170 READ A(I,J)
180 NEXT J
190 NEXT I
200 FOR I = 1 TO M
210 FOR J = 1 TO N
220 B(I,J) = C*A(I,J)
230 NEXT J
240 NEXT I
300 PRINT
310 PRINT "c" " * " "A" " = "
320 PRINT
330 FOR I=1 TO M
340 FOR J = 1 TO N
350 PRINT TAB(1 + 10 * (J - 1)) B(I,J);
360 NEXT J
370 PRINT
380 NEXT I
800 END
900 DATA 2,3,-5
910 DATA 2,-1,5
920 DATA 4,0,-3
```

```
10 'Product of two matrices,  AB
20 '
50 'Enter the number of row of  A,  the number of columns of  A,
60 'and the number of columns of  B  in line 900
70 'Enter  A  by rows starting at line 910
80 'Enter  B  by rows after the entries in  A
90 '
100 READ M,N,P
110 DIM A(M,N),B(N,P),C(M,P)
120 FOR I=1 TO M
125 FOR J=1 TO N
130 READ A(I,J)
140 NEXT J
145 NEXT I
150 FOR J=1 TO N
155 FOR K=1 TO P
160 READ B(J,K)
170 NEXT K
175 NEXT J
200 FOR I = 1 TO M
210 FOR K = 1 TO P
220 SUM = 0
230 FOR J = 1 TO N
240 SUM = SUM + A(I,J) * B(J,K)
250 NEXT J
260 C(I,K) = SUM
270 NEXT K
280 NEXT I
300 PRINT
310 PRINT"The matrix product  AB  equals"
315 PRINT
```

```
320 FOR I = 1 TO M
325 FOR K = 1 TO P
330 PRINT TAB(1 + 8 * (K - 1)) C(I,K);
340 NEXT K
345 PRINT
350 NEXT I
800 END
900 DATA 2,3,4
910 DATA 1,4,3
920 DATA 5,1,-3
930 DATA 2,4,3,-1
950 DATA 1,-2,3,4
960 DATA 2,-1,1,2
```

ANSWERS TO SELECTED EXERCISES

Chapter 1

Section 1.1

1. (1, 2) **3.** (1, 1) **5.** No solution **7.** Yes **9.** No **11.** Yes
13. Yes **15.** Yes **17.** 1 **19.** (−1, 2) **21.** (3, 2, 1, −1) **23.** No
25. Place the mass at $x = 2$

Section 1.2

1. $\left[\begin{array}{cc|c} 1 & -2 & 4 \\ 2 & 1 & -2 \end{array}\right]$ **3.** $\left[\begin{array}{ccc|c} 1 & -3 & 1 & 2 \\ 2 & 0 & -1 & 1 \\ 3 & 1 & -2 & 3 \end{array}\right]$ **5.** $\begin{aligned} 2x + 4y &= 0 \\ x - 2y &= 5 \end{aligned}$

7. $\begin{aligned} x - 2y + 4z &= 4 \\ 2x + 3y - z &= 0 \\ -2x + 5z &= -1 \end{aligned}$ **9.** 3, 1) **11.** (2, 2) **13.** (0, 0) **15.** (3, 2)

17. (1, 2, −2) **19.** (4, 0, −2) **21.** (2, 1, 0, −1) **23.** (2, 1, −1, 1, 2)
39. −6 **41.** −2 **43.** −1 **45.** $((3b - 2a)/4, (2a - b)/4)$
47. $(2a - b, -6a + 4b + c, -5a + 3b + c)$
49. $x = (1/2)x' - (\sqrt{3}/2)y', y = (\sqrt{3}/2)x' + (1/2)y'$

Section 1.3

1. $(4y, y)$ **3.** $(-2y/3, y)$, or $(-2y, 3y)$
5. $(-2x_2 + 3x_3 - 3x_4, x_2, x_3, x_4)$ **7.** $(z/7, -4z/7, z)$, or $(z, -4z, 7z)$
9. $(-z, 4z, 2z)$ **11.** (0, 0, 0) **13.** $(-8z, -5z, 14z)$
15. $(4x_3 + 3x_4, -x_3 + 8x_4, 7x_3, 7x_4)$ **17.** $(10x_3, -3x_3, -x_3, 0)$
19. (0, 0, 0, 0) **21.** No **43.** No **45.** 3 **47.** 2

49. $\begin{bmatrix} 1 & 1/2 \\ 0 & 1 \end{bmatrix}, \begin{bmatrix} 1 & 2 \\ 0 & 1 \end{bmatrix}$ **51.** $\begin{bmatrix} 1 & -1/2 & -1/2 \\ 0 & 1 & 5/3 \\ 0 & 0 & 1 \end{bmatrix}, \begin{bmatrix} 1 & 1 & 2 \\ 0 & 1 & 5/3 \\ 0 & 0 & 1 \end{bmatrix}$

53. $\begin{bmatrix} 1 & 0 \\ 0 & 1 \end{bmatrix}$ **55.** $\begin{bmatrix} 1 & 0 & 0 \\ 0 & 1 & 0 \\ 0 & 0 & 1 \end{bmatrix}$ **57.** $4x - y - 3z = 0$ **59.** −5

Section 1.4

1. Yes **3.** $(7/29, 10/29, -2/29)$ **5.** $(1+z, 1-z, z)$ **7.** $(10, -5, 7)$
9. Inconsistent **11.** $(4z, -2z, z)$ **13.** $(2, 0, 1)$ **15.** $(2y+3, y, -2)$
17. $(x_3 - x_4, 1 - x_4, x_3, x_4)$ **19.** $(4/3 + x_3, 1/3 - x_3 - x_4, x_3, x_4)$
21. $(7z - 6, 5 - 5z, z)$ **23.** $(x_4, 0, 0, x_4)$
25. $(3/2 - 3x_3/2 + x_4, 1/2 - x_3/2, x_3, x_4)$ **29.** Yes **31.** No
33. $a \neq 1/2$ **35.** $5a - 4b + 4 = 0$ **37.** No

Section 1.5

1. $2x - y = 1$ **3.** $3x - 2y = -7$ **5.** The system is inconsistent
7. $y = -x^2 - + 4x + 2$ **9.** $(x-2)^2 + (y+1)^2 = 4$
11. $y = (2x^3 + 2x^2 + 4x - 8)/3$ **13.** $y = x^4 - x^3 - 2x^2 + x + 3$
15. $z = ax - (a+2)y + 4$, a arbitrary
17. The resulting system is inconsistent
19. $(x-1)^2 + (y+1)^2 + (z-2)^2 = 1$ **21.** $(16.63, 16.52, 24.72, 14.72, 22.36)$
23. $A = 0, B = 1/3, C = 0, D = -1/3$ **25.** $A = -1/4, B = 1/2, C = 1/4, D = 1/2$
27. $I_1 = 0.9, I_2 = 2.1, I_3 = 0.3, I_4 = 1.2, I_5 = 1.8$, all in amperes
29. $I_1 = 0.267, I_2 = 2.133, I_3 = 2.400, I_4 = 1.467, I_5 = -0.667, I_6 = 0.933$, all in amperes

Review Exercises

1. $(-y, 2y, 0)$ **3.** $(-z, -2z, 9z)$
5. $((5 - 5x_3 + 2x_4)/4, (1 + 3x_3 + 2x_4)/4, x_3, x_4)$ **7.** $(8, 1)$ **9.** $(0, 0, 0)$
11. Yes **13.** 2 **15.** 4

Chapter 2

Section 2.1

1. 1 **3.** $\sqrt{5}$ **5.** 1 **7.** 3 **15.** $(-3, -4)$ **17.** $(-2, 17, 7, 2)$
19. $(3/2, -1, -1)$ **21.** 7 **23.** -10 **25.** Yes **27.** $(2, 0)$
29. $(5/2, 5/2)$ **31.** $(-1/3, 2/3, 2/3)$ **33.** $\mathbf{v} \cdot (-2, 1) = 0$
37. None exist **39.** $(2, 1, 1)$

Section 2.2

1. $\begin{bmatrix} 3 & 2 \\ 3 & 2 \end{bmatrix}$ **3.** $\begin{bmatrix} 7 & 3 \\ 10 & 6 \end{bmatrix}$ **5.** $\begin{bmatrix} 6 & 14 \\ -12 & -4 \end{bmatrix}$

7. $\begin{bmatrix} 12 & 18 \\ -6 & 0 \end{bmatrix}$ **9.** $\begin{bmatrix} 3/4 & 3/2 \\ -9/4 & 3/2 \end{bmatrix}$

11. $\begin{bmatrix} 3 & 1 \\ 3 & 6 \\ 5 & 6 \end{bmatrix}$ **13.** $\begin{bmatrix} c_1 & 4c_1 \\ 2c_1 & 4c_1 \\ c_1 & 7c_1 \end{bmatrix}$ **15.** $\begin{bmatrix} c_1 + c_2 & 3(c_1 + c_2) \\ -c_1 - c_2 & 2(c_1 + c_2) \\ 3(c_1 + c_2) & 4(c_1 + c_2) \end{bmatrix}$

17. $\begin{bmatrix} a_{11} & a_{12} \\ a_{21} & a_{22} \end{bmatrix}$ **19.** $\begin{bmatrix} 4 & 2 \\ 14 & -3 \end{bmatrix}$ **21.** $\begin{bmatrix} -4 & 10 & 14 \\ 3 & 6 & 13 \end{bmatrix}$

23. $\begin{bmatrix} -2 & 11 & 13 \\ 1 & -8 & -19 \\ 10 & 6 & 0 \end{bmatrix}$ **25.** $\begin{bmatrix} 2 & 11 \\ -11 & 2 \end{bmatrix}$

27. $\begin{bmatrix} 4 & 3 \\ 14 & 3 \end{bmatrix}$ **29.** $\begin{bmatrix} -1 & 5 \\ 1 & 20 \end{bmatrix}$

31. $\begin{bmatrix} -1 & 1 \\ 2 & 4 \end{bmatrix}$ **33.** $\begin{bmatrix} -3 & -2 \\ 0 & 10 \end{bmatrix}$ **35.** $\begin{bmatrix} 4 & 14 \\ 3 & 3 \end{bmatrix}$

37. $\begin{bmatrix} 1 & 16 \\ 2 & 2 \end{bmatrix}$ **39.** $\begin{bmatrix} -4 & -1 \\ -2 & 7 \end{bmatrix}$

41. $\begin{bmatrix} 4 & 6 \\ -2 & 2 \end{bmatrix}$ **43.** $\begin{bmatrix} 5 & 5 \\ 5 & 10 \end{bmatrix}$ **45.** $\begin{bmatrix} 1 & 4 \\ 1 & 5 \end{bmatrix}$

47. Not possible

49. $\begin{bmatrix} 3 & 9 & 3 \\ -3 & 6 & 12 \end{bmatrix}$ **51.** Not possible

53. $\begin{bmatrix} 0 & 0 & 0 \\ 0 & 0 & 0 \end{bmatrix}$ **55.** $\begin{bmatrix} 4 \\ 3 \end{bmatrix}$

57. I, I **59.** $\begin{bmatrix} 2 & 11 & -7 \\ 0 & 8 & -1 \\ 0 & 0 & 6 \end{bmatrix}, \begin{bmatrix} 2 & 4 & 1 \\ 0 & 8 & 0 \\ 0 & 0 & 6 \end{bmatrix}$

61. $\begin{bmatrix} 2 & 0 & 0 \\ 8 & -1 & 3 \\ 1 & 5 & 2 \end{bmatrix}, \begin{bmatrix} 2 & 0 & 0 \\ -1 & -1 & 3 \\ 11 & 5 & 2 \end{bmatrix}$

Section 2.3

1. Both sides: $\begin{bmatrix} 2 & 13 \\ 6 & 11 \end{bmatrix}$ **3.** Both sides: $\begin{bmatrix} 2 & 15 \\ 5 & 15 \end{bmatrix}$

5. Both sides: $\begin{bmatrix} 7 & 30 \\ 0 & 6 \end{bmatrix}$ **7.** No **9.** Yes **11.** No **15.** $\begin{bmatrix} 1 & 2 \\ 2 & 4 \end{bmatrix}$

17. $\begin{bmatrix} 4 & 6 \\ 6 & 13 \end{bmatrix}$ **19.** $A^3 + 3A^2 + 3A + I$ **21.** $A_k^t \ldots A_2^t A_1^t$

23. $\begin{bmatrix} 1 & 0 & 0 \\ 0 & -1 & 0 \\ 0 & 0 & 2 \end{bmatrix}$ **25.** $\begin{bmatrix} 0 & 1 \\ 0 & 0 \end{bmatrix}$ **31.** $\begin{bmatrix} 1 \\ 1 \end{bmatrix}$ **33.** $\begin{bmatrix} 1 & 0 \\ 0 & 1 \end{bmatrix}$

41. No, e.g., $A = \begin{bmatrix} 0 & 1 \\ 0 & 0 \end{bmatrix}, B = \begin{bmatrix} 1 & 0 \\ 0 & 0 \end{bmatrix}$

Section 2.4

1. $\begin{bmatrix} 5 & -4 \\ -1 & 1 \end{bmatrix}$ **3.** $\begin{bmatrix} 1 & -3/2 \\ -1 & 2 \end{bmatrix}$ **5.** $\begin{bmatrix} 1 & 0 \\ -2/5 & 1/5 \end{bmatrix}$ **7.** $(2, -1)$

9. $(-4/7, 9/7)$ **11.** $(-1, -1)$ **13.** $(4, -1, -1)$ **15.** $(-7, -1, 3, 2)$

17. $\begin{bmatrix} 0 & 5 & 1 & 9 \\ 0 & 3 & 1 & 7 \\ 1 & -1 & 1 & -3 \end{bmatrix}$ **19.** $\begin{bmatrix} 1/5 & 1/5 \\ -3/5 & 2/5 \end{bmatrix}$ **21.** $\begin{bmatrix} 7 & -4 & -5 \\ 4 & -2 & -3 \\ -1 & 1 & 1 \end{bmatrix}$

23. $\begin{bmatrix} -9 & 7 & -4 \\ 4 & -3 & 2 \\ 3 & -2 & 1 \end{bmatrix}$ **25.** Not invertible **27.** $\begin{bmatrix} 4/5 & -3/5 \\ -1/5 & 2/5 \end{bmatrix}$

29. Not invertible **31.** $\begin{bmatrix} 1/4 & -1/4 & 1/4 \\ -1/4 & 1/4 & 1/4 \\ -1/2 & -1/2 & 0 \end{bmatrix}$

33. $\begin{bmatrix} 3/4 & -1/4 & -1/4 \\ -1/4 & 3/4 & -1/4 \\ -1/4 & -1/4 & 3/4 \end{bmatrix}$ **35.** $\begin{bmatrix} 5/43 & 2/43 & 6/43 \\ 14/43 & -3/43 & -9/43 \\ -8/43 & 14/43 & -1/43 \end{bmatrix}$

37. $\begin{bmatrix} 17/2 & 7/2 & -3/2 & 1/2 \\ 33/4 & 13/4 & -7/4 & 3/4 \\ 5/2 & 1/2 & -1/2 & 1/2 \\ -13/2 & -5/2 & 3/2 & -1/2 \end{bmatrix}$

39. $\begin{bmatrix} 1 & 1 & 2 & -4 \\ 5/2 & 2 & 7 & -25/2 \\ -3/2 & -1 & -4 & 15/2 \\ 1/2 & 0 & 1 & -3/2 \end{bmatrix}$

Section 2.5

1. $\begin{bmatrix} 0 & 1 \\ 1 & 0 \end{bmatrix}, \begin{bmatrix} 0 & 1 \\ 1 & 0 \end{bmatrix}$ **3.** $\begin{bmatrix} 1 & 0 \\ 0 & -3 \end{bmatrix}, \begin{bmatrix} 1 & 0 \\ 0 & -1/3 \end{bmatrix}$

5. $\begin{bmatrix} 1 & -1 \\ 0 & 1 \end{bmatrix}, \begin{bmatrix} 1 & 1 \\ 0 & 1 \end{bmatrix}$ **7.** $R2 \leftarrow R2 + 2R1$, $\begin{bmatrix} 1 & 0 \\ -2 & 1 \end{bmatrix}$

9. $R1 \leftrightarrow R2$, $\begin{bmatrix} 0 & 1 \\ 1 & 0 \end{bmatrix}$ **11.** $\begin{bmatrix} 0 & 1 & 0 \\ 1 & 0 & 0 \\ 0 & 0 & 1 \end{bmatrix}, \begin{bmatrix} 0 & 1 & 0 \\ 1 & 0 & 0 \\ 0 & 0 & 1 \end{bmatrix}$

13. $\begin{bmatrix} 1 & 0 & 0 \\ -2 & 1 & 0 \\ 0 & 0 & 1 \end{bmatrix}, \begin{bmatrix} 1 & 0 & 0 \\ 2 & 1 & 0 \\ 0 & 0 & 1 \end{bmatrix}$ **15.** $\begin{bmatrix} 1 & 0 & 0 \\ 0 & 1 & 0 \\ 0 & 0 & 4 \end{bmatrix}, \begin{bmatrix} 1 & 0 & 0 \\ 0 & 1 & 0 \\ 0 & 0 & 1/4 \end{bmatrix}$

17. $R2 \leftarrow -3R2$, $\begin{bmatrix} 1 & 0 & 0 \\ 0 & -1/3 & 0 \\ 0 & 0 & 1 \end{bmatrix}$ **19.** $R1 \leftrightarrow R3$, $\begin{bmatrix} 0 & 0 & 1 \\ 0 & 1 & 0 \\ 1 & 0 & 0 \end{bmatrix}$

21. R1 ← R1 − 3R3, $\begin{bmatrix} 1 & 0 & 3 \\ 0 & 1 & 0 \\ 0 & 0 & 1 \end{bmatrix}$

23. $\begin{bmatrix} 1 & 0 \\ 0 & 1/2 \end{bmatrix}\begin{bmatrix} 1 & 0 \\ 3 & 1 \end{bmatrix}\begin{bmatrix} 1 & 0 \\ 0 & -7 \end{bmatrix}\begin{bmatrix} 1 & 1 \\ 0 & 1 \end{bmatrix}\begin{bmatrix} 2 & 0 \\ 0 & 1 \end{bmatrix}$

25. $\begin{bmatrix} 1 & 0 & 0 \\ 1 & 1 & 0 \\ 0 & 0 & 1 \end{bmatrix}\begin{bmatrix} 1 & 0 & 0 \\ 0 & 1 & 0 \\ 2 & 0 & 1 \end{bmatrix}\begin{bmatrix} 1 & 0 & 0 \\ 0 & 1 & 0 \\ 0 & 1 & 1 \end{bmatrix}\begin{bmatrix} 1 & 0 & 0 \\ 0 & -1 & 0 \\ 0 & 0 & 1 \end{bmatrix}$

$\begin{bmatrix} 1 & 0 & 0 \\ 0 & 1 & 0 \\ 0 & 0 & -2 \end{bmatrix}\begin{bmatrix} 1 & 0 & 0 \\ 0 & 1 & -3 \\ 0 & 0 & 1 \end{bmatrix}\begin{bmatrix} 1 & 0 & -1 \\ 0 & 1 & 0 \\ 0 & 0 & 1 \end{bmatrix}$

27. $\begin{bmatrix} 1 & 0 & 0 \\ 2 & 1 & 0 \\ 0 & 0 & 1 \end{bmatrix}\begin{bmatrix} 1 & 0 & 0 \\ 0 & 1 & 0 \\ 1 & 0 & 1 \end{bmatrix}\begin{bmatrix} 1 & 0 & 0 \\ 0 & 1 & 0 \\ 0 & -1 & 1 \end{bmatrix}$

$\begin{bmatrix} 1 & 0 & 0 \\ 0 & 1 & 2 \\ 0 & 0 & 1 \end{bmatrix}\begin{bmatrix} 1 & 2 & 0 \\ 0 & 1 & 0 \\ 0 & 0 & 1 \end{bmatrix}\begin{bmatrix} 1 & 0 & 1 \\ 0 & 1 & 0 \\ 0 & 0 & 1 \end{bmatrix}$ **31.** $\begin{bmatrix} 1 & 0 & 0 \\ 2 & 1 & 0 \\ -4 & 0 & 1 \end{bmatrix}$

Section 2.6

1. $\begin{bmatrix} 1 & 0 \\ 2 & 1 \end{bmatrix}\begin{bmatrix} 1 & 2 \\ 0 & 1 \end{bmatrix}$ **3.** $\begin{bmatrix} 1 & 0 & 0 \\ 2 & 1 & 0 \\ -1 & 2 & 1 \end{bmatrix}\begin{bmatrix} 1 & -2 & 1 \\ 0 & 1 & 3 \\ 0 & 0 & 1 \end{bmatrix}$

5. $\begin{bmatrix} 1 & 0 & 0 & 0 \\ 2 & 1 & 0 & 0 \\ -1 & 2 & 1 & 0 \\ -2 & -2 & 3 & 1 \end{bmatrix}\begin{bmatrix} 1 & 1 & 0 & -2 \\ 0 & 1 & 2 & 2 \\ 0 & 0 & 1 & 3 \\ 0 & 0 & 0 & 1 \end{bmatrix}$

7. $\begin{bmatrix} 1 & 0 & 0 \\ -1 & 1 & 0 \\ -2 & 1 & 1 \end{bmatrix}\begin{bmatrix} 1 & -1 & 3 \\ 0 & 3 & 6 \\ 0 & 0 & 5 \end{bmatrix}$ **9.** $(1/4, 1/4)$ **11.** $(5/4, -3, -3)$

13. $\begin{bmatrix} 1 & 0 \\ 2 & 1 \end{bmatrix}\begin{bmatrix} 1 & 0 \\ 0 & 1 \end{bmatrix}\begin{bmatrix} 1 & 2 \\ 0 & 1 \end{bmatrix}$ **15.** $\begin{bmatrix} 1 & 0 & 0 \\ 1 & 1 & 0 \\ 2 & 0 & 1 \end{bmatrix}\begin{bmatrix} 1 & 0 & 0 \\ 0 & 1 & 0 \\ 0 & 0 & -3 \end{bmatrix}\begin{bmatrix} 1 & 2 & 0 \\ 0 & 1 & 1 \\ 0 & 0 & 1 \end{bmatrix}$

17. $(15/35, 12/35, 8/35)$

19. (a) (31.8, 25.4, 20.8, 17.0, 13.9, 11.1),
(31.8, 25.4, 20.8, 17.0, 13.9, 11.0)

21. (a) (.2917, .4167, .2917), (.2865, .4271, .2865), (.2858, .4284, .2858),
(b) $(2/7, 3/7, 2/7)$

Review Exercises

1. $\begin{bmatrix} 14 & 13 \\ 3 & 1 \end{bmatrix}$ **3.** $\begin{bmatrix} 2 & 8 & 2 \\ 1 & 4 & 1 \\ -2 & -8 & -2 \end{bmatrix}$ **5.** $\begin{bmatrix} -3/7 & 2/7 \\ 5/7 & -1/7 \end{bmatrix}$

7. Not invertible **9.** $\begin{bmatrix} 1 & 2 & -1 \\ 1/2 & -3/2 & 1/2 \\ -1/2 & -1/2 & 1/2 \end{bmatrix}$

11. $\begin{bmatrix} -72 & -19 \\ -26 & -7 \\ 11 & 3 \end{bmatrix}$ **13.** $\begin{bmatrix} 0 & 5 & 3 \\ 1 & 8 & 6 \\ 0 & -3 & -2 \end{bmatrix}$ **15.** $\begin{bmatrix} 2 & 1 & 6 \\ 2 & 0 & 3 \\ -3 & 0 & -5 \end{bmatrix}$

Chapter 3

Section 3.1

1. -4 **3.** 1 **5.** -5 **7.** 0 **9.** 0 **11.** 0 **13.** 2. **15.** 4

17. -44 **19.** $\begin{vmatrix} 8 & 1 \\ 0 & 3 \end{vmatrix}, \begin{vmatrix} 2 & 1 \\ 6 & 3 \end{vmatrix}, \begin{vmatrix} 2 & 8 \\ 6 & 0 \end{vmatrix}, \begin{vmatrix} 5 & 2 \\ 0 & 3 \end{vmatrix},$

$\begin{vmatrix} 4 & 2 \\ 6 & 3 \end{vmatrix}, \begin{vmatrix} 4 & 5 \\ 6 & 0 \end{vmatrix} \begin{vmatrix} 5 & 2 \\ 8 & 1 \end{vmatrix}, \begin{vmatrix} 4 & 2 \\ 2 & 1 \end{vmatrix}, \begin{vmatrix} 4 & 5 \\ 2 & 8 \end{vmatrix}$; $24, 0, -48, -15, 0, 30, -11, 0, 22$

21. 24 **23.** -12 **25.** 12 **27.** $\begin{bmatrix} 1 & 0 \\ 0 & 0 \end{bmatrix}, \begin{bmatrix} 0 & 0 \\ 0 & 1 \end{bmatrix}$

29. Both equal -20 **31.** ± 2 **33.** $0, 1, 2$ **35.** $9, 27$ **37.** $\pm\sqrt{5}$

39. $\det(AB) = \det(BA)$

Section 3.2

1. $\begin{bmatrix} 1/3 & 1/9 \\ -1/3 & 2/9 \end{bmatrix}$ **3.** Not invertible

5. $\begin{bmatrix} -1/2 & -1/2 & 1 \\ 3/2 & 1 & -3/2 \\ -1/2 & 0 & 1/2 \end{bmatrix}$ **7.** $\begin{bmatrix} 1 & 1 & -4 \\ 0 & 1 & -2 \\ 1 & 0 & 1 \end{bmatrix}$

9. $\begin{bmatrix} -1/7 & -4/7 & 2/7 \\ 5/14 & -1/14 & -3/14 \\ -1/7 & 3/7 & 2/7 \end{bmatrix}$ **11.** $(-3, 7/5)$ **13.** $(2/3, 5/3)$

15. $(2, 4, -3)$ **17.** $(7, -3, -9)$ **19.** T

Section 3.4

1. $\begin{vmatrix} x & y & 1 \\ 4 & 1 & 1 \\ 3 & 2 & 1 \end{vmatrix} = 0$ **3.** $\begin{vmatrix} x & y & 1 \\ 1 & -2 & 1 \\ 2 & 1 & 1 \end{vmatrix} = 0$ **5.** $\begin{vmatrix} x^2 & x & y & 1 \\ 1 & 1 & -2 & 1 \\ 4 & 2 & 2 & 1 \\ 4 & -2 & -2 & 1 \end{vmatrix} = 0$

7. $\begin{vmatrix} x^2 & x & y & 1 \\ 1 & -1 & 2 & 1 \\ 4 & 2 & 4 & 1 \\ 9 & 3 & 4 & 1 \end{vmatrix} = 0$ **9.** $3y^2 - 4y - x = 0$

11. $y^2 - 4y - x + 4 = 0$ **13.** $\begin{vmatrix} x & y & z & 1 \\ x_1 & y_1 & z_1 & 1 \\ x_2 & y_2 & z_2 & 1 \\ x_3 & y_3 & z_3 & 1 \end{vmatrix} = 0$

15. $x + y + z = 2$ **17.** $y = 1$

19. $2x^3 + 2x^2 + 4x - 3y - 8 = 0$ **21.** $\begin{vmatrix} x^2 + y^2 & x & y & 1 \\ x_1^2 + y_1^2 & x_1 & y_1 & 1 \\ x_2^2 + y_2^2 & x_2 & y_2 & 1 \\ x_3^2 + y_3^2 & x_3 & y_3 & 1 \end{vmatrix} = 0$

23. $(x + 1)^2 + y^2 = 5$ **25.** 3 **27.** $\sqrt{2}$ **29.** 2 **31.** 16

33. Half the absolute value of the determinant $\begin{vmatrix} 1 & x_1 & y_1 \\ 1 & x_2 & y_2 \\ 1 & x_3 & y_3 \end{vmatrix}$ **35.** 2 **37.** 2

39. 1/6

Review Exercises

1. 0 **3.** 2 **5.** 0 **7.** -8 **9.** $\begin{bmatrix} -3/7 & 2/7 \\ 5/7 & -1/7 \end{bmatrix}$ **11.** Not invertible

13. Not invertible **15.** 0 **17.** $(8/7, -1/7)$

Chapter 4

Section 4.1

5. No **7.** x^2 is in W but $0x^2$ is not **9.** Yes **11.** No **13.** Yes

17. No **19.** Yes **21.** Yes **23.** Yes **25.** No **27.** No

29. $(a_0 + a_1x + a_2x^2 + a_3x^3) + (b_0 + b_1x + b_2x^2 + b_3x^3)$
$= (a_0 + b_0) + (a_1 + b_1)x + (a_2 + b_2)x^2 + (a_3 + b_3)x^3$,
$c(a_0 + a_1x + a_2x^2 + a_3x^3) = ca_0 + (ca_1)x + (ca_2)x^2 + (ca_3)x^3$, Yes

31. No **35.** Yes **37.** No **39.** Yes

Section 4.2

1. $0(1,4) + 0(3,7)$ **3.** No **5.** No **7.** Yes **9.** No **11.** Yes
13. $(3,1)$ **15.** $(2,-4,6)$ **17.** $(2,8,-8)$ **19.** $2(1,0) + 3(0,1)$
21. Not possible **23.** $(-1)(-2,-3) + 0(1,1)$
25. $-(1,0,0) + (0,1,0) + 2(0,0,1)$ **27.** $2(1,2,-1) - (3,3,-4)$
29. $-(1/4)(4,-4,-8)$
31. $(1,0,0,0) + 2(0,1,0,0) - 3(0,0,0,1) - 2(0,0,0,1)$
33. $(1/3)(3,6,-9,-6)$ **35.** $3(1,2,-1,-2) - 2(1,2,0,-2)$
37. $3(1,2,-1) + 4(1,-1,2) - 2(2,0,1)$
39. $(4/3)(1,2,-1) + (5/3)(1,-1,2) - (2,0,1)$
41. Yes **43.** Yes **45.** No **47.** No **49.** Yes **51.** No
53. $(3t, 2t)$ **57.** $(7,-1,1)$ **59.** $s(1,0,-2) + t(2,1,0)$ **61.** $4x + 3y = 0$
63. $(2,4,6)$ **65.** $2x_1 - x_2 = 0$, $3x_1 - x_3 = 0$
69. $s(1,-1,2) + t(1,0,1)$

71. $2\begin{bmatrix}1 & 0\\0 & 0\end{bmatrix} - 3\begin{bmatrix}1 & 1\\0 & 0\end{bmatrix} + 4\begin{bmatrix}1 & 1\\1 & 0\end{bmatrix} + \begin{bmatrix}1 & 1\\1 & 1\end{bmatrix}$

73. Yes **77.** $f = f_2 - 4f_0$ **79.** $h = f_3 - 6f_2 + 12f_1 - 8f_0$ **83.** Yes

Section 4.3

1. No **3.** Yes **5.** No **7.** No **9.** No **11.** Yes **13.** No solution
15. $(-3,2,1)$ **17.** No **19.** No **21.** Yes **23.** ± 6 **25.** 2
27. 2 **29.** S is linearly independent **31.** S is linearly dependent
33. No **35.** Yes **37.** Yes **39.** No **41.** No
43. If S is linearly dependent and T contains S, then T is linearly dependent
45. If S is linearly independent and T is a subset of S, then T is linearly independent
47. Yes **49.** No **51.** No **53.** $4x_1 - x_2 - 9x_3 = 0$
55. $(1,4,0) + 2(2,-1,1)$ and $3(1,4,0) + 2(1,-5,1)$ **59.** Yes **61.** No
63. $B = A^t$

Section 4.4

1. Yes **3.** No **5.** Yes **7.** T **9.** T **11.** T
13. $B = \{(1,1,0,0), (4,0,-1,0), (2,0,0,-1)\}, 3$ **15.** The line $(6t, -2t, t)$

17. The line $(3t, -2t, 0, -t)$ **19.** $\mathbb{R}^3$

21. $\mathbb{R}^3$ **23.** $B = \{(1,0,1,-2),(0,1,2,0),(0,0,1,3)\}, 3$

25. $B = \left\{\begin{bmatrix}1 & 0\\0 & 0\end{bmatrix}, \begin{bmatrix}0 & 1\\0 & 0\end{bmatrix}, \begin{bmatrix}0 & 0\\1 & 0\end{bmatrix}, \begin{bmatrix}0 & 0\\0 & 1\end{bmatrix}\right\}, 4$

27. Yes **29.** No

31. $B = \{(2,-1,0,0,0,0),(3,0,0,-1,0,0),(12,0,-2,0,3,-1)\}, 3$ **33.** Yes

35. $B = \{(1,2,0,-1,1),(2,5,2,-2,0),(1,1,-2,0,5),(-2,-3,2,3,-1)\}, 4$

37. No **39.** Append $(1,0,0,0)$ to the given set

41. $3(1,2,0) - (2,1,-1), (8/3)(1,2,0) - (1/3)(5,1,-3)$

Section 4.5

1. The origin, 0 **3.** $\mathbb{R}^2, 2$ **5.** The span of $\{(1,-3,3),(0,5,-4)\}, 2$

7. The origin, 0 **9.** $\mathbb{R}^2, 2$

11. The span of $\{[1 \quad 2 \quad -2]^t, [0 \quad 1 \quad -3]^t\}, 2$

13. $B = \{(1,0,0,0),(0,1,0,0),(0,0,1,0),(0,0,0,1)\}$

15. $B = \{(1,2,1,3),(0,1,2,5),(0,0,5,12)\}$

17. $B = \{[1 \quad 0 \quad 0 \quad 0]^t, [0 \quad 1 \quad 0 \quad 0]^t, [0 \quad 0 \quad 1 \quad 0]^t, [0 \quad 0 \quad 0 \quad 1]^t\}$

19. $B = \{[1 \quad 2 \quad 2 \quad 5]^t, [0 \quad 1 \quad 3 \quad 5]^t, [0 \quad 0 \quad 0 \quad 1]^t\}$

21. $B = \{(1,-3,3),(2,-1,2)\}$ **23.** $\{(1,2,1,-2),(1,3,0,-2),(2,6,1,-2),(1,-3,3,-3)\}$

25. $1 = 2 \cdot 2 + 1(-3)$
$0 = 3 \cdot 2 + 2(-3)$

27. $(2,3,2) = (3,2,2) - (1,-1,0)$
$(10,0,4) = 2(3,2,2) + 4(1,-1,0)$

29. $(-3,3) = (-3(2,1) + (1,2) + 2(1,2)$
$(5,4) = 2(2,1) + 2(1,2) - (1,2)$

31. $\begin{bmatrix}1\\0\end{bmatrix} = 2\begin{bmatrix}2\\3\end{bmatrix} - 3\begin{bmatrix}1\\2\end{bmatrix}$

33. $\begin{bmatrix}2\\10\end{bmatrix} = 3\begin{bmatrix}1\\2\end{bmatrix} + \begin{bmatrix}-1\\4\end{bmatrix}, \begin{bmatrix}3\\0\end{bmatrix} = 2\begin{bmatrix}1\\2\end{bmatrix} - \begin{bmatrix}-1\\4\end{bmatrix}, \begin{bmatrix}2\\4\end{bmatrix} = 2\begin{bmatrix}1\\2\end{bmatrix}$

35. $\begin{bmatrix}-2\\1\end{bmatrix} = \begin{bmatrix}0\\2\end{bmatrix} + \begin{bmatrix}1\\4\end{bmatrix} + \begin{bmatrix}-3\\-5\end{bmatrix},$

$\begin{bmatrix}-5\\-2\end{bmatrix} = 2\begin{bmatrix}0\\2\end{bmatrix} + \begin{bmatrix}1\\4\end{bmatrix} + 2\begin{bmatrix}-3\\-5\end{bmatrix},$

$\begin{bmatrix}-7\\-1\end{bmatrix} \quad 3\begin{bmatrix}0\\2\end{bmatrix} + 2\begin{bmatrix}1\\4\end{bmatrix} + 3\begin{bmatrix}-3\\-5\end{bmatrix}$

37. $\begin{bmatrix}3 & -4\\2 & 3\end{bmatrix}\begin{bmatrix}0 & 1 & 2\\1 & 2 & 1\end{bmatrix} = \begin{bmatrix}-4 & -5 & 2\\3 & 8 & 7\end{bmatrix}$

39. $\begin{bmatrix} 4 & -1 \\ 2 & 2 \\ 1 & 2 \end{bmatrix}\begin{bmatrix} 2 \\ -1 \end{bmatrix} = \begin{bmatrix} 9 \\ 2 \\ 0 \end{bmatrix}$ **41.** $\begin{bmatrix} -1 & 2 & 1 \\ 1 & 1 & -2 \\ 2 & -1 & 1 \end{bmatrix}\begin{bmatrix} 1 & 3 \\ 1 & 1 \\ 1 & 2 \end{bmatrix} = \begin{bmatrix} 2 & 1 \\ 0 & 0 \\ 2 & 7 \end{bmatrix}$

45. $\begin{vmatrix} 1 & 2 \\ 3 & 1 \end{vmatrix}, \begin{vmatrix} 1 & 1 \\ 3 & 1 \end{vmatrix}, \begin{vmatrix} 1 & 3 \\ 3 & 2 \end{vmatrix}, \begin{vmatrix} 2 & 1 \\ 1 & 1 \end{vmatrix}, \begin{vmatrix} 2 & 3 \\ 1 & 2 \end{vmatrix}, \begin{vmatrix} 1 & 3 \\ 1 & 2 \end{vmatrix}$

Section 4.6

9. The line $(t, -3t, -2t)$

11. The system is consistent by Theorem 4.14a, g applied to A^t

15. The row space of A coincides with $\mathbb{R}^n$ by Theorem 4.13f, k **17.** n

19. By Theorem 4.13, if A is any $n \times n$ matrix for which the sum of each row equals 0, then A is not invertible

25. $\begin{bmatrix} 1 & 2 \\ 1 & 0 \\ 0 & -1 \end{bmatrix}\begin{bmatrix} c_1 \\ c_2 \end{bmatrix} = \begin{bmatrix} b_1 \\ b_2 \\ b_3 \end{bmatrix}$

27. Both coincide with $\mathbb{R}^n$, since $\det(A)\det(B) = \det(AB) \neq 0$

Review Exercises

1. All (x, y, z) with two components zero **3.** No **5.** Yes **7.** No

9. $(2, 4, 3) = -3(1, 1, 0) + 4(1, 2, 1) + (1, -1, -1)$ **11.** $(1, -1, -1) = 0(1, 1, 0) + 0(1, 2, 1) + (1, -1, -1)$ **13.** 2 **15.** 2 **17.** $B = \{(3, 1, -2)\}$

Chapter 5

Section 5.1

1. $3(1, 0) + 2(0, 1)$ **3.** $(17/5)(3/5, 4/5) - (6/5)(-4/5, 3/5)$

5. $\|x\| = \sqrt{13}$

7. $(3/\sqrt{2})(1/\sqrt{2}, 0, -1/\sqrt{2}) - (3/\sqrt{6})(1/\sqrt{6}, -2/\sqrt{6}, 1/\sqrt{6})$

9. $(4/\sqrt{5})(2/\sqrt{5}, 0, -1/\sqrt{5}) - (2/\sqrt{6})(1/\sqrt{6}, 1/\sqrt{6}, 1/\sqrt{6}) - (8/\sqrt{30})(1/\sqrt{30}, -5/\sqrt{30}, 2/\sqrt{30})$

11. $B = \{(1/\sqrt{2}, 1/\sqrt{2}), (1/\sqrt{2}, -1/\sqrt{2})\}$

13. $B = \{3/5, 0, 4/5), (28/\sqrt{1850}, 25/\sqrt{1850}, -21/\sqrt{1850})\}$

15. $B = \{(2/3, 2/3, 1/3), (1/3, -2/3, 2/3), (-2/3, 1/3, 2/3)\}$

17. $B = \{(1/\sqrt{2}, 0, 1/\sqrt{2}, 0), (0, 1/\sqrt{2}, 0, 1/\sqrt{2})\}$

19. $B = \{(1/\sqrt{2}, 1/\sqrt{2}, 0, 0), (-1/\sqrt{6}, 1/\sqrt{6}, 2/\sqrt{6}, 0)$ $(1/\sqrt{12}, -1/\sqrt{12}, 1/\sqrt{12}, 3/\sqrt{12})\}$ **21.** $(1, 1/2, 1/2)$

23. $(1/49)(10, 29, 72)$ **25.** $(2/3, -2/3, 1, 1/3)$ **27.** $(1, 1/2, 1/2, 0)$

29. $(3/5, 1, 6/5)$ **31.** $(0, 1, 0, 1)$

33. A system $AX = B$ is consistent if and only if B lies in the or thogonal complement of the nullspace of A^t. **35.** $B = \{(0, 0, 1), (1/\sqrt{2}, 1/\sqrt{2}, 0)\}$

37. $k = \langle \mathbf{x}, \mathbf{a} \rangle / \langle \mathbf{a}, \mathbf{a} \rangle$

Section 5.2

1. No **3.** No

5. $(x_1y_1 + 2x_2y_2 + 3x_3y_3)^2 \leq (x_1^2 + 2x_2^2 + 3x_3^2)(y_1^2 + 2y_2^2 + 3y_3^2)$ **7.** Yes

9. No **11.** No **13.** Yes **15.** $2x_1y_1 + x_1y_2 + x_2y_1 + 4x_2y_2$

17. $A = I$, the identity **19.** $\|\mathbf{x} + \mathbf{y}\|_A = \sqrt{5}$, $\|\mathbf{x}\|_A = \sqrt{13}$, $\|\mathbf{y}\|_A = \sqrt{10}$

35. Both equal $x_1^2 + 3x_1x_2 + 3x_2^2 + x_2x_3 + 2x_3^2$; the expansions x^tAx and x^tA^tx are equal **37.** $\begin{bmatrix} 2 & -3 \\ -3 & 5 \end{bmatrix}$, yes

Section 5.3

1. $y = .99x + 1.14$ **3.** $y = -0.66x + 8.75$

5. $\mathbf{r} = [.06 \quad -0.13 \quad .08 \quad -.01]^t$, r.m.s. error = .08

9. $y = .29x^2 - 2.9x + 3.93$

11. $\mathbf{r} = [-0.05 \quad .23 \quad -0.37 \quad .27 \quad -0.07]^t$, r.m.s. error = .23

13. $\mathbf{r} = [0 \quad 0.2 \quad -0.06 \quad .06 \quad -0.02]^t$, r.m.s. error = .02

15. $y = 1.33x^3 - 1.36x^2 + 2.17x + 0.11$

17. $\mathbf{r} = [.31 \quad -1.26 \quad 1.89 \quad -1.26 \quad .31]^t$, r.m.s. error = 1.18

19. $y = 1.01e^{-0.13x}$ **21.** $y = 15.7x + 34.5$ **23.** $y = 1.97x + 0.92$

Review Exercises

1. No **3.** $(11/5)(3/5, 4/5) + (2/5)(-4/5, 3/5)$ **5.** $\sqrt{5}$ **7.** $(1, -2, 1)$

9. $B = \{(1/\sqrt{2}, 1/\sqrt{2}, 0), (3/\sqrt{22}, -3/\sqrt{22}, 2/\sqrt{22}), (-1/\sqrt{11}, 1/\sqrt{11}, 3/\sqrt{11})\}$

11. $B = \{(2/\sqrt{5}, 0, -1/\sqrt{5}), (3/\sqrt{70}, 5/\sqrt{70}, 6/\sqrt{70})\}$

Chapter 6

Section 6.1

1. Yes **3.** No **5.** No **7.** $[1 \quad -1 \quad 0]$ **9.** $\begin{bmatrix} 2 & 5 & -1 \\ 3 & -1 & 2 \\ 6 & 5 & 1 \end{bmatrix}$

11. $T(x) = (2x, x)$ **13.** $T(x_1, x_2, x_3) = (-x_1 + 4x_2 + 3x_3, 2x_1 + x_2 - 3x_3)$

15. $T(x_1, x_2, x_3, x_4) = (-4x_1 + x_2 + 3x_4, 2x_2 + x_3 + 5x_4, -2x_1 - 3x_2 - x_3)$

17. $(-3, -1, 1/2)$ **19.** $\begin{bmatrix} 1 & 0 \\ 0 & -1 \end{bmatrix}$ **21.** $\begin{bmatrix} 0 & 1 \\ 1 & 0 \end{bmatrix}$ **23.** $\begin{bmatrix} 2 & 0 \\ 0 & 2 \end{bmatrix}$

25. 1 **27.** The line $(5t, 2t)$ **29.** The line (t, t, t) **31.** $\begin{bmatrix} -3 & 2 \\ 1 & 0 \end{bmatrix}$

39. The range of T consists of all polynomials of degree at most 2

43. $B = \{(1, 0, 0), (0, 1, 0), (0, 0, 1)\}$

Section 6.2

1. $[3 \quad -2]^t$ **3.** $[-1 \quad -2, \quad 0]^t$ **5.** $[4 \quad -3]^t$ **7.** $[-1/5 \quad -1/5]^t$

9. $[-3/2 \quad -3/2 \quad -2]^t$ **11.** $[-11/9 \quad -14/9]$ **13.** $\begin{bmatrix} 2 & -1 & 1 \\ 4 & 1 & -1 \end{bmatrix}$

15. $\begin{bmatrix} 898 & 1288 \\ -624 & -895 \end{bmatrix}$ **17.** $\begin{bmatrix} 4 & 0 & 0 \\ 0 & 1 & 0 \\ 0 & 0 & 5 \end{bmatrix}$ **19.** $\begin{bmatrix} 10 & -18 & -11 \\ -15/2 & 14 & 17/2 \end{bmatrix}$

21. $\begin{bmatrix} 25/3 & -13 \\ -7 & 11 \\ 8/3 & -7/2 \end{bmatrix}$ **23.** $\begin{bmatrix} 1 & 0 & 0 \\ 0 & 1 & 0 \\ 0 & 0 & 0 \end{bmatrix}$

25. For B, $\begin{bmatrix} 0 & 1 & 0 \\ 0 & 0 & 2 \\ 0 & 0 & 0 \end{bmatrix}$; for B', $\begin{bmatrix} 0 & 1 & 0 \\ 0 & 0 & 4 \\ 0 & 0 & 0 \end{bmatrix}$

Section 6.3

1. $B' = \left\{ \begin{bmatrix} 1 \\ 2 \end{bmatrix}, \begin{bmatrix} 1 \\ 3 \end{bmatrix} \right\}$ **3.** $B' = \left\{ \begin{bmatrix} 2 \\ 3 \end{bmatrix}, \begin{bmatrix} 1 \\ 2 \end{bmatrix} \right\}$

5. $B' = \left\{ \begin{bmatrix} 9 \\ 7 \end{bmatrix}, \begin{bmatrix} 1 \\ 1 \end{bmatrix} \right\}$ **7.** $B' = \left\{ \begin{bmatrix} 2 \\ 3 \\ 1 \end{bmatrix}, \begin{bmatrix} 1 \\ 2 \\ -1 \end{bmatrix}, \begin{bmatrix} 2 \\ 2 \\ 2 \end{bmatrix} \right\}$

27. y is in the nullspace of A' if and only if Py is in the nullspace of A

Section 6.4

1. $(201, 74)$ **5.** $\begin{bmatrix} k & 0 & 0 \\ 0 & 1 & 0 \\ 0 & 0 & 1 \end{bmatrix}$ **7.** $\begin{bmatrix} 1 & 2 & 0 \\ 0 & 1/2 & 0 \\ 0 & 0 & 1 \end{bmatrix}$ **17.** $e^x \sin x + c$

19. $xe^x + c$

21. $(2 - x + 3x^2 - 2x^3 + x^4) + c$

25. $-2\cos x + 2x\cos x - x\sin x + c$

27. $4\cos x + 2x\cos x - 4x^2\cos x - \sin x + 7x\sin x + c$

Review Exercises

1. $(-4, -3)$ **3.** $(-9, 4)$ **5.** $[-6 \quad 7]^t$ **7.** $\begin{bmatrix} 3 & 7 \\ 2 & 5 \end{bmatrix}$ **9.** $\begin{bmatrix} 5 & 11 \\ -6 & -13 \end{bmatrix}$

11. $\begin{bmatrix} 2 & 1 \\ 0 & 1 \\ 3 & -4 \end{bmatrix}$ **13.** $\begin{bmatrix} 1 & -1 \\ 1 & 1 \end{bmatrix}$ **15.** $\begin{bmatrix} -28 & -69 \\ 13 & 32 \end{bmatrix}$ **17.** $\begin{bmatrix} -33 & 26 \\ -47 & 37 \end{bmatrix}$

Chapter 7

Section 7.1

5. $\lambda^2 - 9\lambda + 24$ **7.** $\lambda^3 - \lambda$ **9.** $2, \left\{\begin{bmatrix}1\\0\end{bmatrix}, \begin{bmatrix}0\\1\end{bmatrix}\right\}$ **11.** $0, \left\{\begin{bmatrix}1\\0\end{bmatrix}, \begin{bmatrix}0\\1\end{bmatrix}\right\}$

13. $2, \begin{bmatrix}1\\0\end{bmatrix}; 3, \begin{bmatrix}1\\-1\end{bmatrix}$

15. $1, \left\{\begin{bmatrix}1\\-1\end{bmatrix}\right\}$ **17.** $4, \left\{\begin{bmatrix}1\\-1\end{bmatrix}\right\}$ **19.** $1, \left\{\begin{bmatrix}1\\0\\0\end{bmatrix}, \begin{bmatrix}0\\1\\0\end{bmatrix}, \begin{bmatrix}0\\0\\1\end{bmatrix}\right\}$

21. $0, \left\{\begin{bmatrix}1\\0\\0\end{bmatrix}, \begin{bmatrix}0\\1\\0\end{bmatrix}, \begin{bmatrix}0\\0\\1\end{bmatrix}\right\}$

23. $1, \left\{\begin{bmatrix}2\\1\\-1\end{bmatrix}\right\}; (-1 + i\sqrt{3})/2, \left\{\begin{bmatrix}1\\2\\(-1 - i\sqrt{3})/2\end{bmatrix}\right\};$

$(-1 - i\sqrt{3})/2, \left\{\begin{bmatrix}2\\1\\(-1 + i\sqrt{3})/2\end{bmatrix}\right\}$ **25.** $2, \left\{\begin{bmatrix}1\\0\\1\end{bmatrix}\right\}; -2, \left\{\begin{bmatrix}1\\0\\-3\end{bmatrix}\right\}; 1, \left\{\begin{bmatrix}2\\1\\3\end{bmatrix}\right\}$

27. $1, \left\{\begin{bmatrix}0\\1\\1\end{bmatrix}\right\}; -1, \left\{\begin{bmatrix}1\\2\\1\end{bmatrix}\right\}$ **29.** $1, \left\{\begin{bmatrix}1\\0\\-1\end{bmatrix}, \begin{bmatrix}1\\-1\\1\end{bmatrix}\right\}; 5, \left\{\begin{bmatrix}1\\1\\1\end{bmatrix}\right\}$ **35.** No

Section 7.2

1. $\text{diag}(2, 1), P = \begin{bmatrix}4 & 1\\-3 & -1\end{bmatrix}$ **3.** Already in diagonal form

5. Not diagonalizable **7.** $\text{diag}(3, 1), P = \begin{bmatrix}1 & 3\\1 & 4\end{bmatrix}$

9. $\text{diag}((-3 - i\sqrt{15})/2, (-3 + i\sqrt{15})/2), P = \begin{bmatrix}5 & 5\\(5 + i\sqrt{15})/2 & (5 - i\sqrt{15})/2\end{bmatrix}$

11. $\text{diag}(1, i, -i), P = \begin{bmatrix}2 & 2 & 2\\-1 & -1 - i & -1 + i\\-1 & -2 & -2\end{bmatrix}$

13. $\text{diag}(1, 0, -1), P = \begin{bmatrix}1 & 2 & 2\\1 & 2 & 1\\-1 & -3 & -3\end{bmatrix}$

15. $\text{diag}(4, -2, 5), P = \begin{bmatrix}2 & 1 & 2\\1 & -1 & -2\\-1 & 1 & -5\end{bmatrix}$

17. $\text{diag}(1, -3, -3)$, $P = \begin{bmatrix} 1 & 1 & 1 \\ -1 & -4 & 0 \\ 1 & 0 & -4 \end{bmatrix}$

19. $\text{diag}(3, 1, 1, 1)$, $P = \begin{bmatrix} 1 & 1 & 0 & 0 \\ 1 & 0 & 1 & 0 \\ 1 & 1 & 2 & 0 \\ 1 & 0 & 0 & 1 \end{bmatrix}$

23. $A^k = \text{diag}(d_1^k, \ldots, d_n^k)$ **25.** $\frac{1}{3}\begin{bmatrix} 1+2^{51} & 1-2^{50} & 1-2^{50} \\ 1-2^{50} & 1+2^{51} & 1-2^{50} \\ 1-2^{50} & 1-2^{50} & 1+2^{51} \end{bmatrix}$

Section 7.3

1. $\text{diag}(2, 0)$, $P = \begin{bmatrix} 1/\sqrt{2} & 1/\sqrt{2} \\ 1/\sqrt{2} & -1/\sqrt{2} \end{bmatrix}$ **3.** $\text{diag}(3, 1)$, $P = \begin{bmatrix} 1/\sqrt{2} & 1/\sqrt{2} \\ 1/\sqrt{2} & -1/\sqrt{2} \end{bmatrix}$

5. $\text{diag}(2, -3)$, $\begin{bmatrix} 2/\sqrt{5} & 1/\sqrt{5} \\ 1/\sqrt{5} & -2/\sqrt{5} \end{bmatrix}$ **7.** $\text{diag}(3, -1)$, $P = \begin{bmatrix} 1/\sqrt{2} & 1/\sqrt{2} \\ 1/\sqrt{2} & -1/\sqrt{2} \end{bmatrix}$

9. $\text{diag}(-7, 8)$, $P = \begin{bmatrix} 2/\sqrt{5} & 1/\sqrt{5} \\ -1/\sqrt{5} & 2/\sqrt{5} \end{bmatrix}$

11. $\text{diag}(1, 2, 4)$, $P = \begin{bmatrix} 1/\sqrt{3} & 1/\sqrt{2} & -1/\sqrt{6} \\ -1/\sqrt{3} & 1/\sqrt{2} & 1/\sqrt{6} \\ 1/\sqrt{3} & 0 & 2/\sqrt{6} \end{bmatrix}$

13. $\text{diag}(1, 5, 2)$, $P = \begin{bmatrix} 1/\sqrt{2} & 1/\sqrt{6} & 1/\sqrt{3} \\ -1/\sqrt{2} & 1/\sqrt{6} & 1/\sqrt{3} \\ 0 & 2/\sqrt{6} & -1/\sqrt{3} \end{bmatrix}$

15. $\text{diag}(-1, 4, 1)$, $P = \begin{bmatrix} 1/\sqrt{2} & 1/\sqrt{3} & 1/\sqrt{6} \\ -1/\sqrt{2} & 1/\sqrt{3} & 1/\sqrt{6} \\ 0 & 1/\sqrt{3} & -2/\sqrt{6} \end{bmatrix}$

17. $\text{diag}(-2, -2, 4)$, $P = \begin{bmatrix} 1/\sqrt{2} & 1/\sqrt{3} & 1/\sqrt{6} \\ -1/\sqrt{2} & 1/\sqrt{3} & 1/\sqrt{6} \\ 0 & -1/\sqrt{3} & 2/\sqrt{6} \end{bmatrix}$

19. $\text{diag}(-1, 2, 2, 2)$, $P = \begin{bmatrix} 1/\sqrt{3} & 1/\sqrt{2} & 1/\sqrt{6} & 0 \\ 1/\sqrt{3} & -1/\sqrt{2} & 1/\sqrt{6} & 0 \\ -1/\sqrt{3} & 0 & 2/\sqrt{6} & 0 \\ 0 & 0 & 0 & 1 \end{bmatrix}$

21. $\operatorname{diag}(1,1,3,3)$, $P = \begin{bmatrix} 1/\sqrt{2} & 0 & 1/\sqrt{2} & 0 \\ 0 & 1/\sqrt{2} & 0 & 1/\sqrt{2} \\ 1/\sqrt{2} & 0 & -1/\sqrt{2} & 0 \\ 0 & 1/\sqrt{2} & 0 & 1/\sqrt{2} \end{bmatrix}$

23. $B = \begin{bmatrix} 1/\sqrt{2} & \sqrt{2}/\sqrt{3} & \sqrt{5}/\sqrt{6} \\ 0 & -\sqrt{2}/\sqrt{3} & 2\sqrt{5}/\sqrt{6} \\ -1/\sqrt{2} & \sqrt{2}/\sqrt{3} & \sqrt{5}/\sqrt{6} \end{bmatrix}$

Section 7.4

1. Ellipse **3.** Ellipse **5.** Hyperbola **7.** Elliptic cylinder

9. $[.542 \quad .308 \quad .150]^t$ **11.** $[.300 \quad .224 \quad .188 \quad .158 \quad .103 \quad .029]^t$

13. $[1.0000 \quad .3328]^t$ (5th iterate) **15.** $[-0.9375 \quad -0.9375 \quad 1.0000]^t$

17. $[-0.7333 \quad -0.8667 \quad 1.0000]^t$, with $x_0 = [1 \quad -1 \quad 1]^t$

19. $[1.0000 \quad 0.5120 \quad -1.0000]^t$ **21.** $[-0.0448 \quad -0.4776 \quad 1.0000]^t$

23. $[-0.5554 \quad 1.0000]^t$

29. Normal modes correspond to $\begin{bmatrix} 1.000 \\ -0.6180 \end{bmatrix}, \begin{bmatrix} .6180 \\ 1.0000 \end{bmatrix}$

31. Normal modes correspond to $\begin{bmatrix} -0.8019 \\ 1.0000 \\ -0.4450 \end{bmatrix}, \begin{bmatrix} 1.0000 \\ 0.4450 \\ -0.8019 \end{bmatrix}, \begin{bmatrix} 0.4450 \\ 0.8019 \\ 1.0000 \end{bmatrix}$

Review Exercises

1. $\lambda^2 + 2\lambda + 7$ **3.** $-1, \begin{bmatrix} 1 \\ -1 \end{bmatrix}$, not diagonalizable

5. $\operatorname{diag}(2i, -2i)$, $P = \begin{bmatrix} 2 & 2 \\ 1-i & 1+i \end{bmatrix}$

7. $2, \begin{bmatrix} 1 \\ 1 \\ 1 \end{bmatrix}; 1, \begin{bmatrix} 4 \\ 3 \\ 2 \end{bmatrix}$; not diagonalizable

9. $-1, \begin{bmatrix} 3 \\ 2 \\ -1 \end{bmatrix}; 1 \begin{bmatrix} 1 \\ 0 \\ -1 \end{bmatrix}$; not diagonalizable

11. $\operatorname{diag}(1, -1, 0)$, $P = \begin{bmatrix} 2 & 0 & 2 \\ -1 & 1 & -4 \\ 1 & -1 & 3 \end{bmatrix}$

13. $\operatorname{diag}(6, 1, 4)$, $P = \begin{bmatrix} 1/\sqrt{2} & -1/\sqrt{3} & 1/\sqrt{6} \\ 0 & 1/\sqrt{3} & 2/\sqrt{6} \\ 1/\sqrt{2} & 1/\sqrt{3} & -1/\sqrt{6} \end{bmatrix}$

15. diag(8, 2, 2), $P = \begin{bmatrix} -1/\sqrt{6} & 1/\sqrt{2} & 1/\sqrt{3} \\ 1/\sqrt{6} & 1/\sqrt{2} & -1/\sqrt{3} \\ 2/\sqrt{6} & 0 & 1/\sqrt{3} \end{bmatrix}$

Chapter 8

Section 8.1

5. (6, 0), (0, 3) **7.** (0, 4), (0, 5), (8, 7), (5/2, 3/2)
9. Any disk of radius $r \geq 6$ will do
11. Maximum 7 at (3, 4); minimum 0 at (0, 0)
13. Maximum 18 at (6, 0); minimum 0 at (0, 0)
15. Maximum 39 at (7, 4); minimum 7 at (2, 0)
17. Maximum 20 at (5, 0); minimum −24 at (1, 7)
19. Use 1667 gallons of blend 2 and 3333 gallons of blend 1 for the maximum profit of $5667
21. Produce 10.67 of type 1 and 1.67 of type 2 (on the average) for the maximum profit of $747

Section 8.2

1. Maximum 7 at (3, 4) **3.** Maximum 18 at (6, 0) **5.** Maximum 19 at (8, 3)
7. Maximum 35 at (0, 7) **9.** Maximum 23 at (1, 2, 2)
11. Maximum 62 at (0, 8, 6) **13.** Maximum 55 at (4, 6, 5)

Section 8.3

1. Minimum 24 at (3, 4) **3.** Minimum 10 at (0, 10)
5. Minimum 17 at (2, 3)
7. Minimum 17 at (5, 2) **9.** Minimum 18 at (6, 6)
11. Minimum −16 at (7, 9) **13.** Maximum 10 at (4, 3)
15. Maximum −1 at (0, 1)
17. No minimum

Review Exercises

1. Yes **3.** Minimum −6 at (0, 2) **5.** Maximum 30 at (6, 0)
7. Minimum 7 at (3, 4)

INDEX